MW01705309

Chapters from

Introduction to Geography

and

Human Geography

Arthur Getis
Judith Getis
Jerome Fellmann

Wm. C. Brown Publishers
College Division
Dubuque, Iowa 52001

ISBN 0-697-15345-2

Printed in the United States of America by Wm. C. Brown Publishers, 2460 Kerper Boulevard, Dubuque, Iowa 52001.

10 9 8 7 6 5 4 3 2 1

Contents

Chapter 1

Introduction: Some Background Basics

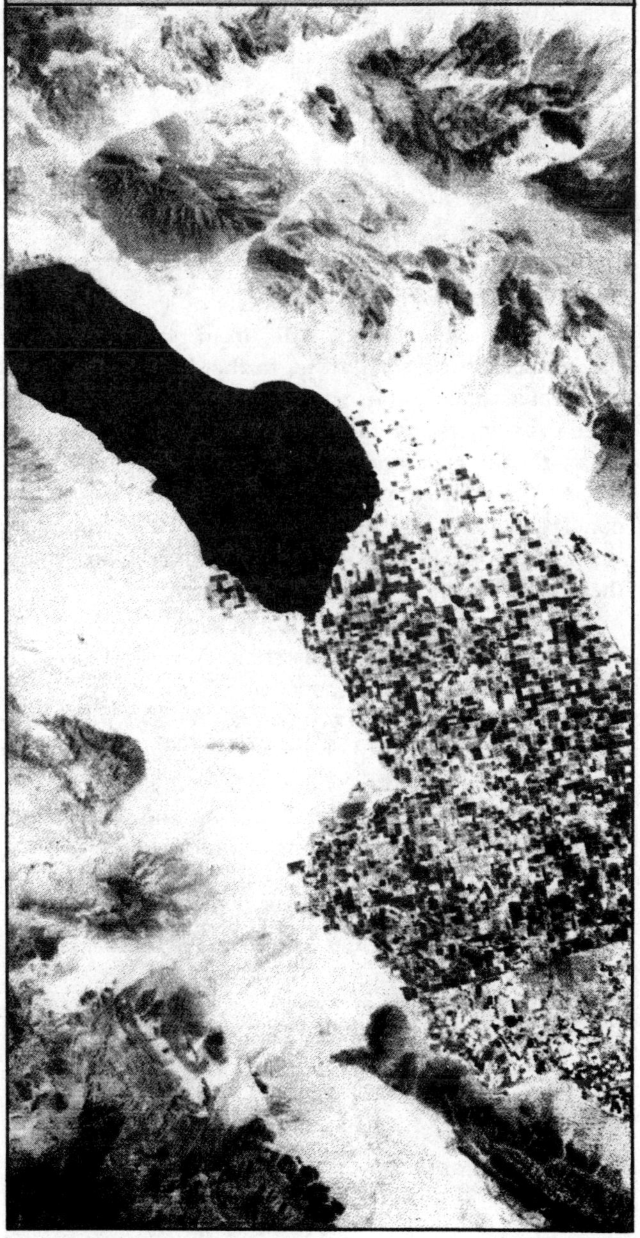

Getting Started

The fundamental question asked by geographers is "Does it make a difference where things are located?" If for any one thing or group of things the answer is "You bet it does," geographers' interest is aroused and geographic inquiry is appropriate. For example, it matters a great deal that languages of a certain kind are spoken in certain places. But knowledge of the location of a specific language group is not of itself particularly significant. Geographic study of a language requires that we attempt to answer questions dealing with the characteristics of the language and the reasons for the distribution of those who speak it. In the process of our study, we would be compelled to discuss such concepts as migration, acculturation, the diffusion of innovation, the effect of physical barriers on communication, and the relationship of language to other aspects of culture. As geographers, we are interested in the areal integration—the spatial system—of things.

Geography is often referred to as the spatial science, that is, the discipline concerned with the use of earth space. In fact, the term *geography* literally means "description of the earth," but describing the earth is really the responsibility of nearly all the sciences. Geography might better be defined as the study of spatial variation, of how—and why—things differ from place to place on the surface of the earth. It is, further, the study of how observable spatial patterns evolved through time. If things were everywhere the same, if there were no spatial variation, the kind of human curiosity that we call "geographic" simply would not exist. Without the certain conviction that in some interesting and important way landscapes, peoples, and opportunities differ across space, there would be no discipline of geography.

But we do not have to deal in such abstract terms. Geography and geographic awareness are vital elements that you consciously or subconsciously recognize and act on in your daily life. You are where you are, doing what you are doing, because of choices you faced and decisions you made. You cannot be here reading this book and simultaneously be somewhere else—working, perhaps, or at the gym. And should you now want to go to work or take an exercise break, the time involved in going from here to there (wherever "there" is) is time not available for other activities in other locations. Of course, the act of going implies knowing where you are now, where "there" is in relation to "here," and the paths or routes you can take to cover the distance.

These are simple examples of the observation that "space matters" in a very personal way. You cannot avoid the implications of geography in your everyday affairs. Your understanding of your hometown, your neighborhood, or your college campus is essentially a geographic understanding. It is based on your awareness of where things are, of their spatial relationships, and of the varying content of the different areas and places you frequent. You carry out your routine activities in particular places and move on your daily rounds within defined geographic areas, following logical paths of connection between different locations.

Just as geography matters in your personal life, so it matters on the larger stage as well. Decisions made by corporations about the locations of manufacturing plants or warehouses in relation to transportation routes and markets are spatially rooted, as are those made by shopping center developers and locators of parks and grade schools. On an even grander scale, judgments about the projection of national power or the claim and recognition of "spheres of influence and interest" among the superpowers are related to the implications of distance and area.

Geography, therefore, is about space and the content of space. We think of and respond to places from the standpoint not only of where they are but, rather more importantly, of what they contain or what we think they contain. Reference to a place or an area usually calls up images about its physical nature or what people do there and often suggests, without conscious thought, how those physical things and activities are related. "Colorado," "mountains," and "skiing" might be a simple example. The content of area, that is, has both physical and cultural aspects, and geography from its inception has been concerned with understanding both (Figure 1.1).

Evolution of the Discipline

Geography's combination of interests was apparent even in the work of the early Greek geographers, who first gave structure to the discipline. Geography's name was reputedly coined by the Greek scientist Eratosthenes over 2200 years ago from the words *geo,* "the earth," and *graphein,* "to write." From the beginning, that writing focused both on the physical structure of the earth and on the nature and activities of the people who inhabited the different lands of the known world. To Strabo (c. 64 B.C.–A.D. 20) the task of geography was to "describe the several parts of the inhabited world . . . to write the assessment of the countries of the world [and] to treat the differences between countries." Greek (and, later, Roman) geographers measured the earth, devised the global grid of latitudes and longitudes, and drew upon this grid surprisingly sophisticated maps (Figure 1.2). Employing nearly modern concepts, they discussed patterns and processes of climates,

Figure 1.1

The ski lift at Lake Louise, Alberta, Canada: physical environmental and human activity patterns in spatial juxtaposition. Climate and terrain have made specialized human use possible. Human development has placed a cultural landscape on and altered the natural environment.

Figure 1.2

World map of the 2nd century A.D. Roman geographer-astronomer Ptolemy. Ptolemy adopted a previously developed map grid of latitude and longitude based on the division of the circle into 360°, permitting a precise mathematical location for every recorded place. Unfortunately, errors of assumption and measurement rendered both the map and its accompanying six-volume gazetteer inaccurate. Many variants of Ptolemy's map were published in the 15th and 16th centuries. The version shown here summarizes the extent and content of the original.

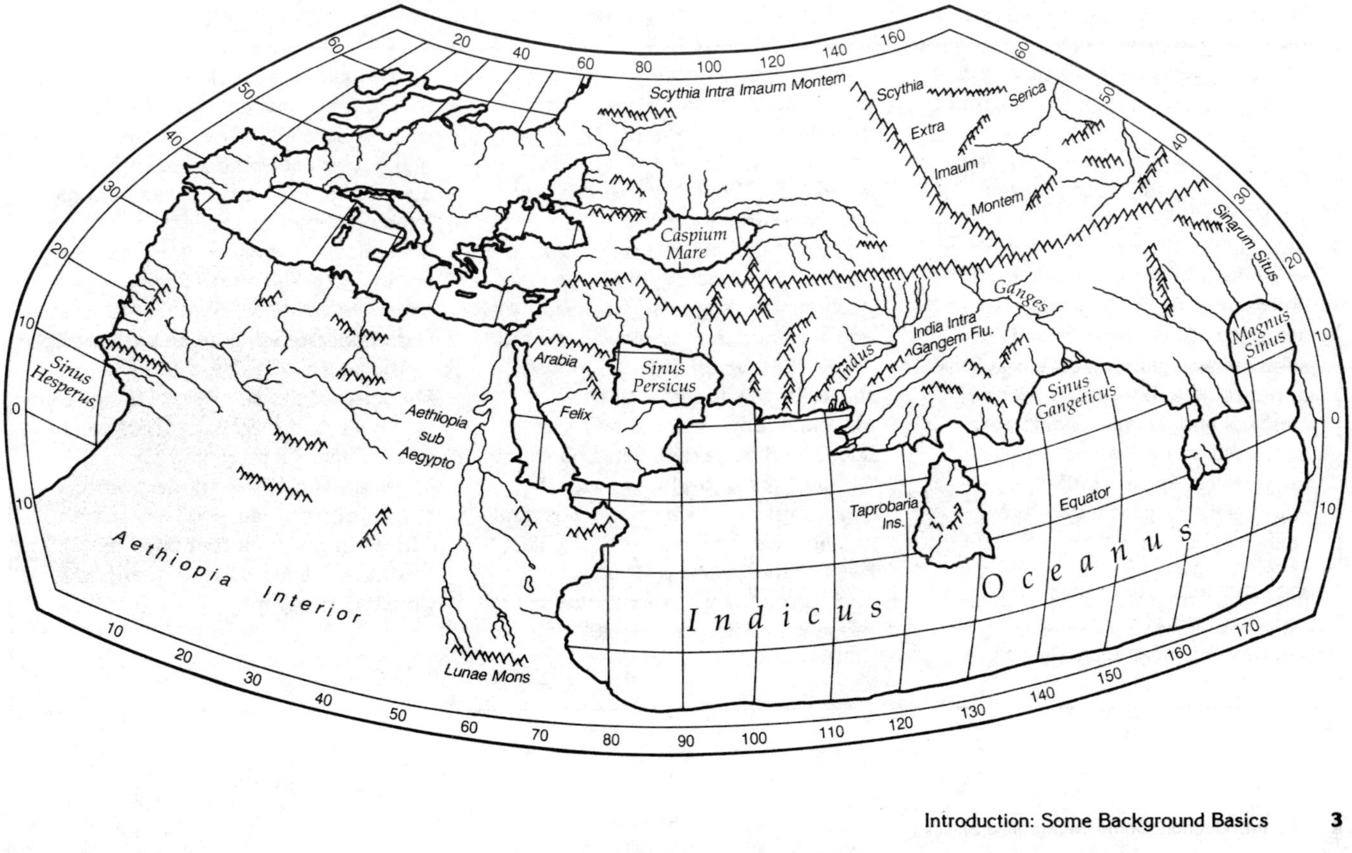

vegetation, and landforms and described areal variations in the natural landscape. Against that physical backdrop, they focused their attention on what humans did in home and distant areas: how they lived; what their distinctive similarities and differences in language, religion, and custom were; and how they used, altered, and perhaps destroyed the lands they inhabited.

These are enduring and universal interests. The ancient Chinese, for example, were as involved in geography as an explanatory viewpoint as were Westerners, though there was no exchange between them. Moreover, as Christian Europe entered the Dark and Middle Ages between A.D. 800 and 1400 and lost their knowledge of Greek and Roman geographic work, Muslim scholars—who retained that knowledge—undertook to describe and analyze their known world in its physical, cultural, and regional variation (see "Roger's Book").

Modern geography had its origins in the surge of scholarly inquiry that, beginning in the 17th century, gave rise to many of the traditional academic disciplines we know today. In its European rebirth, geography from the outset was recognized—as it always had been—as a broadly based integrative study. Patterns and processes of the physical landscape were early interests, as was concern with humans as part of the earth's variation from place to place. The rapid development of geology, botany, zoology, and other natural sciences by the end of the 18th century strengthened regional geographic investigation and increased scholarly and popular awareness of the intricate interconnections of things in space and between places. By that same time, accurate determination of latitude and longitude and scientific mapping of the earth made assignment of place information more reliable and comprehensive. During the 19th century, national censuses, trade statistics, and ethnographic studies gave firmer foundation to human geographic investigation.

By the end of the 19th century, geography had become a distinctive and respected discipline in universities throughout Europe and in other regions of the world where European academic examples were followed. The proliferation of professional geographers and geography programs resulted in the development of a whole series of increasingly narrowed disciplinary specializations.

Geography and Human Geography

Geography's specialized interests are not divisive. The discipline in all its subdivisions is characterized by three dominating interests. The first is in the areal variation of physical and human phenomena on the surface of the earth and in relationships between human societies and the natural environments that they occupy and modify. The second is a focus on the spatial systems that link physical phenomena and human activities and societies in one area of the earth's surface with other areas. Together, these interests lead to a third enduring theme, that of regional analysis, in which the human-environmental—or ecological—relationships and the systems of spatial interaction are studied in specific areal settings. This areal orientation is called *regional geography*.

Roger's Book

The Arab geographer Idrisi, or Edrisi (c. A.D. 1099–1154), a descendant of the Prophet Mohammed, was directed by Roger II, the Christian King of Sicily in whose court he served, to collect all known geographical information and assemble it in a truly accurate representation of the world. An academy of geographers and scholars was gathered to assist Idrisi in the project. Books and maps of classical and Islamic origins were consulted, mariners and travelers interviewed, and scientific expeditions dispatched to foreign lands to observe and record. Data collection took 15 years before the final world map was fabricated on a silver disc some 80 inches in diameter and weighing over 300 pounds. Lost to looters in 1160, the map is survived by "Roger's Book," containing the information amassed by Idrisi's academy and including a world map, 71 part maps, and 70 sectional itinerary maps.

Idrisi's "inhabited earth" is divided into the 7 "climates" of Greek geographers, beginning at the equator and stretching northward to the limit at which, it was supposed, the earth was too cold to be inhabited. Each climate was then subdivided by perpendicular lines into 11 equal parts, beginning with the west coast of Africa on the west and ending with the east coast of Asia. Each of the resulting 77 square compartments was then discussed in sequence in "Roger's Book."

Though Idrisi worked in one of the most prestigious courts of Europe, there is little evidence that his work had any impact on European geographic thought. He was strongly influenced by Ptolemy's work and misconceptions and shared the then common Muslim fear of the unknown western ocean. Yet Idrisi's clear understanding of such scientific truths as the roundness of the earth, his grasp of the scholarly writings of his Greek and Muslim predecessors, and the faithful recording of information on little-known portions of Europe, the Near East, and North Africa set his work far above the mediocre standards of contemporary Christian geography.

Other geographers choose to identify particular classes of things, rather than segments of the earth's surface, for specialized study. These *systematic geographers* may focus their attention on one or a few related aspects of the physical environment or of human populations and societies. In each case, the topic selected for study is examined in its interrelationships with other spatial systems and areal patterns. *Physical geography* directs its attention to the natural environmental side of the human-environment structure. Its concerns are with landforms and their distribution, with atmospheric conditions and climatic patterns, with soils or vegetation associations, and the like. The other systematic branch of geography—and the subject of this book—is *human geography*.

Human Geography

Human geography deals with the world as it is and with the world as it might be made to be. Its emphasis is on people: where they are, what they are like, how they interact over space, and what kinds of landscapes of human use they erect upon the natural landscapes they occupy. It encompasses all those interests and topics of geography that are not directly concerned with the physical environment or, like cartography, technical in orientation. Human geography's content provides integration for all the social sciences, for it gives to those sciences the necessary spatial and systems viewpoint that they otherwise lack. At the same time, human geography draws on other social sciences in the analyses identified with its subfields, such as *behavioral, political, economic,* or *social geography* (Figure 1.3).

Figure 1.3

Some of the subdivisions of human geography and the allied fields to which they are related.

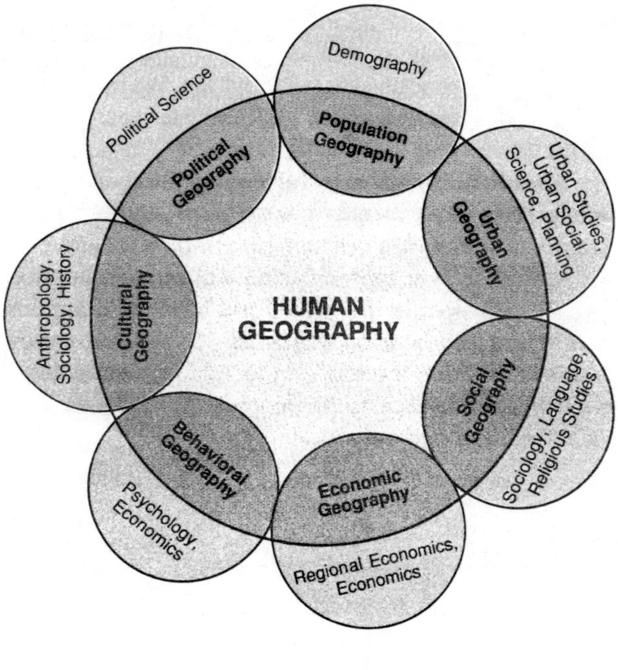

Human geography admirably serves the objectives of a liberal education. It helps us to understand the world we occupy and to appreciate the circumstances affecting peoples and nations other than our own. It clarifies the contrasts in societies and cultures and in the human landscapes they have created in different regions of the earth. Its models and explanations of spatial interaction allow us to better comprehend the economic, social, and political systems within which we all, singly and collectively, live and operate. Its analyses of spatial systems make us more aware of the realities and the prospects of our own society in an increasingly troubled and competitive world. Our study of human geography, therefore, can help make us better informed citizens, more able to understand the important issues facing our communities and our countries and better prepared to contribute to their solution.

The Structure of This Book

By way of getting started, it is useful for you to know how the organization and topics of this text have been structured to help you reach the kinds of understanding we seek.

We begin by exploring the roots and meaning of culture (Chapter 2), establishing the observed ground rules of spatial interaction and spatial behavior (Chapter 3), and examining the areal variations in patterns of population distribution and change (Chapter 4). These set the stage for following separate discussions of spatial patterns of language and religion, ethnic distinctions, and folk and popular culture. These are the principal expressions of unity and diversity and of areal differentiation among the earth's peoples and societies. Understanding their spatial patterns and interrelations goes far toward providing the global view that is our objective.

Beginning with Chapter 8, our focus shifts more to the economic and organizational landscapes humans have created. In turn, we look at economic geography and economic development, urban systems and structures, and patterns of the political ordering of space. Finally, in Chapter 13 dealing with human impacts, we return to the underlying concern of all geographic study: the relationship between human geographic patterns and processes and both the present conditions and the future prospects of the physical and cultural environments we occupy, create, or modify.

To help clarify the connections between the various topics of human geography, the chapters of this book are grouped by common theme and separately introduced. For students new to geography as a subject and for those who want a reminder of its unifying objectives and shared techniques of study, the remainder of this first chapter serves as introduction and review.

Background Basics

Basic Geographic Concepts

The topics included in human geography are diverse. However, that very diversity emphasizes the reality that all geographers—whatever their particular topical or regional specialties—are united by the similar questions they ask and the common set of concepts they employ to consider their answers. Of either a physical or cultural phenomenon they will inquire: What is it? Where is it? How did it come to be what and where it is? Where is it in relation to other things that affect it or are affected by it? How is it part of a functioning whole? How does its location affect people's lives and the content of the area in which it is found?

These questions are spatial in focus and systems analytical in approach. They derive from enduring central themes in geography. In answering them, geographers draw upon a common store of concepts, terms, and methods of study that together form the basic structure and vocabulary of geography. Collectively they reflect the fundamental truths addressed by geography: that things are rationally organized on the earth's surface and that understanding spatial patterns is the necessary starting point for understanding how people live upon and shape the earth's surface.

Geographers use the word *spatial* as an essential modifier in framing their questions and forming their concepts. Geography, they say, is a *spatial science.* It is concerned with *spatial behavior* of people, with the *spatial relationships* that are observed between places on the earth's surface, and with the *spatial processes* that create or maintain those behaviors and relationships. The word *spatial* comes, of course, from *space,* and to geographers it always carries the idea of the way things are distributed, movements occur, or processes proceed over the whole or a defined portion of the surface of the earth. The geographer's space, then, is earth space, the surface area occupied or available to be occupied by humans. Spatial phenomena have locations on that surface, and spatial interactions occur between places, things, and people within the earth area available to them. The need to understand those relationships, interactions, and processes helps frame the questions that geographers ask.

Those questions have their starting point in basic observations about the location and nature of places and about how places are similar to or different from one another. Such observations, though simply stated, are profoundly important to our comprehension of the world we occupy:

Places have location, direction, and distance with respect to other places.
A place may be large or small; scale is important.
A place has both physical structure and cultural content.
The characteristics of places develop and change over time.
Places interact with other places.
The content of places is rationally structured.
Places may be generalized into regions of similarities and differences.

These are basic notions understandable to everyone. They also are the means by which geographers express fundamental observations about the earth spaces they examine and put those observations into a common framework of reference. Each of the concepts is worth further explanation, for they are not quite as simple as they at first seem.

Location, Direction, and Distance

Location, direction, and *distance* are everyday ways of assessing the space around us and identifying our position in relation to other things and places of interest to us. They are also essential to our understanding the processes of spatial interaction that figure so importantly in the study of human geography.

Location

The location of places and things is the starting point of all geographic study as well as of all our personal movements and spatial actions in everyday life. We think of and refer to location in at least two different senses, *absolute* and *relative.*

Absolute location is the identification of place by some precise and accepted system of coordinates; it therefore is sometimes called *mathematical location.* We have several such accepted systems of pinpointing positions. One of them is the global grid of latitude and longitude (discussed later on pages 18–19). With it we can accurately describe the absolute location of any point on the earth by reference to its degrees, minutes, and seconds of *latitude* and *longitude* (Figure 1.4).

Figure 1.4

The latitude and longitude of Singapore is 1° 20′ N, 103° 57′ E. What are the coordinates of Kuala Lumpur?

Figure 1.5

The reality of *relative location* on the globe may be strikingly different from the impressions we form from flat maps. The position of the Soviet Union with respect to North America when viewed from a polar perspective emphasizes that relative location properly viewed is important to our understanding of spatial relationships and interactions between the two world areas.

Other coordinate systems are also in use. Survey systems such as the township, range, and section description of property in much of the United States give mathematical locations on a regional level, while street address precisely defines a building according to the reference system of an individual town. Absolute location is unique to each described place, is independent of any other characteristic or observation about that place, and has obvious utility in legally describing places, in measuring the distance separating places, and in finding directions between places on the earth's surface.

When geographers remark that "location matters," however, their reference is usually not to absolute but to **relative location**—the position of a place in relation to that of other places or activities. Relative location expresses spatial interconnection and interdependence. On an immediate and personal level, we think of the location of the school library not in terms of its street address or room number but where it is relative to our classrooms, or the cafeteria, or some other reference point. On the larger scene, relative location tells us that people, objects, and places exist not in a spatial vacuum but in a world of physical and cultural characteristics that differ from place to place (Figure 1.5).

New York City, for example, may be described in absolute terms as located at (approximately) latitude 40° 43′ N (read as 40 degrees, 43 minutes north) and longitude 73° 58′ W. However, we have a better understanding of the *meaning* of its location when reference is made to its spatial relationships: to the continental interior through the Hudson-Mohawk lowland corridor; to its position on the eastern seaboard of the United States; to its role as a transportation node and multifunctional economic and administrative center of a major industrial nation of the Northern Hemisphere; and to its interconnections, through trade routes, financial ties, and the United Nations, to the international community. Within the city, we gain understanding of the locational significance of, for example, Central Park or the Lower East Side not by reference solely to the street addresses or city blocks they occupy but by their spatial and functional relationships to the total land use, activity, and population patterns of New York City.

In view of these different ways of looking at location, geographers make a distinction between the *site* and the *situation* of a place. **Site** refers to the physical and cultural characteristics and attributes of the place itself. It is more than mathematical location, for it tells us something about the internal features of that place. The site of Pittsburgh, for example, is the valleys of the Allegheny, Monongahela, and Ohio rivers. **Situation,** on the other hand, refers to the external relations of a locale. It is an expression of relative location with particular reference to items of significance to the place in question. The situation of Chicago might be described as at the deepest penetration of the Great Lakes system into the interior of the United States, astride the Great Lakes-Mississippi waterways, and near the western margin of the manufacturing belt, the northern boundary of the corn belt, and the southeastern reaches of a major dairy region. References to railroads, coal deposits, and ore fields amplify its situational characteristics (Figure 1.6).

Direction

Direction is a second universal spatial concept. Like location, it has more than one meaning and can be expressed in absolute or relative terms. **Absolute direction** is based on the cardinal points of north, south, east, and west. These appear uniformly and independently in all cultures, derived from the obvious "givens" of nature: the rising and setting of the sun for east and west, the sky location of the noontime sun and of certain fixed stars for north and south.

We also commonly use **relative** or *relational* **directions.** In the United States, we go "out West," "back East," or "down South"; we worry about conflict in the "Near East" or economic competition from the countries of the "Far East." These directional references are culturally based and locationally variable, despite their reference to cardinal compass points. The Near and the Far East locate parts of Asia from the European perspective; they are retained in the Americas by custom and usage, even though one would normally travel westward across the Pacific, for example, to reach the "Far East" from California, British Columbia, or Chile. For many Americans, "back East" and "out West" are reflections of the migration paths of earlier generations for whom home was in the eastern part of the country, to which they might look back. "Up North" and "down South" reflect our accepted custom of putting north at the top and south at the bottom of our maps.

Figure 1.6

The *situation* of Chicago helps suggest the reasons for its functional diversity.

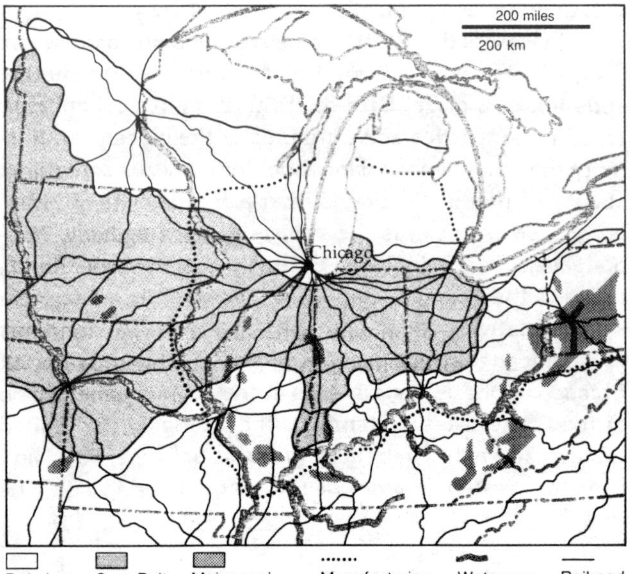

Distance

Distance joins location and direction as a commonly understood term that has dual meanings for geographers. Like its two companion spatial concepts, distance may be viewed in both an absolute and a relative sense.

Absolute distance refers to the spatial separation between two points on the earth's surface measured by some accepted standard unit such as miles or kilometers for widely separated locales and feet or meters for more closely spaced points. **Relative distance** transforms those linear measurements into other units more meaningful for the space relationship in question.

To know that two competing malls are about equidistant in miles from your home is perhaps less important in planning your shopping trip than is knowing that because of street conditions or traffic congestion one is 5 minutes and the other 15 minutes away (Figure 1.7). Most people, in fact, think of time distance rather than linear distance in their daily activities; downtown is 20 minutes by bus, the library is a 5-minute walk. In some instances, money rather than time may be the distance transformation. An urban destination might be estimated to be a $5 cab ride away, information that may affect either the decision to make the trip at all or the choice of travel mode used to get there.

A psychological transformation of linear distance is also frequent. The solitary late-night walk back to the car through an unfamiliar or dangerous neighborhood seems far longer than a daytime stroll of the same distance through familiar and friendly territory. A first-time trip to

Figure 1.7

Lines of equal travel time (*isochrones*) mark off different linear distances from a given starting point, depending on the condition of the route and terrain and changes in the roads and traffic flows over time. On this map, the areas within 30 minutes' travel time from downtown Los Angeles are recorded for the period 1953 to 1971.

Source: Reproduced by permission from *The Los Angeles Metropolitan Experience*, Howard J. Nelson and William A. V. Clark, page 49, Association of American Geographers, 1976.

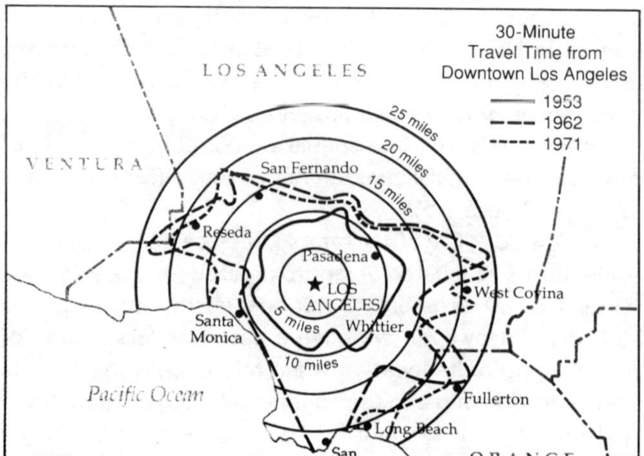

a new destination frequently seems much longer than the return trip over a now-known path. Distance relationships, their measurement, and their meaning for human spatial interaction are fundamental to our understanding of human geography. They are a subject of Chapter 3, and reference to them recurs throughout this book.

Size and Scale

When we say that a place may be large or small, we speak both of the nature of the place itself and of the generalizations one can make about it. In either instance, geographers are concerned with **scale,** though we may use that term in different ways. We can, for example, study a problem—population, say, or agriculture—at the local scale, the regional scale, or on a global scale. Here the reference is purely to the size of unit studied. More technically, scale tells us the relationship between the size of an area on a map and the actual size of the mapped area on the surface of the earth. In this sense, scale is a feature of every map and essential to recognizing the areal meaning of what is shown on that map.

In both senses of the word, *scale* implies the degree of generalization represented (Figure 1.8). Geographic inquiry may be broad or narrow; it occurs at many different size scales. Climate may be an object of study, but research and generalization focused on climates of the world will differ in degree and kind from study of the microclimates of a city. Awareness of scale is of great importance, for in geographic work concepts, relationships, and understandings that are found to have meaning at one scale may not be applicable when the same problem is examined at another scale.

Physical and Cultural Attributes

All places have individual physical and cultural traits that distinguish them from other places. Their attributes give them character, potential, and meaning that set them apart from other earth space. Geographers are concerned with identifying and analyzing the details of those attributes and, particularly, with recognizing the interrelationship between the physical and cultural components of area—the human-environmental interface.

The physical characteristics of a place refer to such natural aspects of a locale as its climate and soil, the presence or absence of water supplies and mineral resources, its terrain features, and the like. These **natural landscape** attributes provide the setting within which human action occurs. They help shape—but do not dictate—how people live. The resource base, for example, is physically determined, though how resources are perceived and used is culturally conditioned.

Figure 1.8

Population density and map scale. "Truth" depends on one's scale of inquiry. Map (*a*) reveals that the maximum population density of midwestern states is no more than 319 people per square mile (123 per km²). From map (*b*), however, we see that population densities in two Illinois counties exceed 1280 people per square mile (494 per km²). Were we to reduce our scale of inquiry even further, examining individual city blocks in Chicago, we would find densities as high as 5000 people per square mile (1930 per km²). Scale matters!

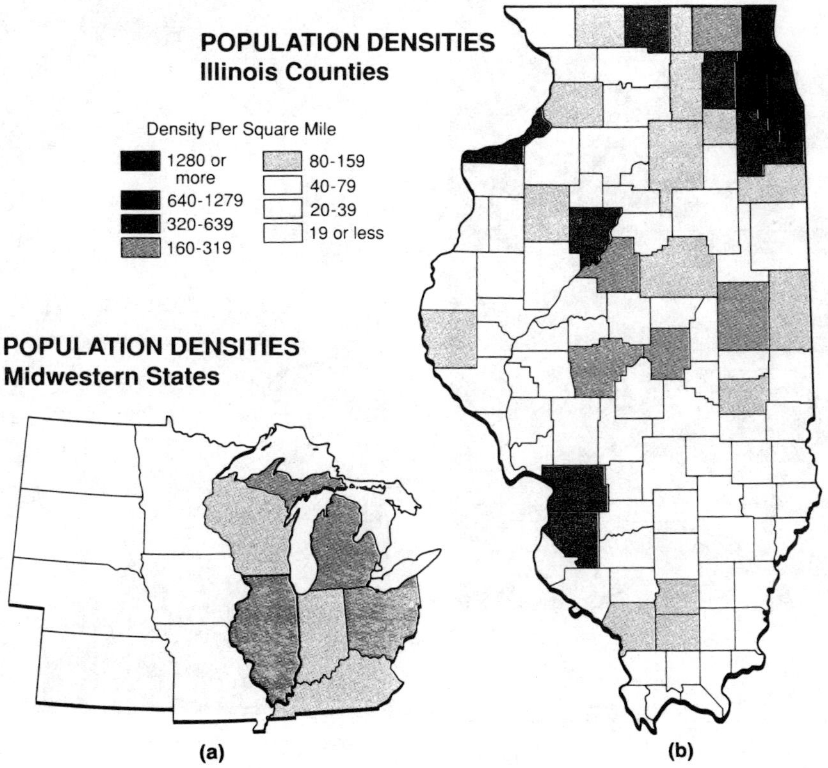

People modify the environmental conditions of a given place simply by occupying it. The existence of the U.S. Environmental Protection Agency (and its counterparts elsewhere) is a reminder that humans are the active and frequently harmful agents in the continuing spatial interplay between the cultural and physical worlds (Figure 1.9). Virtually every human activity leaves its imprint on an area's soils, water, vegetation, animal life, and other resources and on the atmosphere common to all earth space. The visible imprint of that human activity is the **cultural landscape.** It, too, exists at different scales and at different levels of visibility. Differences in agricultural practices and land use between Mexico and Southern California are evident in Figure 1.10, while the signs, structures, and people of Los Angeles's Chinatown leave a smaller, more confined imprint within the larger cultural landscape of the metropolitan area itself.

Although the focus of this book is on the human characteristics of places, geographers are ever aware that the physical content of an area is also important in understanding the activity patterns of people and the interconnections between people and the environments they occupy and modify. Those interconnections and modifications are not static or permanent, however; they are subject to continual change.

Figure 1.9

Sites (and sights) such as this devastation of ruptured barrels and petrochemical contamination near Texas City, Texas, are all-too-frequent reminders of the adverse environmental impacts of humans and their waste products. Many of those impacts are more hidden in the form of soil erosion, water pollution, increased stream sedimentation, plant and animal extinction, deforestation, and the like.

The Changing Attributes of Place

The physical environment surrounding us seems eternal and unchanging, but of course it is not. In the framework of geologic time, change is both continuous and pronounced. Islands form and disappear; mountains rise and are worn low to swampy plains; vast continental glaciers form, move, and melt away, and sea levels fall and rise in response. Geologic time is long, but the forces that give shape to the land are timeless and relentless.

Figure 1.10

This Landsat image reveals contrasting cultural landscapes along the Mexico-California border. Slowly move your eyes from the Salton Sea (the dark patch near the center of the image) to the agricultural land extending to the southeast and to the edge of the picture. Notice how the regularity of the fields and the bright colors (representing growing vegetation) give way to a marked break, where irregularly shaped fields and less prosperous agriculture are evident. Above the break is the Imperial Valley of California; below the border is Mexico.
© NASA.

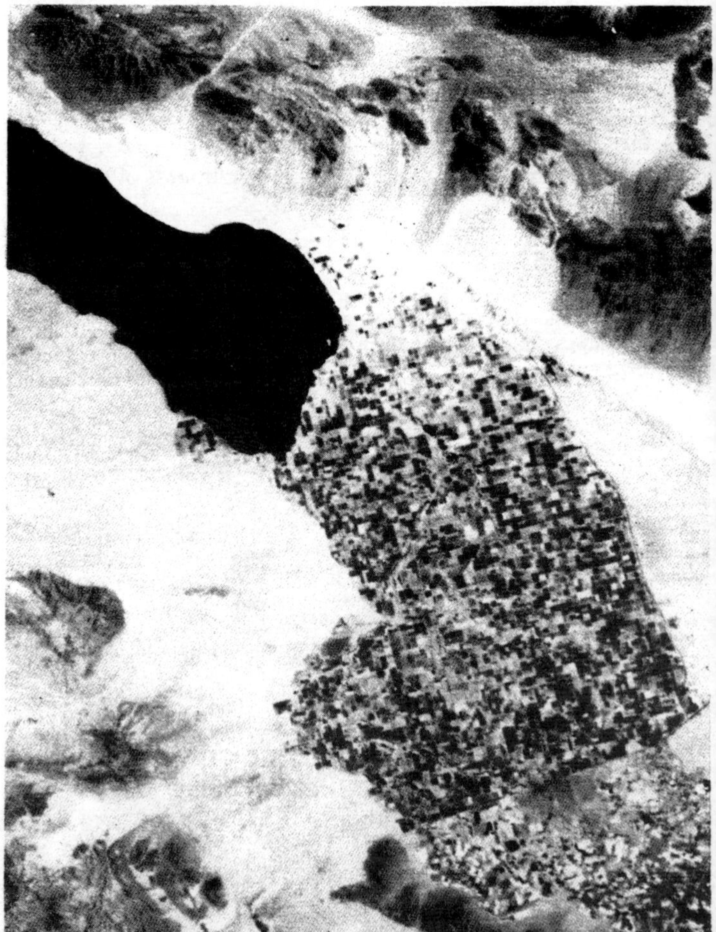

Even within the short period of time since the most recent retreat of continental glaciers—some 10,000 or 11,000 years ago—the environments occupied by humans have been subject to change. Glacial retreat itself marked a period of climatic alteration, extending the area habitable by humans to include vast reaches of northern Eurasia and North America formerly covered by thousands of feet of ice. With moderating climatic conditions came associated changes in vegetation and fauna. On the global scale, these were natural environmental change; humans were as yet too few in number and too limited in technology to alter materially the course of physical events. On the regional scale, however, even early human societies exerted an impact on the environments they occupied. People used fire to clear forest undergrowth, to maintain or extend grassland for grazing animals and to drive them in the hunt, and, later, to clear openings for rudimentary agriculture.

With the dawn of civilizations and the invention and spread of agricultural technologies, humans accelerated their management and alteration of the now no longer "natural" environment. Even the classical Greeks noted how the landscape they occupied differed, for the worse, from its former condition. With growing numbers of people, and particularly with industrialization and the spread of European exploitative technologies throughout the world, the pace of change in the content of area accelerated. The built landscape increasingly replaced the natural landscape. Each new settlement or city, each agricultural assault on forests, each new mine, dam, and factory changed the content of regions and altered the temporarily established spatial interconnections between humans and the environment.

Characteristics of places today are therefore the result of constantly changing past conditions. They are, as well, the forerunners of differing human-environmental balances yet to be struck. Geographers are concerned with places at given moments of time. But to understand fully the nature and development of places, to appreciate the significance of their spatial relationships, and to understand the interplay of their physical and cultural characteristics, geographers must view places as the present result of past operation of distinctive physical and cultural processes (Figure 1.11).

Figure 1.11

The process of change in a cultural landscape. Before the advent of the freeway, this portion of suburban Long Island, New York, was largely devoted to agriculture (left). The construction of the freeway and cloverleaf interchange ramps altered nearby land use patterns (right) to replace farming with new housing developments and new commercial and light industrial activities.

Photo: U.S. Geological Survey.

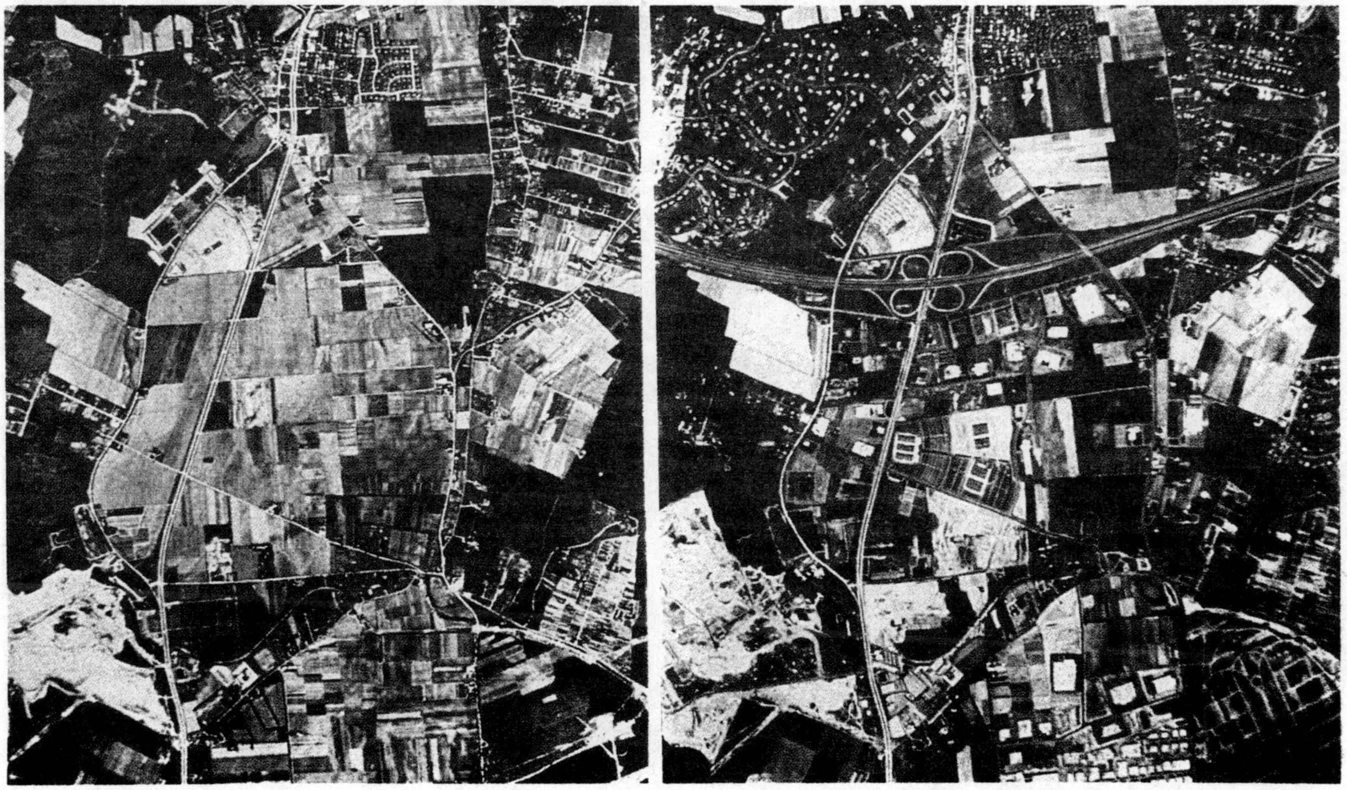

You will recall that one of the questions geographers ask of a place or thing is How did it come to be what and where it is? This is an inquiry about process and about becoming. The forces and events shaping the physical and explaining the cultural environment of places today are an important focus of geography. They are, particularly in their human context, the subjects of most of the separate chapters of this book. To understand them is to appreciate more fully the changing nature of the human spatial order of our contemporary world.

Interaction Among Places

The concepts of relative location and distance that we earlier introduced lead directly to the fundamental spatial reality that places interact with other places in structured and comprehensible ways. In describing the processes and patterns of that **spatial interaction,** geographers add to location and distance the ideas of *accessibility* and *connectivity*.

A basic law of geography tells us that in a spatial sense everything is related to everything else but that relationships are stronger when things are near one another. Our observation, therefore, is that interaction between places diminishes in intensity and frequency as distance between them increases—a statement of the idea of *distance decay* that we explore in Chapter 3.

Consideration of distance implies assessment of **accessibility.** How easy or difficult is it to overcome the "friction of distance"? That is, how easy or difficult is it to surmount the barrier of the time and space separation of places? Distance isolated North America from Europe until the development of ships (and aircraft) that reduced the effective distance between the continents. All parts of the ancient and medieval city were accessible by walking; they were "pedestrian cities," a status lost as cities expanded in area and population with industrialization. Accessibility between city districts could only be maintained by the development of public transit systems whose fixed lines of travel increased ease of movement between connected points and reduced it between areas not on the transit lines themselves.

Accessibility therefore suggests the idea of **connectivity,** a broader concept implying all the tangible and intangible ways in which places are connected: by physical telephone lines, street and road systems, pipelines and sewers; by unrestrained walking across open countryside; by radio and TV broadcasts beamed outward uniformly from a central source. Where routes are fixed and flow is channelized, *networks*—the patterns of routes connecting sets of places—determine the efficiency of movement and the connectedness of points.

There is, inevitably, interchange between connected places. *Spatial diffusion* is the process of dispersion of an idea or an item from a center of origin to more distant points with which it is directly or indirectly connected. The rate and extent of that diffusion are affected by the distance separating the originating center of, say, a new idea or technology and other places where it is eventually adopted. Diffusion rates are also affected by population densities, means of communication, obvious advantages of the innovation, importance or prestige of the originating node, and other conditions. These ideas of diffusion are further explored in Chapter 2.

Geographers study the dynamics of spatial relationships. Movement, connection, and interaction are part of the social and economic processes that give structure to places and regions (Figure 1.12). Geography's interest in those relationships recognizes that spatial interaction is not just an awkward necessity but a fundamental organizing principle of human life on earth.

Figure 1.12

The routes of the 5 million automobile trips made each day in Chicago during the late 1950s are recorded on this light-display map. The boundaries of the region of interaction that they created are clearly marked. Those boundaries (and the dynamic region they defined) were subject to change as residential neighborhoods expanded or developed, as population relocations occurred, and as the road pattern was altered over time.

Source: Chicago Area Transportation Study, *Final Report*, vol. I, p. 44, 1959.

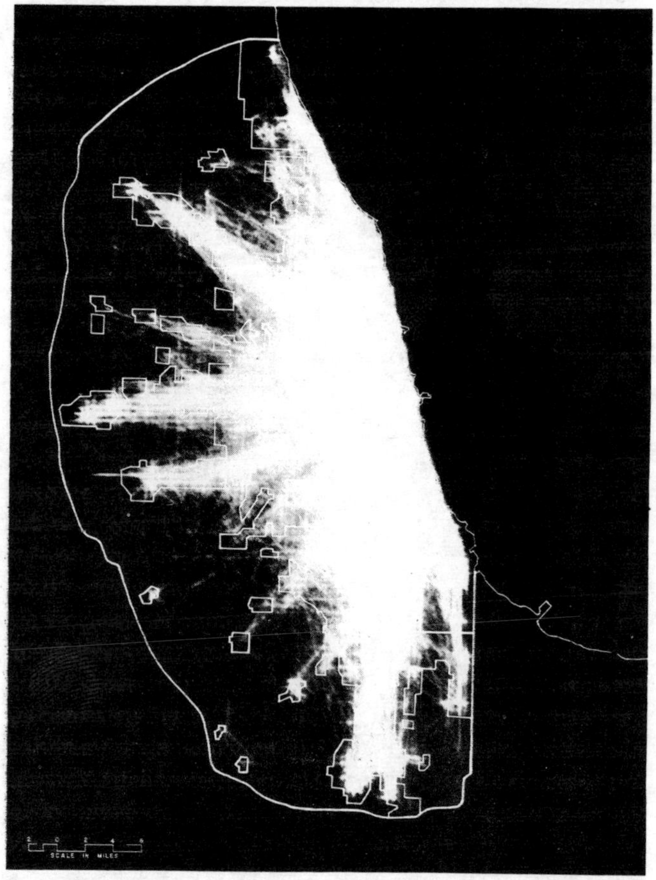

The Structured Content of Place

A starting point for geographic inquiry is how things are distributed in area. That interest distinguishes geography from other sciences, physical or social, and underlies many of the questions geographers ask: Where is a thing located? How is that location related to other items? How did the location we observe come to exist? Such questions carry the conviction that the content of area is comprehensibly arranged or structured. The arrangement of things on the earth's surface is called **spatial distribution** and may be analyzed by the elements common to all spatial distributions: *density, dispersion,* and *pattern.*

Density

The measure of the number or quantity of anything within a defined unit of area is its **density.** It is therefore not simply a count of items but of items in relation to the space in which they are found. When the relationship is absolute, as in population per square kilometer, for example, or dwelling units per acre, we are defining *arithmetic density* (see Figure 1.8). Sometimes it is more meaningful to relate item numbers to a specific kind of area. *Physiological density,* for example, is a measure of the number of people per unit area of arable land. Density defined in population terms is discussed in Chapter 4.

A density figure is a statement of fact but not necessarily one useful in itself. Densities are normally employed comparatively, relative to one another. High or low density implies a comparison with a known standard, with an average, or with a different area. Ohio, with 263 people per square mile (102 per km²) in 1987 might be thought to have a high density compared to neighboring Michigan at 162 people per square mile (63 per km²), and a low one in relation to New Jersey at 1027 (396 per km²).

Figure 1.13

Density and dispersion are independent statements about the location of things in area. *Density* is the number of items or observations within a defined area; it remains the same no matter how the items are distributed. The density of houses per square mile is the same in both (*a*) and (*b*). *Dispersion* is a statement about nearness or separation. The houses in (*a*) are more *dispersed* than those shown *clustered* in (*b*).

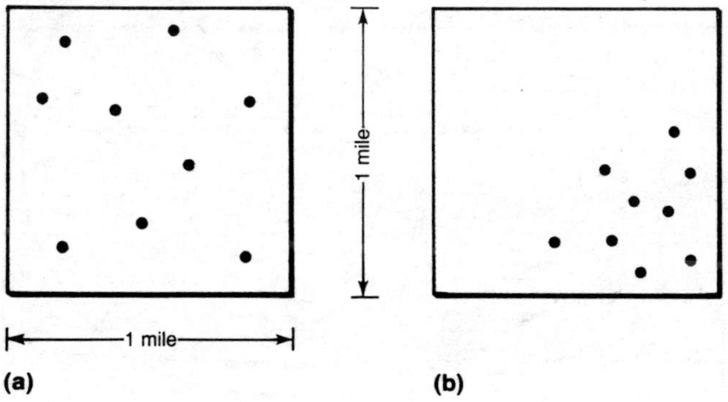

Dispersion

Dispersion (or its opposite, ***concentration***) is a statement of the amount of spread of a phenomenon over an area. It tells us how far things are spread out rather than how many or how much. If they are close together spatially, they are considered *clustered* or *agglomerated.* If they are spread out, they are *dispersed* or *scattered* (Figure 1.13).

If the residents of a metropolitan county were all located within a confined central city, we might say the population was clustered. If, however, that same population redistributed itself, with many city residents moving to the suburbs and occupying a larger portion of the county's territory, it would become more dispersed. In both cases, the *density* of population (number of people in relation to area of the county) would be the same, but the distribution would have changed. Since dispersion deals with separation of things one from another, a distribution that might be described as *clustered* (closely spaced) at one scale of reference might equally well be considered *dispersed* (widely spread) at another scale.

Pattern

The geometric arrangement of things in space is called **pattern.** Like dispersion, pattern refers to distribution, but that reference emphasizes design rather than spacing (Figure 1.14). The distribution of towns along a railroad or houses along a street may be seen as *linear.* A *centralized* pattern may involve items concentrated around a single node. A *random* pattern may be the best description of an unstructured irregular distribution.

The rectangular system of land survey adopted in much of the United States under the Ordinance of 1785 creates a checkerboard rural pattern of "sections" and "quarter sections" of farmland (see Figure 6.28). As a result, in most American cities streets display a *grid* or *rectilinear* pattern. The same is true of cities in Canada, Australia, New Zealand, and South Africa, which adopted similar geometric survey systems. The *hexagonal* pattern

Figure 1.14

Pattern describes spatial arrangement and design. The *linear* pattern of towns in (*a*) perhaps traces the route of a road or railroad or the course of a river. The central city in (*b*) with its nearby suburbs represents a *centralized* pattern, while the dots in (*c*) are *randomly* distributed.

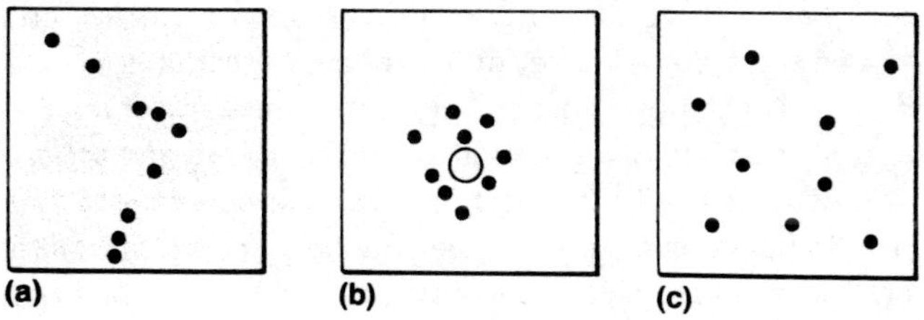

of service areas of farm towns is a mainstay of central place theory, discussed in Chapter 11. These references to the geometry of distribution patterns help us visualize and describe the structured arrangement of things in space. They help us make informed comparisons between areas and use the patterns we discern to ask further questions about the interrelationship of things.

Place Similarity and Regions

The distinctive characteristics of places in content and structure immediately suggest two geographically important ideas. The first is that no two places on the surface of the earth can be *exactly* the same. Not only do they have different absolute locations, but—as in the features of the human face—the precise mix of physical and cultural characteristics of a place is never exactly duplicated.

Since geography is a spatial science, the inevitable uniqueness of place would seem to impose impossible problems of generalizing spatial information. That this is not the case results from the second important idea: The physical and cultural features of an area and the dynamic interconnections of people and places show patterns of spatial similarity. Often the similarities are striking enough for us to conclude that spatial regularities exist. They permit us to recognize and define **regions,** earth areas that display significant elements of internal uniformity and external difference from surrounding territories. Places are therefore both unlike and like other places, creating patterns of areal differences and of coherent spatial similarity.

The problem of the historian and the geographer is similar: Each must generalize about items of study that are essentially unique. The historian creates arbitrary but meaningful and useful historical periods for reference and study. The "Roaring Twenties" and the "Victorian Era" are shorthand summary names for specific time spans, internally quite complex and varied but significantly distinct from what went before or followed after. The region is the geographer's equivalent of the historian's epoch: a device of areal generalization, an attempt to separate into recognizable component parts the otherwise overwhelming diversity and complexity of the earth's surface. In both the time and the space need for generalization, attention is focused on key unifying elements or similarities of the era or area selected for study. In both the historical and geographical cases, the names assigned to those times and places serve to identify the time span or region and to convey a complex set of interrelated attributes.

All of us have a general idea of the meaning of region, and all of us refer to regions in everyday speech and action. We visit "the old neighborhood" or "go downtown"; we plan to vacation or retire in the "Sunbelt," or we speculate on the effects of weather conditions in the "corn belt" on next year's food prices. In each instance, we have a mental image of the area mentioned, and in each we have engaged in an informal place classification to pass along complex spatial, organizational, or content ideas. We have applied the **regional concept** to bring order to the immense diversity of the earth's surface.

What we do informally as individuals, geography attempts to do formally as a discipline—define and explain regions (Figure 1.15). The purpose is clear: to make the infinitely varying world around us understandable through spatial summaries. That world is only rarely subdivided into neat, unmistakable "packages" of uniformity. Neither the environment nor human areal actions present us with a compartmentalized order, any more than the sweep of human history has predetermined "eras" or all plant specimens come labeled in nature with species names. We all must classify to understand, and the geographer classifies in regional terms.

Regions are spatial expressions of ideas or summaries useful to the analysis of the problem at hand. Although as many possible regions exist as there are physical, cultural, and organizational attributes of area, the geographer selects for study those areal variables that contribute to the understanding of a specific topic or areal problem. All other variables are disregarded as irrelevant. Regional boundaries are assumed to be marked where the region's internal unifying characteristics change so materially that different regional summaries are required.

Figure 1.15

The Middle West as seen by different professional geographers. Agreement upon the need to recognize spatial order and to define regional units does not imply unanimity in the selection of boundary criteria. All of the sources concur in the significance of the Middle West as a regional entity in the spatial structure of the United States and agree upon its core area. The sources differ, however, in their assessment of its limiting characteristics.

Sources: (*a*) John H. Garland, editor *The North American Midwest*. Copyright © 1955 John Wiley & Sons, Inc., New York, NY; (*b*) John R. Borchert and Jane McGuigan, *Geography of the New World*. Copyright © 1961 Rand McNally, Chicago, IL; and (*c*) Otis P. Starkey and J. Lewis Robinson, *The Anglo-American Realm*. Copyright © 1969 McGraw-Hill Publishing Company, New York, NY.

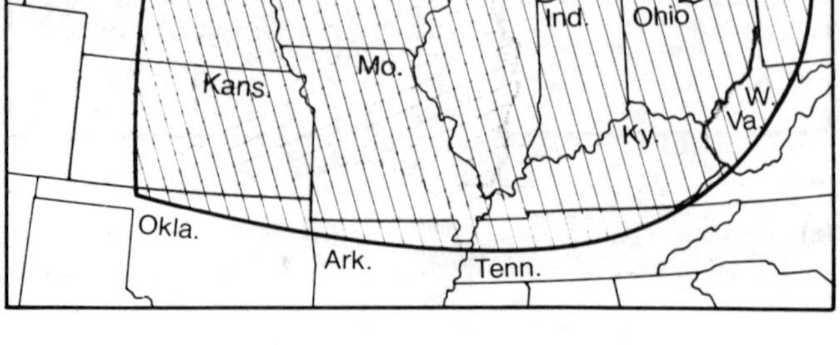

The Characteristics of Regions

The regional concept tells us that all regions share certain common characteristics related to earth space:

Regions have *location,* often expressed in the regional names selected, such as the Middle West, the Near East, North Africa, and the like. This form of regional name underscores the importance of *relative location.*

Regions have *spatial extent.* By definition, they recognize territories over which defined characteristics or generalizations about physical, cultural, or organizational content are constant.

Regions have *boundaries* based on the areal spread of the features selected for study. Since regions are the recognition of the features defining them, their boundaries are drawn where those features no longer occur or dominate. Regional boundaries are rarely as sharply defined as those suggested by Figure 1.16 or by the regional maps in this and other geography texts. More frequently, there exist broad zones of transition from one distinctive core area to another as the dominance of the defining regional features gradually diminishes outward from the core to the regional periphery. Linear boundaries are arbitrary divisions made necessary by the scale of world regional maps and by the summary character of most regional discussions.

Regions are *hierarchically arranged.* Although regions vary in scale, type, and degree of generalization, none stands alone as the ultimate key to areal understanding. Each defines a part of spatial reality (Figure 1.17) and at the same time exists as part of a larger, equally valid regional unit.

Figure 1.16

Aachen in 1649. The regional units of contemporary concern are rarely as precisely and visibly demarked as the walled medieval city. Its sprawling modern counterpart may be more difficult to define, but the boundary significance of the concept of *urban* remains.

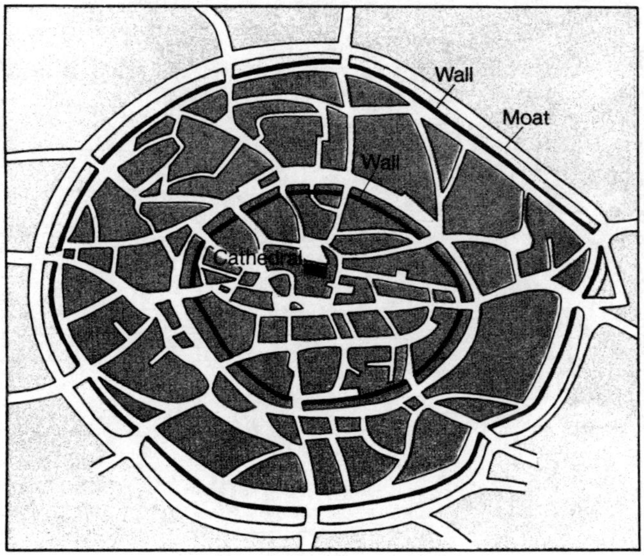

Types of Regions

Regions may be either *formal, functional,* or *perceptual.* **Formal** or **uniform regions** are nonoverlapping areas of essential uniformity in one or a limited combination of physical or cultural features. Your home state is a precisely bounded formal political region within which uniformity of law and administration is found. Later in this book we encounter formal (homogeneous) cultural regions in which standardized characteristics of language, religion, ethnicity, or economy exist. The endpaper maps of landform regions and national units show other formal regional patterns. Whatever the basis of its definition, the formal region is the largest area over which a valid generalization of attribute uniformity may be made. Whatever is stated about one part of it holds true for its remainder.

The **functional** or **nodal region,** in contrast, is a spatial system defined by the interactions and connections that give it a dynamic, organizational basis (Figure 1.18). Its boundaries remain constant only as long as the interchanges establishing it remain unaltered.

Figure 1.17

A hierarchy of regions. One possible nesting of regions within a regional hierarchy defined by differing criteria. Each regional unit has internal coherence. The recognition of each adds comprehension of the nature of its subdivisions. On a formal regional scale of size progression, the Delmarva Peninsula of the eastern United States may be seen as part of the Atlantic Coastal Plain, which is in turn a portion of the eastern North American humid continental climatic region, a hierarchy that changes the basis of regional recognition as the level and purpose of generalization alter.

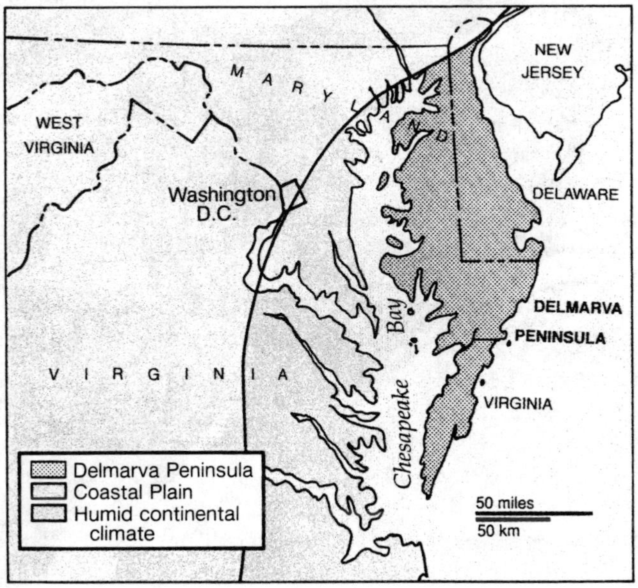

Perceptual regions are less rigorously structured than the formal and functional regions geographers devise. They reflect feelings and images rather than objective data, and because of that they may be more meaningful in the lives and actions of those who recognize them than are the more abstract regions of geographers.

Ordinary people have a clear idea of spatial variation and employ the regional concept to distinguish between territorial entities. People individually and collectively agree on where they live. They occupy a type of perceptual region called *vernacular regions,* areas that have reality in their minds and that are reflected in regionally based names used in businesses, by sports teams, and in advertising slogans. The frequency of references to "Dixie" in the southeastern United States represents that kind of regional consensus and awareness. Such vernacular regions reflect the way people view space, assign their loyalties, and interpret their world. At a different scale, such urban ethnic enclaves (see Chapter 6) as "Little Italy" or "Chinatown" have comparable regional identity in the minds of their inhabitants. Less clearly perceived by outsiders but unmistakable to their inhabitants are the "turfs" of urban clubs or gangs. Their boundaries are sharp, and the perceived distinctions between them are paramount in the daily lives and activities of their occupants.

Maps

Maps are tools we use to identify regions and to analyze their content. The spatial distributions, patterns, and relations of interest to geographers usually cannot easily be observed or interpreted in the landscape itself. Many, such as landform or agricultural regions or major cities, are so extensive spatially that they cannot be seen or studied in their totality from one or a few vantage points. Many, such as regions of language usage or religious belief, are spatial phenomena but are not tangible or visible. Many interactions, flows, and exchanges imparting the dynamic quality

Figure 1.18

***Functional regions* based on correspondent banking linkages** between large central cities and smaller towns. The regions marked on this map suggest one form of *connectivity* between principal cities and locales beyond their own immediate metropolitan area.

Source: Reproduced by permission from the *Annals of the Association of American Geographers*, John R. Borchert, Volume 62, page 358, Association of American Geographers, 1972.

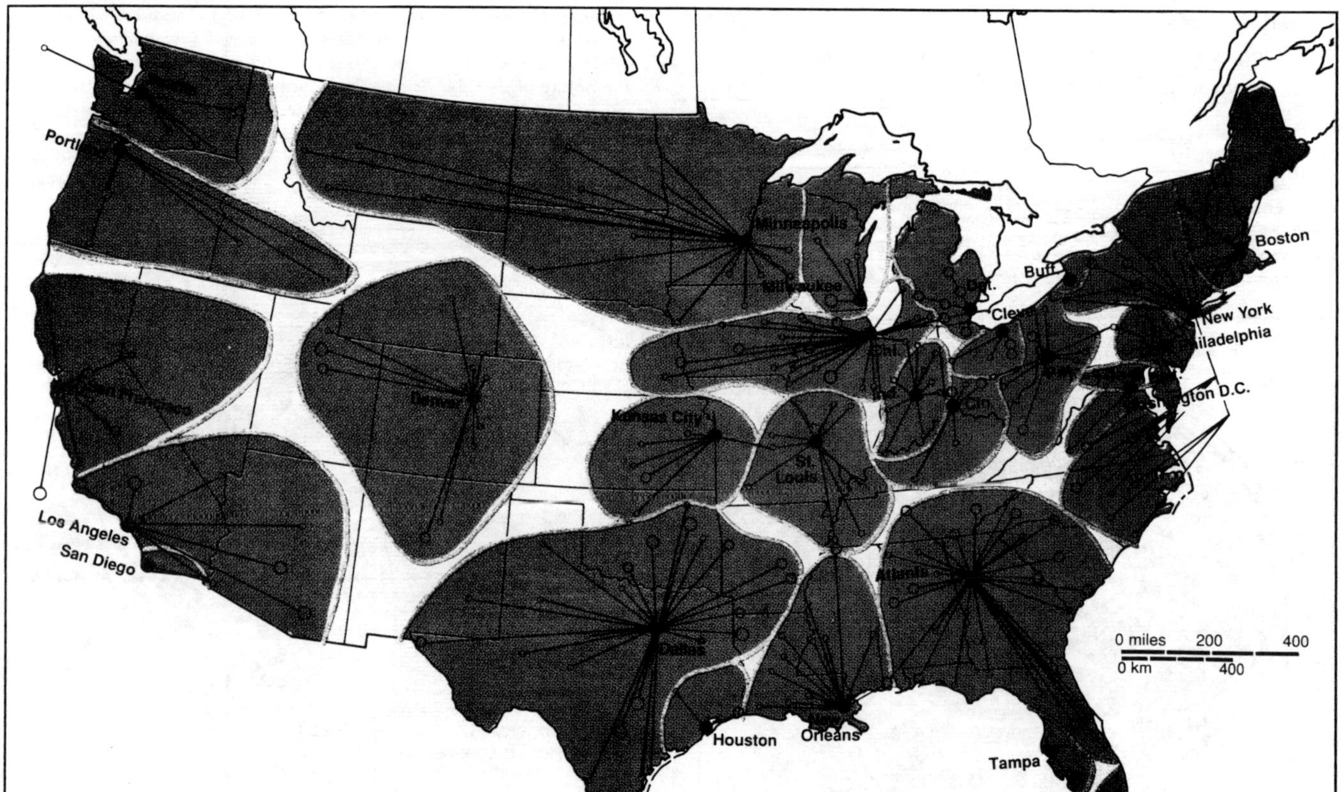

to spatial interaction may not be directly observable at all. And even if all things of geographic interest could be seen and measured through field examination, the infinite variety of tangible and intangible content of area would make it nearly impossible to isolate for study and interpretation the few matters of regional interest selected for special investigation.

Therefore, the map has become the essential and distinctive tool of geographers. Only through the map can spatial distributions and interactions of whatever nature be reduced to an observable scale, isolated for individual study, and combined or recombined to reveal relationships not directly measurable in the landscape itself. But maps can serve their purpose only if their users have a clear idea of their strengths, limitations, and diversity and of the conventions observed in their preparation and interpretation.

Map Scale

We have already seen (page 9) that scale is a vital element of every map. Because it is a much reduced version of the reality it summarizes, a map generalizes the data it displays. *Scale*—the relationship between size or length of a feature on the map and the same item on the earth's surface—determines the amount of that generalization (Figure 1.19). The smaller the scale of the map, the larger is the area it covers and the more generalized are the data it portrays. The larger the scale, the smaller is the depicted area and the more accurately can its content be represented. It may seem backward, but large-scale maps show small areas, and small-scale maps show large areas.

Map scale is selected according to the amount of generalization of data that is acceptable and the size of area that must be depicted. The user must consider map

Figure 1.19

The effect of scale on area and detail. The larger the scale, the greater the number and kinds of features that can be included. Scale may be reported to the map user in one (or more) of three ways. A *verbal* scale is given in words ("1 inch to 1 mile"). A *representative fraction* (such as that placed at the left, below each of the four maps shown here) is a statement of how many linear units on the earth's surface are represented by one unit on the map. A *graphic* scale (such as that placed at the right, below each of these maps) is a line or bar marked off in map units but labeled in ground units.

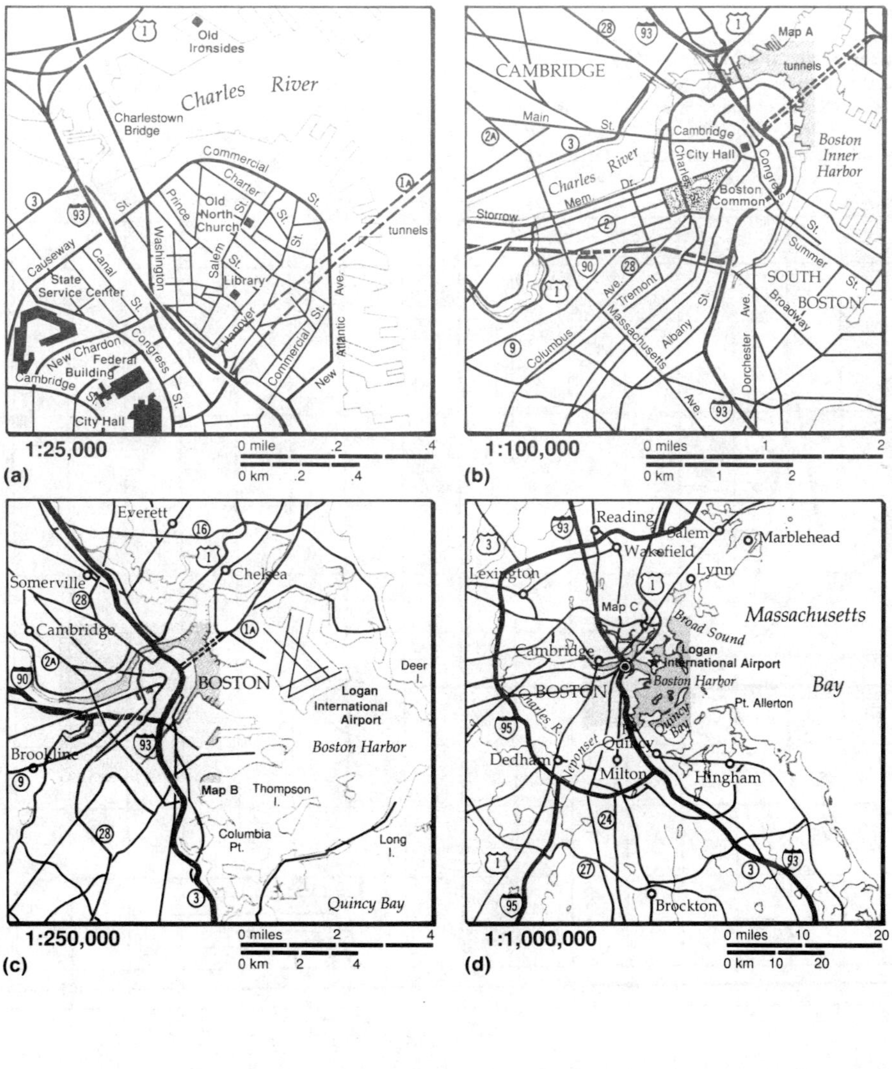

scale in evaluating the reliability of the spatial data that is presented. Regional boundary lines drawn on the world maps in this and other textbooks and atlases would cover many kilometers or miles on the earth's surface. They obviously distort the reality they are meant to define, and on small-scale maps major distortion is inevitable.

Projection

A general rule of thumb is that the larger the earth area shown on a map, the greater is the distortion built into the map. This is so because a map has to depict the curved surface of the three-dimensional earth on a two-dimensional sheet of paper. The term **projection** designates the method chosen to represent the earth's curved surface as a flat map. Because absolutely accurate representation is impossible, all projections inevitably distort. Specific projections may be selected, however, to minimize the distortion of at least one of the four main map properties: area, shape, distance, and direction.

Equal area or *equivalent* projections show the areas of regions in correct or constant proportion to earth reality, but the shape of the portrayed area is inevitably distorted. A square on the earth, for example, may become a rectangle on the map, but that rectangle has the correct area. A *conformal* map, in contrast, shows true shapes for *small* areas (no projection can provide correct shapes for large regions) but distorts areal relationships (Figure 1.20). No map can be both equivalent and conformal.

Distance relationships are nearly always distorted on a map, but some projections do maintain true distances in one direction or along certain selected lines. Others, called *equidistant* projections, show true distance in all directions, but only from one or two central points. As is true of distances, directional relationships are inevitably distorted on almost all maps.

The Globe Grid

All spatial analysis starts with locations, and all locations are related to the global grid of latitude and longitude. Since these lines of reference are drawn on the spherical earth, their projection onto a map distorts their grid relationships. The extent of variance between the globe grid and a map grid helps tell us the kind and degree of distortion the map will contain.

Figure 1.20

Distortion on the Mercator projection. A perfect five-pointed star was drawn on a globe, and the latitude and longitude of the points of the star were transferred to the Mercator map shown here. The manner in which the star is distorted reflects the way the projection distorts land areas, greatly enlarging the apparent size of territories in upper latitudes. Whole generations of school children—exposed for years to this projection created for mariners, not makers of wall maps—have grown to maturity convinced that the state of Alaska is nearly the size of all the lower 48 states put together. Mercator maps should be accompanied, as here, by a scale diagram showing the linear values of degrees of longitude at different latitudes.

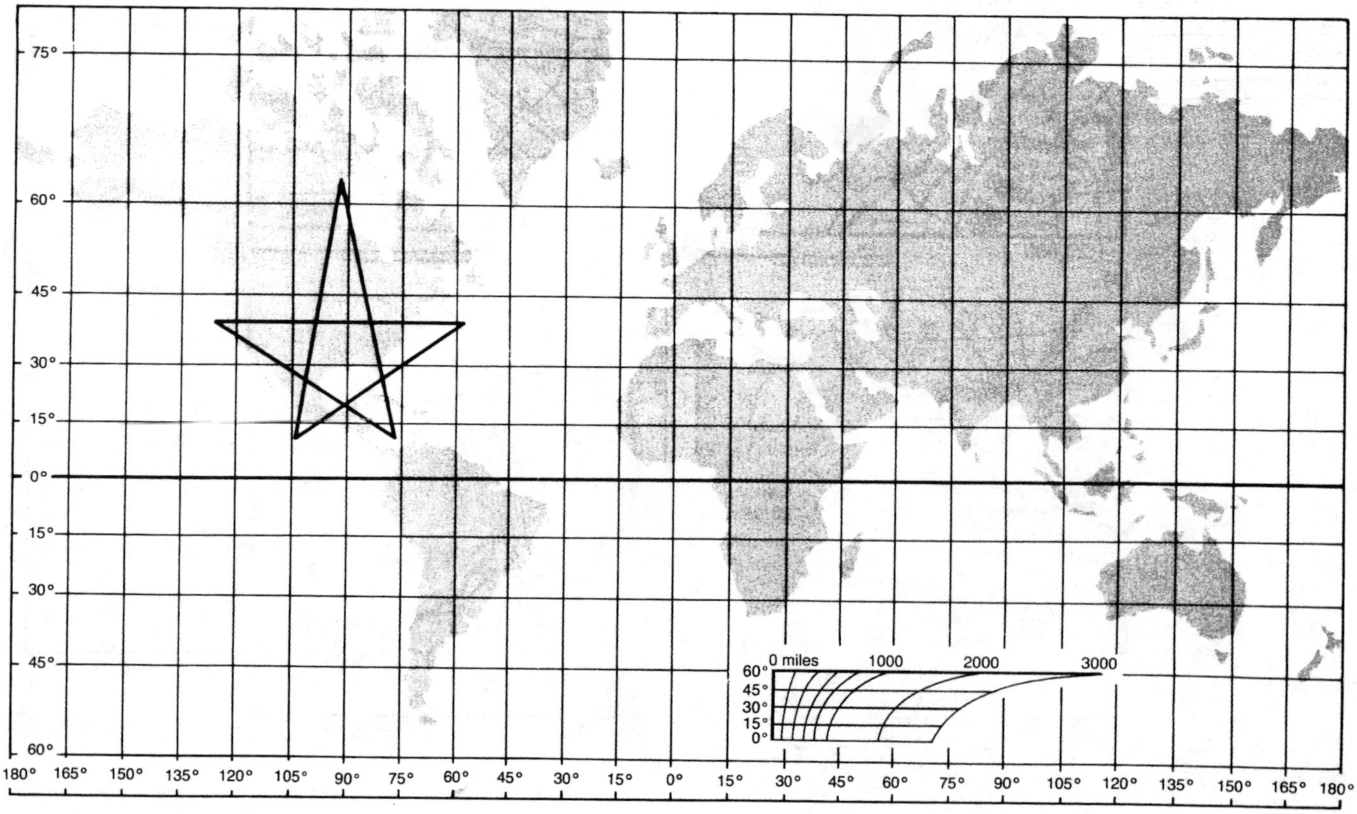

The key reference points in the *grid system* are the North and South poles and the equator, which are given in nature, and the *prime meridian,* which is agreed upon by cartographers. Because a circle contains 360 degrees, the distance between the poles is 180 degrees and between the equator and each pole, 90 degrees (Figure 1.21). *Latitude* measures distance north and south of the equator (0°), and *parallels* of latitude run due east-west. *Longitude* is the angular distance east or west of the prime meridian and is depicted by north-south lines (called *meridians*) that converge at the poles. The properties of the globe grid the mapmaker tries to retain and the map user should look for are as follows:

1. All meridians are of equal length; each is one-half the length of the equator.
2. All meridians converge at the poles and are true north-south lines.
3. All lines of latitude (parallels) are parallel to the equator and to each other.
4. Parallels decrease in length as one nears the poles.
5. Meridians and parallels intersect at right angles.
6. The scale on the surface of the globe is everywhere the same in every direction.

Only the globe grid itself retains all these characteristics. To project it onto a surface that can be laid flat is to distort some or all of these properties and consequently to distort the reality the map attempts to portray.

How Maps Show Data

The properties of the globe grid and of alternate projections are the concern of the cartographer. The geographer is more interested in the depiction of spatial data and in the analysis of the patterns and interrelationships those data present. Out of the myriad items comprising the content of an area, the geographer must, first, select those that are of concern to the problem at hand and, second, decide on how best to display them for study or demonstration.

A *thematic map* is the general term applied to a map of any scale that presents a specific spatial distribution or a single category of data. The way the information is shown on such a map may vary according to the type of information to be conveyed and the level of generalization desired. A *statistical map* records the actual numbers or occurrences of the mapped item per established unit area or location (Figure 1.22). The actual count of each state's

Figure 1.21

The grid system of parallels of latitude and meridians of longitude. Since the meridians converge at the poles, parallels become increasingly shorter away from the equator. On the globe, the 60th parallel is only one-half as long as the equator, and a degree of longitude along it measures only about 35 miles compared to about 70 miles at the equator (0°).

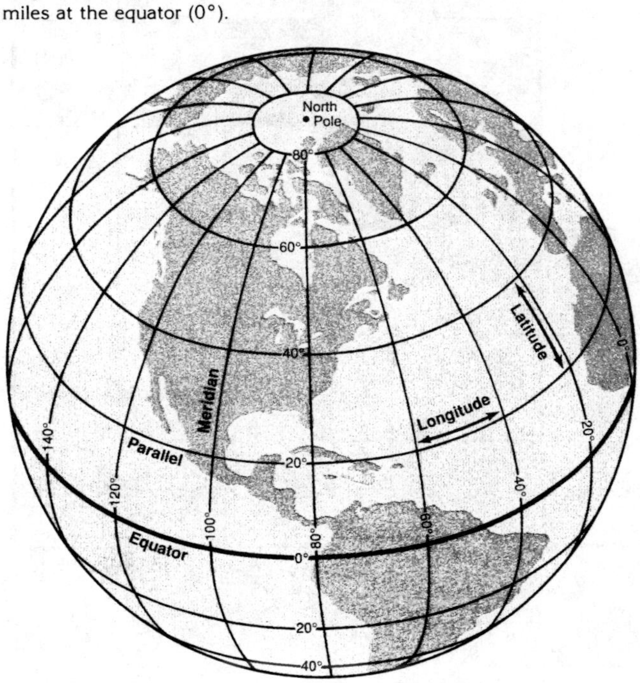

Figure 1.22

The starting point of quantitative thematic maps is the location and magnitude of things or events. Here observations are assigned to their areas of occurrence. The spatial arrangement of events is recorded, but their pattern of occurrence is difficult to see.

Source: McCarty/Lindberg, *A Preface to Economic Geography,* © 1966, pp. 31, 33, 34, 35. Reprinted by permission of Prentice-Hall, Inc., Englewood Cliffs, NJ.

4	9	13	11	9	4
8	13	15	17	12	10
6	19	18	12	7	3
9	15	12	10	8	3
8	9	10	8	5	2
5	6	7	6	1	1

Figure 1.23

A cartogram with states drawn in proportion to news coverage of the 1984 presidential nomination contests.

Source: William C. Adams, "As New Hampshire Goes . . ." in *Media and Momentum: The New Hampshire Primary and Nomination Politics,* ed. Gary R. Orren and Nelson W. Polsby (Chatham, N.J.: Chatham House, 1987), 43.

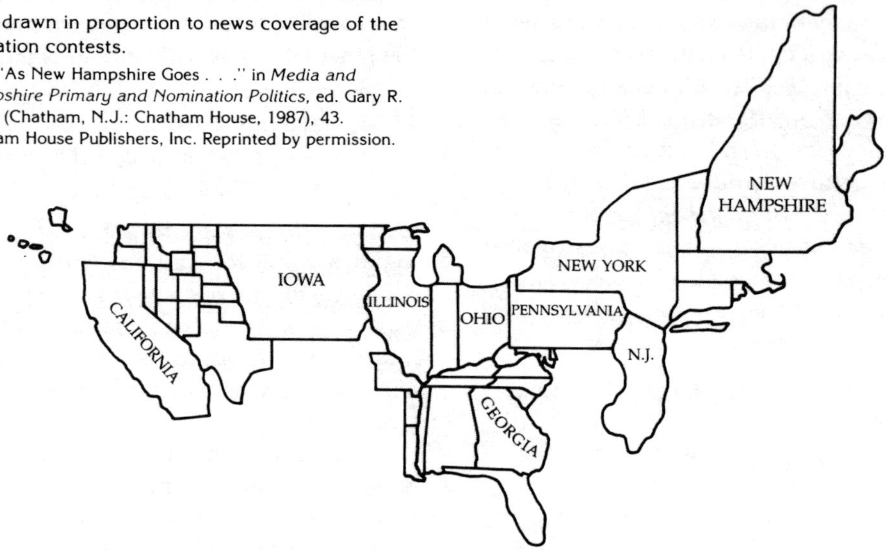

Figure 1.24

A dot map example. In this dot map the data of Figure 1.22 are represented by dots placed randomly within the areal statistical units.

Source: McCarty/Lindberg, *A Preface to Economic Geography,* © 1966, pp. 31, 33, 34, 35. Reprinted by permission of Prentice-Hall, Inc., Englewood Cliffs, NJ.

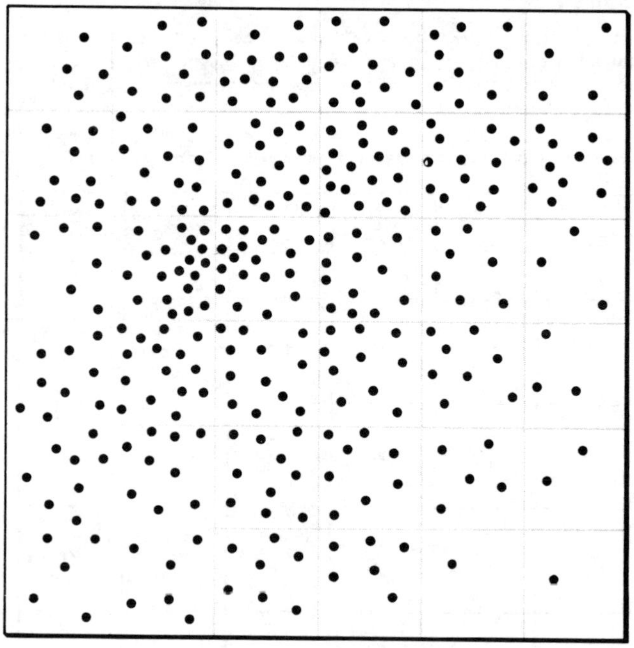

Figure 1.25

A choropleth map example. This choropleth map is derived from Figure 1.22. Data are assigned to class intervals, and colors are used to distinguish magnitude of occurrences. The selection of data intervals and choice of colors can alter the message and impact of the map.

Source: McCarty/Lindberg, *A Preface to Economic Geography,* © 1966, pp. 31, 33, 34, 35. Reprinted by permission of Prentice-Hall, Inc., Englewood Cliffs, NJ.

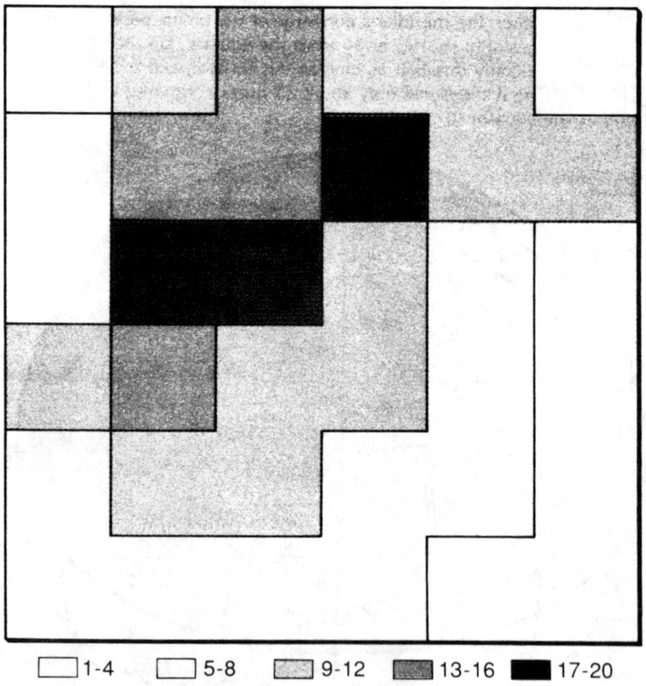

colleges and universities shown on an outline map of the United States and the number of traffic accidents at each street intersection within a city are examples of statistical maps. A *cartogram* uses such statistical data to transform territorial space so that the largest areal unit on the map is the one showing the greatest statistical value (Figure 1.23).

A *dot map* gives a different view of data by representing quantities or occurrences by a dot located on the map in the approximate location of the occurrence or, perhaps, uniformly distributed within the unit area of occurrence (Figure 1.24). The numerical value of a single dot is determined by the compiler and recorded in the map legend. The dot map serves not only to record data but to suggest their spatial pattern, distribution, and dispersion. A *choropleth map* presents average value of the data studied per unit area—dwelling unit values by city block, for example, or (in th population densities by individual townships ties. Each unit area on the map is then shade to suggest the magnitude of the event or item fou its borders (Figure 1.25).

An *isoline map* features lines (*isolines*) that conn points registering equal values of the item mapped (*iso* means "equal"). The *isotherms* shown on a daily weather map connect points recording the same temperature at the same moment of time or the same average temperature during the day. Identical elevations above sea level may be shown by a form of isoline called a *contour line*. Were you to walk along a mapped contour line you would never go up or down hill but remain always on the level, eventually returning to your starting point (Figure 1.26).

Figure 1.26

A portion of the Santa Barbara, California, topographic quadrangle of the U.S. Geological Survey 1:24,000 series. Where contour lines are closely spaced in the upper-right quadrant, they indicate a rapid change in elevation—that is, a steep slope. Widely spaced contours tell us the slope (elevation change) is gradual. The practiced user of topographic maps can visualize terrain with the same ease as if viewing a three-dimensional model. Color and symbols on topographic maps also tell us much about the landscape. Here the pink tint indicates built-up areas, in which only landmark buildings are shown. The light purple tint is used to indicate extensions of urban areas. The scale of the original map no longer applies to this photographic reduction.

Source: U.S. Geological Survey.

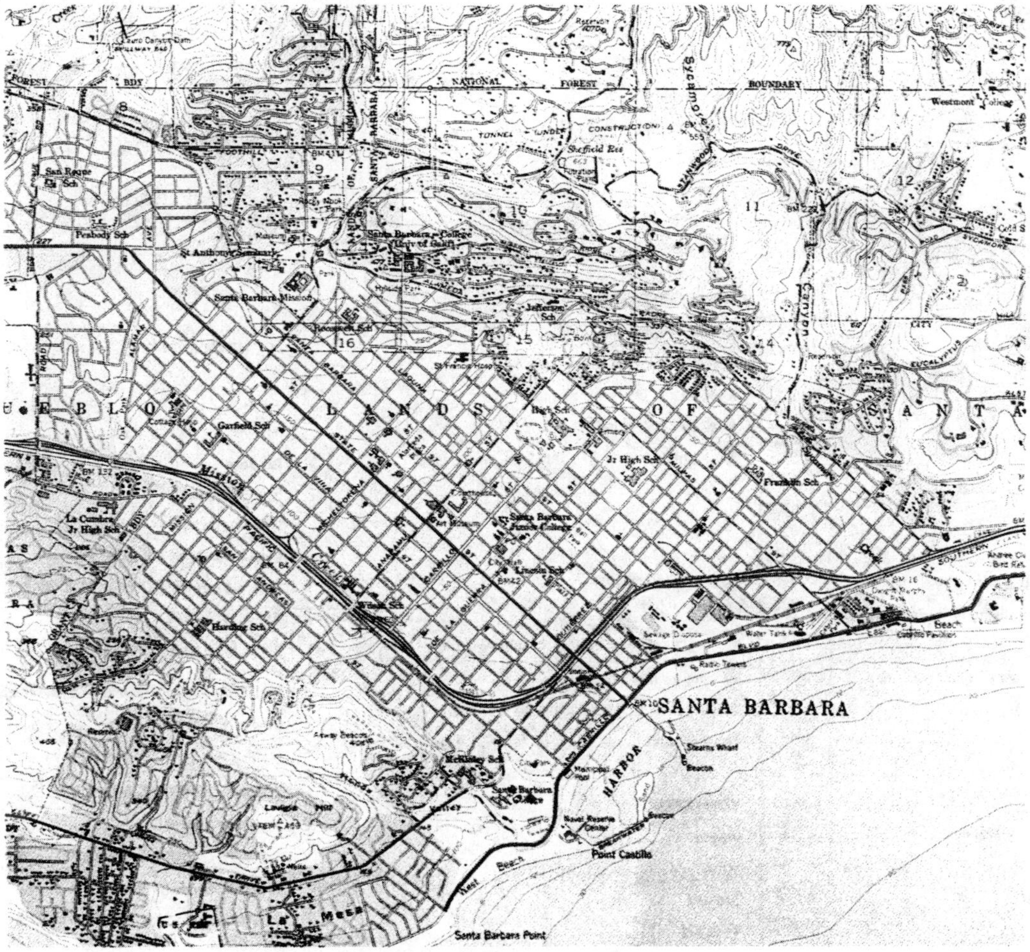

Figure 1.27

Standard U.S. Geological Survey topographic map symbols.

Source: U.S. Geological Survey.

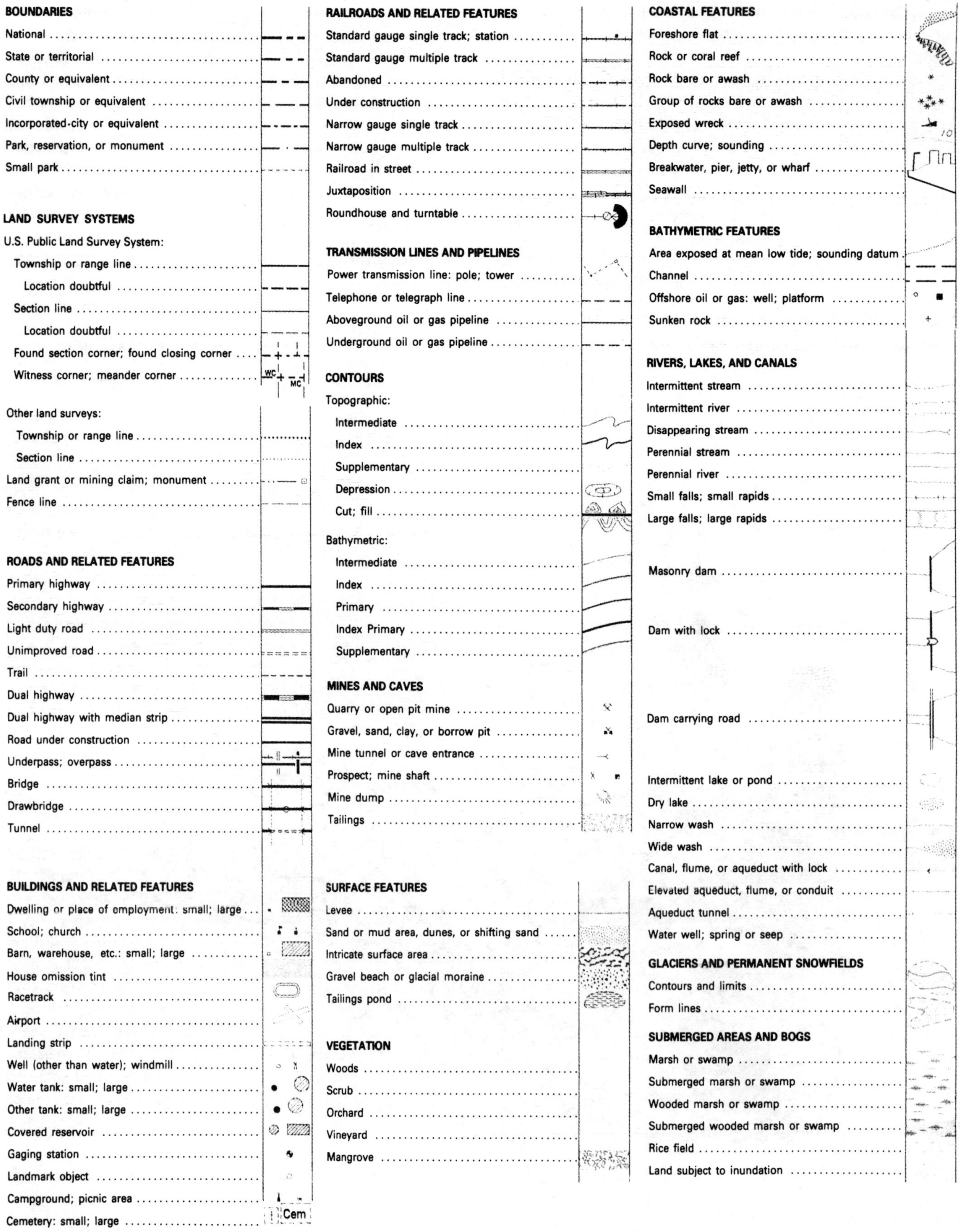

U.S. Geological Survey topographic maps show a great deal more than topography and contour lines, as Figures 1.26 and 1.27 document.

On many maps, isolines are drawn to connect points showing identical calculated values, such as percentage of total possible sunlight received at a location. On other maps, the calculation refers not to a point but to an areal statistic—for example, people per square kilometer or average percentage of cropland in corn—and the isoline connects average values for unit areas. For emphasis, the area enclosed by isolines may be shaded to indicate approximately uniform occurrence of the thing mapped, and the isoline itself may be treated as the boundary of a uniform region (Figure 1.28).

Mental Maps

Maps that shape our understanding of distributions and locations or influence our perceptions of the world around us are not always drawn on paper. We carry about with us *mental maps* that in some ways are more accurate in reflecting our view of spatial reality than the formal maps created by geographers or cartographers. **Mental maps** are images about an area or an environment developed by an individual on the basis of information or impressions received, interpreted, and stored. We use this information—this mental map—in organizing our daily activities: selecting our destinations and the sequence in which we will visit them, deciding on our routes of travel, recognizing where we are in relation to where we wish to be. Such maps are every bit as real to their creators (and we all have them) as are the street maps or highway maps commercially available, and a great deal more immediate in their impact on our spatial decisions. We may choose routes or avoid neighborhoods not on objective grounds but on emotional or perceptual ones. Whole sections of a community may be voids on our mental maps, as unknown as the interiors of Africa or South America were two centuries ago to western Europeans. Our areas of awareness generally increase with the increasing mobility that comes with age (Figure 1.29), affluence, and education and may be enlarged or restricted for different social groups within the city (Figure 1.30).

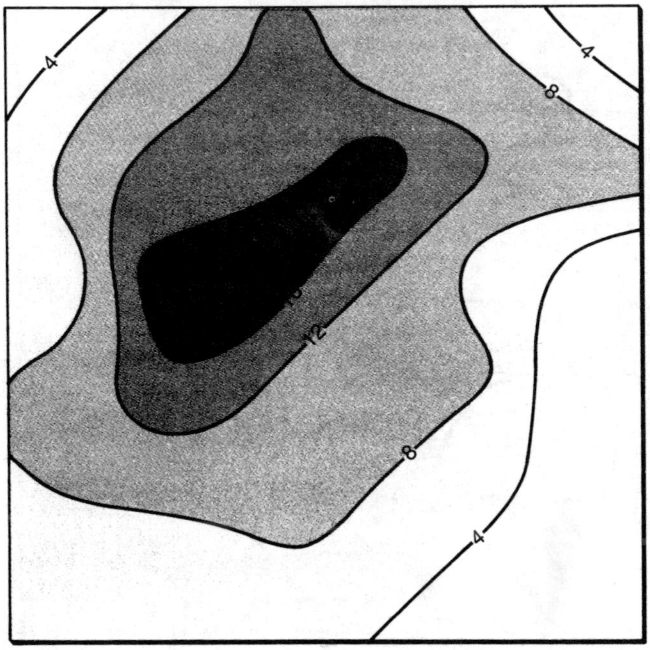

Figure 1.28

An isoline generalization of areal data. The basic assumptions of such a map are (1) that events occur at the central point of the statistical unit they represent and (2) events occur continuously across the map in smooth gradients. The isolines are drawn by interpolating values between actual numerical amounts. The shadings applied to areas between the isolines on this map are those of the Figure 1.25 class intervals.

Source: McCarty/Lindberg, *A Preface to Economic Geography*, © 1966, pp. 31, 33, 34, 35. Reprinted by permission of Prentice-Hall, Inc., Englewood Cliffs, NJ.

Systems, Maps, and Models

The content of area is interrelated and constitutes a **spatial system** that, in common with all systems, functions as a unit because its component parts are interdependent. Only rarely do individual elements of area operate in isolation, and to treat them as if they did is to lose touch with spatial reality. The systems of geographic concern are those in which the functionally important variables are spatial: location, distance, direction, density, and the other basic concepts we have reviewed. The systems that they define are not the same as regions, though spatial systems may be the basis for regional identification.

Systems have components, and the analysis of the role of components helps reveal the operation of the system as a whole. To conduct that analysis, individual system elements must be isolated for separate identification and, perhaps, manipulated to see their function within the structure of the system or subsystem. Maps and models are the devices geographers use to achieve that isolation and separate study.

Maps, as we have seen, are effective to the degree that they can segregate at an appropriate level of generalization those system elements selected for examination. By compressing, simplifying, and abstracting reality, maps record in manageable dimension the real-world conditions of interest. A **model** is a simplified abstraction of reality, structured to clarify causal relationships. Maps are a kind of model. They represent reality in an idealized form so that certain of its properties may be more clearly seen. They are a special form of model, of course, for their abstractions are rendered visually and at a reduced scale so that they may be displayed, for example, on the pages of this book.

Figure 1.29

Three young people, aged 6, 10, and 13, who lived in the same house, were asked to draw maps of their neighborhood. They received no further instructions. Notice how perspectives broaden and neighborhoods expand with age. (*a*) For the 6-year-old, the "neighborhood" consists of the houses on either side of her own. (*b*) The square block on which she lives is the neighborhood for the 10-year-old. (*c*) The wider horizons of the 13-year-old are reflected in this drawing. The square block that the 10-year-old drew is shaded in this sketch.

(a)
(b)
(c)

Figure 1.30

Four mental maps of Los Angeles. The upper-middle-income residents of Northridge and Westwood have expansive views of the metropolis, reflecting their mobility and area of travel. Residents of Boyle Heights and Avalon, both minority districts, have a much more

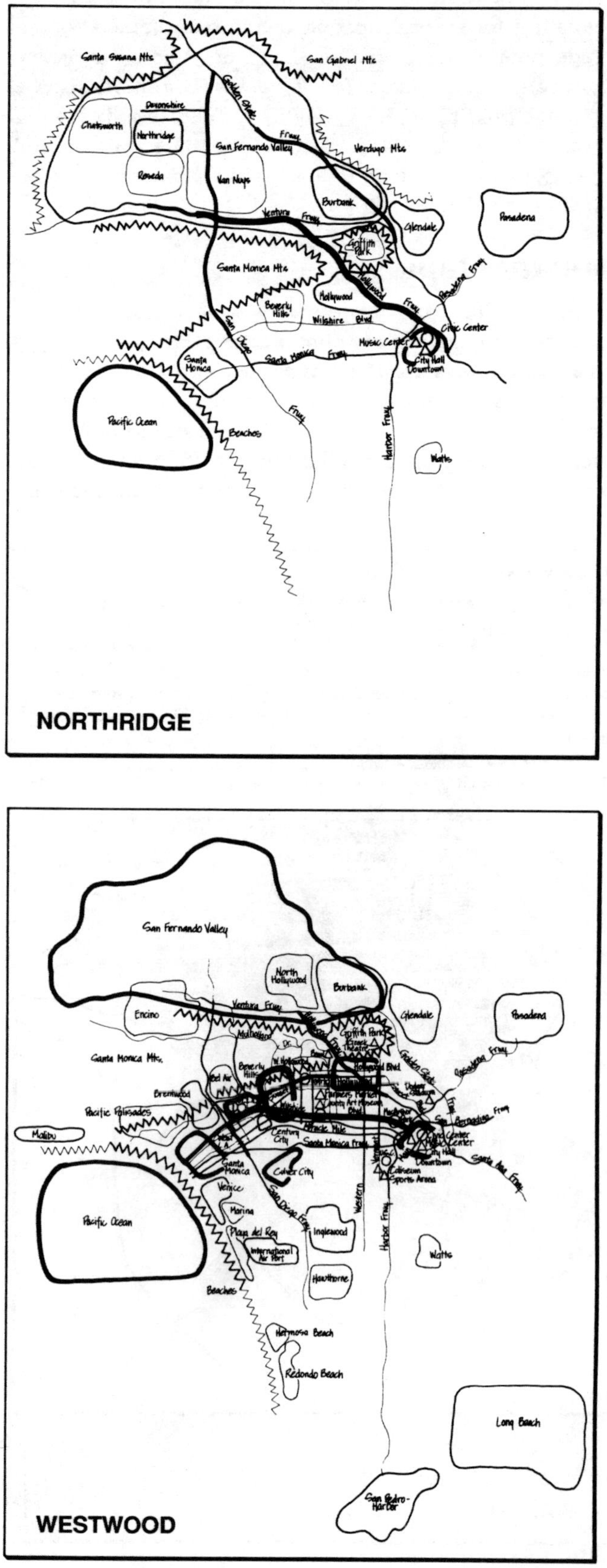

restricted and incomplete mental image of the city. Their limited mental maps reflect and reinforce their spatial isolation within the metropolitan area.

Source: From Los Angeles City Planning Department, *The Visual Environment of Los Angeles*, 1971. Reprinted by permission.

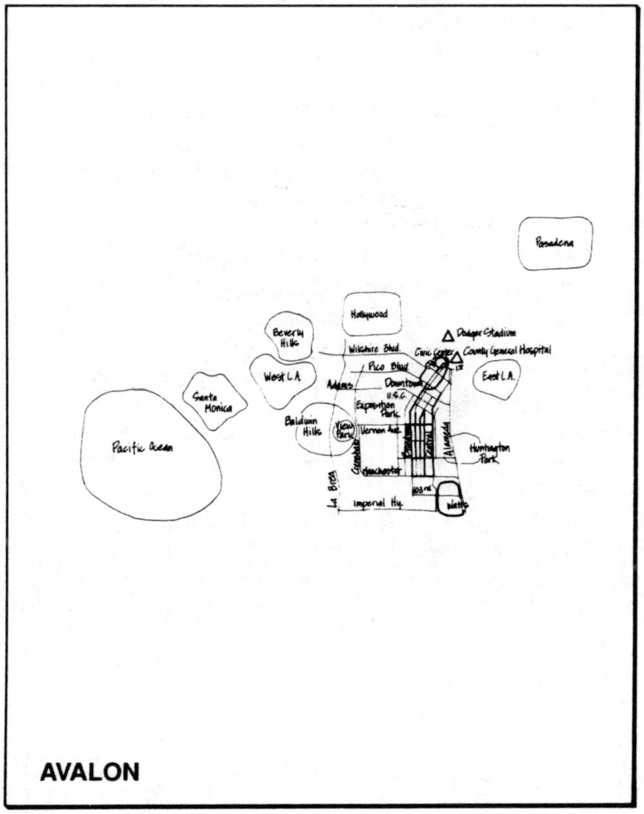

The complexities of spatial systems analysis—and the opportunities for quantitative analysis of systems made possible by computers and sophisticated statistical techniques—have led geographers to use other kinds of models in their work. Model building is the technique social scientists employ to simplify complex situations, to eliminate (as does the map) unimportant details, and to isolate for special study and analysis the role of one or more interacting elements in a total system.

An interaction model discussed in Chapter 3, for instance, suggests that the amount of exchange expected between two places depends on the distance separating them and on their population size. The model indicates that the larger the places and the closer their distance, the greater is the amount of interaction. Such a model helps us to isolate the important components of the spatial system, to manipulate them separately, and to reach conclusions concerning their relative importance. When a model satisfactorily predicts the volume of intercity interaction in the majority of cases, the lack of agreement between what is observed and what is expected in a particular case leads to an examination of the circumstances contributing to the disparity. The quality of connecting roads, political barriers, or other variables may affect the specific places examined, and these causative elements may be isolated for further study.

Indeed, the steady pursuit of more refined and definitive analysis of human geographic questions—the "further study" that continues to add to our understanding of how people occupy and utilize the earth, interact with each other, and organize and alter earth space—has led to the remarkably diversified yet coherent field of modern human geography. With the content of this introductory chapter as background to the nature, traditions, and tools of geography, we are ready to begin its exploration.

Summary

Geography is about earth space and its physical and cultural content. Throughout its long history, geography has remained consistent in its focus on human-environmental interactions, the interrelatedness of places, and the similarities and differences in physical and cultural content of area that exist from place to place. The collective interests of geographers are summarized by the spatial and systems analytical questions they ask. The responses to those questions are interpreted through basic concepts of location, distance, direction, content evolution, spatial interaction, and regional organization.

Geographers employ maps and models to abstract the complex reality of space and to isolate its components for separate study. Maps are imperfect renderings of the three-dimensional earth and its parts on a two-dimensional surface. In that rendering, some or all of the characteristics of the globe grid are distorted, but convenience and data manageability are gained. Spatial information may be depicted visually in a number of ways, each designed to simplify and clarify the infinite complexity of spatial content. Geographers also use verbal and mathematical models for the same purpose, to abstract and analyze.

In their pursuit of the study of the earth's surface as the occupied and altered space within which humans operate, geographers may concentrate on the integration of physical and cultural phenomena in a specific earth area (regional geography). They may, instead, emphasize systematic geography through study of the earth's physical systems of spatial and human concern or, as here, devote primary attention to people. This is a text in *human geography.* Its focus is on human interactions, both with the physical environments people occupy and alter and with the cultural environments they have created. We are concerned with the ways people perceive the landscapes and regions they occupy, act within and between them, make choices about them, and organize them according to the varying cultural, political, and economic interests of human societies. This is a text clearly within the social sciences, but like all geography, its background is the physical earth as the home of humans. As a human geography, its concern is with how that home has been altered by societies and cultures. Culture is the starting point, and we begin with an inquiry about the roots and nature of culture in the next chapter.

Key Words

absolute location 6
accessibility 12
connectivity 12
cultural landscape 10
formal region 15
functional region 15
mental map 23
model 23
natural landscape 9
perceptual region 16
projection 18
region 14
relative location 7
scale 9
site 7
situation 7
spatial interaction 12

For Review

1. In what two meanings and for what different purposes do we refer to *location?*
2. Describe the *site* and the *situation* of the town where you live, work, or go to school.
3. What kinds of distance transformations are suggested by the term *relative distance?* How is the concept of *psychological distance* related to relative distance?
4. What are the common elements of *spatial distribution?* What different aspects of the spatial arrangement of things do they address?
5. What are the common characteristics of *regions?* How are *formal* and *functional* regions different in concept and definition? What is a *perceptual region?*
6. List at least four properties of the globe grid. What do we mean by *conformal?* By *equivalent?* By *equidistant?*
7. What does *prime meridian* mean? What happens to the length of a degree of longitude as one approaches the poles?
8. What different ways of displaying statistical data on maps can you name and describe?

Suggested Readings

American Cartographic Association. Committee on Map Projections. *Choosing a World Map: Attributes, Distortions, Classes, Aspects.* Special Publication No. 2. Falls Church, Va. American Congress on Surveying and Mapping, 1988.

American Cartographic Association. Committee on Map Projections. *Which Map Is Best? Projections for World Maps.* Special Publication No. 1. Falls Church, Va. American Congress on Surveying and Mapping, 1986.

Broek, J. O. M. *Geography: Its Scope and Spirit.* Columbus, Ohio: Charles E. Merrill, 1965.

Buttimer, Anne, and David Seamon, eds. *The Human Experience of Space and Place.* New York: St. Martin's Press, 1980.

Dent, Borden D. *Principles of Thematic Map Design.* Reading, Mass.: Addison-Wesley, 1985.

Dickinson, Robert E., and Osbert J. R. Howarth. *The Making of Geography.* Oxford: Clarendon Press, 1933.

Downs, Roger M., and David Stea. *Maps in Minds: Reflections on Cognitive Mapping.* New York: Harper & Row, 1977.

Freeman, T. W. *A Hundred Years of Geography.* Chicago: Aldine, 1961.

Gold, John R. *An Introduction to Behavioral Geography.* New York: Oxford University Press, 1980.

Goodall, Brian. *The Facts on File Dictionary of Human Geography.* New York: Facts on File Publications, 1987.

Gould, Peter, and Rodney White. *Mental Maps.* 2d ed. Winchester, Mass.: Allen & Unwin, 1986.

Gritzner, Charles F., Jr. "The Scope of Cultural Geography." *Journal of Geography* 65 (1966):4–11.

Holt-Jensen, Arild. *Geography: Its History and Concepts.* Totowa, N.J.: Barnes & Noble Books, 1980.

James, Preston E., and Geoffrey J. Martin. *All Possible Worlds: A History of Geographical Ideas.* 2d ed. New York: John Wiley & Sons, 1981.

Johnston, R. J. *Philosophy and Human Geography.* 2d ed. London: Methuen, 1986.

Johnston, R. J., Derek Gregory, and David M. Smith. *The Dictionary of Human Geography.* 2d ed. Oxford: Basil Blackwell, 1986.

Lanegran, David A., and Risa Palm. *An Invitation to Geography.* 2d ed. New York: McGraw-Hill, 1978.

Larkin, Robert P., and Gary L. Peters. *Dictionary of Concepts in Human Geography*. Westport, Conn.: Greenwood Press, 1983.

Ley, David. "Cultural/Humanistic Geography." *Progress in Human Geography* 5 (1981):249–257; 7 (1983):267–275.

Ley, David, and Marwyn S. Samuels, eds. *Humanistic Geography: Prospects and Problems*. Chicago: Maaroufa Press, 1978.

Lynch, Kevin. *The Image of the City*. Cambridge, Mass.: M.I.T. Press, 1960.

McDonald, James R. "The Region: Its Conception, Design, and Limitations." *Annals of the Association of American Geographers* 56 (1966):516–528.

McDonald, James R. *A Geography of Regions*. Dubuque, Iowa: Wm. C. Brown Publishers, 1972.

Minshull, Roger. *Regional Geography: Theory and Practice*. Chicago: Aldine, 1967.

Morrill, Richard L. *The Spatial Organization of Society*. 2d ed. North Scituate, Mass.: Duxbury Press, 1974.

Muehrcke, Phillip C. *Thematic Cartography*. Association of American Geographers, Commission on College Geography. *Resource Paper* No. 19. Washington, D.C.: Association of American Geographers, 1972.

Muehrcke, Phillip C. *Map Use: Reading, Analysis, and Interpretation*. 2d ed. Madison, Wis.: J. P. Publications, 1986.

Murphey, Rhoads. *The Scope of Geography*. 3d ed. New York: Methuen, 1982.

Robinson, Arthur H., Randall D. Sale, Joel L. Morrison, and Phillip C. Muehrcke. *Elements of Cartography*. 5th ed. New York: John Wiley & Sons, 1984.

Rowntree, Lester. "Cultural/Humanistic Geography." *Progress in Human Geography* 10 (1986):580–586.

Thomson, J(ames) Oliver. *History of Ancient Geography*. New York: Biblo and Tannen, 1965.

Tuan, Yi-fu. "Humanistic Geography." *Annals of the Association of American Geographers* 66 (1976):266–276.

Warntz, William, and Peter Wolff. *Breakthroughs in Geography*. New York: New American Library, 1971.

Whittlesey, Derwent. "The Regional Concept and the Regional Method." Chap. II in *American Geography: Inventory and Prospect,* edited by Preston E. James and Clarence F. Jones. Syracuse, N.Y.: Syracuse University Press, 1954.

Chapter 2

Roots and Meaning of Culture

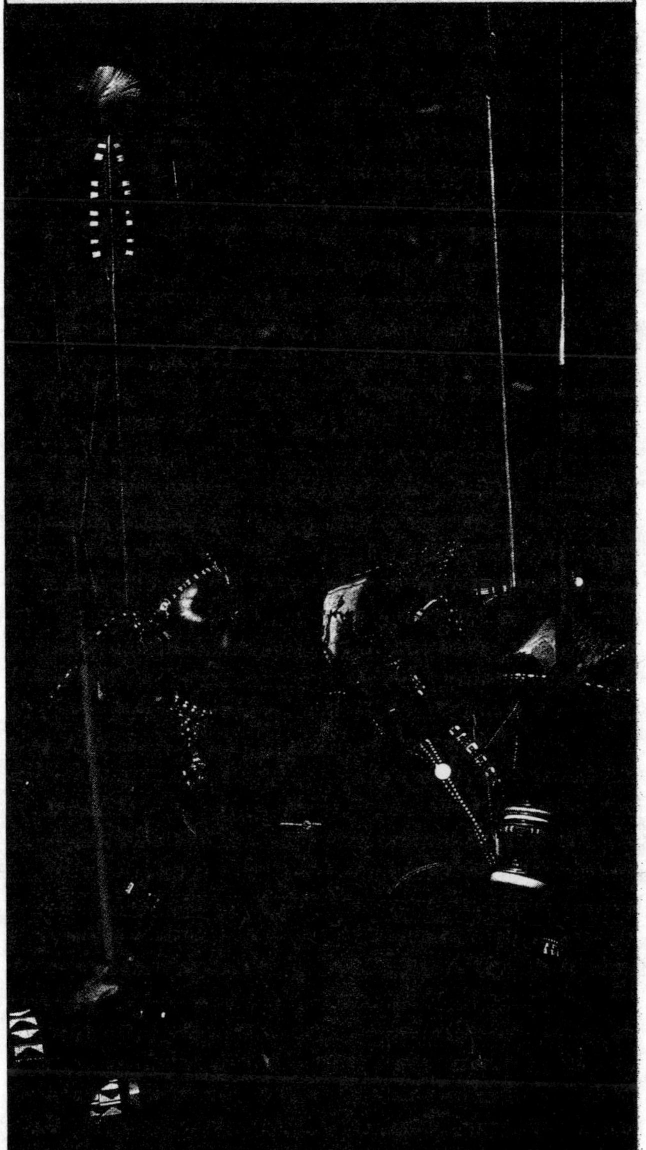

They buried him there in the cave where they were working, less than 4 miles from the edge of the ice sheet. Outside stretched the tundra, summer feeding grounds for the mammoths whose ivory they had come so far to collect. Inside, near where they dug his grave, was stacked the ivory they had gathered and were cutting and shaping. They prepared the body carefully and dusted it with red ochre, then buried it with tundra flowers and offerings of food in an elaborate grave, placing a bracelet on its arm, a pendant about its throat, and 40 to 50 polished rods of ivory by its side. It rested there, in modern Wales, undisturbed for some 18,000 years until discovered early in the 19th century. The 25-year-old hunter had died far from the group's home some 400 miles away near present-day Paris, France. He had been part of a routine annual summer expedition overland from the forested south across the as yet unflooded English Channel to the mammoths' grazing grounds at the edge of the glacier.

As always, they were well prepared for the trip. Their boots were carefully made. Their sewn skin leggings and tunics served well for travel and work; heavier fur parkas warded off the evening chill. They carried emergency food, fire-making equipment, and braided cord, which they could fashion into nets, fishing lines, ropes, or thread. They traveled by reference to sun and stars, recognizing landmarks from past journeys and occasionally consulting a crude map etched on bone. Although the hunters returned bearing the sad news of their companion's death, they also brought the ivory to be carved and traded among the scattered peoples of Europe from the Atlantic Ocean to the Ural Mountains.

As shown by their tools and equipment, their behaviors and beliefs, these Stone Age travelers displayed highly developed and distinctive characteristics, primitive only from the vantage point of our own differently expressed technologies and customs. They represented the culmination of a long history of development of skills, invention of tools, and creation of life-styles that set them apart from peoples elsewhere in Europe, Asia, or Africa, who were the products of still different cultural heritages.

To writers in newspapers and the popular press, "culture" means the arts (literature, painting, music, and the like). To a social scientist, **culture** is the specialized behavioral patterns, understandings, adaptations, and social systems that summarize a group of people's learned way of life. In this broader sense, culture is an ever-present part of the regional differences that are the essence of human geography. The visible and invisible evidences of culture—buildings and farming patterns, language, political organization, and ways of earning a living, for example—are all parts of the spatial diversity human geographers study. Cultural differences over time may present contrasts as great as those between the Stone Age ivory hunters and modern urban Americans. Cultural differences in areas result in human landscapes with variations as subtle as the differing "feel" of urban Paris, Moscow, or New York or as obvious as the sharp contrasts of rural Sri Lanka and the Prairie Provinces of Canada (Figure 2.1).

Figure 2.1

Visible cultural contrasts are clearly evident between the rice paddies of central Sri Lanka and the broad regularity of extensively farmed fields of the Canadian prairies.

Since such tangible and intangible cultural differences exist and have existed in various forms for thousands of years, human geography addresses the question of Why? Why, since humankind constitutes a single species, are cultures so varied? What and where were the origins of the different culture regions we now observe? How, from whatever limited areas individual culture traits developed, were they diffused over a wider portion of the globe? How did people who had roughly similar origins come to display significant areal differences in technology, social structure, ideology, and the other myriad details and amalgams of human geographic diversity? In what ways, and why, are there distinctive cultural variations even in "melting pot" societies such as the United States and Canada or in the outwardly homogeneous long-established countries of Europe? Part of the answer to these questions lies in the way separate human groups developed techniques to solve regionally varied problems of securing food, clothing, and shelter and, in the process, created areally distinctive customs and ways of life.

Components of Culture

Culture is transmitted within a society to succeeding generations by imitation, instruction, and example. In short, it is learned, not biological; it has nothing to do with instinct or with genes. As members of a social group, individuals acquire integrated and cohesive sets of behavioral patterns, environmental and social perceptions, and knowledge of existing technologies. Of necessity, each of us learns the culture in which we are born and reared; but we need not—indeed, cannot—learn its totality. Age, sex, status, or occupation may dictate the aspects of the cultural whole in which an individual becomes fully indoctrinated. A culture, that is, despite overall generalized and identifying characteristics and even an outward appearance of uniformity and conformity, displays a social structure—a framework of roles and interrelationships of individuals and established groups. Each individual learns and adheres to the rules and conventions not only of the culture as a whole but also of those specific to the subgroup to which he or she belongs, a subgroup that may have its own recognized social structure (Figure 2.2).

Culture is an interlocked web of such complexity and pervasiveness that it cannot be grasped and, in fact, may be misunderstood by being glimpsed or generalized from limited, obvious traits. Distinctive eating utensils, the use of gestures, or the ritual of religious ceremony may summarize and characterize a culture for the casual observer. These are, however, only individually insignificant parts of a totality that one can appreciate only when one experiences the whole.

The richness and intricacy of human life compel us to seek those more fundamental cultural variables that give order to societies and to summarize them in meaningful conceptual and spatial terms. We begin with *culture traits,* the smallest distinctive item of culture. **Culture traits** are units of learned behavior ranging from the language spoken to the tools used or to the games played. A trait may be an object (a fishhook, for example), a technique (weaving and knotting of a fishnet), a belief (in the spirits resident in water bodies), or an attitude (a conviction that fish is superior to other animal protein). Such traits are the most elementary expressions of culture, the building blocks of the complex behavioral patterns of distinctive groups of peoples.

Figure 2.2

Both the traditional rice farmer of rural Japan and the harried commuter of Tokyo are part of a common Japanese culture. They occupy vastly different positions in its social structure.

Individual cultural traits that are functionally interrelated comprise a **culture complex.** The existence of such complexes is universal. Keeping cattle was a culture trait of the Masai of Kenya and Tanzania. Related traits included the measurement of personal wealth by the number of cattle owned, a diet containing milk and the blood of cattle, and disdain for labor unrelated to herding. The assemblage of these and other related traits yielded a culture complex descriptive of one aspect of Masai society (Figure 2.3). In exactly analogous ways, religious complexes, business behavior complexes, sports complexes, and others can easily be recognized in American or any other society.

Culture traits and complexes have areal extent. When they are plotted on maps, the regional character of the components of culture is revealed. Although human geographers are interested in the spatial distribution of these individual elements of culture, their usual concern is with the **culture region,** a portion of the earth's surface occupied by populations sharing recognizable and distinctive cultural characteristics that summarize their collective attributes or activities. Examples include the political organizations societies devise, the religions they espouse, the form of economy they pursue, and even the type of clothing they wear, eating utensils they use, or kind of housing they occupy. There are as many culture regions as there are culture traits and complexes identified for population groups. Their recognition will be particularly important in discussions of ethnic, folk, and popular cultures in later chapters of this book.

Finally, a set of culture regions showing related culture complexes and landscapes may be grouped to form a **culture realm.** The term recognizes a large segment of the earth's surface having an assumed fundamental uniformity in its cultural characteristics and showing a significant difference in them from adjacent realms. Culture realms are, in a sense, culture regions at the broadest scale of generalization. In fact, the scale is so broad and the diversity within the recognized realms is so great that the very concept of realm may mislead more than it informs. One of many possible divisions of human cultures into realms is offered in Figure 2.4. The spatial pattern and characteristics of these generalized realms will help us place the discussions and examples of human geography of later chapters in their regional context.

Interaction of People and Environment

Culture develops in a physical environment that, in its way, contributes to differences among people. In primitive societies, the acquisition of food, shelter, and clothing, all parts of culture, depends on the utilization of the natural resources at hand. The interrelations of people with the environment of a given area, their perceptions and utili-

Figure 2.3

The formerly migratory Masai of Kenya are now largely sedentary, partially urbanized, and frequently owners of fenced farms. Today, traditional costumes are more representative of the tourist trade than of the culture complex they formerly signified.

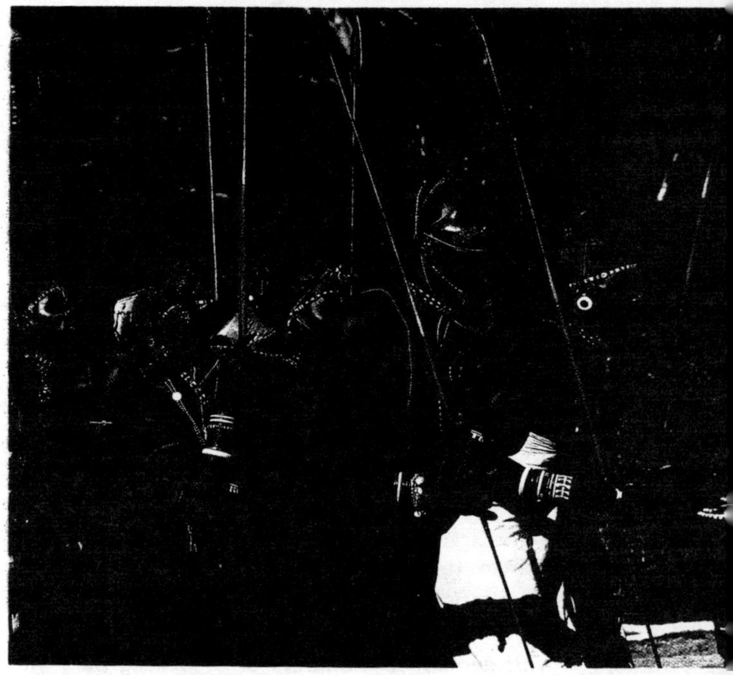

zation of it, and their impact on it are interwoven themes of **cultural ecology**—the study of the relationship between a culture group and the natural environment it occupies.

Environments as Controls

Geographers have long dismissed as intellectually limiting and demonstrably invalid the ideas of **environmental determinism,** the belief that the physical environment exclusively shapes humans, their actions, and their thoughts. Environmental factors alone cannot account for the cultural variations that occur around the world. Levels of technology, systems of organization, and ideas about what is true and right have no obvious relationship to environmental circumstances.

The environment does place certain limitations on the human use of territory, although such limitations must be seen not as absolute, enduring restrictions but as relative to technologies, cost considerations, national aspirations, and linkages with the larger world. Human choices in the use of landscapes are affected by group perception of the feasibility and desirability of their occupance and exploitation. These are not circumstances inherent in the land. Mines, factories, and cities were (and are being) created in the formerly nearly unpopulated tundra and forests of Siberia as a reflection of Soviet developmental programs, not in response to recent environmental improvement. **Possibilism** is the viewpoint that people, not

Figure 2.4

Culture realms of the modern world.

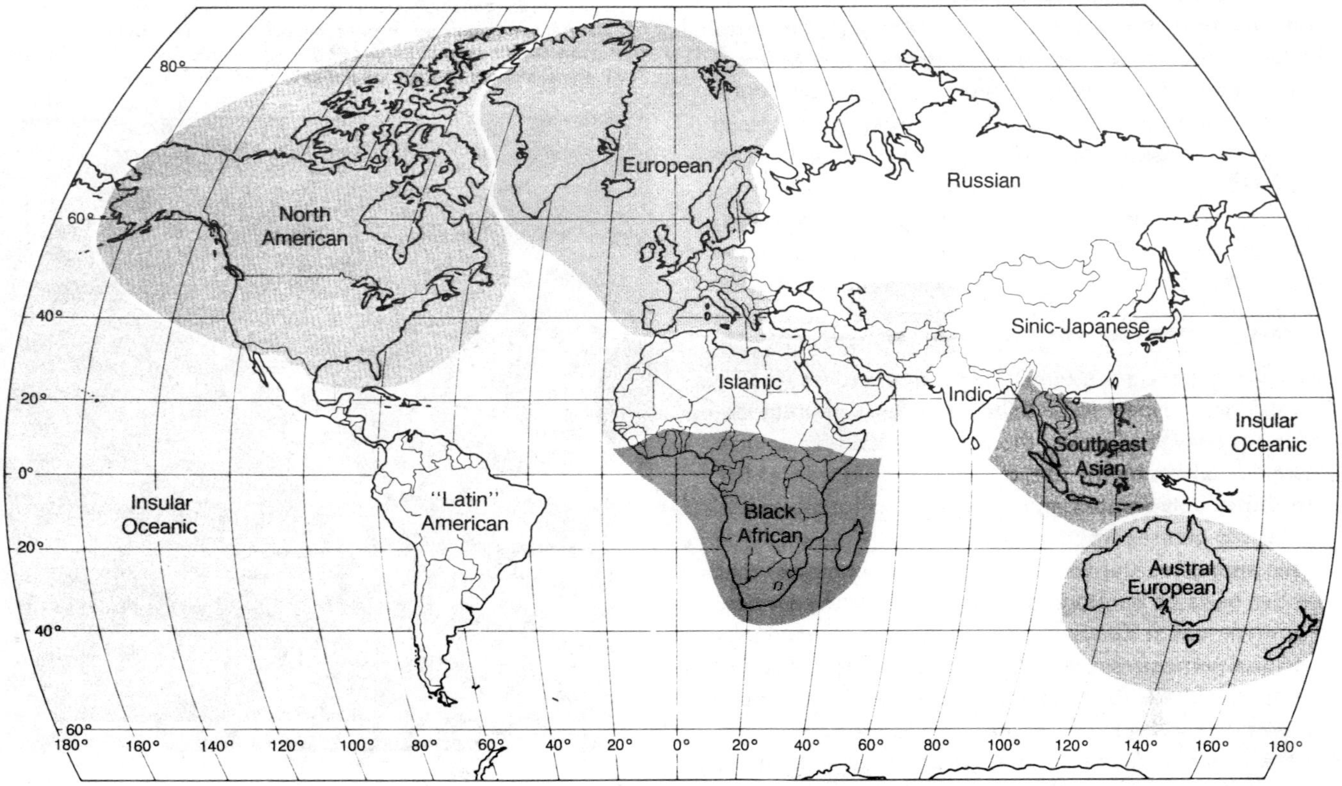

environments, are the dynamic forces of cultural development. The needs, traditions, and level of technology of a culture affect how that culture assesses the possibilities of an area and shape what choices the culture makes regarding them. Each society uses natural resources in accordance with its culture. Changes in a group's technical abilities or objectives bring about changes in its perceptions of the usefulness of the land.

Map evidence suggests the nature of some environmental limitations on use of area. The vast majority of the world's population is differentially concentrated on less than one-half of the earth's land surface, as Figure 4.20 suggests. Areas with relatively mild climates that offer a supply of fresh water, fertile soil, and abundant mineral resources are densely settled, reflecting in part the different potentials of the land under earlier technologies to support population. Even today, the polar regions, high and rugged mountains, deserts, and some hot and humid areas contain very few people. If resources for feeding, clothing, or housing ourselves within an area are lacking, or if we do not recognize them there, there is no inducement for people to occupy the territory.

Environments that do contain such resources provide the framework within which a culture operates. Coal, oil, and natural gas have been in their present locations throughout human history, but they were of no use to preindustrial cultures and did not impart any recognized advantage to their sites of occurrence. Not until the Industrial Revolution did these deposits gain importance and come to influence the location of such great industrial complexes as the Midlands in England, the Ruhr in Germany, and the steel-making districts formerly so important in parts of the northeastern United States. American Indians made one use of the environment around Pittsburgh, whereas 19th-century industrialists made quite another.

Human Impacts

People are also able to modify their environment, and this is the other half of the human-environment relationship of geographic concern. Geography, including cultural geography, examines the reactions of people to the physical environment and also their impact on that environment. We modify our environment through the material objects we place on the landscape: cities, farms, roads, and so on (Figure 2.5). The form these take is the product of the kind of culture group in which we live. The **cultural landscape,** the earth's surface as modified by human action, is the tangible physical record of a given culture. House types, transportation networks, parks and cemeteries, and the size and distribution of settlements are among the indicators of the use that humans have made of the land.

Human actions, both deliberate and inadvertent, modifying or even destroying the environment are perhaps as old as humankind itself. People have used, altered, and replaced the vegetation in vast areas of the tropics and midlatitudes. They have hunted to extinction vast herds and whole species of animals. They have, through overuse and abuse of the earth and its resources, rendered sterile and unpopulated formerly productive and attractive regions.

Fire has been called the first great tool of humans, and the impact of its early and continuing use is found on nearly every continent. Poleward of the great rain forests of equatorial regions of South America, Africa, and South Asia lies the *tropical savanna* of extensive grassy vegetation separating scattered trees and forest groves (Figure 2.6). The trees appear to be the remnants of naturally occurring tropical dry forests, thorn forests, and scrub now largely obliterated by the use, over many millennia, of fire to remove the unwanted and unproductive forest and to clear off old grasses for more nutritious new growth. The grasses supported the vast herds of grazing animals that were the basis of hunting societies. After independence, the government of Kenya in East Africa sought to protect its national game preserves by prohibiting the periodic use of fire. It quickly found that the immense herds of gazelles, zebras, antelope, and other grazers (and the lions and other predators that fed upon them) that tourists came to see were being replaced by less-appealing browsing species—rhinos, hippos,and elephants—as, with fire prohibited, the forests began to reclaim their natural environment.

The same form of vegetation replacement occurred in midlatitudes. The grasslands of North America were greatly extended by Native Americans, who burned the forest margin to extend grazing areas and to drive animals in the hunt. The control of fire in modern times has resulted in the advance of the forest once again in formerly grassy areas ("parks") of Colorado, northern Arizona, and other parts of the United States West.

Examples abound. The *Pleistocene overkill*—the Stone Age loss of whole species of large animals on all inhabited continents—is often ascribed to the unrestricted hunting to extinction carried on by societies familiar with fire to drive animals and hafted weapons to slaughter them. With the use of these, according to one estimate, about 40% of African large-animal genera passed to extinction. In North America, most of the original large-animal species had disappeared by 10,000 years ago under pressure from the hunters migrating to and spreading across the continent. Although some have suggested that climatic changes were at least partially responsible, human action seems the more likely explanation for the abrupt faunal changes. No uncertainty exists in the record of faunal destruction by the Maoris of New Zealand or of Polynesians who had exterminated some 80% to 90% of South Pacific bird species by the time Captain Cook arrived in the 18th century.

Figure 2.5

The physical and cultural landscapes in juxtaposition. Advanced societies are capable of so altering the circumstances of nature that the cultural landscapes they create become the controlling environment. The twin cities of Juarez, Mexico and El Paso, Texas are "built environments" largely unrelated to their physical surroundings.

Figure 2.6

The parklike landscape of grasses and trees, characteristic of the tropical savanna, is seen in this view from Kenya, Africa.

Not only destruction of animals but of the life-supporting environment itself has been a frequent consequence of human misuse of area (see "Chaco Canyon Desolation"). North Africa, the "granary of Rome" during the empire, became wasted and sterile in part because of mismanagement. Roman roads standing high above the surrounding barren wastes give testimony to the erosive

Chaco Canyon Desolation

It is not certain when they first came, but by A.D. 1000 the Anasazi people were building a flourishing civilization in present-day Arizona and New Mexico. In the Chaco Canyon alone, they erected as many as 75 towns, all centered around pueblos, huge stone-and-adobe apartment buildings as tall as five stories and with as many as 800 rooms. These were the largest and tallest buildings of North America prior to the construction of "cloudscrapers" in major cities at the end of the 19th century. An elaborate network of roads and irrigation canals connected and supported the pueblos. About A.D. 1200, the settlements were abruptly abandoned. The Anasazi, advanced in their skills of agriculture and communal dwelling, were forced to move on by the ecological disaster their pressures had brought to a fragile environment.

They needed forests for fuel and for the hundreds of thousands of logs used as beams and bulwarks in their dwellings. The pinyon-juniper woodland of the canyon was quickly depleted. For larger timbers needed for construction, the Anasazi first harvested stands of ponderosa pine found some 25 miles away. As early as A.D. 1030, these, too, were exhausted, and the community switched to spruce and Douglas fir from mountaintops surrounding the canyon. When they were gone by 1200, the Anasazi fate was sealed—not only by the loss of forest but by the irreversible ecological changes deforestation and agriculture had occasioned. With forest loss came erosion that destroyed the topsoil. The surface water channels that had been built for irrigation were deepened by accelerated erosion, converting them into enlarging arroyos, useless for agriculture.

The material roots of their culture destroyed, the Anasazi turned upon themselves; warfare convulsed the region. Smaller groups sought refuge elsewhere, recreating on reduced scale their pueblo way of life but now in nearly inaccessible, highly defensible mesa and cliff locations. The destruction they had wrought destroyed the Anasazi in turn.

power of wind and water when natural vegetation is unwisely removed and farming techniques are inappropriate. Easter Island in the South Pacific was covered lushly with palms and other trees when Polynesians settled there about A.D. 400. By the beginning of the 18th century, Easter Island had become the barren wasteland it remains today. Deforestation increased soil erosion, removed the supply of timbers needed for the vital dugout fishing canoes, and made it impossible to move the massive stone statues that were significant in the islanders' religion (Figure 2.7). With the loss of livelihood resources and the collapse of religion, warfare broke out and the population was decimated. A similar tragic sequence is occurring on Madagascar in the Indian Ocean today. Despite current romantic notions, not all primitive societies lived in harmony with their environment.

The more technologically advanced and complex the culture, the more apparent is its impact on the environment. In sprawling urban-industrial societies, the cultural

Figure 2.7

Now treeless, Easter Island once was lushly forested. The statues (some weighing up to 85 tons) that dot the island were rolled to their locations and lifted into place with logs.

landscape has come to outweigh the natural physical environment in its impact on people's daily lives. It interposes itself between "nature" and humans, and residents of the cities of such societies—living and working in climate-controlled buildings, driving to enclosed shopping malls—can go through life with very little contact with or concern about the physical environment.

Roots of Culture

Earlier humans found the physical environment more immediate and controlling than we do today. Some 11,000 years ago, the massive glaciers—moving ice sheets of great depth—that had covered much of the land and water of the Northern Hemisphere (Figure 2.8) began to retreat. Animal, plant, and human populations that had been spatially confined by the ice margin and the harsh climates of middle-latitude regions began to spread, colonizing newly opened territories. The name *Paleolithic* (Old Stone Age) is used to describe the period near the end of glaciation during which small and scattered groups such as the ivory hunters at this chapter's start began to develop regional variations in culture, that is, in their ways of life and livelihood.

All were **hunter-gatherers,** preagricultural people dependent on the year-round availability of plant and animal foodstuffs they could secure with the limited variety of rudimentary stone tools and weapons at their disposal. Even during the height of the Ice Age, the unglaciated sections of western, central, and northeastern Europe (the continent with the best-documented evidence of Paleolithic culture) were covered with tundra vegetation, the mosses, lichens, and low shrubs typical of areas too cold to support forests. Southeastern Europe and southern Russia had forest, tundra, and grasslands, and the Mediterranean areas had forest cover (Figure 2.9). Gigantic herds of herbivores—reindeer, bison, mammoth, and horses—browsed, bred, and migrated throughout the tundra and the grasslands. An abundant animal life filled the forests.

Human advancement northward into present-day Sweden, Finland, and the USSR demanded a much more elaborate set of tools and provision for shelter and clothing than had previously been required. It necessitated the crossing of a number of ecological barriers and the occupation of previously avoided difficult environments. By the end of the Paleolithic period, humans had spread to all the continents but Antarctica, carrying with them their common hunting-gathering culture and social organization. The settlement of the lands bordering the Pacific Ocean is shown in Figure 2.10.

While spreading, the total population also increased. But hunting and foraging bands require considerable territory to support a relatively small number of individuals. There were contacts between groups and, apparently, even planned gatherings for trade, socializing, and selecting spouses from outside the home group. Nevertheless, the bands tended to live in isolation. Estimates place the Paleolithic population of all of the island of Great Britain, which was on the northern margin of habitation, at only some 400 to 500 people living in widely separated families of 20 to 40 members. Total world population at about 9000 B.C. probably ranged from 5 to 10 million. Variations in the types of tools characteristic of different population groups steadily increased as people migrated and encountered new environmental problems (Figure 2.11). Improved tool technology greatly extended the range of possibilities in the use of locally available materials; the result was more efficient and extensive exploitation of the physical environment than earlier had been possible. At the same time, regional contrasts in plant and animal life and in environmental conditions accelerated the differentiation of culture between isolated groups who under earlier less varied conditions had shared common characteristics.

Within many environments, even harsh ones, the hunting and foraging process was not particularly demanding of either time or energy. Recent studies of South African San people (Bushmen), for example, indicate that such bands survive well on the equivalent of a 2½-day workweek. Time was available for developing skills in working flint and bone for tools, in developing regionally distinctive art and sculpture, and in making decorative beads and shells for personal adornment and trade. By the end of the Ice Age (about 11,000 years ago), language, religion, long-distance trade, permanent settlements, and

Figure 2.8

Maximum extent of glaciation. In their fullest development, glaciers of the most recent Ice Age covered large parts of Eurasia and North America. Even areas not covered by ice were affected as ocean levels dropped and rose and climate and vegetation regions changed with glacial advance and retreat.

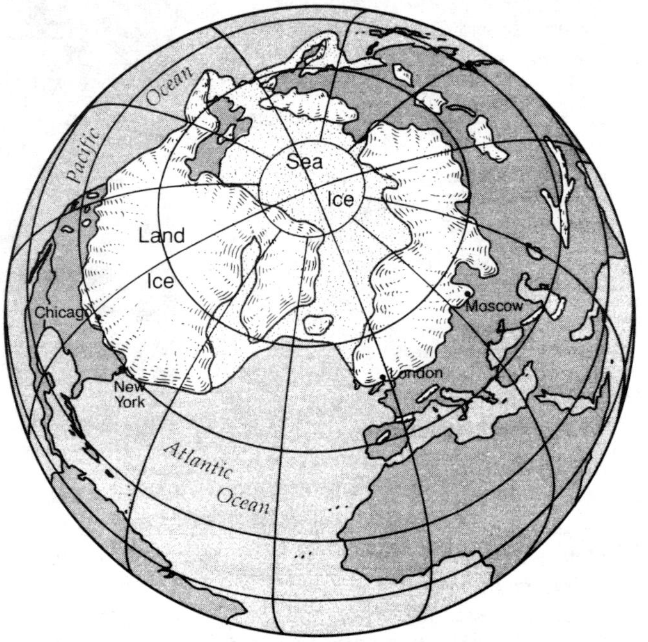

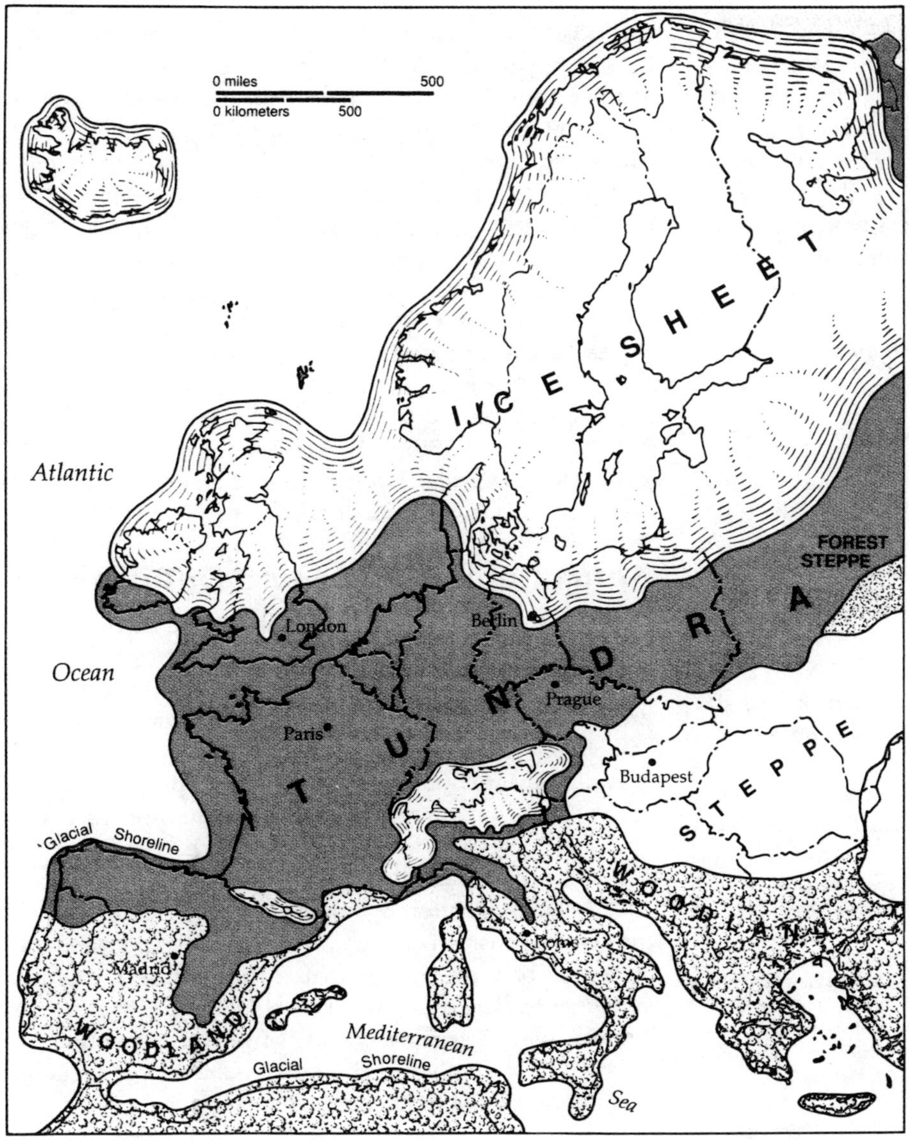

Figure 2.9

Late Paleolithic environments of Europe. During the late Paleolithic period new food-gathering, shelter, and clothing strategies were developed to cope with harsh and changing environments, so different from those in Europe today.

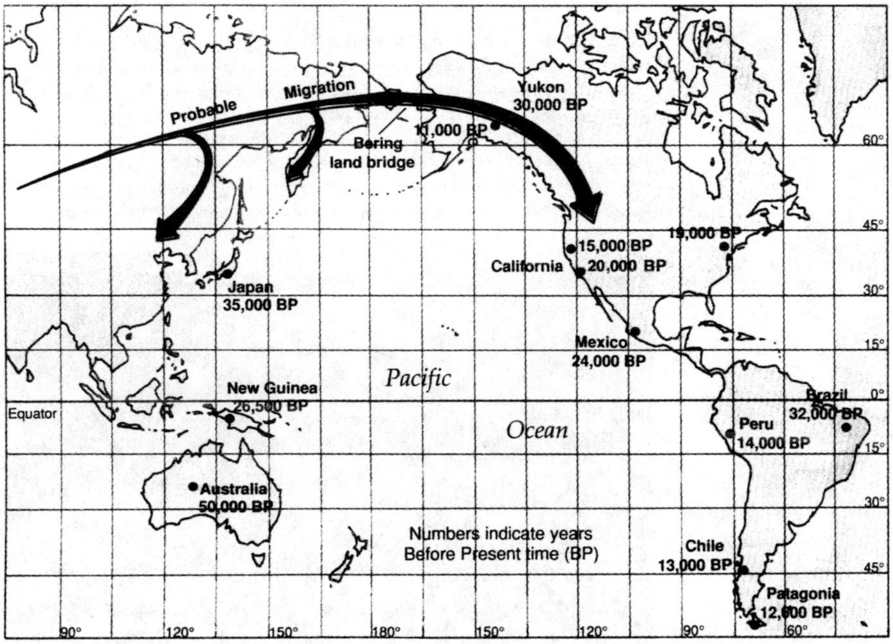

Figure 2.10

Settlement of the Americas and the Pacific basin. The date of arrival of humans in the Americas is in dispute. Geological evidence indicates that the Bering Strait land bridge—over which, it is speculated, migrants from northern China and northeastern Siberia passed in perhaps three different migration waves—disappeared about 14,000 years ago. At that time, glacier melt released sufficient water to raise sea levels several hundred feet worldwide. Radiocarbon dates from northeastern Brazil suggest cave dwellers were there as early as 32,000 years ago. A Pennsylvania rock shelter is thought to have been occupied some 19,000 years ago.

social stratification within groups appear to have been well developed in many European culture areas. What was learned and created was transmitted within the cultural group. The increasing variety of adaptive strategies and technologies and the diversity of noneconomic creations in art, religion, language, and custom meant an inevitable cultural variation of humankind. That diversification began to replace the rough cultural uniformity among hunting and gathering people that had been based on their similar livelihood patterns, informal leadership structures, small-band organizations based on kinship groups, and the like (Figure 2.12).

Seeds of Change

The retreat of the last glaciers marked the end of the Paleolithic and the beginning of successive periods of cultural evolution leading from primitive hunting and gathering economies at the outset through the development of agriculture and animal husbandry to, ultimately, the urbanization and industrialization of modern societies and economies. Since not all cultures passed through all stages at the same time, or even at all, **cultural divergence** between human groups became evident.

Figure 2.11

Representative tools of the Paleolithic period in Europe. The basis of many Stone Age tools was the stone blade, formed by new techniques of flaking stone that permitted efficient manufacture of special-purpose implements. Among these was the *burin,* a chisel-edged blade for working wood, bone, and antlers. The wooden or bone spear thrower (called the *atlal* after its Aztec name) permitted more reliable harvesting of large game. Development of bone and antler fishhooks and harpoons made possible a sedentary life based on fishing. Clothing and tents were fashioned with bone awls, needles, and fasteners.

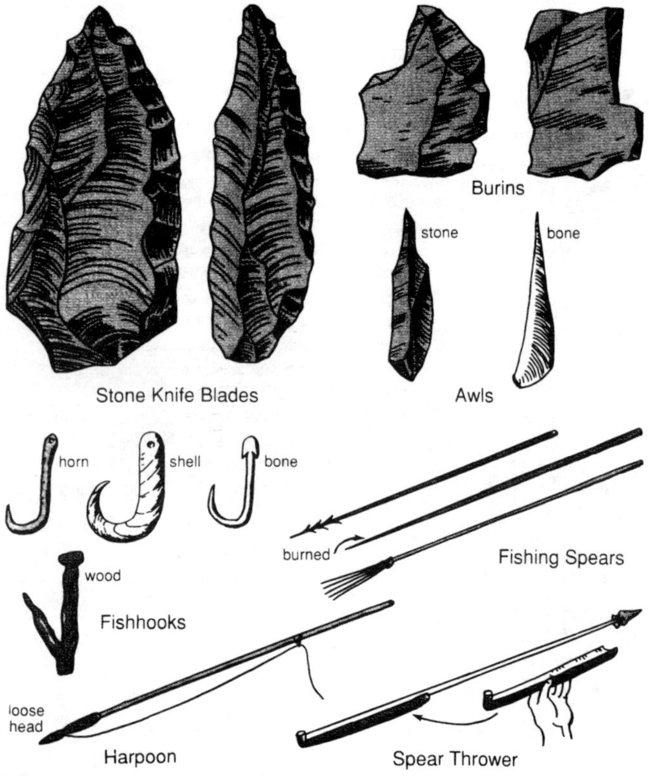

Glacial recession brought new ecological conditions to which people had to adapt. The weather became warmer, forests began to appear on the open plains and tundras of Europe and northern China. In the Middle East, where plant and animal domestication would later occur, savanna (grassland) vegetation replaced more arid landscapes. Populations grew and through hunting depleted the large herds of grazing animals already retiring northward with the retreating glacial front. Further population growth demanded new food bases and production techniques, for the **carrying capacity**—the number of people supportable within a given area by the technologies at their disposal—of the earth for hunter-gatherers is low. The *Mesolithic* (Middle Stone Age) period, from about 11,000 to 5000 B.C. in Europe, marked the transition from the collection of food to its production.

Agricultural Origins and Diffusions

As the population of hunter-gatherers rose slowly at the end of the glacial period, domestication of plants and animals began to occur independently in more than one world area. There is no agreement on whether the domestication of animals preceded or followed that of plants. The sequence may well have been different in different areas. What appears certain is that animal domestication, the successful breeding of species that are dependent on human beings, began during the Mesolithic not as a conscious economic effort by humans but as an outgrowth of the keeping of small or young wild animals as pets and the attraction of scavenger animals to the refuse of human settlements. The assignment of religious significance to certain animals and the docility of others to herding by hunters all strengthened the human-animal connections that ultimately led to full domestication.

Figure 2.12

Hunter-gatherers practiced the most enduring life-style in human history, trading it for the more arduous life of farmers under the necessity to provide larger quantities of less diversified foodstuffs for a growing population. Unlike their settled farmer rivals and successors, among hunter-gatherers age and sex differences, not caste or economic status, were and are the primary basis for the division of labor and of interpersonal relations. Here a San (Bushman) hunter of Botswana, Africa, stalks his prey. Men also help contribute to the gathered food that constitutes 80% of the San diet.

Radiocarbon dates suggest the domestication of goats in the Near East as early as 8000 B.C., of sheep in Turkey by about 7500 B.C., and of cattle and pigs in both Greece and the Near East about 7000 B.C. North Africa, India, and southeastern Asia were other Old World domestication sources, as were—less successfully—Meso-America and the Andean Uplands. Although the concept of animal domestication diffused from limited source regions, once its advantages were learned numerous additional domestications were accomplished elsewhere. The widespread natural occurrence of species able to be domesticated made that certain. Cattle of different varieties, for example, were domesticated in India, north-central Eurasia, Southeast Asia, and Africa. Pigs and various domestic fowl are other examples.

The domestication of plants, like that of animals, appears to have occurred independently in more than one world region over a time span of between 10,000 and perhaps as much as 20,000 years ago. Clear evidence exists that African peoples were raising crops of wheat, barley, dates, lentils, and chickpeas on the floodplains of the Nile River as early as 18,500 years ago. In other world regions, farming began more recently.

Familiarity with plants of desirable characteristics is universal among hunter-gatherers. In those societies, females were assigned the primary food-gathering role and thus developed the greatest familiarity with nutritive plants. Their fundamental role in initiating crop production to replace less reliable food gathering seems certain. Indeed, women's major contributions as innovators of technology—in food preparation and clothing production, for example—or as inventors of such useful and important items as baskets and other containers, baby slings, yokes for carrying burdens, and the like are unquestioned.

Agriculture itself, however, seems most likely to have been not an "invention" but the logical extension to food species of plant selection and nurturing habits developed for nonfood varieties. Plant poisons applied to hunting arrows or spread on lakes or streams to stun fish made food gathering easier and more certain. Plant dyes and pigments were universally collected or prepared for personal adornment or article decoration. Medicinal and mood-altering plants and derivatives were known, gathered, protected, and cultivated by all primitive cultures. Indeed, persuasive evidence exists to suggest that early gathering and cultivation of grains was not for grinding and baking as bread but for brewing as beer, a beverage that became so important in some cultures for religious and nutritional reasons that it may well have been a first and continuing reason for sedentary agricultural activities.

Nevertheless, full-scale domestication of food plants, like that of animals, can be traced to a limited number of origin areas from which its techniques spread (Figure 2.13).

Figure 2.13

Chief centers of plant and animal domestication. The southern and southeastern Asian center was characterized by the domestication of plants such as taro, which are propagated by the division and replanting of existing plants (vegetative reproduction). Reproduction by the planting of seeds (e.g., maize and wheat) was more characteristic of Meso-America and the Near East. The African and Andean areas developed crops reproduced by both methods. The lists of crops and livestock associated with the separate origin areas are selective, not exhaustive.

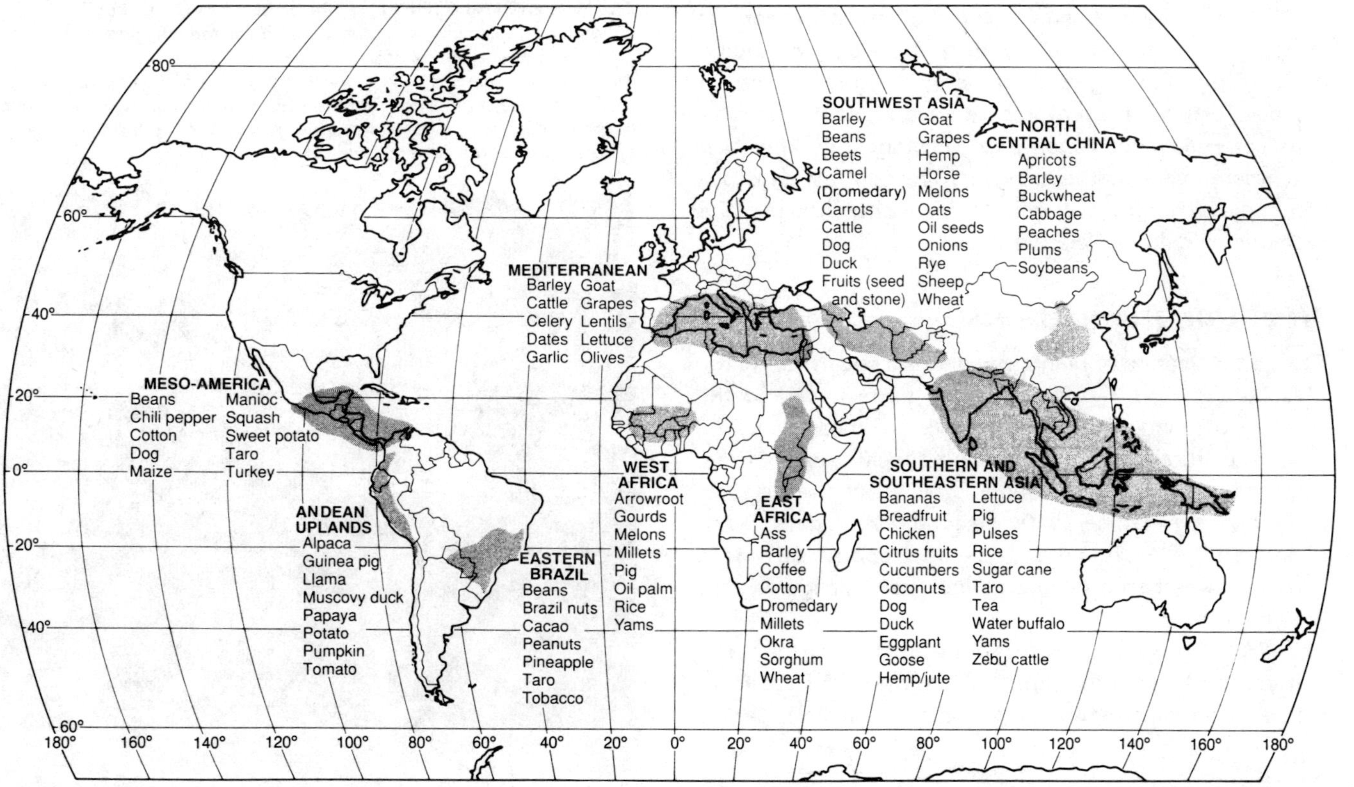

Although source regions were several, certain uniformities united them. In each, domestication focused on plant species selected apparently for their capability of providing large quantities of storable calories or protein. In each, there was a population already well fed and able to devote time to the selection, propagation, and improvement of plants available from a diversified vegetation. In the tropics and humid subtropics, selected plants were apt to be those that reproduced vegetatively—from roots, tubers, or cuttings. Outside those regions, wild plants reproducing from seeds were more common and the objects of domestication. Although there was some duplication, each of the origin areas developed crop complexes characteristic of itself alone, as Figure 2.13 summarizes. From each, there was dispersion of crop plants to other areas, slowly at first under primitive systems of population movement and communication, more rapidly and extensively with the onset of European exploration and colonization after A.D. 1500.

While adapting wild plant stock to agricultural purposes, the human cultivators, too, adapted. They assumed sedentary residence to protect the planted areas from animal, insect, and human predators. They developed labor specializations and created more formalized and expansive religious structures in which fertility and harvest rites became important elements. The regional contrasts between hunter-gatherer and sedentary agricultural societies increased. Where the two groups came in contact, farmers were the victors and hunter-gatherers the losers in competition for territorial control. The contest continued into modern times. During the past 200 years, European expansion totally dominated the hunting and gathering cultures it encountered in large parts of the world such as North America and Australia. Even today, in the forests of Zaire in central Africa, Bantu farmers put continuing pressure on hunting and gathering Pygmies; and in southern Africa, Hottentot herders and Bantu farmers constantly advance upon the territories of the San (Bushmen) hunter-gatherer bands. The contrast and conflict between hunter-gatherers and agriculturalists provide dramatic evidence of cultural divergence.

Neolithic Innovations

The domestication of plants and animals began during the Mesolithic period, but in its refined form it marked the onset of the *Neolithic* (New Stone Age), which is more a stage of cultural development than a specific span of time. The term implies the creation of an advanced set of tools and technologies to deal with the conditions and needs encountered by an expanding, sedentary population whose economy was based on the agricultural management of the environment (Figure 2.14).

Not all peoples in all areas of the earth made the same cultural transition at the same time. In the Near East, from which most of our knowledge of this late prehistoric period comes, the Neolithic lasted from approximately 8000 to 3500 B.C. There, as elsewhere, it brought complex and revolutionary changes in human life. Culture began to alter at an accelerating pace, and change itself became a way of life. In an interconnected adaptive web, technological and social innovations came with a speed and genius surpassing all previous periods. Humans learned the arts of spinning and weaving plant and animal fibers. They learned to use the potter's wheel and to fire clay and make utensils. They developed techniques of brick making, mortaring, and construction, and they discovered the skills of mining, smelting, and casting metals. On the foundation of such technical advancements, a more complex exploitative culture appeared, a stratified society to replace the rough equality of adults in nomadic hunting and gathering economies. Special local advantages in resources or products promoted the development of long-distance trading connections, which the invention of the sailboat helped to maintain.

By the end of the Neolithic period, certain spatially restricted groups, having created a food-producing rather than a foraging society, undertook the purposeful restructuring of their environment. They began to modify plant and animal species; to manage soil, terrain, water, and mineral resources; and to utilize animal energy to supplement that of humans. They used metal to make refined tools and superior weapons—first pure copper and later the alloy of tin and copper that produced the harder, more durable bronze. Humans had moved from adopting and shaping to the art of creating.

Figure 2.14

The Mediterranean scratch plow, the earliest form of this basic agricultural implement, was essentially an enlarged digging stick dragged by an ass, an ox or pair of oxen, or—as here, near El-Djem, Tunisia—by a camel. The first application of nonhuman power to agriculture, the scratch plow represented a technological breakthrough of great importance. It is depicted in Egyptian tomb drawings and in art preserved in the ancient Middle East.

As people gathered together in larger communities, new and more formalized rules of conduct and control emerged, especially important where the use of land was involved. We see the beginnings of governments to enforce laws and specify punishments for wrongdoers. The protection of private property, so much greater in amount and variety than that carried by the nomad, demanded more complex legal codes, as did the enforcement of the rules of societies increasingly stratified by social privileges and economic status.

Religions became more formalized. For the hunter, religion could be individualistic, and his worship was concerned with personal health and safety. The collective concerns of farmers were based on the calendar: the cycle of rainfall, the seasons of planting and harvesting, the rise and fall of waters to irrigate the crops. Religions responsive to those concerns developed rituals appropriate to seasons of planting, irrigation, harvesting, and thanksgiving. An established priesthood was required, one that stood not only as intermediary between people and the forces of nature but also as authenticator of the timing and structure of the needed rituals.

In daily life, occupations became increasingly specialized. Metalworkers, potters, sailors, priests, merchants, scribes, and in some areas, warriors complemented the work of farmers and hunters.

Culture Hearths

The social and technical revolutions that began in and characterized the Neolithic period were initially spatially confined. The new technologies, the new ways of life, and the new social structures diffused from those points of origin and were selectively adopted by people who were not a party to their creation. The term **culture hearth** is used to describe such centers of innovation and invention from which key culture traits and elements moved to exert an influence on surrounding regions. The hearth may be viewed as the "cradle" of a culture group whose developed systems of livelihood and life created distinctive cultural landscapes. All hearth areas developed the trappings of *civilizations*. The definition of that term is not precise, but indicators of its achievement are commonly assumed to be writing, metallurgy, long-distance trade connections, astronomy and mathematics, social stratification and labor specialization, formalized governmental systems, and a structured urban culture.

Several major culture hearths emerged in the Neolithic period. Prominent centers of early creativity were found in Egypt, Mesopotamia, the Indus Valley of the Indian subcontinent, northern China, southeastern Asia, and in several locations in Africa, the Americas, and elsewhere (Figure 2.15). They arose in widely separated areas of the

Figure 2.15

Early culture hearths of the Old World and the Americas.
Generalized hearth characteristics are summarized in Table 2.1.

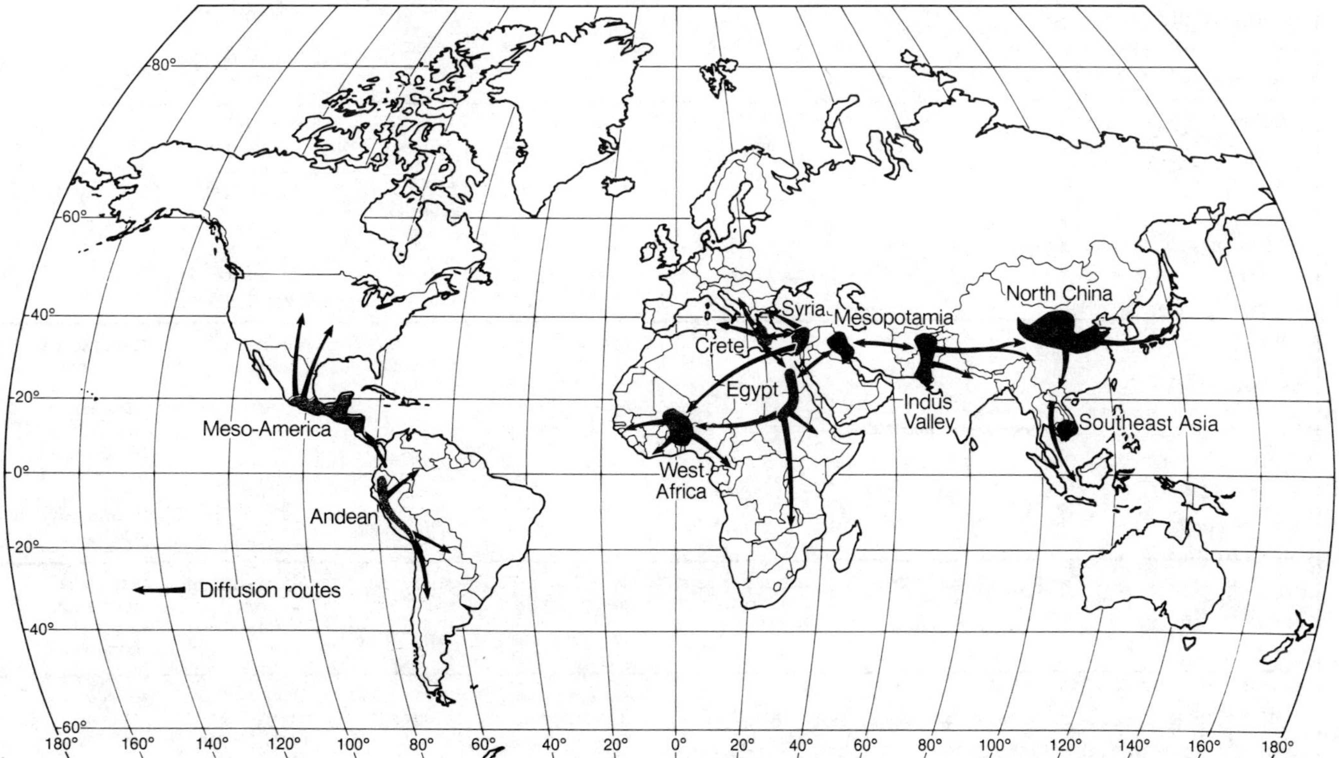

Table 2.1 Emerging Culture Hearths: Periods and Features of Development

Hearth Region	Before 10,000 B.C.	10,000–8000 B.C.	8000–6000 B.C.	6000–4000 B.C.	4000–2000 B.C.	2000 B.C.–A.D. 1	A.D. 1–1000
Near East	Earliest domestication of dog Early hunter-gatherer villages	Domestication of sheep and goats First permanent communities Long-distance trade	First farming villages Wheat, barley, legumes Sophisticated houses Metalworking Pottery, textiles	First irrigation Early records on clay tablets Potter's wheel	First cities Earliest writing Wheeled vehicles First legal codes Bronze Age Plow	First alphabet Glass Iron smelting Birth of Christ	Birth of Mohammed (A.D. 570) Arab/Muslim expansion
Nile Valley	Evidence of cultivation of wheat, barley, lentils, dates		Pottery Fishing villages	Domestication of cattle Metalworking Farming villages	Cattle herding Farming Cloth Sailing ships Cities Writing Long-distance trade		
Indus Valley					Village farming Rise of cities Long-distance trade	End of Indus Valley cities (1600 B.C.)	
East Asia			Cultivation of rice; root crops, beans, millet Pottery	Settled villages Wide range of crops, domestic animals Plow Irrigation	Metalworking (bronze) First Chinese city	Chinese walled cities Buddha (563–483 B.C.) Confucius (551–479 B.C.) Ideographic script Iron working (China)	First Southeast Asian cities
Europe	Cave art Ivory, stone figurines	Long-distance trade	First farming in Greece and Aegean	Megalithic tombs	Olive, grape domestication First European cities Copper working	Minoan civilization on Crete Mycenaean culture in Greece "Golden Age" of Greece and Rome	Fall of Roman Empire Dark Ages
West Africa				Pottery	Yam cultivation, Farming	Village clusters	Ceramic art First cities Well-developed agriculture Iron smelting Long-distance trade Empire of Ghana
Andean America		Roots, tubers (potato) as food crops	Beans, peppers, other plant domestications		Pottery	Metalworking Ceramics Textiles	City formation City-state conquests
Meso-America			Maize domestication (Mexico)	Beans, squash, chili peppers	First farming villages	Olmec culture Early cities Early Mayan culture Astronomy Writing Raised field agriculture	Apex of Mayan culture

From John E. Pfeiffer, *The Emergence of Society: A Prehistory of the Establishment*.

world, at different times, and under differing ecological circumstances. Each displayed its own unique mix of culture traits and amalgams, some aspects of which are summarized in Table 2.1.

All were urban centered, the indisputable mark of civilization first encountered in Mesopotamia 5500 to 6000 years ago, but the urbanization of each was differently arrived at and expressed (Figure 2.16). In some hearth areas, such as Mesopotamia and Egypt, the transition from settled agricultural village to urban form was gradual and prolonged. In Minoan Crete, urban life was less explicitly developed than in the Indus Valley, where early trade contacts with the Near East suggest the importance of exchange in fostering urban growth (see "Cities Brought Low"). Trade seems particularly important in the development of West African culture hearths, such as Ghana and Kanem. Coming later (from the 8th to the 10th centuries) than the Nile or Mesopotamian centers, their numerous stone-built towns seem to have been supported by both an extensive agriculture whose origins were probably as early as those of the Middle East and, particularly, by long-distance trade across the Sahara. The Shang kingdom on the middle course of the Huang He (Yellow River) on the North China Plain had walled cities containing wattle-and-daub buildings but no monumental architecture.

Figure 2.16

Urbanization was invariably a characteristic of culture hearths of both the Old and the New worlds. Pictured is the Sun Temple and Street of the Dead at Teotihuacán, a city that at its height between A.D. 300 and 700 spread over nearly 7 square miles (18 km²). Located some 30 miles northeast of Mexico City in the Valley of Mexico, the planned city of Teotihuacán featured broad, straight avenues and an enormous pyramid complex. The Street of the Dead, bordered with low stone-faced buildings, was nearly 2 miles in length.

Each culture hearth showed a rigorous organization of agriculture resulting in local productivity sufficient to enable a significant number of people to engage in nonfarm activities. Therefore, each hearth region saw the creation of a stratified society that included artisans, warriors, merchants, scholars, priests, and administrators. Each also developed or adopted astronomy, mathematics, and the all-essential calendar. Each, while advancing in cultural diversity and complexity, exported technologies, skills, and learned behaviors far beyond its own boundaries.

Writing appeared first in Mesopotamia and Egypt at least 5000 years ago, as cuneiform in the former and as hieroglyphics in the latter. The separate forms of writing have suggested to some that they arose independently in separate hearths; others maintain that the idea of writing originated in Mesopotamia and diffused outward to Egypt, the Indus Valley, Crete, and perhaps even China. The systems of record keeping developed in New World hearths were not related to those of the Old, but once developed, they were widely spread in areas under the influence of Andean and Meso-American hearths. Skill in working iron, so important in Near Eastern kingdoms, was an export of sub-Saharan African hearths.

The anthropologist Julian Steward (1902–1972) proposed the concept of **multilinear evolution** to explain the common characteristics of widely separated cultures developed under similar ecological circumstances.

Cities Brought Low

Just as the activities of humans may alter or destroy their environment, so may spontaneous changes in the environment bring low the works of organized society. Recent research suggests that one of the world's great early cultures, that of the Indus Valley of Pakistan, was destroyed not by fire or sword but by drowning in a sea of mud.

The great cities of the civilization, among them Harappa and Mohenjo-Daro, underwent an abrupt downfall about 2000 B.C. Excavations showing deep accumulations of mud, the erosional collapse of city walls, and recurring attempts to rebuild city structures document a losing battle against rising water backed up behind a dam created by earthquakes, rock shifts, and mud slides. The Indus was converted from a free-flowing stream to a huge, swampy lake, trapping mud formerly carried to the sea. As the lake rose, it destroyed forests and ruined agriculture. Although people erected dikes and brick walls to shield the cities, their food bases and very life were choked. The dam was cut through by the river and replugged at least five times, each flood recurrence taking its toll of urban life. The degeneration of pottery and other evidences of cultural decline record the struggle with an altering environment.

He suggested that each major environmental zone—arid, high altitude, midlatitude steppe, tropical forest, and so on—tends to induce common adaptive traits in the cultures of those who exploit it. Those traits were, at base, founded upon the development of agriculture and the emergence of similar cultural and administrative structures in the several culture hearths. But *similar* does not imply *identical.* Steward simply suggested that since comparable sequences of developmental events cannot always or even often be explained on the basis of borrowing or exporting of ideas and techniques (because of time and space differences in cultures sharing them), they must be regarded as evidence of parallel creations based on similar ecologies. From similar origins, but through separate adaptations and innovations, distinctive cultures emerged.

Nonetheless, the common characteristics deriving from multilinear evolution and the diffusion of specific culture traits and complexes contained the roots of **cultural convergence.** That term describes the sharing of technologies, organizational structures, and even cultural traits and artifacts that is so evident among widely separated societies in a modern world united by instantaneous communication and efficient transportation.

The Structure of Culture

Even such fundamental culture *traits* as writing and metalworking or culture *complexes* of agricultural techniques or urban life represented only a fraction of the total social systems developed in the hearth regions. Each culture trait or complex is but a minor part of a whole that can be understood only when that whole is envisaged, perhaps an impossibility for anyone who is not part of the culture itself. The anthropologist Leslie White (1900–1975) suggested that for analytical purposes, a culture could be viewed as a three-part structure composed of subsystems that he termed *ideological, technological,* and *sociological.* In a similar classification, the biologist Julian Huxley (1887–1975) identified three components of culture: *mentifacts, artifacts,* and *sociofacts.* Together, according to these interpretations, the subsystems—identified by their separate components—comprise the system of culture as a whole. But they are integrated; each reacts on the others and is affected by them in turn.

The **ideological subsystem** consists of ideas, beliefs, and knowledge of a culture and of the ways in which these things are expressed in speech or other forms of communication. Mythologies and theologies, legend, literature, philosophy, and folk wisdom make up this category. Passed on from generation to generation, these abstract belief systems, or **mentifacts,** tell us what we ought to believe, what we should value, and how we ought to act. Beliefs form the basis of the socialization process (see "The Gauda's Son"). Often we know—or think we know—what tne beliefs of a group are from their oral or written statements. Sometimes, however, we must depend on the actions or objectives of a group to tell us what its true ideas and values are. "Actions speak louder than words" and "Do as I say, not as I do" are commonplace

The Gauda's Son

The traditional belief system of an Indian village was embedded in—but of course only partially revealed by—the special treatment given to the son of the Gauda (the village headman). How many different elements of the technological, sociological, and ideological subsystems of the village can you detect in the following extract?

The Gauda's son is eighteen months old. Every morning, a boy employed by the Gauda carries the Gauda's son through the streets of Gopalpur. The Gauda's son is clean; his clothing is elegant. When he is carried along the street, the old women stop their ceaseless grinding and pounding of grain and gather around. If the child wants something to play with, he is given it. If he cries, there is consternation. If he plays with another boy, watchful adults make sure that the other boy does nothing to annoy the Gauda's son.

Shielded by servants, protected and comforted by virtually everyone in the village, the Gauda's son soon learns that tears and rage will produce anything he wants. At the same time, he begins to learn that the same superiority which gives him license to direct others and to demand their services places him in a state of danger. The green mangoes eaten by all of the other children in the village will give him a fever; coarse and chewy substances are likely to give him a stomachache. While other children clothe themselves in mud and dirt, he finds himself constantly being washed. As a Brahmin [a religious leader], he is taught to avoid all forms of pollution and to carry out complicated daily rituals of bathing, eating, sleeping, and all other normal processes of life.

In time, the Gauda's son will enter school. He will sit motionless for hours, memorizing long passages from Sanskrit holy books and long poems in English and Urdu. He will learn to perform the rituals that are the duty of every Brahmin. He will bathe daily in the cold water of the private family well, reciting prayers and following a strict procedure. The gods in his house are major deities who must be worshipped every day, at length and with great care.

recognitions of the fact that actions, values, and words do not always coincide. Two basic strands of the ideological subsystem—language and religion—are the subject of Chapter 5.

The **technological subsystem** is composed of the material objects, together with the techniques of their use, by means of which people are able to live. Such objects are the tools and other instruments that enable us to feed, clothe, house, defend, transport, and amuse ourselves. We must have food, we must be protected from the elements, and we must be able to defend ourselves. Huxley termed the material objects we use to fill these basic needs **artifacts** (Figure 2.17). In Chapter 10 we will examine the relationship between technological subsystems and regional patterns of economic development.

The **sociological subsystem** of a culture is the sum of the expected and accepted patterns of interpersonal relations that find their outlet in economic, political, military, religious, kinship, and other associations. These **sociofacts** define the social organization of a culture. They regulate how the individual functions relative to the group, whether it be family, church, or state. There are no "givens" as far as the patterns of interaction in any of these associations are concerned, except that most cultures possess a variety of formal and informal ways of structuring behavior. Differing patterns of behavior are learned and transmitted from one generation to the next (Figure 2.18).

Figure 2.17

(*a*) This Balinese farmer working with draft animals uses tools typical of the lower technological levels of subsistence economies. (*b*) Cultures with advanced technological subsystems use complex machinery to harness inanimate energy for productive use.

(a) **(b)**

Figure 2.18

All societies prepare their children for membership in the culture group. In each of these settings, certain values, beliefs, skills, and proper ways of acting are being transmitted to the youngsters.

Classifications are of necessity arbitrary, and these classifications of the subsystems and components of culture are no exception. The three-part categorization of culture, while helping us to appreciate its structure and complexity, can simultaneously obscure the many-sided nature of individual elements of culture. A dwelling, for example, is an artifact providing shelter for its occupants. It is, simultaneously, a sociofact reflecting the nature of the family or kinship group it is designed to house and a mentifact summarizing a culture group's convictions about appropriate design, orientation, and building materials of dwelling units. In the same vein, clothing serves as an artifact of bodily protection appropriate to climatic conditions, available materials and techniques, or the activity in which the wearer is engaged. But garments also may be sociofacts, identifying an individual's role in the social structure of the community or culture, and mentifacts, evoking larger community value systems (Figure 2.19).

Nothing in a culture stands totally alone. Changes in the ideas that a society holds may affect the sociological and technological systems just as changes in technology force changes in the social system. The abrupt alteration of the ideological structure of Russia following the 1917 Communist Revolution from a monarchical, agrarian, capitalistic system to an industrialized, communistic society involved sudden, interrelated alteration of all facets of that country's culture system. The interlocking nature of all aspects of a culture is termed **cultural integration.**

Culture Change

The recurring theme of cultural geography is change. No culture is, or has been, characterized by a permanently fixed set of material objects, systems of organization, or even ideologies, although all of these may be long enduring within a society at equilibrium with its resource base and totally isolated from a need or an influence for change. Such isolation and stagnation have always been rare. On the whole, while cultures are essentially conservative, they are simultaneously constantly changing. They are always in a state of flux. Some changes are major and pervasive. The transition from hunter-gatherer to sedentary farmer, as we have seen, affected markedly every facet of the cultures experiencing that change. Profound, too, has been the impact of the Industrial Revolution and its associated urbanization on all societies it has touched.

Change may, of course, differently affect different segments of the same culture group. The Masai of East Africa, mentioned earlier in this chapter, have undergone great changes in their transition from primitive herders to citizens of modern nations. Once, dominating other populations by their spears, they ranged with their herds unchecked over the East African uplands until European conquerors ended their raiding and claimed their lands. Barely tolerated by the sedentary farmers whom they once terrorized, the Masai were in the middle 20th century picturesque oddities—with ochre dreadlocks, lean-bodied from their diet of worm-infested raw meat, carrying spears no longer needed for defense or pillage, herding cattle as gaunt as themselves. After Kenya gained independence in 1963, the Masai of that country benefited from available medication and improved diet. Breed improvement has fattened the herds. Private landholding, wells and water retention ponds, and participation in a commercial economy have fundamentally altered the Kenyan Masai appearance and way of life. Tanzanian Masai, however, under a different set of economic, social, and political controls, have retained many of their traditional adornments and ways of life, less affected by the revolutionary changes embraced by their Kenyan kin.

Figure 2.19

(*a*) When clothing serves primarily to cover, protect, or assist in activities, it is an *artifact.* (*b*) Some garments are *sociofacts,* identifying a role or position within the social structure: the distinctive "uniforms" of the soldier, the cleric, or the beribboned ambassador immediately proclaim their respective roles in a culture's social organizations. (*c*) The mandatory chadors of Iranian women are *mentifacts,* indicative not specifically of the role of the wearer but of the values of the culture the wearer represents.

(a)

(b)

(c)

Not all change is so pervasive as that following the introduction of agriculture or the Industrial Revolution, or so quickly apparent as the Kenyan Masai move to modern life. Many changes are so slight individually as to go almost unnoticed at their inception, though cumulatively they may substantially alter the affected culture. Think of how the culture of the United States differs today from what you know it to have been in 1940—not in essentials, perhaps, but in the innumerable electrical, electronic, and transportational devices that have been introduced and in the social, behavioral, and recreational changes they have wrought. Such cumulative changes occur because the cultural traits of any group are not independent; they are clustered in a coherent and integrated pattern. Change in an apparently minor and limited fashion will have wide repercussions as associated traits arrive at accommodation with the adopted adjustment. Change, both major and minor, within cultures is induced by *innovation* and *diffusion.*

Innovation

Innovation implies changes to a culture that result from ideas created within the social group itself. The novelty may be an invented improvement in material technology, such as the bow and arrow or the jet engine. It may involve the development of nonmaterial forms of social structure and interaction: feudalism, for example, or Christianity.

Primitive and traditional societies characteristically are not innovative. In societies at equilibrium with their environment and with no unmet needs, change has no adaptive value and no reason to occur. Indeed, all societies have an innate resistance to change. Complaints about youthful fads or the glorification of times past are familiar cases in point. However, when a social group is inappropriately unresponsive—mentally, psychologically, or economically—to changing circumstances and innovation, it is said to exhibit **cultural lag.**

Innovation, frequently under stress, has characterized the history of humankind. As we have seen, growing populations at the end of the Ice Age necessitated an expanded food base. In response, domestication of plants and animals appears to have occurred independently in more than one world area. Indeed, a striking fact about early agriculture is the universality of its development or adoption within a very short span of human history. In 10,000 B.C., the world population of no more than 10 million was exclusively hunter-gatherers. By A.D. 1500, only 1% of the world's 350 million people still followed that way of life. The revolution in food production affected every facet of the threefold subsystems of culture of every society accepting it. All innovation has a radiating impact on the web of culture. The more basic the innovation, the more pervasive its consequences.

In most modern societies, innovative change has become common, expected, and inevitable. The rate of invention, at least as measured by the number of patents granted, has steadily increased, and the period between idea conception and product availability has been decreasing. A general axiom is that the more ideas available and the more minds able to exploit and combine them, the greater the rate of innovation. The spatial implication is that larger urban centers of advanced technologies tend to be centers of innovation. This is not just because of their size but because of the number of ideas interchanged. Ideas not only stimulate new ideas but also create circumstances in which the society must develop new solutions to maintain its forward momentum (Figure 2.20).

Diffusion

Diffusion is the process by which an idea or innovation is transmitted from one individual or group to another across space. Diffusion may assume a variety of forms, each different in its impact on social groups, but basically two processes are involved: either people move, for any of a number of reasons, to a new area and take their culture with them (e.g., immigrants to the American colonies), or information about an innovation (e.g., barbed wire or hybrid corn) may spread throughout a society (Figure 2.21). The former is known as **relocation diffusion,** the latter as **expansion diffusion** (Figure 2.22).

Expansion diffusion involves the spread of an item or idea from one place to others, but in the process the thing diffused also remains—and is frequently intensified—in the origin area. When that expansion affects nearly uniformly all individuals and areas outward from the source region, it is termed **contagious diffusion.** The term implies the importance of direct contact between those who developed or have adopted the innovation and those who

Figure 2.20

The rate of innovation through human history. Hunter-gatherers, living in easy equilibrium with their environment and their resources base, had little need for innovation and no need for cultural change. Increased population pressures led to the development of agriculture and the diffusion of the ideas and techniques of domestication, urbanization, and trade. With the Industrial Revolution, dramatic increases in innovation began to alter cultures throughout the world.

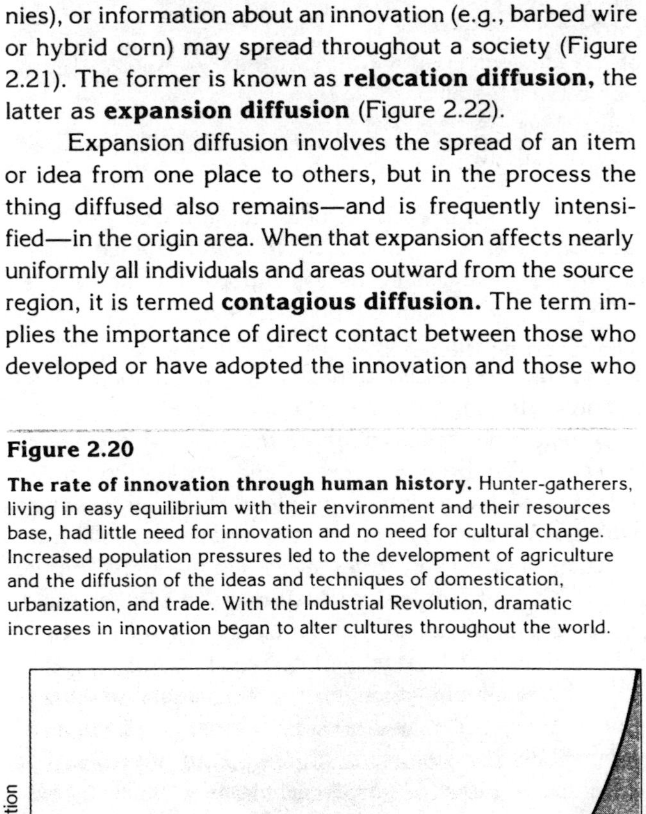

Figure 2.21

A street scene in Guatemala City. In modern society, advertising is a potent force for diffusion. Advertisements over radio and television, in newspapers and magazines, and on billboards and signs communicate information about many different products and innovations.

Figure 2.22

Patterns of diffusion. (*a*) In *relocation diffusion,* innovations or ideas are transported to new areas by carriers who permanently leave the home locale. The process is spatially selective. By accident or design, some areas or regions are settled by the innovators while other potential locations are bypassed. (*b*) In *expansion diffusion,* a phenomenon spreads from one place to neighboring locations, but in the process it remains and is often intensified in the place of origin.

Source: Reproduced by permission from *Spatial Diffusion,* Peter R. Gould, page 4, Association of American Geographers, 1969.

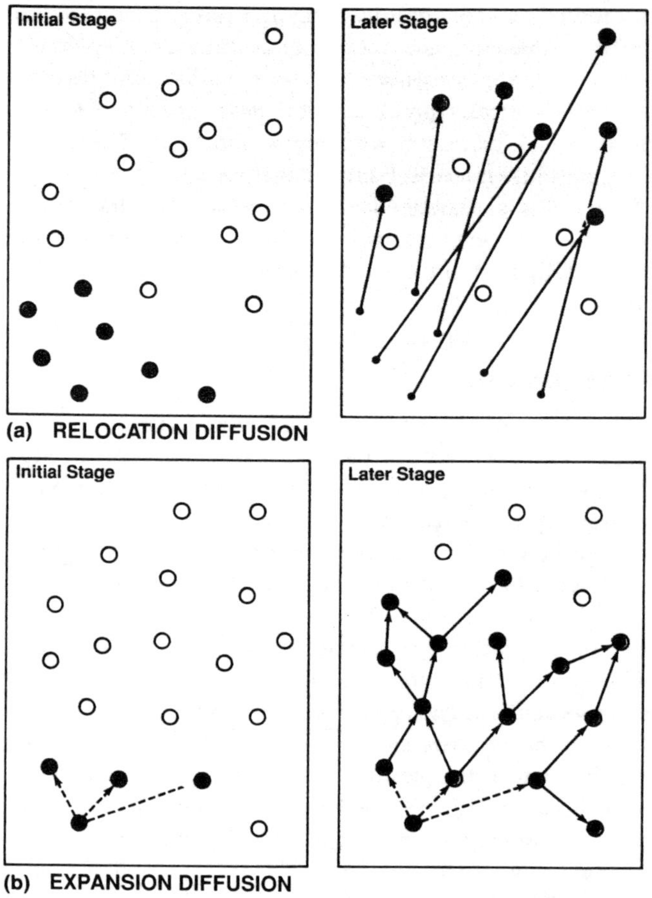

newly encounter it, and is reminiscent of the course of infectious diseases (Figure 2.23*a*). In some instances, however, geographic distance is less important in the transfer of ideas than is communication between major centers or important people. News of new clothing styles, for example, quickly spreads internationally between major cities and only later filters down irregularly to smaller towns and rural areas. The process of transferring ideas first between larger places or prominent people and only later to smaller or less important points or people is known as **cascade diffusion.** It always involves a systematic dispersal from higher to successively lower levels of a hierarchy (Figure 2.23*b*). In **hierarchical diffusion,** on the other hand, movement or dispersal may be either upward or downward through the hierarchy of classes or centers.

In relocation diffusion, the innovation or idea is physically carried to new areas by migrating individuals or populations that possess it (Figure 2.22*a*). Mentifacts or artifacts are therefore introduced into new locales by new settlers who become part of populations not themselves associated or in contact with the origin area of the innovation. The spread of religions by settlers or conquerors is a clear example of relocation diffusion. Christian Europeans brought their faiths to areas of colonization or economic penetration throughout the world; Islam was dispersed by conquest and settlement from its Near Eastern hearth across North Africa and into central and southern Asia. At the world scale, massive and pervasive relocation diffusion resulted from the European colonization and economic penetration that began in the 16th century. More localized relocation diffusion continues today as Asian refugees or foreign "guest workers" bring their cultural traits to their new areas of settlement in Europe or North America.

It is not always possible, of course, to determine the precise point of origin or the routes of diffusion of innovations now widely adopted (see "Documenting Diffusion"). Nor is it always certain whether the existence of a cultural trait in two different areas is the result of diffusion or of **independent** (or *parallel*) **invention.** Cultural similarities do not necessarily prove that diffusion has occurred. The pyramids of Egypt and of the Central American Maya civilization, pictured in Figure 2.24, most likely were separately conceived and are not evidence, as some have proposed, of pre-Columbian voyages from the Mediterranean to the Americas. A monument-building culture,

Figure 2.23

Two forms of expansion diffusion. (*a*) The process of *contagious diffusion* is sensitive to both time and distance, as suggested by the diffusion pathways of the European influenza pandemic of 1781. The pattern there was a wavelike radiation from a Russian nodal origin area. (*b*) The process of *cascade diffusion*, a special case of hierarchical diffusion, is one of transfer between the large and important before dispersal to the smaller or less important. New discoveries are shared among scientists at leading universities before they appear in textbooks or become general knowledge through the public press. Prosperous and locally important farmers in any region are the first to adopt new agricultural techniques that only later are understood and taken up by their less adventurous neighbors.

Sources: (*a*) Based on Gerald F. Pyle and K. David Patterson, *Ecology of Disease*, vol. 2, no. 3, 1984. p. 179; (*b*) Reproduced by permission from *Spatial Diffusion*, Peter R. Gould, page 6, Association of American Geographers, 1969.

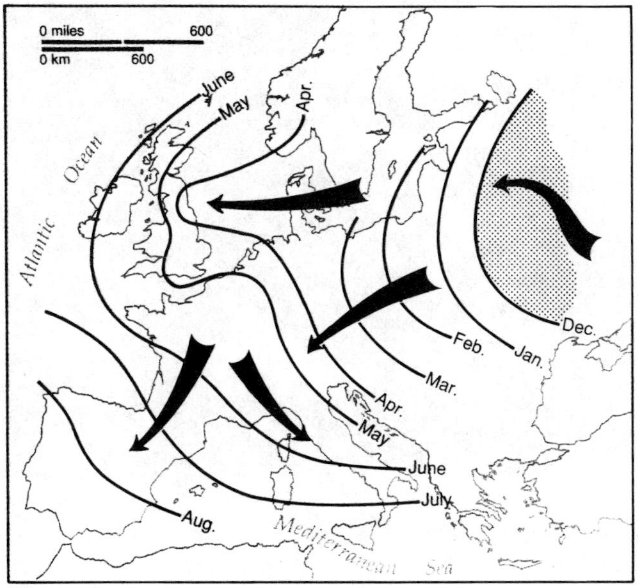

(a)

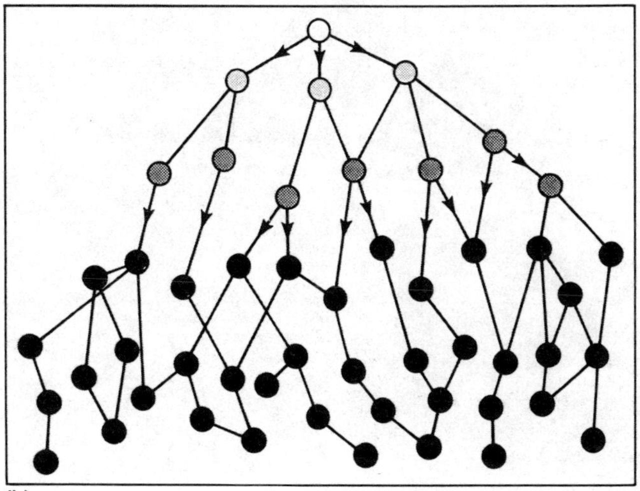

(b)

Documenting Diffusion

The places of origin of many ideas, items, and technologies important in contemporary cultures are only dimly known or supposed, and their routes of diffusion are at best speculative. Gunpowder, printing, and spaghetti are presumed to be the products of Chinese inventiveness; the lateen sail has been traced to the Near Eastern culture world. The moldboard plow is ascribed to 6th-century Slavs of northeastern Europe. The sequence and routes of the diffusion of these innovations has not been documented.

In other cases, such documentation exists, and the process of diffusion is open to analysis. For example, hybrid corn was originally adopted by imaginative farmers of northern Illinois and eastern Iowa in the mid-1930s. By the late 1930s and early 1940s the new seeds were being planted as far east as Ohio and north to Minnesota, Wisconsin, and northern Michigan. By the late 1940s, all commercial corn-growing districts of the United States and southern Canada were cultivating hybrid corn varieties.

Clearly marked as well is the diffusion path of the custom of smoking tobacco, a practice that originated among Amerindians. Sir Walter Raleigh's Virginia colonists, returning home in 1586, introduced smoking in English court circles, and the habit very quickly spread among the general populace. England became the source region of the new custom for northern Europe. Smoking was introduced to Holland by English medical students in 1590. Dutch and English together spread the habit by sea to the Baltic and Scandinavian areas and overland through Germany to Russia, where by 1634, laws were being passed to curb the practice of smoking. The innovation continued its eastward diffusion, and within a hundred years, tobacco had spread across Siberia and was, in the 1740s, reintroduced to the American continent at Alaska as both a habit and an item of trade carried by Russian fur traders. A second route of diffusion for tobacco smoking can be traced from Spain, where the custom was introduced in 1558, and from which it spread more slowly through the Mediterranean area into Africa, the Near East, and Southeast Asia.

Figure 2.24

Outward appearance is not the only similarity between Egyptian (left) and Mayan (right) pyramids. In both societies, the pyramids were royal burial sites, goods were placed inside the tombs to accompany the soul on its journey to eternity, and pains were taken to hide the entrance to the tomb.

after all, has only a limited number of shapes from which to choose. Historical examples of independent, parallel invention are numerous: logarithms by Napier (1614) and Burgi (1620), the calculus by Newton (1672) and Leibnitz (1675), the telephone by Elisha Gray and Alexander Graham Bell (1876) are commonly cited examples. It appears beyond doubt that agriculture was independently invented not only in both the New World and the Old but also in more than one culture hearth in each of the hemispheres.

Contact Between Regions

All cultures are amalgams of innumerable innovations spread spatially from their points of origin and integrated into the structure of the receiving societies. It has been estimated that no more than 10% of the cultural items of any society are traceable to innovations created by its members and that the other 90% come to the society through diffusion (see "A Homemade Culture"). Since, as we have seen, the pace of innovation is affected strongly by the mixing of ideas among alert, responsive people and is increased by exposure to a variety of cultures, the most active and innovative historical hearths of culture were those at crossroad locations and those deeply involved in distant trade and colonization. Ancient Mesopotamia and classical Greece and Rome had such locations and involvements, as did the West African culture hearth after the 5th century and, much later, England during the Industrial Revolution and the spread of its empire.

Recent changes in technology permit us to travel farther than ever before, with greater safety and speed, and to communicate without physical contact more easily and completely than previously possible. This intensification of contact has resulted in an acceleration of innovation and in the rapid spread of goods and ideas. Several millennia ago, innovations such as smelting of metals took hundreds of years to diffuse. Today, worldwide diffusion may be almost instantaneous.

Obstacles do exist, of course. **Diffusion barriers** are any conditions that hinder either the flow of information or the movement of people and thus retard or prevent the acceptance of an innovation. Because of the *friction of distance,* generally the farther two areas are from each other the less likely is interaction to occur. The concept of *distance decay* tells us that, all else being equal, the amount of interaction decreases as the distance between two areas increases. Distance as a factor in spatial interaction is further explored in Chapter 3. For now it is sufficient to note that distance may be an *absorbing barrier,* halting the spread of an innovation.

Interregional contact can also be hindered by the physical environment and by a lack of receptivity by a contacted culture. Oceans and rugged terrain can, and have, acted as physical *interrupting barriers,* delaying or deflecting the path of diffusion. Cultural obstacles that are equally impenetrable may also exist. For example, for at least 1500 years, most California Indians were in contact with cultures utilizing both maize and pottery, yet they failed to accept either innovation. Should such reluctant adopters intervene between hearths and receptive cultures, the spread of an innovation can be slowed. It can

A Homemade Culture

Reflecting on an average morning in the life of a "100% American," Ralph Linton noted the following:

Our solid American citizen awakens in a bed built on a pattern which originated in the Near East but which was modified in Northern Europe before it was transmitted to America. He throws back covers made from cotton, domesticated in India, or linen, domesticated in the Near East, or wool from sheep, also domesticated in the Near East, or silk, the use of which was discovered in China. All of these materials have been spun and woven by processes invented in the Near East. . . . He takes off his pajamas, a garment invented in India, and washes with soap invented by the ancient Gauls. . . .

Returning to the bedroom, . . . he puts on garments whose form originally derived from the skin clothing of the nomads of the Asiatic steppes [and] puts on shoes made from skins tanned by a process invented in ancient Egypt and cut to a pattern derived from the classical civilizations of the Mediterranean. . . . Before going out for breakfast he glances through the window, made of glass invented in Egypt, and if it is raining puts on overshoes made of rubber discovered by the Central American Indians and takes an umbrella invented in southeastern Asia. . . .

[At breakfast] a whole new series of borrowed elements confronts him. His plate is made of a form of pottery invented in China. His knife is of steel, an alloy first made in southern India, his fork a medieval Italian invention, and his spoon a derivative of a Roman original. He begins breakfast with an orange, from the eastern Mediterranean, a canteloupe from Persia, or perhaps a piece of African watermelon. With this he has coffee, an Abyssinian plant. . . . [H]e may have the egg of a species of bird domesticated in Indo-China, or thin strips of flesh of an animal domesticated in Eastern Asia which have been salted and smoked by a process developed in northern Europe.

When our friend has finished eating, . . . he reads the news of the day, imprinted in characters invented by the ancient Semites upon a material invented in China by a process invented in Germany. As he absorbs the accounts of foreign troubles he will, if he is a good conservative citizen, thank a Hebrew deity in an Indo-European language that he is 100 per cent American.

Ralph Linton, *The Study of Man: An Introduction.*

also be delayed when cultural contact is overtly impeded by governments that interfere with radio reception, control the flow of foreign literature, and discourage contact between their citizens and foreign nationals.

More commonly, barriers are at least partially *permeable*; they permit passage (acceptance) of at least some innovations encountering them. The more similar two cultural areas are to each other, the greater is the likelihood of the adoption of an innovation, for diffusion is a selective process. The receiver culture may adopt some goods or ideas from the donor society and reject others. The decision to adopt is governed by the receiving group's own culture. Political restrictions, religious taboos, and other social customs are cultural barriers to diffusion. The French Canadians, although close geographically to many centers of diffusion, such as Toronto, New York, and Boston, are only minimally influenced by such centers. Both their language and culture complex govern their selective acceptance of Anglo influences. Traditional groups, perhaps controlled by firm religious conviction, may very largely reject culture traits and technologies of the larger society in whose midst they live (Figure 7.2).

Adopting cultures do not usually accept intact items originating outside the receiving society. Diffused ideas and artifacts commonly undergo some alteration of meaning or form that makes them acceptable to a borrowing group. The process of the fusion of the old and new is called **syncretism** and is a major feature of culture change. It can be seen in alterations to religious ritual and dogma made by convert societies seeking acceptable conformity between old and new beliefs; the mixture of Catholic rites and voodooism in Haiti is an example. On a more familiar level, syncretism is reflected in subtle or blatant alterations of imported cuisines to make them conform to the demands of America's fast-food franchises (Figure 2.25).

Figure 2.25

Foreign foods modified for American tastes and American palates accustomed to dishes from all cultures together represent *syncretism* in action.

Cultural Modification and Adoption

A culture group may undergo major modifications in its own identifying traits by adopting some or all of the characteristics of another, dominant culture group. Such is the case in *acculturation*—discussed at greater length in Chapter 6 (pp. 171–172)—as immigrant populations take on the values, attitudes, customs, and speech of the receiving society. A different form of contact and subsequent cultural alteration may occur in a conquered or colonized region where the subordinate or subject population is either forced to adopt the culture, introduced through relocation diffusion, of the new ruling group or does so voluntarily, overwhelmed by the superiority in numbers or the technical level of the conqueror. Tribal Europeans in areas of Roman conquest, native populations in the wake of Slavic occupation of Siberia, and Native Americans stripped of their lands following European settlement of North America experienced this kind of cultural modification or adoption.

In many instances, close contact between two different groups may involve adjustments of the original cultural patterns of both. For example, changes in Japanese political organization and philosophy were imposed by occupying Americans after World War II, and the Japanese voluntarily adopted some more frivolous aspects of American life (Figure 2.26). In turn, American life was greatly enriched by the selective importation of Japanese cuisine, architecture, and philosophy, demonstrating the two-way nature of cultural diffusion.

Figure 2.26

Baseball, an import from America, is one of the most popular sports in Japan, attracting millions of spectators annually.

Summary

The web of culture is composed of many strands. Together, culture traits and complexes in their spatial patterns create human landscapes, define culture regions, and distinguish culture groups. Those landscapes, regions, and group characteristics change through time as human societies interact with their environment, develop for themselves new solutions to collective needs, or are altered through innovations adopted from outside the group itself. The cultural uniformity of a preagricultural world composed solely of hunter-gatherers was lost as domestication of plants and animals in many world areas led to the emergence of culture hearths of wide-ranging innovation and to a cultural divergence between farmers and gatherers. Innovations spread outward from their origin points, carried by migrants through relocation diffusion or adopted by others through a variety of expansion diffusion processes. Although diffusion barriers exist, most successful or advantageous innovations find adopters, and both cultural modification and cultural convergence of different societies result.The details of the technological, sociological, and ideological subsystems of culture define the differences that still exist between world areas.

The ivory hunters who opened our chapter showed how varied and complex the culture of even a primitive group can be. Their artifacts of clothing, fire making, hunting and fishing displayed diversity and ingenuity. They were part of a structured kinship system and engaged in organized production and trade. Their artistic efforts and ritual burial customs speak of a sophisticated set of abstract beliefs and philosophies. Their culture complex did not develop in isolation; it reflected at least in part their contacts with other groups, even those far distant from their Paris Basin homeland. As have culture groups always and everywhere, the hunters carried on their own pursuits and interacted with others in spatial settings. They exhibited and benefited from structured *spatial behavior*, the topic to which we next turn our attention.

Key Words

artifact 47
cultural convergence 46
cultural divergence 40
cultural integration 48
cultural landscape 35
culture 32
culture complex 34
culture hearth 43
culture realm 34
culture trait 33
diffusion 49
diffusion barrier 52
ideological subsystem 46
innovation 49
mentifact 46
sociofact 47
sociological subsystem 47
technological subsystem 47

For Review

1. What is included in the concept of *culture?* How is culture transmitted? What personal characteristics affect the aspects of culture that any single individual acquires or fully masters?
2. What do we mean by *domestication?* When and where did the domestication of plants and animals occur? What impact upon culture and population numbers did plant domestication have?
3. What is a *culture hearth?* What new traits of culture characterized the early hearths? Identify and locate some of the major culture hearths that emerged at the close of the Neolithic period.
4. What do we mean by *innovation?* By *diffusion?* What different patterns of diffusion can you describe? Discuss the role played by innovation and diffusion in altering the cultural structure in which you are a participant from that experienced by your great-grandparents.
5. Differentiate between *culture traits* and *culture complexes.* Between *environmental determinism* and *possibilism.*
6. What are the components or subsystems of the three-part system of culture? What are characteristics—the aspects of culture—included in each of the subsystems?

Suggested Readings

Brown, Lawrence A. *Innovation Diffusion: A New Perspective.* London and New York: Methuen, 1981.

Brown, Lawrence A., and Kevin R. Cox. "Empirical Regularities in the Diffusion of Innovation." *Annals of the Association of American Geographers* 61 (1971): 551–559.

Bryan, Alan L., ed. *New Evidence for the Pleistocene Peopling of the Americas.* Orono: University of Maine, Center for the Study of Early Man, 1986.

Butzer, Karl. *Early Hydraulic Civilization in Egypt: A Study in Cultural Ecology.* Chicago: University of Chicago Press, 1976.

Chen, Allan. "Unravelling Another Mayan Mystery." *Discover* 8, no.6 (June 1987): 40–49.

Coe, Michael, Dean Snow, and Elizabeth Benson. *Atlas of Ancient America.* New York: Facts on File Publications, 1986.

Crosby, Alfred W. *Ecological Imperialism: The Biological Expansion of Europe 900–1900.* Cambridge: Cambridge University Press, 1986.

Deane, Phyllis. *The First Industrial Revolution.* 2d ed. New York: Cambridge University Press, 1979.

Frankfort, Henri. *The Birth of Civilization in the Near East.* Garden City, N.Y.: Doubleday, 1956.

Glacken, Clarence. *Traces on the Rhodian Shore: Nature and Culture in Western Thought from Ancient Times to the End of the Eighteenth Century.* Berkeley: University of California Press, 1967.

Gould, Peter. *Spatial Diffusion.* Association of American Geographers, Commission on College Geography. *Resource Paper* No. 4. Washington, D.C.: Association of American Geographers, 1969.

Gowlett, John. *Ascent to Civilization.* London: Collins, 1984.

Henderson, John S. *The World of the Ancient Maya.* Ithaca, N.Y.: Cornell University Press, 1981.

Isaac, Erich. *Geography of Domestication.* Englewood Cliffs, N.J.: Prentice-Hall, 1970.

Keatinge, Richard W., ed. *Peruvian Prehistory.* New York: Cambridge University Press, 1987.

Kroeber, Alfred L., and Clyde Kluckhohn. "Culture: A Critical Review of Concepts and Definitions." Harvard University. *Papers of the Peabody Museum of American Archaeology and Ethnology* 47, No. 2 (1952).

"The Peopling of the Earth." *National Geographic* (October 1988): 434–503.

Pfeiffer, John E. *The Emergence of Society: A Prehistory of the Establishment.* New York: McGraw-Hill, 1977.

Pfeiffer, John E. *The Creative Explosion.* New York: Harper & Row, 1982.

Platt, Robert S. "Determinism in Geography." *Annals of the Association of American Geographers* 38 (1948): 126–132.

Redman, Charles L. *The Rise of Civilization: From Early Farmers to Urban Society in the Ancient Near East.* San Francisco: W. H. Freeman, 1978.

Reed, Charles A., ed. *Origins of Agriculture.* The Hague: Mouton, 1977.

Rogers, Everett M. *Diffusion of Innovations.* 3d ed. New York: Free Press, 1983.

Sauer, Carl. *Agricultural Origins and Dispersals.* New York: American Geographical Society, 1952.

Sjoberg, Gideon. "The Origin and Evolution of Cities." *Scientific American* 213 (1965): 54–63.

Spencer, Joseph E. and William L. Thomas, Jr. *Introducing Cultural Geography.* 2d ed. New York: John Wiley & Sons, 1978.

Steward, Julian H. *Theory of Culture Change.* Urbana: University of Illinois Press, 1955.

Thomas, William L., Jr., ed. *Man's Role in Changing the Face of the Earth.* Chicago: University of Chicago Press, 1956.

Ucko, Peter J., and G. W. Dimbleby. *The Domestication and Exploitation of Plants and Animals.* Research Seminar in Archaeology and Related Subjects, London University. London: Duckworth, 1969.

Wagner, Philip L., and Marvin W. Mikesell, *Readings in Cultural Geography.* Chicago: University of Chicago Press, 1962.

White, Leslie A. *The Science of Culture: A Study of Man and Civilization.* New York: Farrar, Straus & Giroux, 1969.

White, Randall. *Dark Caves, Bright Visions: Life in Ice Age Europe.* New York: American Museum of Natural History in Association with W. W. Norton & Company, 1986.

C H A P T E R

Behavioral Geography

Commuters in Tokyo, Japan.

EARLY in January of 1849 we first thought of migrating to California. It was a period of National hard times . . . and we longed to go to the new El Dorado and "pick up" gold enough with which to return and pay off our debts. Our discontent and restlessness were enhanced by the fact that my health was not good. . . . The physician advised an entire change of climate thus to avoid the intense cold of Iowa, and recommended a sea voyage, but finally approved of our contemplated trip across the plains in a "prairie schooner." Full of the energy and enthusiasm of youth, the prospects of so hazardous an undertaking had no terror for us, indeed, as we had been married but a few months, it appealed to us as a romantic wedding tour.*

So begins Catherine Haun's account of her journey from Clinton, Iowa, to California in 1849, a trip that was to last 9 months and cover 2400 miles (3900 km). The Hauns were just two of the 250,000 people who traveled across the continent on the Overland Trail in one of the world's great migrations. The dangers inherent in such a trip were numerous; thousands were to die en route, and many others stopped or turned back short of their goal. The migrants faced at least 6 months, and often more, of grueling travel over badly marked routes that crossed swollen rivers, deserts, and mountains. The weather was often foul, with hailstorms, drenching rains, and summer temperatures that could exceed 110°F (43°C) inside the covered wagons. Wagon breakdowns were frequent. Graves along the route were a silent testimony to the lives claimed by buffalo stampedes, Indian skirmishes, cholera epidemics, and other disasters.

What inducements were so great as to make emigrants leave behind all that was familiar and risk their lives on an uncertain venture? Catherine Haun's account is unusual in that it gives the reasons for their trip. She alludes to economic hard times; the depression that swept the United States in 1837 inaugurated a prolonged period of bank closures, depressed prices for agricultural goods, and high rates of unemployment. The Hauns hoped to strike it rich by mining gold. Other migrants were attracted by reports of free land and rich soil in the Oregon and California territories, of productive fishing grounds, and of ample furs for trapping. Like other migrants, the Hauns were also attracted by the climate in the West, which was said to be always sunny and free of disease. Finally, like most who undertook the trip along the Overland Trail, the Hauns were young, moved by restlessness and a sense of adventure.

*Source: Catherine Haun, "A Woman's Trip Across the Plains in 1849," as quoted in Lillian Schlissel, *Women's Diaries of the Westward Journey*, Schocken Books, New York, 1982.

Catherine Haun's story is unique only in its particulars. As did her predecessors back to the beginnings of humankind, she and her family acted in space and over space on the basis of acquired information and awareness of opportunity (Figure 9.1). Her story summarizes the content of this chapter, a survey of how individuals make spatial behavioral decisions and how those separate decisions may be summarized by models and generalizations to explain collective actions.

Behavioral geography is a relatively new field of the discipline of geography. In the last 30 years, geographers have become more interested in studying culture from the viewpoint of the individual. In previous chapters in this section we outlined what might be called the general patterns of cultural geography. In this chapter we pose a question basic to the themes that we have been exploring in preceding chapters: What considerations influence how

Figure 9.1
Cross-country movement was slow, arduous, and dangerous early in the 19th century, and the price of long-distance spatial interaction was far higher in time and risks than a comparable journey today.

individual human beings use space and act within it? Implicit in this question are subsidiary analytical concerns: How do individuals view (perceive) their environment? How is information transmitted through space and acted on? How might all the separate decisions of many individuals be summarized so that we may understand the order that underlies the seeming randomness of individual action?

We will begin with a discussion of human environmental perception. This leads to a consideration of how people define the space within which they carry out their activities. Then we turn to the way ideas are transmitted over space. Finally, we bring together all of these concepts in a section devoted to the migration process.

Perception of the Environment

The term **environmental perception** refers to our awareness, as individuals, of home and distant places and the beliefs we have about them. It involves our feelings, reasoned or irrational, about the complex of natural and cultural characteristics of an area. Whether our view accords with that of others or truly reflects the "real" world seen in abstract descriptive terms is not the major concern. Our perceptions are our reality. The decisions people make about the use of their lives are based not necessarily on reality but on their perceptions of reality.

Geographers interested in determining how we arrive at our environmental understanding employ the concept taken from psychology termed **cognition**. Cognition refers to the way individuals give mental meaning to the information they receive. Perceptions of the environment are sent to the brain where they are stored together with our previously accumulated environmental knowledge. The mental or cognitive structures influence the way we process and recall our perceptions. We might perceive the college classroom by physically being there, but our later recollections of the classroom depend on the way we mentally have organized our perceptions.

The fact that people have had different experiences has much to do with the way they perceive their environment. As people order new information, they store it within their own developing mental structures. Thus, it is not unusual for people to interpret similar phenomena in different ways. The more uniform the life experiences of a group of people, the greater the chance that they will have similar perceptions.

In technologically advanced societies, television and radio, magazines and newspapers, books and lectures, travel brochures and hearsay all combine to help develop a mental picture of unfamiliar places. The most effectively transmitted information seems to come from word-of-mouth reports. These may be in the form of letters or conversations with relatives, friends, and associates. Probably the strongest lines of attachment to relatively unknown regions, whether nearby or far away, develop through the information supplied by family members and friends.

Of course, our knowledge of close places is greater than our knowledge of far places. But barriers to information flow give rise to *directional biases*. Not having friends or relatives in one part of a country may represent a barrier to individuals, so that interest in and knowledge of the area beyond the "unknown" region are sketchy. In the United States, both northerners and southerners tend to be less well informed about each other's areas than about the western part of the country. Traditional communication lines in the United States follow an east-west rather than a north-south direction, which is the result of early migration patterns, business connections, and the pattern of the development of principal cities.

Mental Maps

When information about a place is sketchy, blurred pictures develop. These influence the impression we have of places and cannot be discounted. We might say that each individual has a **mental map** of the world. No single person, of course, has a true and complete image of the world; therefore there can be no completely accurate mental map. In fact the best mental map that most individuals have is that of their own residential neighborhood, the place where they spend the most time.

No one can reproduce on paper an exact replica of the mental image that he or she might have of an area. The study of mental maps must by necessity be indirect. If we want to know how particular people envisage their town, we must either ask them questions about the town or ask them to draw sketch maps. Although the result will not be a completely accurate picture of what they have in mind, it will be suggestive of the mental map.

Whenever individuals think about a place or how to get to a place, they produce a mental map. What are believed to be unnecessary details are left out, and only the important elements are incorporated (Figure 9.2). Those elements usually include awareness that the object or the destination does indeed exist, some conception of the distances separating the starting point and the named object(s), and a feeling for the directional relationships between points. A mental route map might also include reference points to be encountered on the chosen path of connection or on alternate lines of travel. Although mental maps are highly personalized, people with similar experiences tend to give similar answers to questions about the environment and to produce roughly comparable sketch maps.

Awareness of places is usually accompanied by opinions about them, but there is no necessary relationship between the depth of knowledge and the perceptions held. In general, the more familiar we are with a locale, the more sound will be the factual basis of our mental

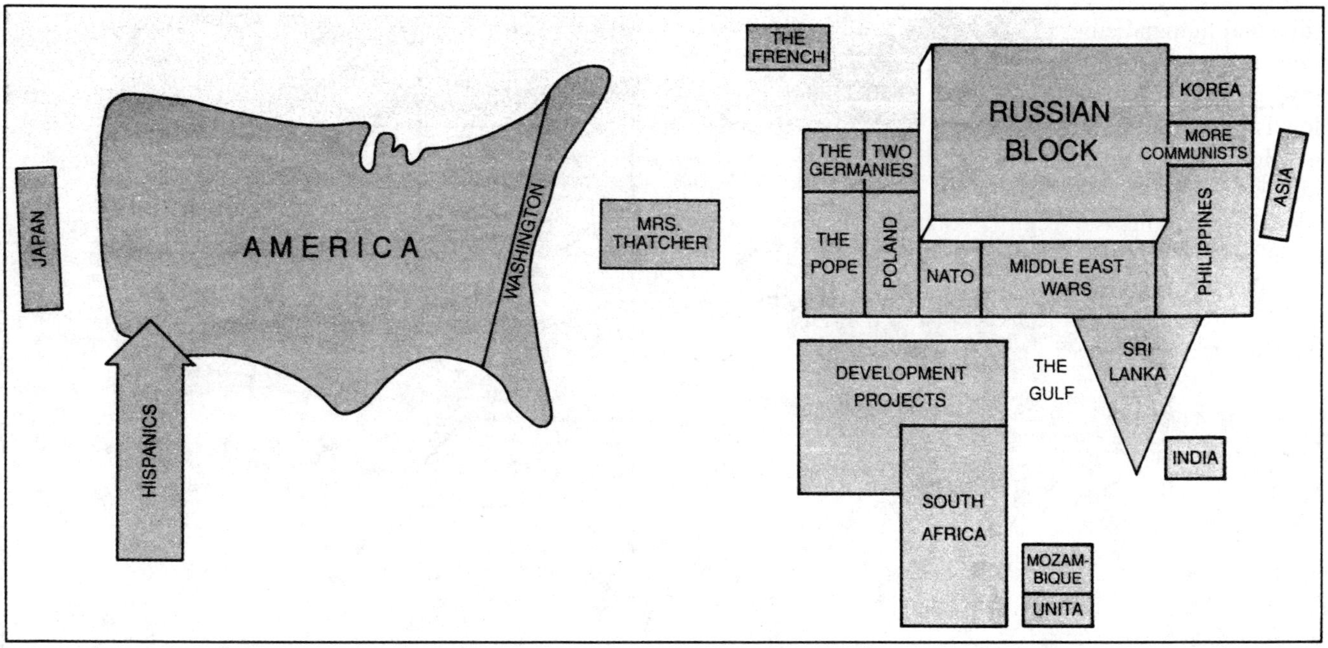

Figure 9.2
A mental map of the world as fashioned from newscasts heard by a Peace Corps volunteer working in the Central African Republic. The map is based on several months of listening to the Africa service of the Voice of America during 1987. Both areas and topics shown were those in the news and selected for emphasis by the broadcasters, whose choices helped form the mental maps and awareness patterns of their listeners. The "Russian Block" took on a three-dimensional reality not intended by the VOA. (Courtesy of Hilary Hope Getis.)

image of it. But individuals form firm impressions of places totally unknown by them personally, and these may color travel or migration decisions.

One way to ascertain how individuals envisage the environment is to ask them what they think of various places. For instance, they may be asked to rate places according to desirability—perhaps residential desirability—or to make a list of the best and worst places in a region such as the United States. Certain regularities appear in such studies. Figure 9.3 presents some residential desirability data as elicited from college students in three provinces of Canada. These and comparable mental maps suggest that near places are preferred to far places unless much information is available about the far places. Places with similar cultural forms are preferred, as are places with high standards of living. Individuals tend to be indifferent to unfamiliar places and to dislike unfamiliar areas that have competing cultural interests (such as disliked political and military activities) or a physical environment known to be unpleasant.

People tend mentally to increase the size of the familiar and to decrease the size of all else. They tend to place their own location in a central position and to increase the size of things nearby. The regional study, "The Yurok World View" (Chapter 13), demonstrates these conclusions in the collective mental map of a primitive society. Figure 9.4 gives two examples of the way people are affected by the familiar. Also, as we grow older our perspectives change, as shown in Figure 9.5.

Perception of Natural Hazards

Mental maps of home areas do not generally include as an overriding concern an acknowledgment of potential natural dangers. An intriguing area of geographical research deals with how people perceive *natural hazards*, defined as processes or events in the physical environment that are not caused by humans but that have consequences harmful to them. Most climatic (hurricanes, tornados, blizzards) and geological (earthquakes, volcanic eruptions) hazards cannot be prevented, and their consequences may be disastrous. The hurricanes that struck the delta area of Bangladesh in 1970 and again in 1985 left dead at least 500,000 people; the 1976 earthquake in the Tangshan area of China devastated a major urban-industrial complex with casualties reported between 700,000 and 1 million. These were major and exceptional natural hazards, but more common occurrences are experienced and are apparently discounted by those affected. Johnstown, Pennsylvania, has suffered recurrent floods, and yet its residents rebuild; violent storms strike the Gulf and East coasts of the United States, and people remain or return.

Why do people choose to settle in high-hazard areas, in spite of the potential threat to their lives and property? Why do hundreds of thousands of people live along the San Andreas fault in California, build houses in Pacific coastal areas known to experience severe erosion during storms (Figure 9.6), or farm in flood-prone areas adjacent

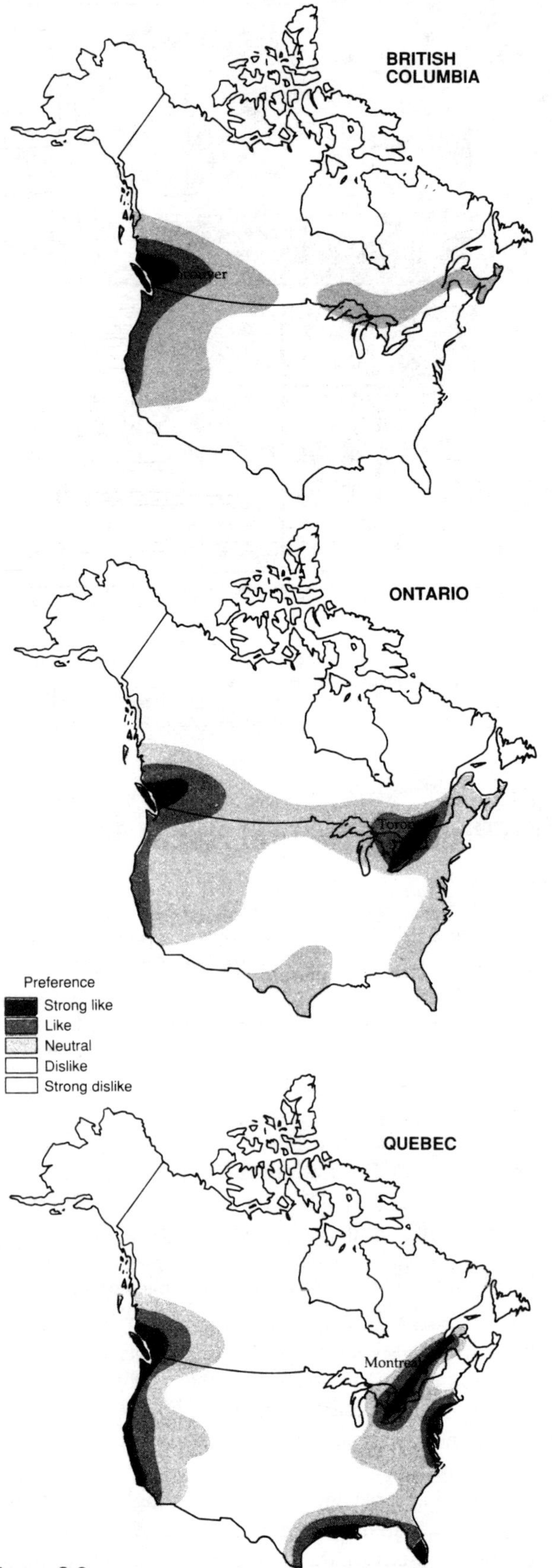

Figure 9.3

Each of these maps shows the residential preference of a sampled group of Canadians from the provinces of British Columbia, Ontario, and Quebec, respectively. Note that each group of respondents prefers its own area but that all like the Canadian and U.S. west coasts.

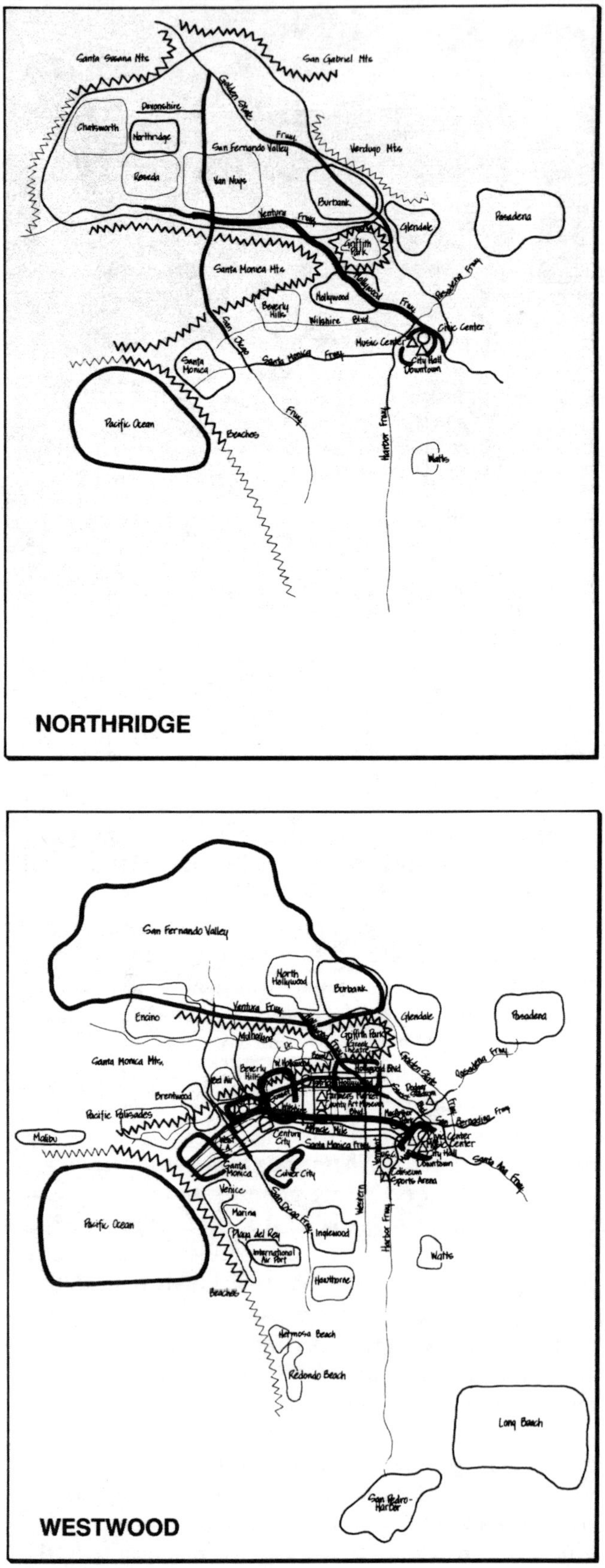

Figure 9.4

Four mental maps of Los Angeles. The upper-middle-income residents of Northridge and Westwood have expansive views of the metropolis, reflecting their mobility and area of travel.

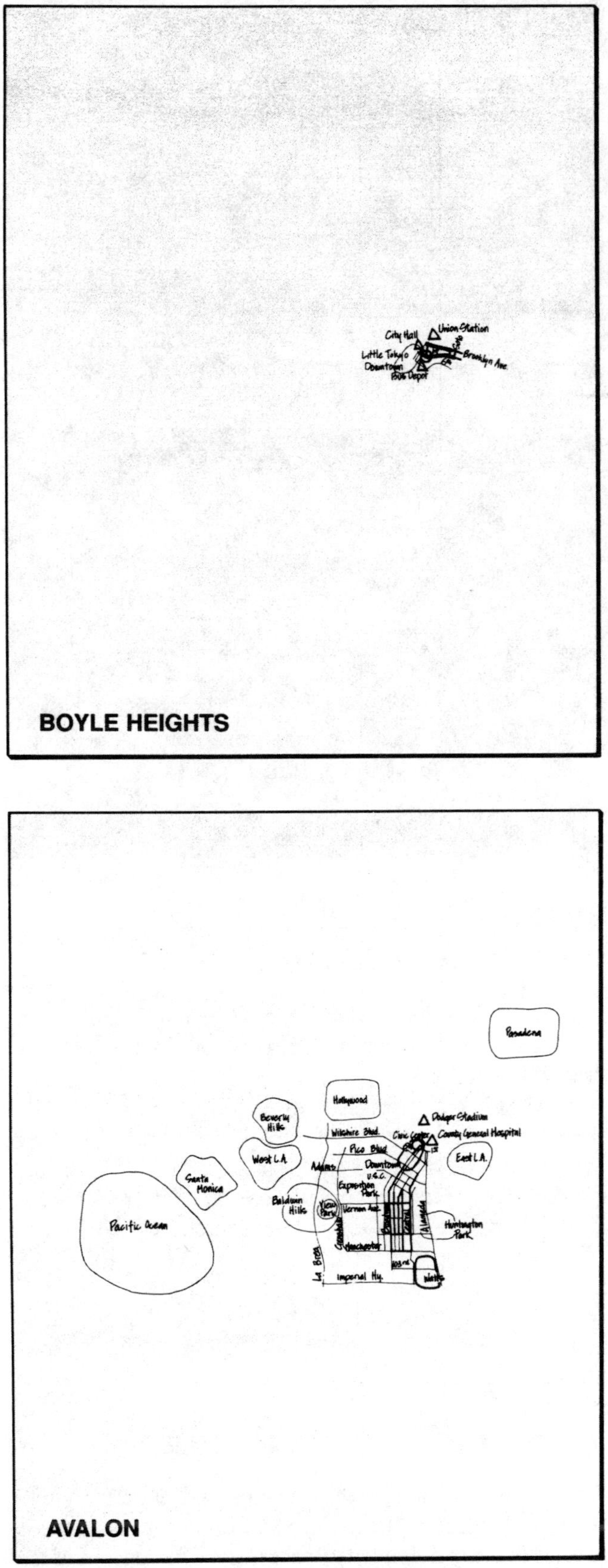

Figure 9.4 continued. Residents of Boyle Heights and Avalon, both minority districts, have a much more restricted and incomplete mental image of the city. Their limited mental maps reflect and reinforce their spatial isolation within the metropolitan area.

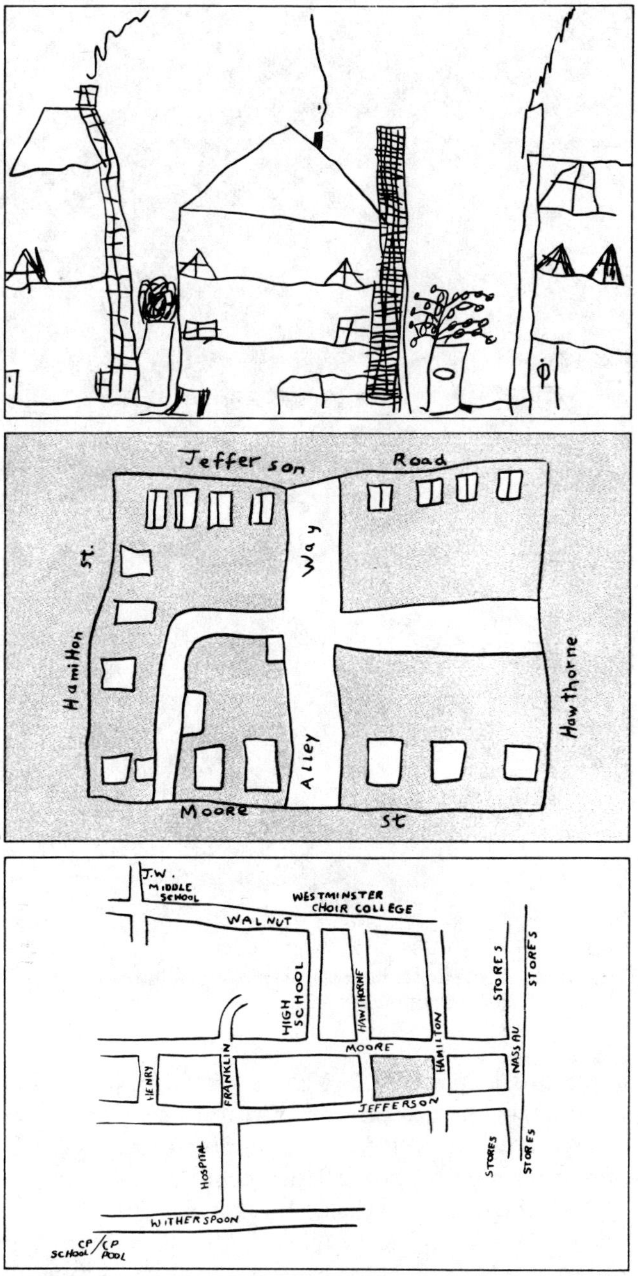

Figure 9.5

Three children, aged 6, 10, and 13, who lived in the same house, were asked to draw maps of their neighborhood. No further instructions were given. Notice how perspectives broaden and neighborhoods expand with age. For the 6-year-old, the neighborhood consisted of the houses on either side of her own. The square block on which she lived was the neighborhood for the 10-year-old. The wider activity space of the 13-year-old is also evident. The square block that the 10-year-old drew is shaded in the 13-year-old's sketch.

Figure 9.6
The aftermath of Hurricane Hugo, which struck the coast of South Carolina in September 1989.

to the Mississippi River? What is it that makes the risk worth taking? There are many reasons that hazardous areas are perceived differently by people and thus that some people choose to settle in them. Of major importance is the fact that specific hazards are relatively rare occurrences. Many people think that the likelihood of an earthquake, flood, or other natural calamity is sufficiently remote that it is not economically feasible to protect themselves against it. They are also influenced by the fact that the scientists who study such hazards may themselves differ on the probability of an event or on the damage that it may inflict. And, in fact, the prediction of hazards is not an exact science, being based on the calculation of the probability of occurrence of what are uncommon events.

People are also influenced by their past experiences in high-hazard areas. If they have not suffered much damage in the past, they may be optimistic about the future. If, on the other hand, past damage has been great, they may think that the probability of similar occurrences in the future is low (Table 9.1). People's memories can be short. In the years following an earthquake, for example, a sense of security grows, building codes or their interpretation are relaxed, and population in the area increases.

High-hazard areas are often sought out not because they pose risks but because they possess desirable topography or scenic views, as do, for instance, the Atlantic and Pacific coasts. Once people have purchased property in a known hazard area, they may be unable to sell it for a reasonable price even if they so desire. They think that they have no choice but to remain and protect their investment. The cultural hazard—loss of livelihood and investment—appears to be more serious than whatever natural hazards there may be.

Individual Activity Space

We saw in Chapter 8 that groups and countries draw boundaries around themselves and divide space into territories that are defended, if necessary. The concept of **territoriality**—the emotional attachment to and the defense of home ground—has been seen by some as a root explanation of much of human action and response. It is true that some collective activity appears to be governed by

Table 9.1
Common Responses to the Uncertainty of Natural Hazards

Eliminate the Hazard	
Deny or denigrate its existence	*Deny or denigrate its recurrence*
"We have no floods here, only high water."	"Lightning never strikes twice in the same place."
"It can't happen here."	"It's a freak of nature."
Eliminate the Uncertainty	
Make it determinate and knowable	*Transfer uncertainty to a higher power*
"Seven years of great plenty. . . . After them seven years of famine."	"It's in the hands of God."
"Floods come every five years."	"The government is taking care of it."

territorial defense responses: the conflict between street groups in claiming and protecting their "turf" (and their fear for their lives when venturing beyond it) and the sometimes violent rejection by ethnic urban neighborhoods of an encroaching black, Hispanic, or other population group.

But for most, our personal sense of territoriality is a tempered one. Homes and property are regarded as defensible private domains but are opened to innocent visitors, known or unknown, or to those on private or official business. Nor do we confine our activities so exclusively within controlled home territories as street-gang members do within theirs. Rather, we have a more or less extended home range, an **activity space** within which we move freely on our rounds of regular activity, sharing that space with others who are also about their daily affairs. Figure 9.7 suggests a probable activity space for a suburban family of five for one day. Note that the activity space for each individual for one day is rather limited, even though two members of the family use automobiles. If one week's activity were shown, more paths would have to be added to the map, and in a year's time, several long trips would probably have to be noted. Because long trips are taken irregularly, we will confine our idea of activity space to often-visited places.

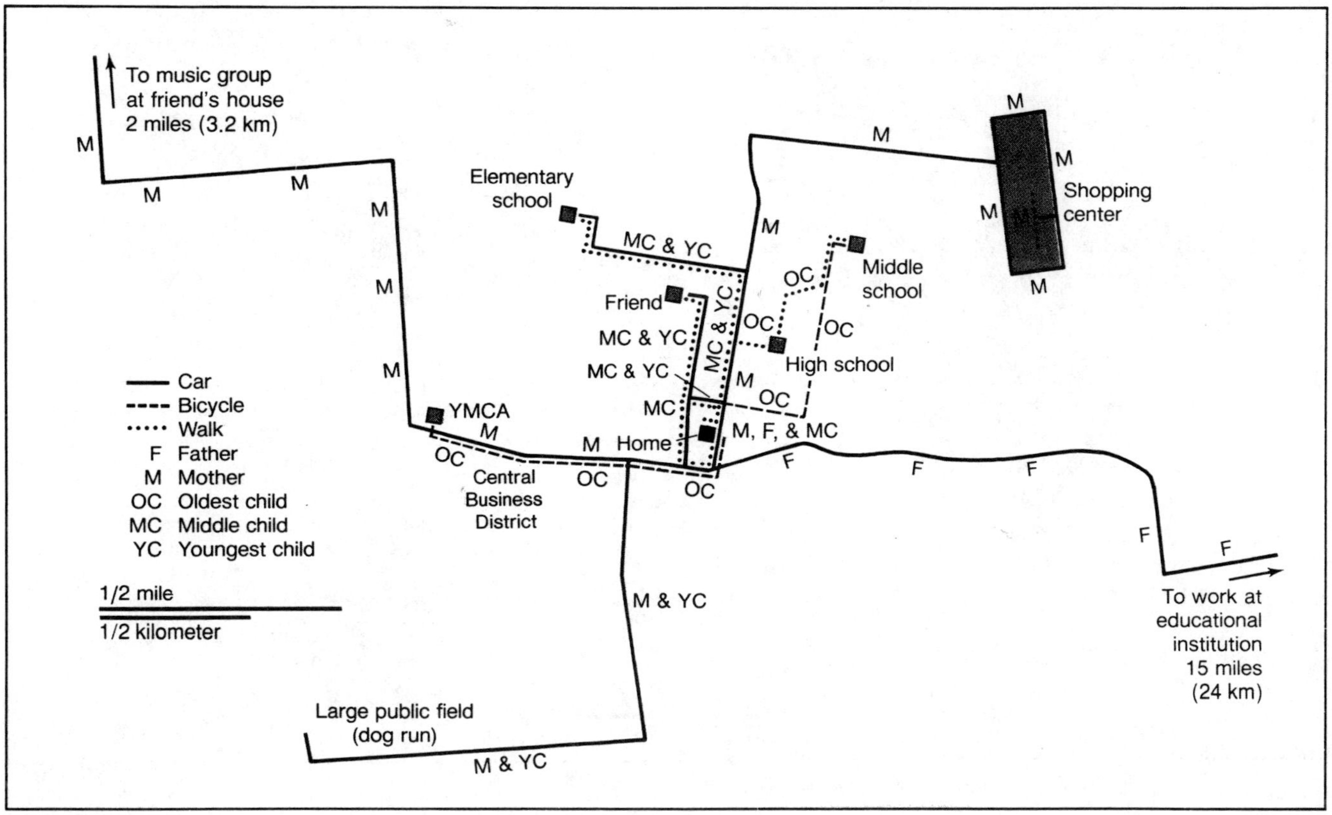

Figure 9.7
Activity space for each member of a family of five for a typical weekday. Routes of regular movement and areas recurrently visited help to foster a sense of territoriality and to color one's perceptions of space.

The kind of activities individuals engage in can be classified according to type of trip: journeys to work, to school, to shops, for recreation, and so on. People in nearly all parts of the world make these same types of journeys, though the spatially variable requirements of culture and economy dictate their frequency, duration, and significance in the time budget of an individual. Figures 9.8, 9.9, and 9.10 illustrate this point. Figure 9.8 depicts variations in travel patterns in two different culture groups in rural midwestern Canada. It suggests that "modern" rural Canadians, who want to take advantage of the variety of goods offered in the regional capital, are willing to travel longer distances than are people of a traditionalist culture who have different tastes in clothing and consumer goods and whose demands are satisfied in local settlements. Also, in this case the traditionalists do not own cars, thus limiting their spatial range. Figures 9.9 and 9.10 suggest the importance of the journey to work among urban populations in two different cities.

The types of trips that individuals make and thus the extent of their activity space depend on at least three variables: their stage in the life cycle, the means of mobility at their command, and the demands or opportunities implicit in their daily activities. The first, *stage in the life cycle*, refers to membership in specific age groups. Stages include preschooler, school-age, young adult, adult, and elderly. Preschoolers stay close to home unless they accompany their parents. School-age children usually travel short distances to lower schools and longer distances to upper-level schools. After-school activities tend to be limited to walking or bicycle trips to nearby locations. High-school students are usually more mobile and take part in more activities than do younger children. Adults responsible for household duties make shopping trips and trips related to child care as well as journeys away from home for social, cultural, or recreational purposes. Wage-earning adults usually travel farther from home than other family members. Elderly people normally do not find it feasible or desirable to have extended activity spaces.

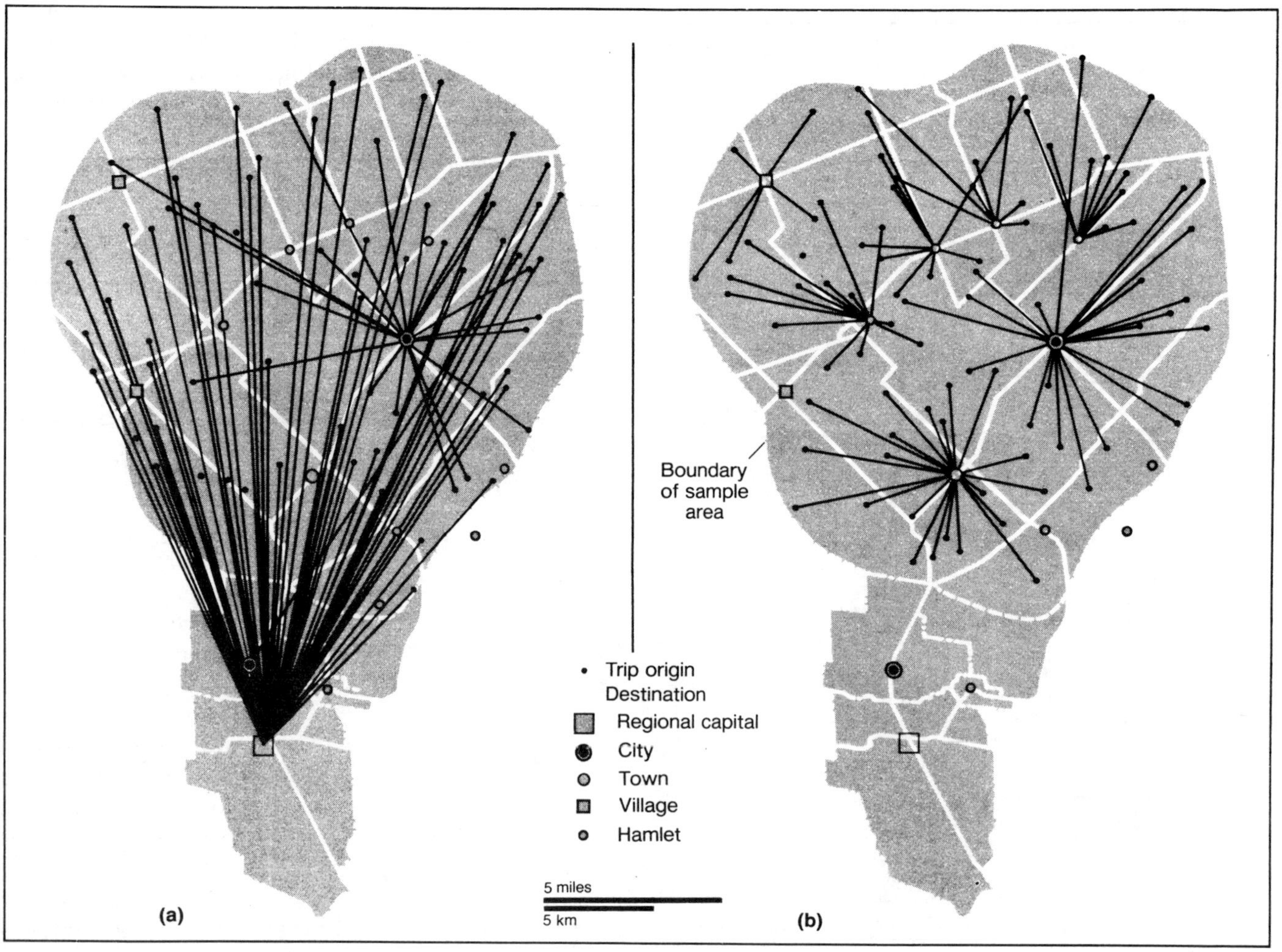

Figure 9.8

Travel patterns for purchase of clothing and yard goods of (a) rural cash-economy Canadians and (b) Canadians of the old-order Mennonite sect. These strikingly different travel behaviors demonstrate the great differences that may exist in the action spaces of different culture groups occupying the same territory.

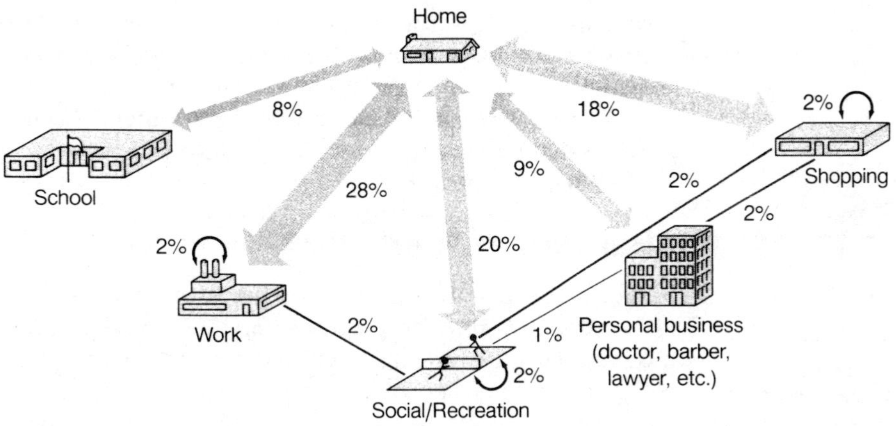

Figure 9.9

Chicago travel patterns. The numbers are the percentages of all urban trips taken in Chicago. The greatest single movement is the journey to and from work. Over 96% of all trips are represented on the diagram.

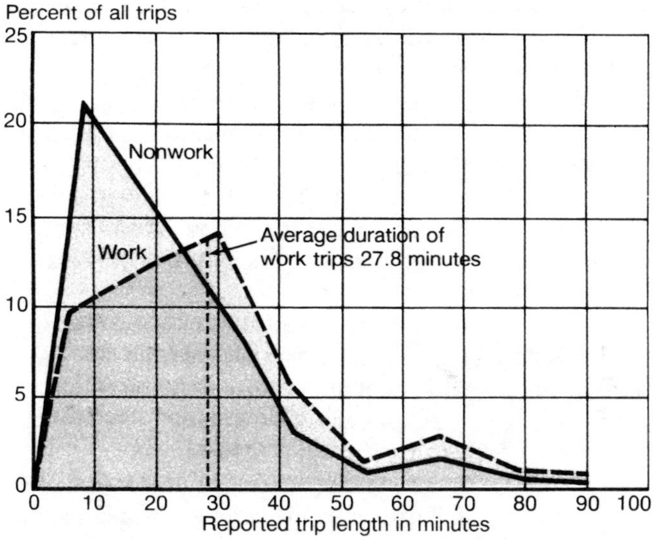

Figure 9.10

The frequency distribution of work and nonwork trip lengths in minutes in Toronto. Work trips are usually longer than other internal city journeys. For North America, 20 minutes seems to be the upper limit of commuting time before people begin to consider time seriously in their choice of homesite. Most nonwork trips are very short.

The second variable that affects the extent of activity space is *mobility*, or the ability to travel. An informal consideration of the cost and effort required to overcome the friction of distance is implicit. Where incomes are high, automobiles are available, and the cost of fuel is a minor item in the family budget, mobility may be great and individual action space large. In societies where cars are not a standard means of personal conveyance, the daily activity space may be limited to the shorter range afforded by the bicycle or by walking. Obviously, both intensity of purpose and the condition of the roadway affect the execution of movement decisions.

The mobility of individuals in countries or in sections of countries with high incomes is relatively great; their activity space horizons are broad. These horizons, however, are not limitless. There are a fixed number of hours in a day, most of them consumed in performing work, preparing and eating food, and sleeping. In addition, there are a fixed number of road, rail, and air routes, so that even the most mobile individuals are constrained in the amount of activity space that they can use. No one can easily claim the world as his or her activity space.

A third factor limiting activity space is the individual assessment of the availability of possible activities or *opportunities*. In the subsistence economies discussed in Chapter 10, the needs of daily life are satisfied at home, so the impetus for journeys away from the residence is minimal. If there are no stores, schools, factories, or roads, one's expectations and opportunities are limited, and one's activity space is therefore reduced. In impoverished nations or neighborhoods, low incomes limit the inducements, opportunities, destinations, and necessity of travel.

Distance and Spatial Interaction

People make many more short-distance trips than long ones. In the words of the behavioral geographer, there is greater human interaction over short distances than long distances. If we drew a boundary line around our activity space, it would be evident that trips to the boundary are taken much less often than short-distance trips near the home. Think of activity space as more intensively used (greater spatial interaction) near one's home place or base and as declining in use with increasing distance from the base. This is the principle of **distance decay** (introduced briefly in Chapter 1), the exponential decline of an activity, function, or amount of interaction with increasing distance from the point of origin. The tendency is for the frequency of trips to fall off very rapidly beyond an individual's **critical distance**. Figure 9.11 illustrates this principle with regard to journeys from the homesite.

The critical distance is the distance beyond which cost, effort, and perception play an overriding role in our willingness to travel. A small child, for example, will make many trips up and down its block but is inhibited by parental admonitions from crossing the street. Different but

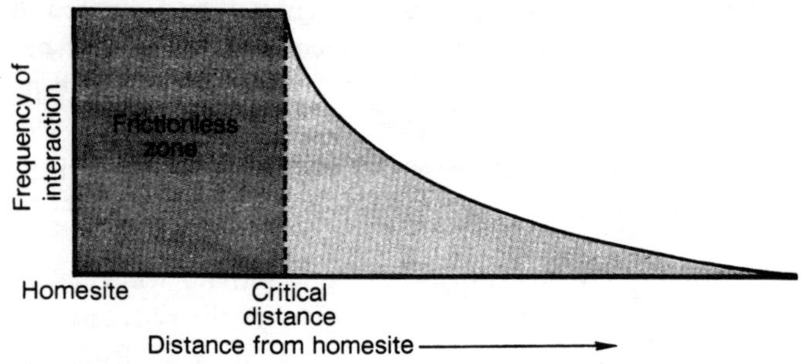

Figure 9.11

This general diagram indicates how distance is observed by most people. For each activity, there is a distance beyond which the intensity of contact declines. This is called the *critical distance*, if distance alone is being considered, or the *critical isochrone*, if time is the measuring rod. For the distance up to the critical distance, a frictionless zone is identified in which time or distance considerations do not effectively figure in the trip decision.

equally effective constraints control adult behavior. Daily or weekly shopping may be within the critical distance of an individual, and little thought may be given to the cost or the effort involved, but shopping for special goods is relegated to infrequent trips, and cost and effort are considered.

Effort may be measured in terms of *time-distance*, that is, the time required to complete the trip. For the journey to work, often time rather than cost plays the major role in the determination of the critical distance. When there are significant differences between our cognition of distance and real distance, we use the term *psychological distance* to describe our perception of distance. A number of studies show that people tend to psychologically consider known places as nearer than they really are and little-known places as farther than true distance.

Spatial Interaction and the Accumulation of Information

The critical distance is different for each person. The variables of life-cycle stage, mobility, and opportunity, together with an individual's interests and demands, help to define how much and how far a person will travel. On the basis of these variables, we can make inferences about the likelihood that a person will gain more or less information about his or her activity space and the space beyond.

We gain information about the world from many sources. Although information obtained from radio, television, and newspapers is important to us, face-to-face contact is assumed to be the most effective means of communication. If we combine the ideas of activity space and distance decay, we see that as the distance away from the home place increases, the number of possible face-to-face contacts usually decreases. We expect more human interactions at short distances than at long distances. Where population densities are high, such as in cities (particularly central business districts during business hours), the spatial interaction between individuals can be at a very high level, which is one reason that these centers of commerce and entertainment are often also centers for the development of new ideas.

Barriers to Interaction

Recent changes in technology permit us to travel farther than ever before, with greater safety and speed, and to communicate without physical contact more easily and completely than previously possible. This intensification of contact has resulted in an acceleration of innovation and in rapid spread of goods and ideas. Several millennia ago, innovations such as smelting of metals took hundreds of years to spread. Today, worldwide diffusion may be almost instantaneous.

The fact that the possible number of interactions is high, however, does not necessarily mean that the effective occurrence of interactions will be high. That is, there are a number of barriers to interaction. Such barriers are any conditions that hinder either the flow of information or the movement of people, and thus retard or prevent the acceptance of an innovation.

Distance itself is a barrier to interaction. Generally, the farther two areas are from one another, the less likely is interaction. The concept of *distance decay* says that, all else being equal, the amount of interaction decreases as the distance between two areas increases.

Cost represents another barrier to interaction. Relatives, friends, and associates living long distances apart may find it difficult to afford to interact. The frequency and time allocated to telephone communication, a relatively inexpensive form of interaction, is very much a function of the rate structure—which, of course, favors short-distance interaction.

Interregional contact can also be hindered by the physical environment and by the cultural barriers of differing religions, languages, ideologies, and political systems. Mountains and deserts, oceans and rivers can, and have, acted as physical barriers slowing or preventing diffusion. Cultural barriers may be equally impenetrable. For at least 1500 years, for example, most American Indians in California were in contact with cultures utilizing both maize and pottery, yet they failed to accept either innovation. Should such nonadopters intervene between hearths and receptive cultures, the spread of an innovation can be delayed. It can also be delayed when cultural contact is overtly impeded by governments that interfere with radio reception, control the flow of foreign literature, and discourage contact between their citizens and foreign nationals.

Even when some members of a given population are receptive to change, there will be others who will not innovate, that is, who will not accept new ideas. In some societies—particularly traditional, dispersed agricultural ones—the number of nonadopters for any innovation may be very high indeed. In other societies, where innovation and change are given high status, and where the innovations are not risky, ideas spread much more rapidly.

Finally, in crowded areas, people commonly set psychological barriers around themselves so that only a limited number of interactions take place. The barriers are raised in defense against information overload and for psychological well-being. We must have a sense of privacy in order to filter out the information that does not directly concern us. As a result, individuals tend to reduce their interests to a narrow range when they find themselves in crowded situations, allowing their wider interests to be satisfied by use of the communications media.

Spatial Interaction and Innovation

The probability that new ideas will be generated out of old ideas is a function of the number of available old ideas in contact with one another. People who specialize in a particular field of interest seek out others with whom they wish to interact. Crowded central cities are characteristically composed of specialists in very narrow fields of interest. Consequently, under short-distance, high-density circumstances, the old ideas are given a hearing and new ideas are generated by the interaction. New inventions and new social movements usually arise in circumstances of high spatial interaction. An exception, of course, is the case of intensely traditional societies—Japan in the 17th and 18th centuries, for example—where the culture rejects innovation and clings steadfastly to customary ideas and methods.

The culture hearths (discussed in Chapter 7) of an earlier day were the most densely settled, high-interaction centers of the world. At the present time, the great national and regional capital cities attract people who want or need to interact with others in special-interest fields. The association of population concentrations and the expression of human ingenuity has long been noted. The home addresses recorded for patent applicants by the U.S. Patent Office over the last century indicate that the inventors were typically residents of major urban centers, presumably closely in contact and exchanging ideas with those in shared fields of interest. It still appears that the metropolitan centers of the world attract those who are young and ambitious, and that face-to-face or word-of-mouth contact is important in the creation of new ideas and products. The recent revolution in communications that now allows for inexpensive interaction by a variety of telephone services and computer equipment have suggested to some that the traditional importance of cities as collectors of creative talent may decline in the future.

Diffusion and Innovation

As we noted in Chapter 7, **diffusion** is the process by which a concept, practice, or substance spreads from its point of origin to new territories. This concept will be discussed in more detail here since it is at the conceptual heart of behavioral geography. When such aspects of a culture as ideas, behaviors, or substances are transferred from one place to another through the migration of people, the process is called *relocation diffusion*. It is discussed later in this chapter under the heading "Migration."

Ideas generated in a center of activity will remain there unless some process is available for their spread. *Innovations*, the changes to a culture that result from the adoption of new ideas, spread in various ways. Some new inventions are so obviously advantageous that they are quickly put to use by those who can afford and profit from them. A new development in petroleum extraction may promise such material reward as to lead to its quick adoption by all of the major petroleum companies, irrespective of their distance from the point of introduction. The new strains of wheat and rice that were part of the Green Revolution were quickly made known to agronomists in all cereal-producing countries, but they were more slowly taken up in poor countries, which could benefit most from them, partly because the farmers had difficulty paying for them.

Many innovations are of little consequence by themselves, but sometimes the widespread adoption of seemingly inconsequential innovations brings about large changes when viewed over a period of time. A new musical tune, "adopted" by a few people, may lead many individuals to fancy that tune plus others of a similar sound, which in turn may have a bearing on dance routines, which in turn may bear on clothes selection, which in turn may affect retailers' advertising campaigns and consumers' patterns of expenditure. Eventually a new cultural form will be identified that may have an important impact on the thinking processes of the adopters and on those who come into contact with the adopters. Notice that a broad definition of innovation is used, but notice also that what is important is whether or not innovations are adopted.

In spatial terms, we may identify a number of processes for the diffusion of innovations. Each is based on the manner in which innovations spread from person to person and, therefore, from place to place. These are discussed below.

Contagious Diffusion

Let us suppose that a scientist develops a gasoline additive that noticeably improves the performance of his or her car. Suppose further that the person shows friends and associates the invention and that they in turn tell others. This process is similar to the spread of a contagious disease. The innovation will continue to diffuse until barriers are met (that is, people not interested in adopting the new idea) or until the area is saturated (that is, all available people have adopted the innovation). This **contagious diffusion** process follows the rules of distance decay at each step. Short-distance contacts are more likely than long-distance contacts, but over time the idea may have spread far from the original site. Figure 9.12 illustrates how a theoretical contagious diffusion process differs from relocation diffusion.

A number of characteristics of this kind of diffusion are worth noting. If the idea has merit in the eyes of potential adopters, the number of contacts of adopters with potential adopters will compound. Consequently, the innovation will spread slowly at first and then more and more rapidly, until there is saturation or a barrier is reached. The incidence of adoption is represented by the S-shaped curve in Figure 9.13 and shown for an actual case in Figure 9.14. In a similar manner, the area in which the adopters are located will at first be small, and then will enlarge at a faster and faster rate. The spreading process will slow as the available areas and/or people decrease.

If an inventor's idea fell into the hands of a commercial distributor, the diffusion process might follow a somewhat different course than that discussed above. The distributor might "force" the idea into the minds of individuals by using the mass media. If the media were local in impact, such as newspapers, then the pattern of adoptions would be similar to that described earlier (Figure 9.15). If, however, a nationwide television, newspaper, or magazine advertising campaign were undertaken, the innovation would become known in numbers roughly corresponding to the population density. Where more people live, there would, of course, be more potential adopters. Economic or other barriers may also affect the diffusion. One immediately sees, however, why large TV markets are so valuable and why national advertising is so expensive.

This type of contagious diffusion process may act together with the distance-decay process. Many of those who accept the innovation after learning of it in the mass media will tell others, so that a locally contagious effect will begin to take over soon after the original contact is made. Each

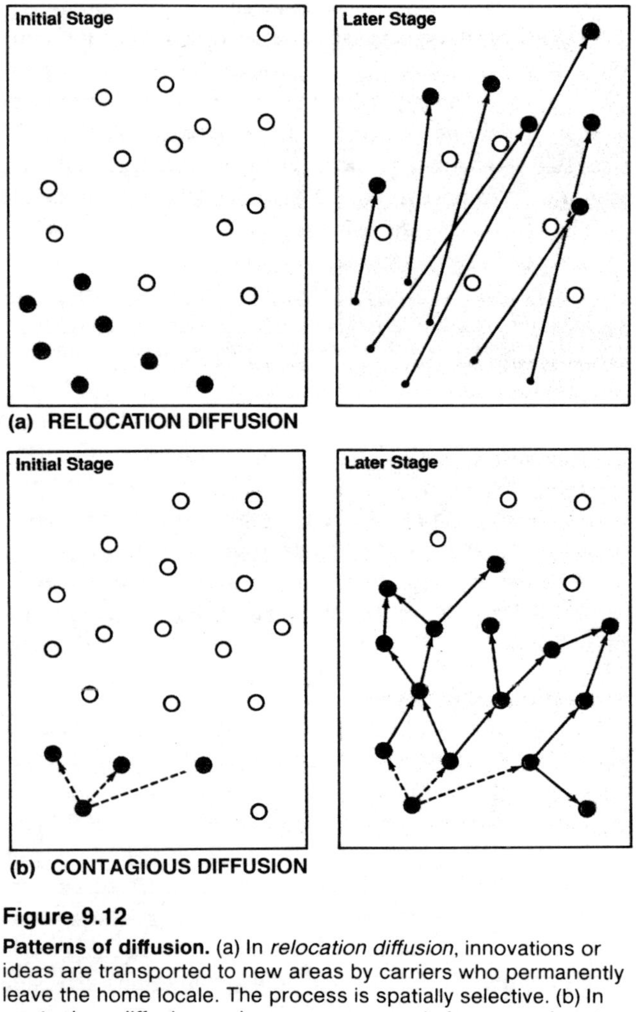

Figure 9.12

Patterns of diffusion. (a) In *relocation diffusion*, innovations or ideas are transported to new areas by carriers who permanently leave the home locale. The process is spatially selective. (b) In *contagious diffusion*, a phenomenon spreads from one place to neighboring locations, but in the process it remains and is often intensified in the place of origin.

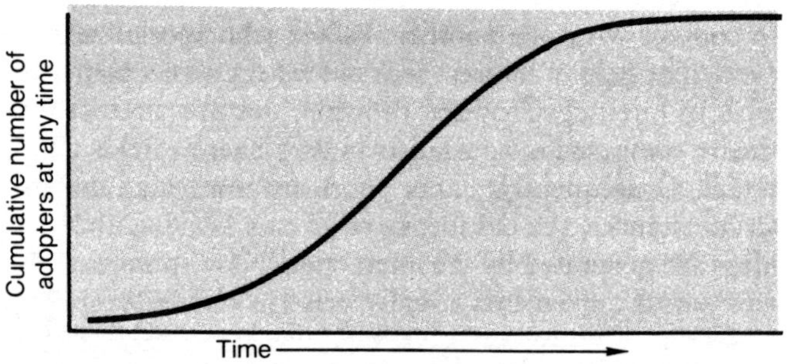

Figure 9.13

The diffusion of innovations over time. The number of adopters of an innovation rises at an increasing rate until the point at which about one-half of the total who ultimately decide to adopt the innovation have made the decision. At this point, the number of adopters increases at a decreasing rate.

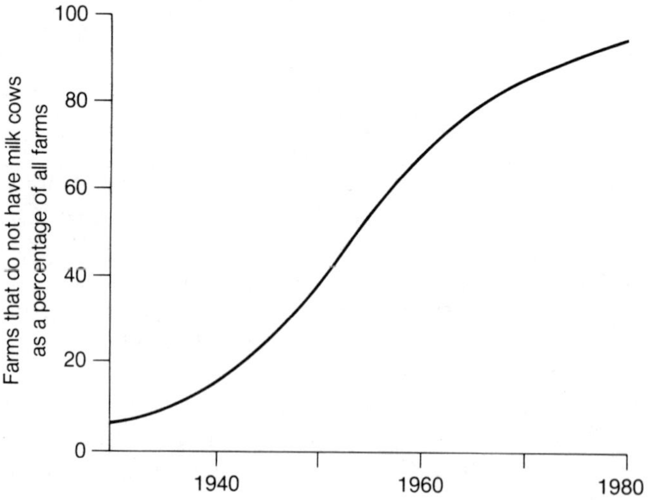

Figure 9.14

Farmers in Illinois found it disadvantageous to herd dairy cattle during the period 1930 to 1978. The idea of discontinuing having milk cows was adopted slowly but increased rapidly from 1940 to 1960. Compare this real example with the theoretical diagram shown in Figure 9.13.

type of medium has its own level of effectiveness. Advertisers have found that they must repeat messages time and again before they are accepted as important information. This fact says something about the effectiveness of the mass media as opposed to, say, face-to-face contact.

Hierarchical Diffusion

A second way innovations are spread combines some aspects of contagious diffusion with a new element: the hierarchy. A *hierarchy* is a classification of objects into categories so that each category is increasingly complex or has increasingly higher status. Hierarchies are found in many systems of organization, such as government offices (the organization chart), universities (instructors, professors, deans, president), and cities (villages, towns, regional centers, metropolises). By **hierarchical diffusion** we mean the spread of innovation up or down a hierarchy of places.

As an example, let us suppose that a new way of processing passengers is adopted at an airport in a major city. Information on the innovation is spread, but only officials at comparably-sized airports are in a position to accept the idea at first. It may be that the quality of information diffused to the larger airports is better or that the larger airports are more financially able to adopt the idea than smaller airports. Eventually, the innovation is adopted at the airports of somewhat smaller cities, and so on down the hierarchy as it becomes better known or as it becomes financially feasible. A hypothetical scheme showing how a four-level hierarchy might be connected in the flow of information is presented in Figure 9.16. Note that the lowest-level centers are connected to higher-level centers but not to each other. Note, too, that connections might bypass intermediate levels and link only with the highest-level center.

Many times, hierarchical diffusion takes place simultaneously with contagious diffusion. One might expect variations when the density of high-level centers is great and when the distances between centers are short. A quick and inexpensive way to spread an idea is to communicate information about it at high-order hierarchical levels. Then the three types of diffusion processes may be used most effectively; even while an idea is diffusing through a high level in the hierarchy, it is also spreading outward from the high-level centers. Consequently, low-level centers that are a short distance from high-level centers may be apprised of the innovation before more distant medium-level centers. People living in suburbs and small towns near a large city are privy to much that is new in the large city, as are individuals in other large cities half a continent away. Figures 9.17 and 9.18 show these patterns for a case taken from Japan.

These forms of diffusion operate in the spread of culture. The consequences are the spatial interaction and innovation discussed earlier in this chapter. We should also recall from Chapter 7 that migration, invasions, selective cultural adoptions, and cultural transference aid the diffusion of innovation. These broader movements and exchanges represent interactions of people beyond their usual activity spaces.

Figure 9.15
A street scene in Guatemala City. In modern society advertising is a potent force for diffusion. Advertisements over radio and television, in newspapers and magazines, and on billboards and signs communicate information about many different products and innovations.

Migration

An important aspect of human history has been the migration of peoples, the evolution of their separate cultures, and the relocation diffusion of those cultures. Portions of that story have been touched on in Chapter 6. The settlement of North America, Australia, and New Zealand involved great long-distance movements of peoples. The flight of refugees from past and recent wars, the settlement of Jews in Israel, the movement south in Africa from the drought-stricken Sahel, the current migration of workers to the United States from overpopulated Mexico,

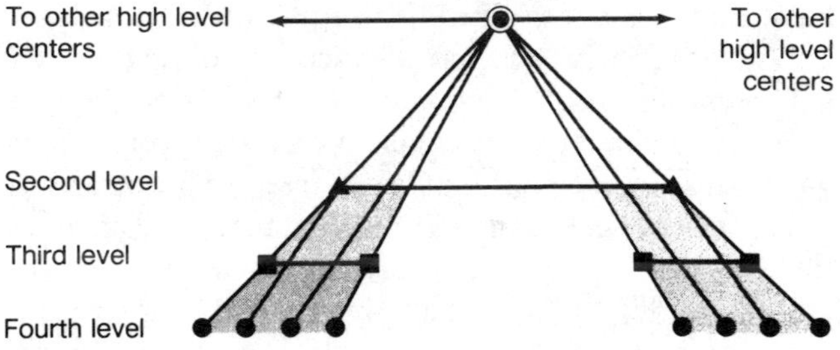

Figure 9.16
A four-level communication hierarchy.

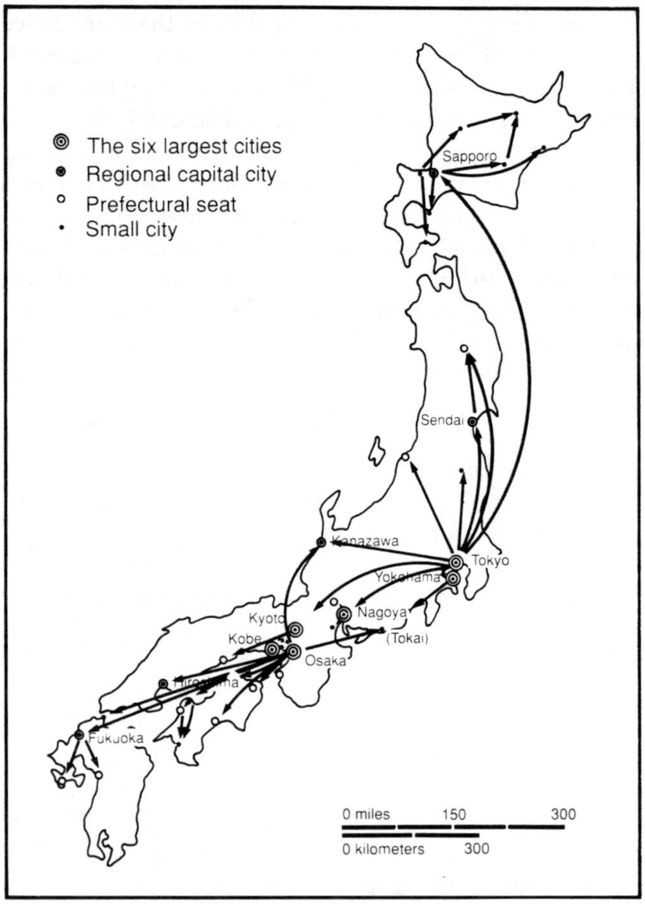

Figure 9.17
Rotary Clubs, members of the international service association, were established in the large cities of Japan during the 1920s. New clubs were established under the sponsorship of the original clubs. This map shows the pattern of diffusion. Note the regional and spatial effects of the diffusion process.

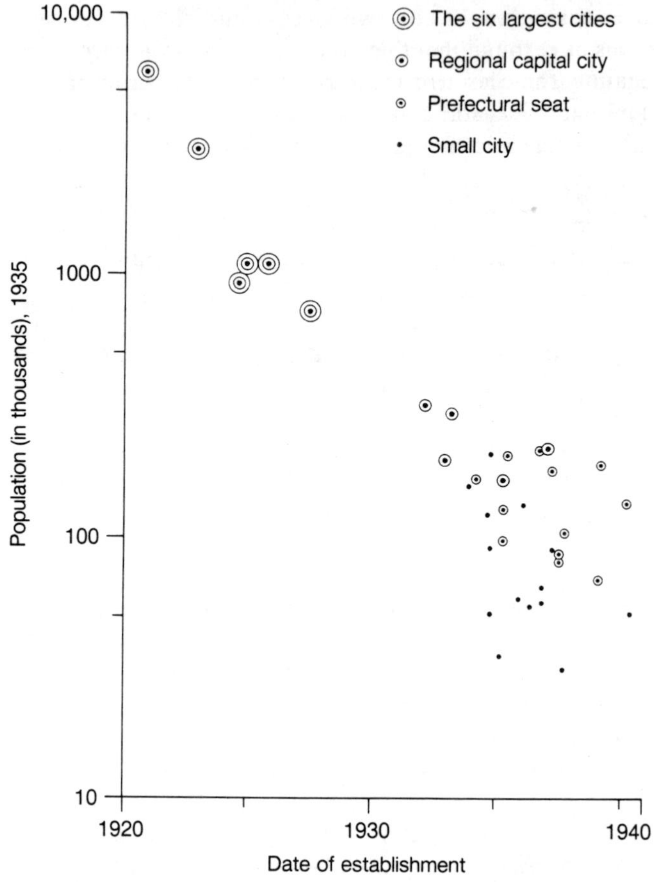

Figure 9.18
This diagram shows the hierarchical diffusion of Rotary Clubs in Japan. The largest cities were the first centers of Rotary Club activity, followed by cities at lower and lower levels of urban population and city function.

and innumerable other examples of mass movement come quickly to mind. In all cases, societies transplanted their cultures to the new areas, their cultures therefore diffused and intermixed, and history was altered. The vast numbers who suffered the hardships of movement to the American West, like Catherine Haun, were the predecessors of the regionally relocating job seekers and retirees of today.

The *planned two-way trip* is a long-distance movement that is not considered to be a migration. It includes the exciting, perhaps historically decisive, daring journey of exploration about which we read in history, or the more common contemporary vacation, business, or social trip in which modern societies indulge so freely. The latter sort of trip, of course, enhances the mental maps and enlarges the awareness space of the participants and may be the prelude to migration. It may also contribute to the diffusion of information through cultural contact, but its individual impact is small because it is so transitory.

Much more important is **migration**: a relocation of both residential environment and activity space. Naturally, the length of the move and its degree of disruption of normal household activities raise distinctions important in the study of migration. A change of residence from the central city to the suburbs certainly changes the activity space of schoolchildren and adults in many of their nonworking activities, but the workers may still retain the city—indeed, the same place of employment there—as an action space. On the other hand, immigration from Europe to the United States and the massive farm-to-city movements of rural Americans around the turn of the century meant a total change of all aspects of behavioral patterns.

The Decision to Migrate

The decision to move is a cultural and temporal variable. Nomads fleeing the famine and the spreading deserts in the Sahel obviously are motivated by different considerations from those of the executive receiving a job transfer to Chicago, the resident of Appalachia seeking factory employment in the city, or the retired couple searching for sun and sand. In general, people who decide to migrate are seeking better economic, political, or cultural conditions or certain amenities. Of course, as in the case of the Hauns, the reasons for migration are frequently a combination of several of these.

Economic causes have impelled more migrations than any other single category. If migrants face unsatisfactory conditions at home (unemployment or famine) and believe that the economic opportunities are better elsewhere, they will be attracted to the thought of a move. To a nomadic herdsman, better opportunities might be abundant grass and water; to a western emigrant, rich soil; to a modern worker, the promise of a high-paying job in a Los Angeles suburb. Attractions such as these are called **pull factors**, and those that contribute to dissatisfaction at home are called **push factors**. Very often migration is a result of both push and pull factors (Figure 9.19).

The desire to escape war and persecution at home and to pursue the promise of freedom in a new location is a political incentive for migration (Figure 9.20). Americans are familiar with the history of settlers who emigrated to North America seeking religious and political freedom. In more recent times, the United States has received hundreds of thousands of refugees from countries such as Hungary (following the uprising of 1956), Cuba

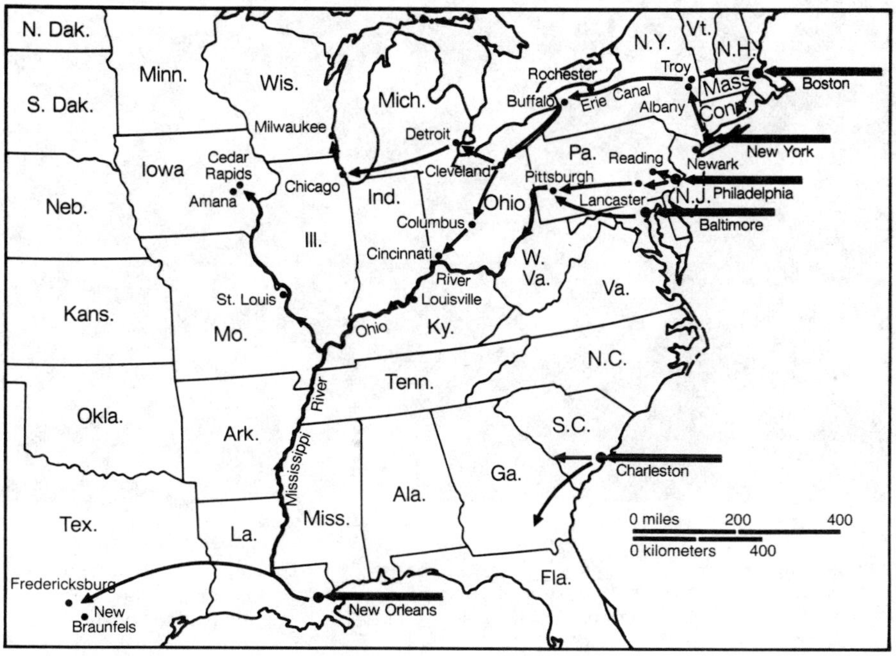

Figure 9.19
The major paths of the early migration of Germans to America constitute a relocation diffusion process. Most emigrants left Germany because of religious and political persecution. They chose the United States not only because immigrants were made welcome, but also because labor was in demand and farmland was available. The first immigrants landed, and many settled, in Boston, New York, Philadelphia, Baltimore, and Charleston. As is characteristic of a relocation diffusion process, the migrants carried with them such aspects of their culture as religion, language, and food preferences.

Documenting Diffusion

The places of origin of many ideas, items, and technologies important in contemporary cultures are only dimly known or supposed, and their routes of diffusion are at best speculative. Gunpowder, printing, and spaghetti are presumed to be the products of Chinese inventiveness; the lateen sail has been traced to the Near Eastern culture world. The moldboard plow is ascribed to 6th century Slavs of northeastern Europe. The sequence and routes of the diffusion of these innovations have not been documented. In other cases, such documentation exists, and the process of diffusion is open to analysis. For example, hybrid corn was originally adopted by imaginative farmers of northern Illinois and eastern Iowa in the mid-1930s. By the late 1930s and early 1940s the new seeds were being planted as far east as Ohio and north to Minnesota, Wisconsin, and northern Michigan. By the late 1940s, all commercial corn-growing districts of the United States and southern Canada were cultivating hybrid varieties.

Also clearly marked is the diffusion path of the custom of smoking tobacco, a practice that originated with Amerindians. Sir Walter Raleigh's Virginia colonists, returning home in 1586, introduced smoking in English court circles, and the habit very quickly spread among the general populace. England became the source region of the new custom for northern Europe; smoking was introduced to Holland by English medical students in 1590. Dutch and English together spread the habit to the Baltic and Scandinavian areas and through Germany to Russia. The innovation continued its eastward diffusion, and within a hundred years, tobacco had spread across Siberia and was, in the 1740s, reintroduced to the American continent at Alaska by Russian fur traders. A second route of diffusion for tobacco smoking can be traced from Spain through the Mediterranean area into Africa, the Near East, and Southeast Asia.

Figure 9.20

Some of the approximately 350,000 Vietnamese "boat people" who fled their country after 1975 to seek a new life in a new location. The United Nations High Commissioner for Refugees in 1988 estimated that worldwide there were over 13 million refugees, people who had left their country because of the well-founded fear of being persecuted for their political views, economic activities, or membership in certain social or religious groups.

(after its takeover by Fidel Castro), and Vietnam (after the fall of South Vietnam). The massive movements of Hindus and Muslims across the Indian subcontinent in 1947, when Pakistan and India were established as governing entities, and the exodus of Jews fleeing persecution in Nazi Germany in the 1930s are other examples of politically inspired moves. Figure 9.21 identifies recent major migrations.

Migration normally involves a hierarchy of decisions. Once people have decided to move and have selected a general destination (e.g., America or the West), they must still choose a particular site at which to settle. At this scale, cultural variables can be important pull factors. Migrants tend to be attracted to areas where the language, the religion, and the racial or ethnic background of the inhabitants is similar to their own. This similarity can help the migrants feel at home when they arrive at their destination and may make it easier to find a job and become assimilated into the new culture. The Chinatowns and Little Italys of large cities attest to the drawing power of cultural factors, as we saw in Chapter 7.

Another set of inducements is grouped under the heading *amenities*, the particularly attractive or agreeable features characteristic of a place. Amenities may be natural (mountains, oceans, climate, and the like) or cultural (the arts and music opportunities available in large cities). They are particularly important to relatively affluent people seeking "the good life." They help account for the attractiveness of the U.S. Sunbelt states for retirees; a similar movement to the south coast has also been observed in countries such as the United Kingdom and France.

The significance of the various factors also varies according to the age, sex, education, and economic status of the migrants. For the modern American, reasons to migrate have been summarized into a limited number of categories that are not mutually exclusive. They include:

1. changes in life cycle (e.g., getting married, having children, getting a divorce, or needing less dwelling space when the children leave home)

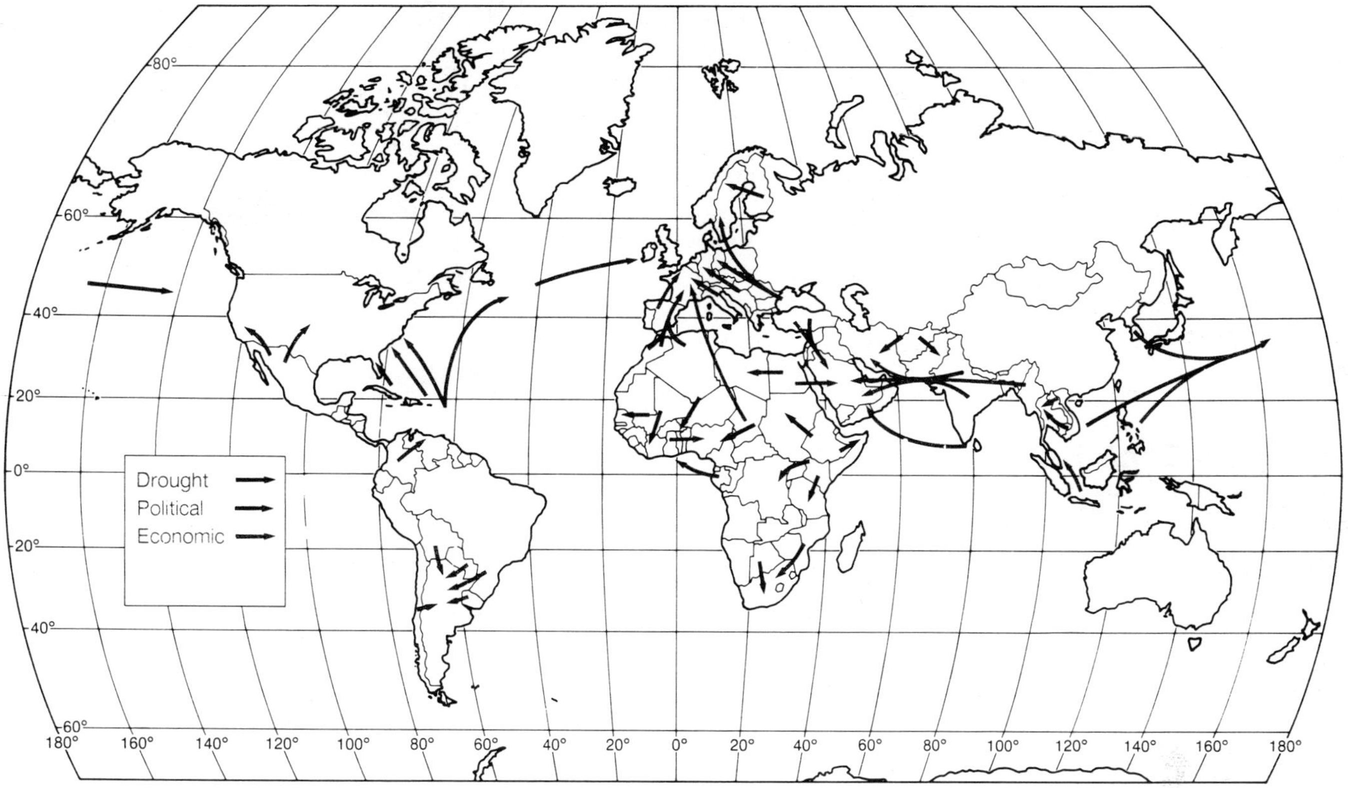

Figure 9.21

Major recent international migration streams. Most of those shown are economic, that is, people moving to the location of available jobs. In the case of the Vietnamese, Afghan, and Sahel migrations, however, politics or drought is the fundamental cause.

2. changes in the career cycle (e.g., leaving college, getting a first job or a promotion, receiving a career transfer, or retiring)
3. forced migrations associated with urban development, construction projects, and the like
4. neighborhood changes from which there is flight, perhaps pressures from new and unwelcome ethnic groups, building deterioration, street gangs, and similar rejected alterations in activity space
5. changes of residence associated with individual personality

Some people simply seem to move often for no easily discernible reason, whereas others, *stayers*, settle into a community permanently. Of course, for a country such as the Soviet Union, with its limitations on emigration, a tightly controlled system of internal passports, restrictions on job changes, and severe housing shortages, a totally different set of summary migration factors would be present.

The factors that contribute to mobility tend to change over time. There is a group, however, that in most societies has always been the most mobile: young adults (Figure 9.22). They are the members of society who are launching careers and making their first decisions about

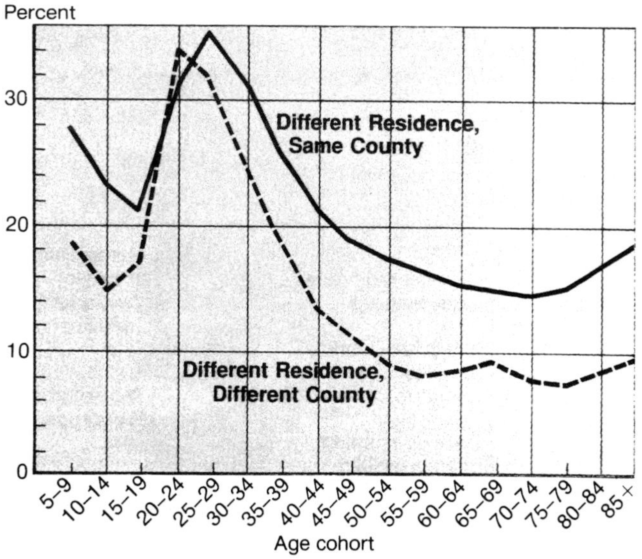

Figure 9.22

Percentage of 1970 population more than 5 years of age with different residence in 1965. Young adults figured most prominently in both short- and long-distance moves in the United States between 1965 and 1970. Movement data through the middle 1980s recorded the same age-related pattern of mobility.

occupation and location. They have the fewest responsibilities of all adults and thus are not as strongly tied to family and institutions as older people. Most of the major voluntary migrations have been composed primarily of young people who suffered from lack of opportunities in the home area and who were easily able to take advantage of opportunities elsewhere.

The concept of **place utility** helps us to understand the decision-making process that potential migrants undergo. It is the value that an individual places on each known, potential migration site. The decision to migrate is a reflection of the appraisal by a potential migrant of the current homesite as opposed to other sites of which something is known. The individual may adjust to conditions at the homesite and thus decide not to migrate.

In the evaluation of the place utility of each potential site, the decision maker considers not only present relative place utility but also expected place utility. The evaluations are matched with the individual's aspiration level, that is, the level of accomplishment or ambition that the individual sees for himself or herself. Aspirations tend to be adjusted to what an individual considers attainable. If one is satisfied with present circumstances, search behavior is not initiated. If, on the other hand, there is dissatisfaction, a utility is assigned to each of the possible

Rational Decision Making: An Example

An example of a structure developed for making a rational decision whether or not to migrate can be found in the case of the selection by a hypothetical unmarried male who is a recent college graduate. This person partitioned his consideration into four major aspects: monetary compensation, geographic location, travel requirements, and the nature of the work. Each of these aspects was subdivided, and some were subdivided again, as illustrated in the table.

Each category was given a value on a scale from 0 to 1.0, according to the importance that the potential migrant placed on it. The restriction that the total of the values must add up to 1.0 enabled the graduate to multiply across any path to find the true relative worth of the 15 categories. In the table, we see that the graduate placed the greatest emphasis on starting salary and the kind of work to be done at the outset. Retirement and insurance benefits were least important.

Once the graduate had a clearer idea of his values, he could take each of his job opportunities and evaluate them according to the best information available about each subcategory. For each possible job, he assigned a value to every category according to the system 90–100 is excellent, 80–89 is good, 70–79 is fair, and so on. By multiplying each of these by the relative worth of each category and summing for each column, the graduate was able to make an objective decision. For example, for job opportunity No. 1, 95 x 0.040 = 3.80; 80 x 0.059 = 4.72; and so on. For job opportunity No. 1, the score was 71.8; for No. 2, the value was 71.2; and for No. 3, it was 77.3.

This decision was based on the subject's own value system. Each person's categories and weighting systems would be different. In the case illustrated, the graduate finally ranked the local opportunity ahead of the national and regional opportunities. The ranking led to the decision not to migrate at the present time.

				Relative worth placed on each category	Ratings of job opportunities: 1 Large national company	2 Regional office	3 Local job
Nature of work	Current and future work features		Management training program	0.040	× 95 = 3.80	80 = 3.20	50 = 2.00
			Variety of work	0.059	× 80 = 4.72	75 = 4.43	70 = 4.13
			Technical challenge	0.049	× 80 = 3.92	80 = 3.92	95 = 4.66
	Immediate work features			0.132	× 70 − 9.24	80 − 10.56	90 = 11.88
Travel requirements	Long trips away from office		Trip lengths	0.082	× 90 = 7.38	70 = 5.74	60 = 4.92
			Proportion time away	0.054	× 80 = 4.32	70 = 3.78	60 = 3.24
	Daily commuting characteristics			0.034	× 50 = 1.70	70 = 2.38	100 = 3.40
Geographic location			Climate	0.034	× 80 = 2.72	70 = 2.38	60 = 2.04
			Proximity to leisure-time activities	0.068	× 90 = 6.12	80 = 5.44	60 = 4.08
			Proximity to relatives	0.068	× 60 = 4.08	70 = 4.76	100 = 6.80
Monetary compensation	Future salary prospects		10-year increase	0.035	× 90 = 3.15	80 = 2.80	80 = 2.80
			3-year increase	0.064	× 60 = 3.84	70 = 4.48	80 = 5.12
	Immediate prospects	Fringe benefits	Retirement	0.009	× 95 = 0.86	70 = 0.63	60 = 0.54
			Insurance	0.014	× 95 = 1.33	70 = 0.98	60 = 0.84
		Starting salary		0.209	× 70 = 14.63	75 = 15.68	100 = 20.90
				1.000	71.81	71.16	77.35

new sites. The utility is based on past or expected future rewards at the various sites. Because the new places are unfamiliar to the individual, the information received about them acts as a substitute for the personal experience of the homesite. The decision maker can do no more than sample information about the new sites, and, of course, there may be sampling errors.

One goal of the potential migrant is to minimize uncertainty. Most decision makers either elect not to migrate or postpone the decision unless uncertainty can be lowered sufficiently. Most migrants reduce uncertainty by imitating the successful procedures followed by others. We see that the decision to migrate is not a perfunctory, spur-of-the-moment reaction to information. It is usually a long, drawn-out process based on a great deal of sifting and evaluation of data.

An example of some of these observations and generalizations in action can be seen in the case of the large numbers of young Mexicans who have migrated both legally and illegally to the United States over the last 20 years. Faced with rural poverty and overpopulation at home, they regard the place utility there as minimal. Their space-searching ability is, however, limited by lack of money and by lack of alternatives in the land of their birth. With a willingness to work and with aspirations for success—perhaps wealth—in the United States, they are eager listeners to friends and relatives who tell of numerous job opportunities in the United States. Hundreds of thousands, presented with the glittering prospect of a certain job, low-paying though it might be, quickly place high utility on perhaps a temporary relocation (maybe 5 or 10 years) to the United States. Many know that dangerous risks are involved if they attempt to enter the country illegally, but the rewards are worth the risk. Their arrival indicates their assignment of higher utility to the new site than to the old.

In the 20th century, nearly all countries have experienced a movement of population to the cities from their agricultural areas. The migration has presumably paralleled the number of perceived opportunities within the cities and convictions of absence of place utility in the rural districts. Perceptions, of course, do not necessarily accord with reality. The enormous influx of rural folk to major urban areas in the developing countries and the economic destitution of many of those in-migrants suggest recurrent gross misperceptions, faulty information, and sampling error. In Chapter 12, we discuss the reasons for this phenomenon.

Barriers to Migration

Paralleling the incentives to migration is a set of disincentives, or barriers, to migration. They help to account for the fact that many people do not choose to move even when conditions are bad at home and are known to be better elsewhere. Migration depends on a knowledge of the opportunities in other areas. People with a limited

Problems in the Sunbelt

Since 1970, there has been a net in-migration of over 12 million people to the region known as the Sunbelt, the tier of southern states stretching from the Atlantic to the Pacific Coast. This region captures the bulk of the country's migrants and immigrants from other countries. Arizona's population grew to 3.4 million in 1987 from 1.8 million in 1970. Florida's growth rate for the period was over 50%. Ironically, the rapid growth of the Sunbelt causes problems that threaten some of the very qualities that make the area attractive.

Low energy costs, low wages, and low taxes attracted businesses and industries from the North to the Sunbelt. Both blue- and white-collar workers from the North followed the companies, drawn by the promise of jobs and the particular attractions of the region: warmth and sunshine, unspoiled land, recreational opportunities, and other amenities. Although the Sunbelt continues to attract retirees, companies, and workers, its advantages are eroding rapidly, and growth itself is the reason.

With economic growth have come higher wages. In some parts of the region, taxes have risen already, and they will rise more as the states and cities try to cope with the services that population and economic growth demand. Across the Sunbelt, the rapid influx of people has strained the capacity of sewage-treatment plants. Depending on which city is considered, poor roads, inadequate fire and police protection, and limited mass transit are problems. Florida and the Southwest face water shortages as well.

Many Sunbelt cities, although prosperous, exhibit the kinds of problems that we associate with northern and eastern cities: crowded freeways and traffic jams, noise, urban sprawl, rising crime rates, unemployment, poor inner cities, and air pollution. The clean air that attracted people with allergies, asthma, and other respiratory problems to Arizona is being polluted not only by cars and mining operations but also by the foliage that the migrants planted to remind them of home. For example, the mulberry, a fast-growing tree that provides good shade, produces pollen levels four times as great as those of other plants. Overall, Arizona has experienced a tenfold increase in airborne pollen since 1960. The problems caused by growth tend to inhibit further growth. The high crime rate in Florida, for example, hurts the huge tourist industry, and higher wages in the Southeast deter new companies from moving there.

A city does not create urban problems independently of the people who live there. When a substantial number of those people move to another city, the potential for the reappearance in the new city of the problems that they hoped to leave behind is considerable.

knowledge of the opportunities elsewhere are less likely to migrate than are better-informed individuals. Other barriers include physical features, the costs of moving, ties to individuals and institutions in the original activity space, and government regulations.

Physical barriers to travel include seas, mountains, swamps, deserts, and other natural features. In prehistoric times, they played an especially significant role in limiting movement. Thus, the spread of the ice sheets across most of Europe in Pleistocene times was a barrier to both migration and human habitation. Physical barriers to movement have probably assumed less importance only within the last 400 years. The developments that made possible the great age of exploration, beginning about A.D. 1500, and the technological advances associated with industrialization have enabled people to conquer space more easily. With industrialization came improved forms of transportation that made travel faster, easier, and cheaper. Still, as the account of conditions on the Overland Trail illustrated, only a century and a half ago travel could be arduous, and in some parts of the world, it remains so today.

Economic barriers to migration include the cost of travel and establishing a residence elsewhere. Frequently the additional expense of maintaining contact with those left behind is also a cost factor that must be considered. Normally all these costs increase with the distance traveled and are a more significant barrier to travel for the poor than for the rich. Many immigrants to this country were married men who came alone; when they had acquired enough money, they sent for their families to join them. This phenomenon is still evident among recent immigrants from the Caribbean area to the United States, and among Turks, Yugoslavs, and West Indians who have settled in Northern and Western European countries.

The *cost factor* limits long-distance movement, but the larger the differential between present circumstances and perceived opportunities, the more individuals are willing to spend on moving. Figure 9.23 is an example of the effect of distance on relocation. For many people, especially older people, the differentials must be extraordinarily high for movement to take place.

Cultural factors also contribute to decisions not to migrate. Family, religious, ethnic, and community relationships defy the principle of differential opportunities. Many people will not migrate under any but the most pressing of circumstances. The fear of change and human inertia—the fact that it is easier not to move than to do so—may be so great that people consider but reject a move. Ties to one's own country, cultural group, neighborhood, or family may be so strong as to compensate for the disadvantages of the home location. Returning migrants may convince potential leavers that the opportunities are not in fact better elsewhere—or, even if better, that they are not worth entering an alien culture or sacrificing home and family.

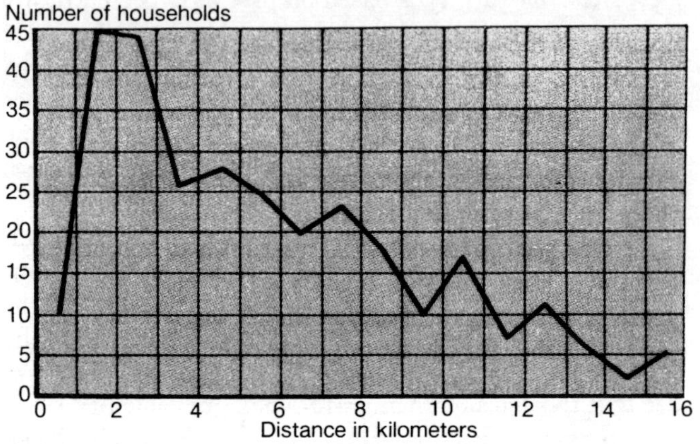

Figure 9.23

Distance between old and new residences in the Asby area of Sweden. Notice how the number of movers decreased with increasing distance.

Restrictions on immigration and emigration constitute *political barriers* to migration. Many governments frown on movements into or outside their own borders. Recognizing that immigration might be economically or politically disadvantageous, many nations restrict outmigration. These restrictions may make it impossible for potential migrants to leave, and they certainly limit the number who can do so.

On the other hand, countries suffering from an excess of workers often encourage emigration. The huge migration of people to the Americas in the late 19th and early 20th centuries is a good example of perceived opportunities for economic gain far greater than in the home country. Many European countries were overpopulated, and their political and economic systems stifled domestic economic opportunity at a time when people were needed by American entrepreneurs hoping to increase their wealth in untapped resource-rich areas.

Countries where per capita incomes are high or are perceived as high are generally the most desired international destinations. In order to protect themselves against overwhelming migration streams, such countries have restrictive policies on immigration. In addition to absolute quotas on the number of immigrants (usually classified by country of origin), a country may impose other requirements, such as the possession of a labor permit or sponsorship by a recognized association.

Patterns of Migration

Several geographic concepts deal with patterns of migration. The first of these is the **migration field**. For any single place, the origin of its in-migrants and the destination of its out-migrants remain fairly stable spatially over time. Areas that dominate a locale's in- and out-migration patterns constitute the migration fields for the place in question, as shown in Figure 9.24. As we would expect, areas

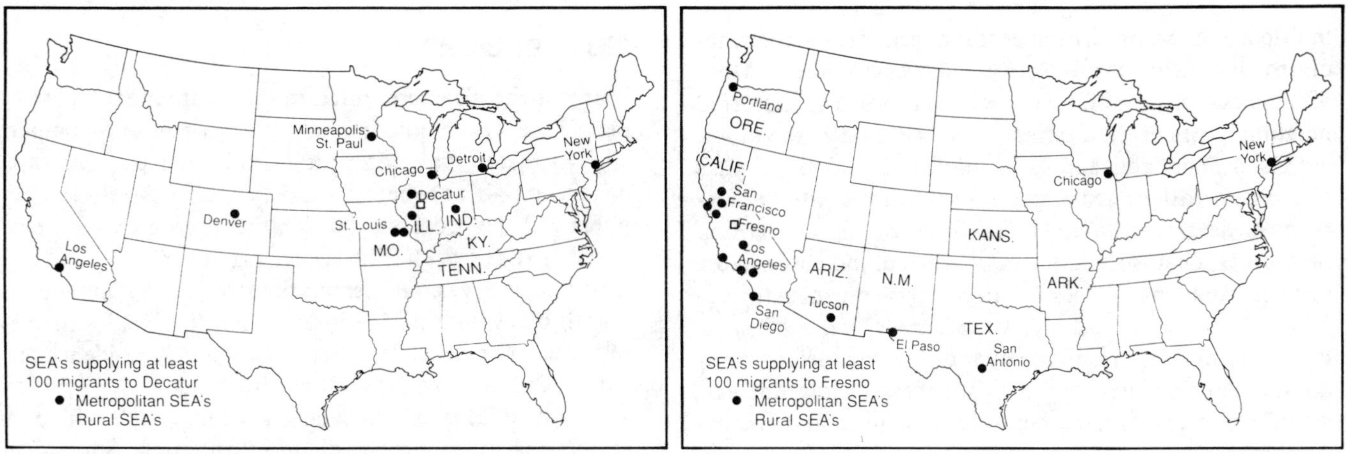

Figure 9.24

The in-migration fields of Decatur, Illinois, and Fresno, California, by State Economic Areas (SEA). For Decatur, as expected, most of the in-migrants originate in nearby rural areas and cities. For Fresno, the pattern is similar with an important addition—rural areas in the Southwest and Kansas. This secondary pattern represents the constant westward movement of population in the United States.

near the point of origin make up the largest part of the migration field. However, places far away, especially large cities, may also be prominent. These characteristics of migration fields are functions of the hierarchical movement to larger places (as we shall see) and the fact that there are so many people in large metropolitan areas that one might expect some migration into and out of them from most areas within a country.

Migration fields do not conform exactly to the diffusion concepts mentioned earlier. As shown by Figure 9.25, some migration fields reveal a distinctly **channelized** pattern of flow. The channels link areas that are socially and economically tied to one another by past migration patterns, by economic trade considerations, or by some other affinity. As a result, flows of migration along these channels are greater than would otherwise be the case. The movements of blacks from the southern United States to the North, of Scandinavians to Minnesota and Wisconsin, of Mexicans to such border states as California, Texas, and New Mexico, and of retirees to Florida and Arizona are all examples of channelized flows.

Sometimes the channelized migration is specific to occupational as well as to ethnic groups. For example, nearly all newspaper vendors in New Delhi, in the north of India, come from one small district in Tamil Nadu, in the south of India. Most construction workers in New Delhi come either from Orissa, in the east of India, or Rajasthan, in the northwest, and taxicab drivers originate in the Punjab. The diamond trade of Bombay, India, is dominated by a network of about 250 related families who come from a small town several hundred miles to the north.

Return migration is a term used to refer to those who, soon after migrating, decide to return to their point of origin. If freedom of movement is not restricted, it is not unusual for as many as 25% of all migrants to return to

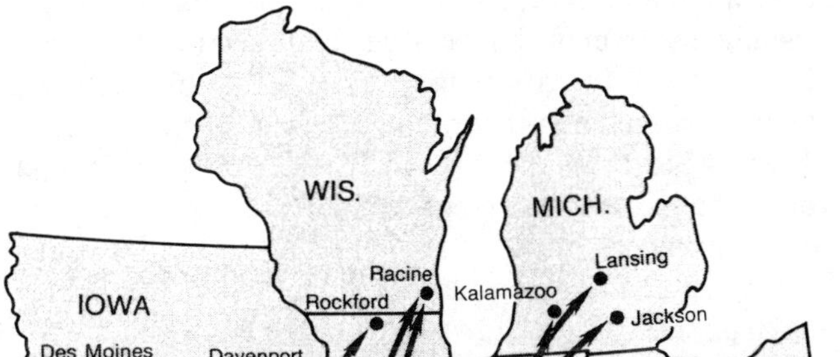

Figure 9.25

Channelized migration flows from the rural South to midwestern cities of medium size. Distance is not necessarily the main determinant of flow direction. Perhaps through family and friendship links, the rural southern areas are tied to particular midwestern destinations.

their place of origin. Unsuccessful migration is sometimes due to an inability to adjust to the new environment. More often, it is a result of false expectations based on distorted mental images of the destination at the time of the move. Myths, secondhand and false information, and people's own exaggerations contribute to what turn out to be mistaken decisions to move. Although return migration often represents the unsuccessful adjustment of individuals to a new environment, it does not necessarily mean that negative information about a place returns with the migrant. It usually means a reinforcement of the channel, as communication lines between the unsuccessful migrant and would-be migrants take on added meaning and understanding.

In addition to channelization, the influence of large cities causes migration fields to deviate from the distance-decay pattern. The concept of **hierarchical migration** assists in understanding the nature of migration fields. Earlier we noted that sometimes information diffuses according to a hierarchical rule, that is, from city to city at the highest level in the hierarchy and then to lower levels. Hierarchical migration, in a sense, is a response to that flow. The tendency is for individuals in domestic relocations to move up the level in the hierarchy from small places to larger ones. Very often, levels are skipped on the way up; only in periods of general economic decline is there considerable movement down the hierarchy. The suburbs of large cities are considered part of the metropolitan area, so that movement from a town to a small suburb would be considered moving up the hierarchy. From this pattern, we can envisage information flowing to a place via a hierarchical routing. Once the information is digested, some respond by migrating to the area from which the information came.

Spatial Search

Many of the ideas presented in this chapter are related to people's decision making in a geographic environment. That is, after evaluating alternatives, people come to a conclusion about where to satisfy their desires. Geographers call the process by which the alternatives are evaluated **spatial search**. The seeker may or may not actually travel to the various destinations; one may pursue the search by evaluating the information that comes to one's residence. For example, searching for a new residence may entail reading newspaper advertisements.

The quality of the decision is very much a function of the quality of the available information. Some decisions, like the decision to buy a new home, are so important that individuals take months or even years to gather information. And, of course, in such circumstances, one is usually obliged to visit each of the possible homesites (Figure 9.26).

Decision making is based on a relationship between information availability and the preferences and motivations of the individual decision maker. Unfortunately, many decisions are made with inadequate or distorted information. Gathering information is a time-consuming activity. Many people are unable or unwilling to use their time for information gathering, and many disappointments are based on insufficient information. Of course, even if the time were available and people were willing to gather information, the needed information might just be unavailable. Many of the return migrants discussed above were responding to an inadequate understanding of the migration location.

Decision makers evaluate information by assessing the utility of each of the possible outcomes of the various courses of action. They depend on their image (perhaps a

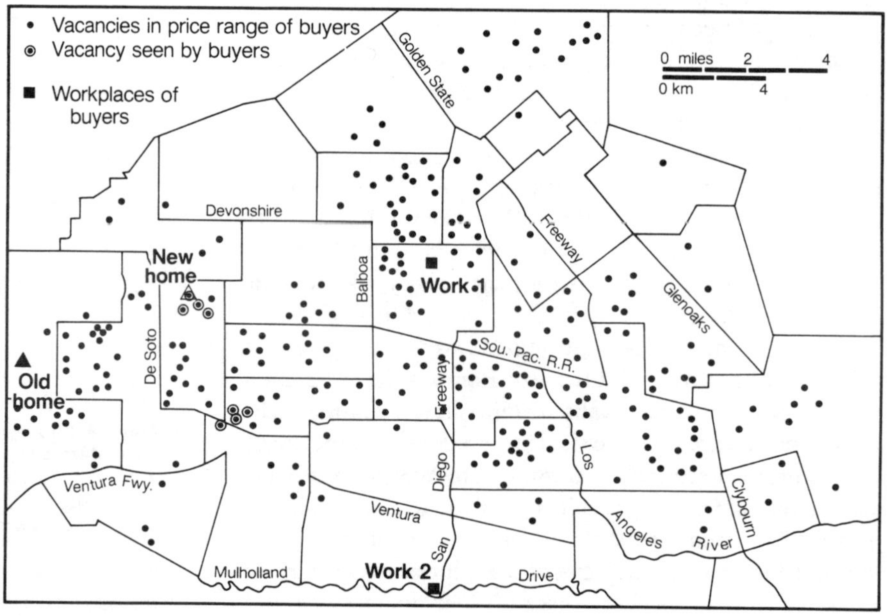

Figure 9.26

An example of spatial search in the San Fernando Valley area of Los Angeles. The dots represent the house vacancies in the price range of a sample family. Note (1) the relationship of the new house location to the workplaces of the married couple, (2) the relationship of the old house location to the new house location, and (3) the limited spatial search space. This example is typical of intraurban moves.

Third World International Migration

The young merchant assessed his options. If he remained at home, there was little chance that his life would improve materially. But if he were willing to take the risk, he could travel surreptiously to the border, slip across it at night, and enter one of the richest countries in the region . . . a country whose high rate of economic growth was marked by modern office buildings, highways, and well-stocked stores. If he weren't deported, he could expect to earn many times the annual wages he received at home.

The story of a Mexican immigrant to the United States? Not at all. Marcel Amao is Malian, and his destination was the Ivory Coast (Côte d'Ivoire), which along with Nigeria is the richest of the West African countries. In recent years, immigrants have flooded into the Ivory Coast at such a rate that they now constitute about one-third of the country's population of 12 million (1989). By the thousands, they come not only from Mali but also from Benin, Burkina Faso, Ghana, and Senegal.

That migration echoes many of the themes explored in this chapter. Most migrants leave their home for economic reasons; "pushed" by poverty, they are "pulled" by the chance to better their lot in the Ivory Coast. The migration is not strictly distance-affected; emigrants from Benin and Senegal bypass neighboring countries and cross two or three national borders before entering the Ivory Coast. The movement is somewhat channelized. In the capital city of Abidjan, people from Benin tend to be cooks and cabinetmakers, while immigrants from the impoverished Sahelian nations of Burkina Faso and Mali find work as street sweepers and night-watchmen, laundrymen, and sellers of cloth. Finally, migrants are both "stayers" and "leavers." In the words of its president, the Ivory Coast "is a country without a passport: one comes, one leaves, one stays, but more often one stays."

mental map) of the thing or place being considered, and they use some type of scaling or preference criteria, keeping in mind the uncertainties that exist. Their motivations are based on their belief and value systems. For example, does the decision maker expect instant gratification or is she or he more inclined to expect gratification in several weeks, months, or years? Does one think that family satisfaction is more important than individual satisfaction? And so on.

In a study of the search for housing and residential choice in Toronto, many of the variables and factors involved in spatial search were evaluated. Among other things, it was found that those looking for apartments spent much less time looking than did potential home owners. The difference was ascribed to the importance people place on buying as opposed to renting. It was found, however, that apartment dwellers, even though they may complain about noise, inadequate space, and so on, are not much less satisfied with their decision than home owners. Most young apartment dwellers think of their quarters as temporary and therefore set lower standards of comfort in order to take advantage of lower-cost housing, more convenience, and less responsibility.

In terms of the search itself, people spend longer in deciding to look for new housing than they do in actually inspecting and choosing it. Most home buyers look at seven or more housing units, compared to only about three for those who eventually choose apartments. A Los Angeles study concluded that those home buyers not leaving the metropolitan area altogether search for homes within a short distance of both their present home place and their workplace. The limiting factor, it was determined, is the income level of the decision maker.

Summary

In this discussion of behavioral geography, we have emphasized the factors that influence how individuals view and use space. At the beginning of the chapter, we posed a number of questions used as our theme. In considering how individuals view their environment, the concepts of spatial perception and spatial cognition were helpful. We spoke of the nature of the information available and of the age of people, their past experiences, and their values. The question of differences in the extent of space used led us to the concept of activity space. The age of an individual, his or her degree of mobility, and the availability of opportunities all play a role in defining the limits of individual activity space. We saw, however, that as distance increases, familiarity with the environment decreases. The concept of critical distance identifies the distance beyond which the decrease in familiarity with places begins to be significant.

How space is used is a function of all of the factors mentioned, but certain factors having to do with the diffusion of innovations indicate what opportunities exist for individuals living in various places. Contagious and hierarchical diffusion influence the geographic direction that cultural change will take. Of course, there are barriers to diffusion, such as effort and cost. A special type of human interaction is migration. When strong enough the various push and pull forces motivate a long-distance, permanent move. Migration fosters the spread of culture by means of a relocation diffusion process.

Finally, we examined how the ability of many to make well-reasoned, meaningful decisions is a function of the utility that they assign to places and the opportunities at those places.

Behavioral geographers view individual action in a special way. There is strong emphasis on information flow and the effect of distance decay. In many respects, it is an interdisciplinary field in that the combined ideas of geographers, psychologists, and other social scientists are crucial to an understanding of spatial behavior. Such behavior has an economic component, and to that we next turn our attention.

Key Words

activity space
channelized migration
cognition
contagious diffusion
critical distance
diffusion
distance decay
environmental perception
hierarchical diffusion
hierarchical migration
mental map
migration
migration field
place utility
push and pull factors
return migration
spatial search
territoriality

For Review

1. What is the difference in meaning between *environmental perception* and *cognition?* Give an example of each.
2. On a blank piece of paper, and without any maps to guide you, draw a map of the United States, putting in state boundaries wherever possible; this is your mental map of the country. Compare it with a standard atlas map. What conclusions can you reach?
3. What is meant by *activity space?* What factors affect the areal extent of the activity space of an individual?
4. Recall the places that you have visited in the past week. In your movements, were the distance-decay and critical-distance rules operative? What variables affect an individual's critical distance?
5. Briefly distinguish between *contagious diffusion* and *hierarchical diffusion.* In what ways, if any, were these forms of diffusion in operation in the culture hearths discussed in Chapter 7?
6. What considerations affect a decision to migrate? What is *place utility,* and how does its perception induce or inhibit migration?
7. What common barriers to migration exist? Why do most people migrate within their own country?
8. Define the term *migration field.* Some migration fields show a channelized flow of people. Select a particular channelized migration flow (such as the movement of Scandinavians to the United States, or people from the Great Plains to California, or southern blacks to the North) and explain why a channelized flow developed.

Selected References

Brown, Lawrence A. *Innovation Diffusion: A New Perspective.* New York: Methuen, 1981.

Cox, Kevin R., and Reginald G. Golledge. *Behavioral Problems in Geography Revisited.* New York: Methuen, 1982.

Downs, Roger M., and David Stea. *Cognitive Mapping and Spatial Behavior.* Chicago: Aldine, 1974.

Fligstein, Neil. *Going North: Migration of Whites and Blacks from the South, 1900–1950.* New York: Academic Press, 1981.

Gold, John R. *An Introduction to Behavioral Geography.* New York: Oxford University Press, 1980.

Golledge, Reginald G., and Robert J. Stimson. *Analytical Behavioural Geography.* New York: Croom Helm, 1987.

Gould, P. R., and Rodney White. *Mental Maps,* 2d ed. Boston: Allen & Unwin, 1986.

Hägerstrand, Torsten. *Innovation Diffusion as a Spatial Process.* Chicago: University of Chicago Press, 1967.

Leis, G. J. *Human Migration: A Geographical Perspective.* London: Croom Helm, 1982.

Lynch, Kevin. *The Image of the City.* Cambridge: MIT Press, 1960.

Newman, Oscar. *Defensible Space*. New York: Macmillan, 1972.

Porteous, J. Douglas. *Environment and Behavior: Planning and Everyday Urban Life*. Reading, Mass.: Addison-Wesley, 1977.

Roseman, Curtis C. *Changing Migration Patterns within the United States*. Resource Papers for College Geography, No. 77–2. Washington, D.C.: Association of American Geographers, 1977.

Rushton, Gerard, and R. G. Golledge, eds. *Spatial Choice and Spatial Behavior*. Columbus: Ohio State University Press, 1976.

Saarinen, Thomas F. *Environmental Planning: Perception and Behavior*. Boston: Houghton Mifflin, 1976.

Chapter 11

Urban Systems and Urban Structures

In the 1930s Mexico City was described as perhaps the handsomest city in North America and the most exotic capital city of the hemisphere, essentially unchanged over the years and timeless in its atmosphere. It was praised as beautifully laid out, with wide streets and avenues, still "a city of palaces" that Baron von Humboldt called it in the 19th century. The 200-foot-wide Paseo de la Refoma, often noted as "one of the most beautiful avenues in the world," was shaded by a double row of trees and lined with luxurious residences.

By the 1950s, with a population of over 2 million and an area of 20 square miles (52 km²), Mexico City was no longer unchanged. The old, rich families who formerly resided along the Paseo de la Reforma had fled from the noise and crowding; their "palaces" were being replaced by tall blocks of apartments and hotels. Industry was expanding and multiplying, tens of thousands of peasants were flocking in from the countryside every year. By the late 1980s, with its population estimated at 20 million and its area at over 390 square miles (1000 km²), metropolitan Mexico City was adding a half-million immigrants each year as well as growing prodigiously by its own 3% birth rate. The toll exacted by that growth is heavy. Each year the city pours more than 5 million tons of pollutants into its air, 80% coming from its estimated 3 million motor vehicles (up from 55,000 in 1950), the rest from some 35,000 industrial plants. More than 4 million people citywide have no access to tap water; in some squatter neighborhoods less than 50% do. Some 3 million residents have no access to the sewage system. Approximately 30% of all families—and they average five people—live in but a single room, and that room generally is in a hovel in one of the largest slums in the world.

Figure 11.1

Surrounded by mountains, Mexico City endures frequent atmospheric inversions—most commonly on winter mornings—when cold air is trapped beneath a layer of warm air. The pollutants surging into the air each day are therefore unable to rise and drift away from the city. The result is frequent deadly smog as visibility drops to less than 500 yards (450 m) and citizens are afflicted with respiratory problems and eye irritation.

The changes in Mexico City since the 1930s have been profound (Figure 11.1). Already the world's most populous metropolitan area, Greater Mexico City is expected to swell further, to at least 31 million by the year 2000. And Mexico City is but a worst case scenario of an urban explosion that sees an increasing proportion of the world's population housed within a growing number of immense cities.

The Urbanizing Century

Figure 11.2 documents the observation that this has been the century of cities. The more than 280 metropolitan areas with over 1 million inhabitants in 1990 far exceeded in number and population the 13 in that size class at the beginning of the century. At least 14 metropolises (Beijing, Bombay, Buenos Aires, Cairo, Calcutta, Los Angeles, London, Mexico City, New York/northeastern New Jersey, Rio de Janeiro, São Paulo, Seoul, Shanghai, and Tokyo/Yokohama,) have (1990) over 10 million people. In 1900, there were no metropolitan areas of that size. Of course, as we saw in Chapter 4, the world's total population has greatly increased, so that we would also expect the urban component to increase. But the fact remains that urbanization and metropolitanization have increased more rapidly than the growth of total population. The amount of urban growth differs from continent to continent and from country to country (Figure 11.3), but nearly all regions have two things in common: the proportion of their people living in cities is rising, and the cities themselves are large and growing.

Table 11.1 shows world urban population by region with projections to 2000. Note that the most industrialized parts of the world, North America, Western Europe, and East Asia (including Japan, Taiwan, South Korea, and Hong Kong) are the most urbanized—that is, have the highest percentages of their populations living in cities—while Africa and the rest of Asia have the lowest proportions of urban population (Figure 4.24). What is evident is that the

Figure 11.2

Trends of world urbanization.

Courtesy of Population Reference Bureau, Washington, DC.

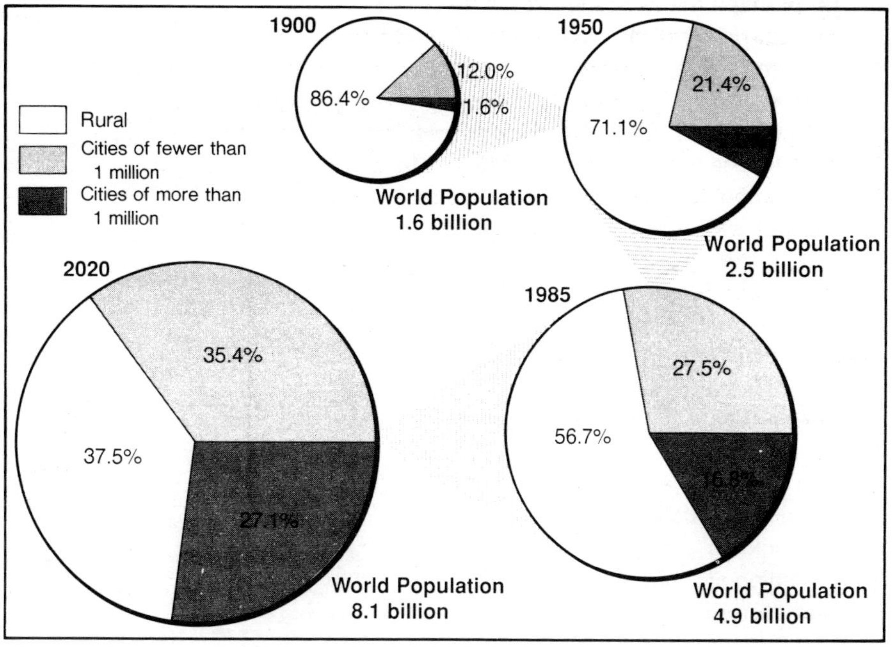

Figure 11.3

Annual average urban population growth rates, 1965–1985. Annual average percentage increases in urban population range from less than 1% to more than 15% (United Arab Emirates). In general, developing countries show the highest percentage increases in their urban populations, and the already highly urbanized and industrialized countries have the lowest—less than 1% per annum in parts of Europe.

Source: Data from World Bank, *World Development Report 1988*.

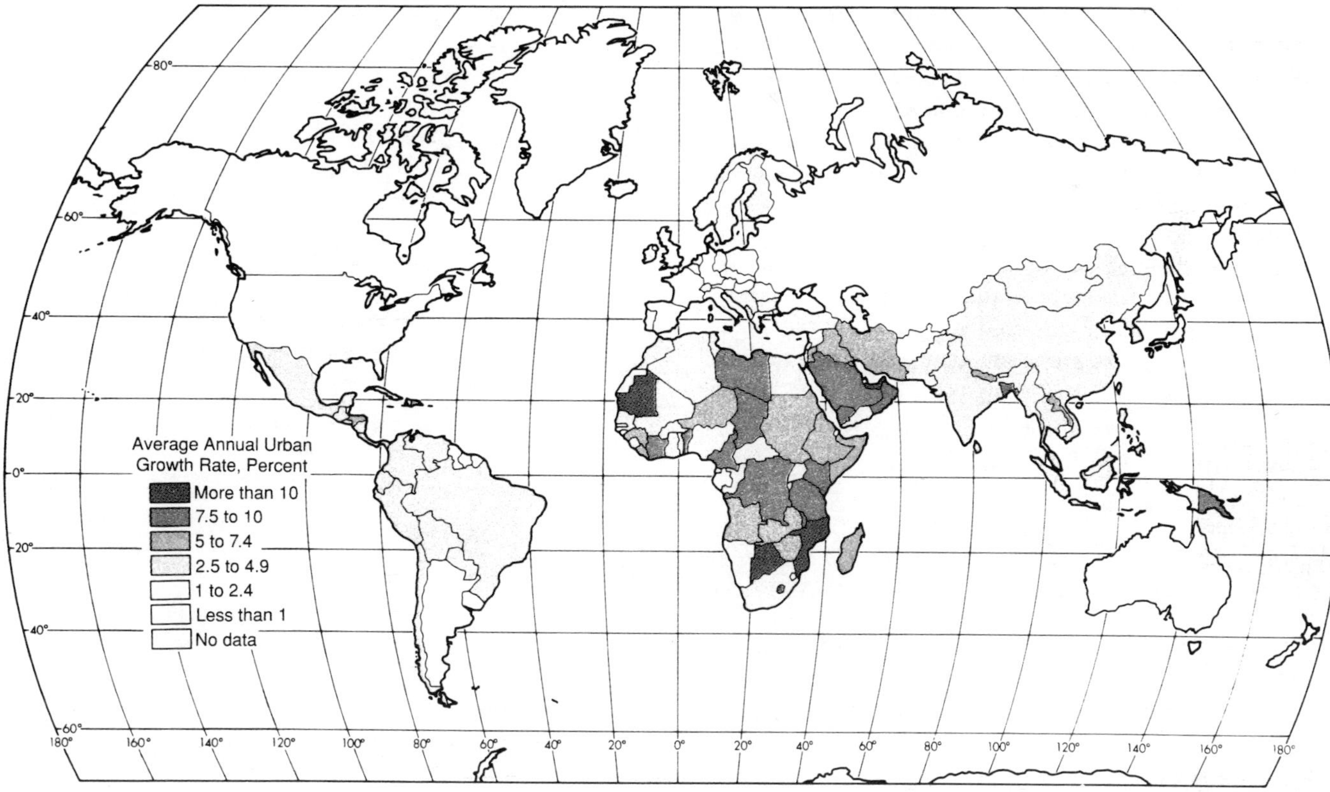

Table 11.1 Estimated Urban Share of Total Population, 1950 and 1988, with Projections to 2000

Region	1950	1988	2000
North America	64%	74%	78%
Europe	56	75	79
Soviet Union	39	65	74
East Asia (except China)	43	73	79
China	12	41	50
Southeast Asia	—	25	35
South Asia	15	25	39
Latin America	41	68	77
Africa	15	30	44
Oceania	61	70	74
World	29	45	51

Sources: Lester R. Brown and Jodi L. Jacobson, *The Future of Urbanization*. Worldwatch Paper 77 (Washington, D.C.: Worldwatch Institute, 1987), p. 8; Population Reference Bureau, *1988 World Population Data Sheet;* and composite projections.

industrialization process has also been one of urbanization. As the world, especially the Third World, continues to industrialize, we can expect further large increases of city dwellers. Even now, although dominantly agricultural Africa, China, and Southeast and South Asia have relatively low proportions of people in cities, their large and growing total populations give them immense urban numbers. For them, even a small percentage increase translates into millions of people. It sometimes escapes us that there are many large cities throughout parts of the world where subsistence agriculture still engages most people.

Although major cities appear on all continents, their world distribution, as displayed in Figure 11.4, shows differential patterns of concentration and dispersion. Four major world regions of clustered supercities in the multimillion size categories are easily identified. The most extensive of them stretches from the British Isles across western and central Europe through the Mediterranean basin into the Middle East. The cultural and economic boundaries it traverses clearly indicate that major urban development is no longer exclusively confined to the Western industrial nations. That observation is reinforced by a second megacity concentration, in South Asia from Pakistan through India to Bangladesh and beyond. Although Table 11.1 tells us urban populations are in the minority there, the sheer numbers of people involved assure cities of massive size. Industrialization and also population pressures figure in the third major urban concentration, that of East Asia from Manchuria (Dongbei) and Korea through developed Japan and industrializing coastal China southward into Indonesia. The fourth region—really a two-part district—is North American. Its eastern component

Figure 11.4

World metropolitan areas of 1 million or more, 1989. Massive metropolitan areas, conurbations, and urbanized districts are no longer characteristic only of the industrialized, developed nations. They are now found on every continent, in all latitudes, as part of most economies and societies. Among the largest metropolitan areas are those in developing nations least able to provide the public facilities, housing, and employment opportunities their expanding urban populations require.

Source: List of metropolitan areas prepared by Richard L. Forstall for Rand McNally and Company's *The International Atlas*, 1989. Used by permission.

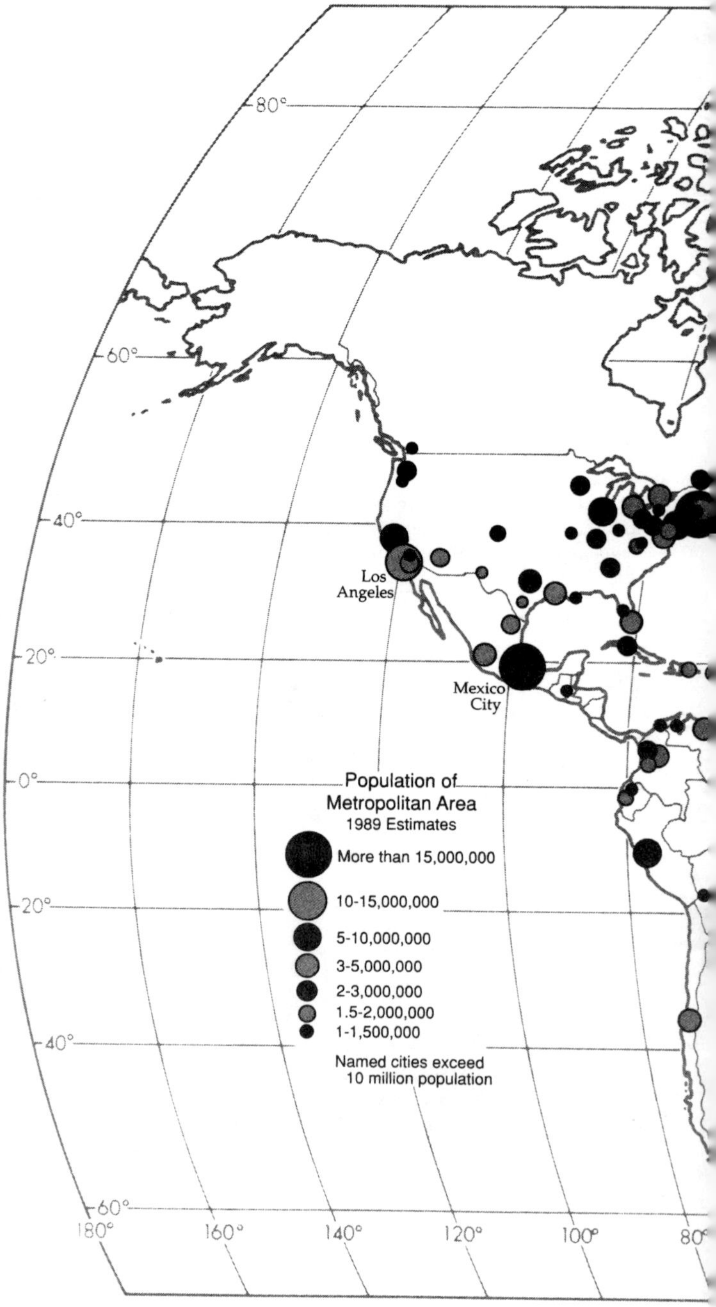

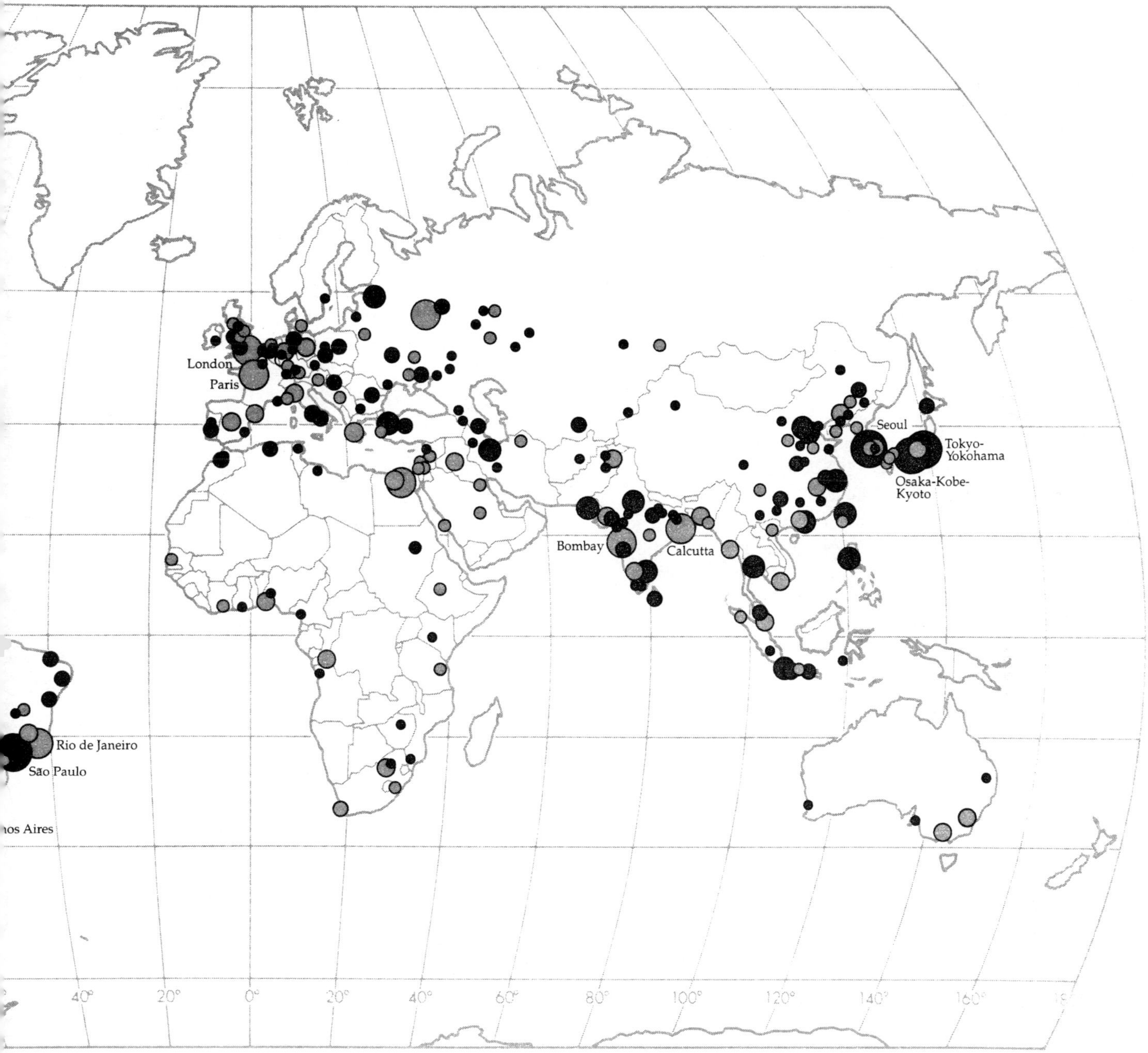
London
Paris
Seoul
Tokyo-
Yokohama
Osaka-Kobe-
Kyoto
Bombay
Calcutta
Rio de Janeiro
São Paulo
nos Aires
40°
20°
0°
20°
40°
60°
80°
100°
120°
140°
160°

extends from Montreal and Minneapolis on the north southward through Houston and Dallas to Monterrey and Mexico City. The western outlier extends from Vancouver, British Columbia, southward through the coastal states to San Diego/Tijuana and inland to Phoenix, Arizona. Outside the great world urban regions, the pattern of large cities is peripheral in Africa and South America, rimming the two continents but not yet part of their interiors. In Australia, the southeast coast alone stands out.

When separate major metropolitan complexes expand along the superior transportation facilities connecting them, they may eventually meet, coalesce at their outer margins, and create extensive metropolitan regions or **conurbations.** *Megalopolis,* already encountered in Chapter 9 (p. 293), is the major conurbation of North America, a nearly continuous urban string that stretches from north of Boston to south of Washington, D.C. Other North American present or emerging conurbations include the Southern Great Lakes region stretching from north of Milwaukee through Chicago and eastward to Detroit, Cleveland, and Pittsburgh; the Coastal California zone of San Francisco–Los Angeles–San Diego–Tijuana, Mexico; and the Canadian "core region" conurbation from Montreal to Windsor, opposite Detroit, Michigan, where it connects with the Southern Great Lakes region (Figure 11.5). Outside North America, examples of conurbations are numerous and growing, still primarily in the most industrialized European and East Asian (Japanese) districts, but hatching as well in the other major world urban regions. Prominent are those of the north of England, including Liverpool, Manchester, Leeds, Sheffield, and the several dozen industrial towns in between; the Ruhr of West Germany, with a dozen large connected cities such as Essen and Düsseldorf; and the Tokyo region of Japan. Were it not for the mountains between Tokyo and Nagoya, and between Nagoya and Osaka-Kobe-Kyoto, there is no doubt that these huge metropolises would by now be one megalopolitan area.

Settlement Roots

The major cities of today had humbler origins, their roots lying in the clustered dwellings that everywhere have been the rule of human settlement. People are gregarious and cooperative. Even Stone Age hunters and gatherers pursued their life-style and economy in groups, not as single individuals or isolated families. Primitive cultures are communal for protection, cooperative effort, sharing of tasks by age and sex, and for more subtle psychological and social reasons. Communal dwelling became the nearly universal rule with the advent of sedentary agriculture wherever it developed, and the village became the norm of human society.

In most of the world still, most rural people live in nucleated settlements, that is, in villages or hamlets rather than in dispersed dwellings or isolated farmsteads. Only in North America, parts of northern and western Europe, and in Australia and New Zealand do rural folk tend to live apart, with houses and farm buildings located on land that is individually worked. In those regions, farmsteads tend to be spatially separate one from another, and the farm village is a much less common settlement form. It was not, of course, unknown in North America; Mormon Utah, Mennonite Manitoba, and other districts of cluster migration (see p. 174) were frequently village centered, as were such cooperative and utopian communities as Oneida, New

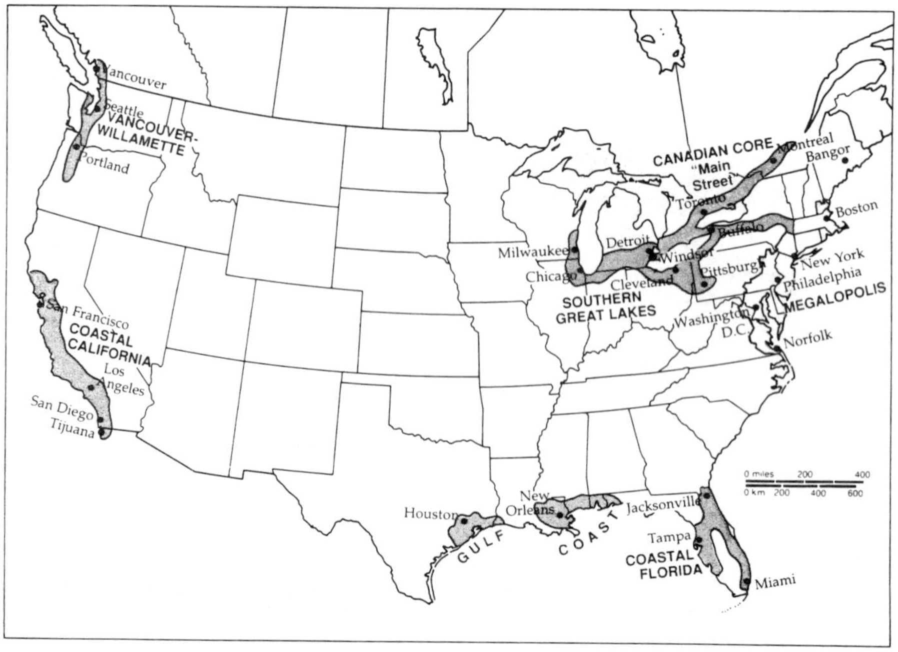

Figure 11.5

Megalopolis and developing North American conurbations. *Megalopolis*, a 600-mile-(965-km-) long continuous series of primarily metropolitan counties stretching from Bangor, Maine, southward to Norfolk, Virginia, contains the economic, political, and administrative core of the United States. A Canadian counterpart core region—dubbed *Main Street* by the geographer Maurice Yeates—connects with other U.S. conurbations through Buffalo, New York, and Detroit, Michigan. For some of their extent, conurbations fulfill their classic definition of continuous built-up urban areas. In other portions, they are more statistical than landscape entities, marked by counties that qualify as "urban" or "metropolitan" even though land uses may appear dominantly rural.

York; Amana, Iowa; Harmony, Indiana; the various Shaker settlements, and others of the 19th century. Elsewhere in the world, villages and hamlets were the settlement norm, though with size and form that varied by region and culture. Intensity of agricultural land use, density of population, complexity and specialization of life and livelihood, and addition of functions other than the purely residential affected the size, distribution, external form, and internal structure of settlements (Figures 11.6 and 11.7).

Figure 11.6

Basic settlement forms.

From Joseph E. Spencer and William L. Thomas, *Introducing Cultural Geography*, 2d ed. Copyright © 1978 John Wiley & Sons, Inc., New York, NY. Reprinted by permission of John Wiley & Sons, Inc.

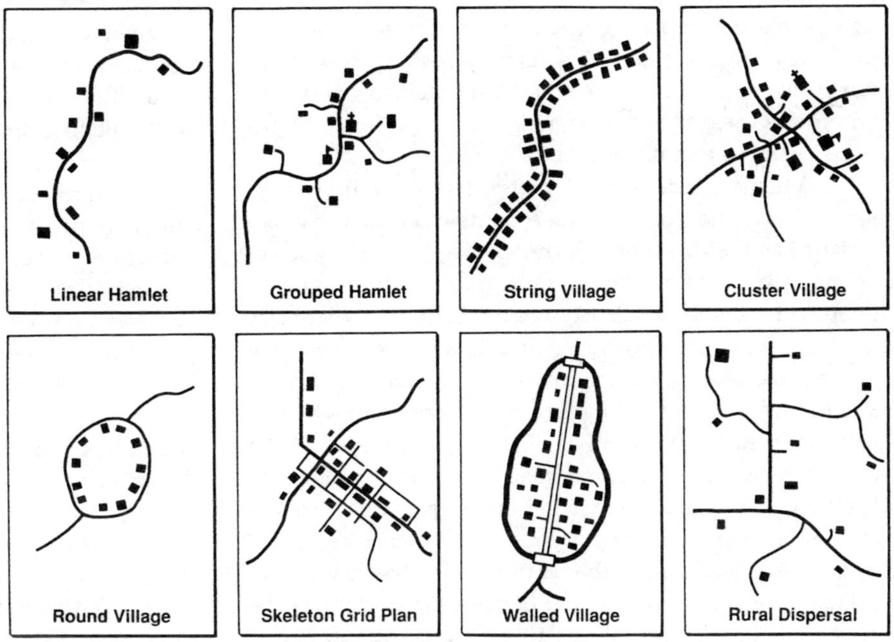

Figure 11.7

Rural settlements in largely subsistence economies vary from the rather low populations characteristic of compact African communities, such as the Zulu village, or kraal, in South Africa shown in (*a*), to more dispersed and populous settlements, such as the Nepalese mountain mountain village of Konar seen in (*b*), to the very large, densely populated Indian rural communities like that seen in (*c*).

Source: (*b*) Nepalese village of Konar courtesy of Colin Thorn.

(a)

(b)

(c)

Rural settlements in developing countries are often considered as expressions of subsistence economic systems in which farming and fishing cultures produce no more than their individual families can consume. That clearly is not always the case. Even in the poorest farm settlements of India or Bangladesh, for example, there is a good deal of trading, buying, and selling of farm goods and family crafts for other needed commodities, and at least some village land is used for other than residential purposes (Figure 11.8). The farm or fishing settlement itself, however, may be nearly self-contained, with little commercial exchange with neighboring villages or between villages and distant cities.

When trade does develop between two or more rural settlements, they begin to take on new physical characteristics as their inhabitants engage in additional types of occupations. The villages lose the purely social and residential character of subsistence agricultural settlements and assume urban features. There is a tendency for the houses to cluster along the main road or roads, leading to a linear, cross, or starlike pattern. No longer are the settlements nearly completely self-contained; they become part of a system of communities. The beginnings of urbanization are seen in the types of buildings that are erected and in the heightened importance of the main streets and of the roads leading to other settlements. The location of villages relative to one another becomes important as the once self-contained rural settlements become towns and cities engaged in urban activities.

The Nature of Cities

Cities are among the oldest marks of civilization. Dating from 4000 to 6000 or more years ago, they originated in—or diffused from—the culture hearths that first developed sedentary agriculture (Table 2.1, page 44). They are as well among the newest experiences of a growing share of the world's population, as Table 11.1 suggests. In all their embodiments, ancient and modern, recurring themes and regularities appropriate to their time and place of existence are evident.

First, all cities perform functions—have an economic base—generating the income necessary to support themselves and to sustain their contained population. Second, none exists in a vacuum; all are part of a larger urban and rural society and economy with which they have essential reciprocal connections. That is, each is a unit in a system of cities and a focus for a surrounding nonurban area. Third, each has a more or less orderly internal arrangement of land uses, social groups, and economic functions. These arrangements may be partially planned and controlled and partially determined by individual decisions, competition for sites, and operational efficiencies. Finally, all cities, large or small, ancient or modern, have or have had maladjustments of land use, recurring and serious social problems and conflicts, environmental inadequacies, and concerns of administration and development. Yet cities, though flawed, remain the capstone of our cultures, the organizing centers of modern societies and economies, the magnet of people everywhere.

Whatever their size, age, or location, urban settlements exist for the efficient performance of functions required by the society that creates them, functions that cannot be adequately carried out in dispersed locations. They reflect the saving of time, energy, and money that the agglomeration of people and activities implies. The more accessible the producer to the consumer, the worker to the workplace, the worshiper to the church, or the lawyer or doctor to the client, the more efficient is the

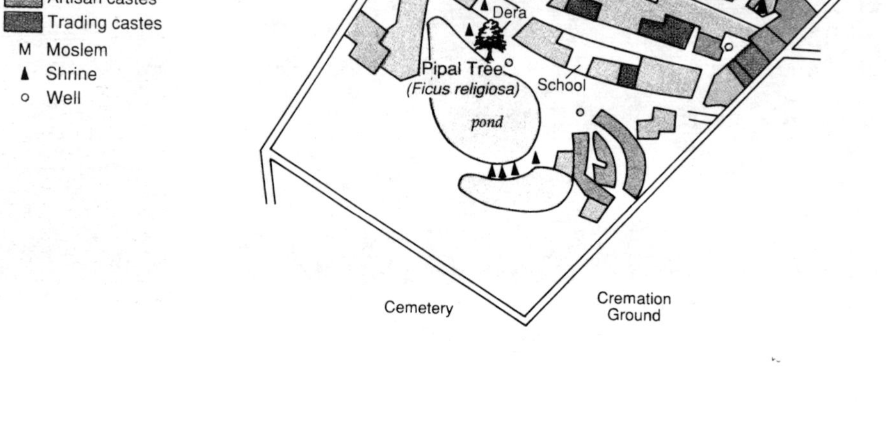

Figure 11.8

A village in the Punjab region, India. In the 1960s, Kunran village had some 1000 inhabitants of several different occupational castes. Most numerous were the Sikh Jat (landowners: 76 households), Ramdasia (peasants and casual laborers: 27 households), and Mazhbi Sikh (agricultural laborers: 12 households). Other castes (and occupations) included Tarkhan (carpenter), Bazigar (acrobat), Jhiwar (water carrier), Sunar (goldsmith), Nai (barber), and Bania (shopkeeper). The trades, crafts, and services they (and others) pursued created a more complex land use pattern than is implied by the generalized village forms depicted in Figure 11.6. The Kunran village plat was "based on a sketch in *Census of India*, vol. 13, part 6, 1963."

From Jan O. M. Broek and John M. Webb, *A Geography of Mankind*. Copyright © 1968 McGraw-Hill Publishing Company, New York, NY. Reprinted by permission.

performance of the separate activities of each, and the more effective is the integration of urban functions.

Urban areas may provide all or some of the following types of functions: retailing, wholesaling, manufacturing, professional and personal services, entertainment, religious observance, governmental administration, military defense, and transportation and communication services. Because all urban functions and people cannot be located at a single point, cities themselves must take up space, and land uses and populations must have room within them. Because interconnection is essential, the nature of the transportation system will have an enormous bearing on the total number of services that can be performed and the efficiency with which they can be carried out. The totality of people and functions of a city constitutes a distinctive cultural landscape whose similarities and differences from place to place are the subjects for urban geographic analysis.

Some Definitions

Urban units are not of a single type, structure, or size. What they have in common—what the term *urban* means—is that they are nucleated, nonagricultural settlements. At one end of their size scale, urban areas are hamlets or small towns with at most a single short main street of shops. At the opposite end, they are complex multifunctional metropolitan areas, supercities, or "megalopolises." The word *urban* is often used in place of such terms as town, city, suburb, and metropolitan area, but it is a general term and does not specify a particular type or size of settlement (Figure 11.9). Although the terms designating the different types of urban settlement, such as *city,* are employed in common speech, not everyone uses them in the same way. What is recognized as a city by a resident of rural Vermont or West Virginia might not at all be afforded that name and status by an inhabitant of California or New Jersey. It is necessary in this chapter to define the meanings of terms commonly employed but varyingly interpreted.

The words *city* and *town* denote nucleated settlements, multifunctional in character, including an established central business district and both residential and nonresidential land uses. Towns have smaller size and less functional complexity than cities but still have a nuclear business concentration. **Suburb** implies a subsidiary area, a functionally specialized segment of a larger urban complex. Suburbs may be dominantly or exclusively residential, industrial, or commercial, but by the specialization of their land uses and functions, they are not individually self-sufficient. They are part of—and correlated with—urbanized areas outside their boundaries. They can, however, be independent political entities. For large cities having many suburbs, it is common to call that part of the urban area contained within the official boundaries of the main city around which the suburbs have been built the **central city.** When separately incorporated, urban places of all types are often referred to as *legal* cities.

Some or all of these urban types may be spatially and functionally associated into larger landscape units. The **urbanized area** refers to a continuously built-up landscape defined by building and population densities without reference to the political boundaries that limit the legal city of which it is the extension. It may be viewed as the *physical* city and may contain a central city and many contiguous cities, towns, suburbs, and other urban tracts. A **metropolitan area,** on the other hand, refers to a large-scale *functional* entity, perhaps containing several urbanized areas, discontinuously built up but nonetheless operating as a coherent and integrated economic whole (see "The Definition of *Metropolitan* in the United States"). Figure 11.10 shows such areas in a hypothetical American county.

Figure 11.9

The differences in size, density, and land use complexity between New York City and an American small town are immediately apparent. Clearly, one is a city and one is a town, but both are *urban*.

The Definition of *Metropolitan* in the United States

Every national jurisdiction has its own definition of *urban* and its own census decisions of the minimum size, legal status, and other criteria of settlements that are considered urban. In Canada, for example, *urban* includes populations living in incorporated places with a population of 1000 or more or in unincorporated places with a density of at least 1000 per square mile. In the United States, *urban places* for many years have been defined as composed of 2500 people resident within an incorporated legal entity, although in the past their official minimum size ranged as high as 8000 residents. A *central city,* the nucleus of a defined *metropolitan area* has been recognized in the United States as a legal city having 50,000 or more people.

The U.S. Bureau of the Census has refined and redefined the concept of metropolitan from time to time to summarize the realities of the changing population, physical size, and functions of urban regions. *Metropolitan districts* were first recognized in 1910. In 1950, the bureau established the *Standard Metropolitan Area (SMA)* and in 1959 renamed it the *Standard Metropolitan Statistical Area (SMSA)* to be replaced after the 1980 census by the *Metropolitan Statistical Area (MSA).* The metropolitan areas were all based on building blocks of one or more functionally integrated counties focusing on a central city of at least 50,000 inhabitants.

In 1983, MSAs were subdivided and the minimum size requirement for central cities was dropped. Central city status was then determined by other qualities—whether, for example, a city was an employment center surrounded by bedroom community suburbs. Automatically, the number of central cities (and metropolitan areas) increased. The statistical structure of urban America was altered as individual communities exercised their right to withdraw from former metropolitan affiliations or opted to join neighboring cities in new ones. The subdivision of MSAs gave rise to still other units whose recognition was deemed necessary by the changing realities of American urban structure. *Primary Metropolitan Statistical Areas (PMSAs)* are those counties that are part of MSAs that have less than 50% resident workers working in a different county. *Consolidated Metropolitan Statistical Areas* are MSAs containing 1 million or more people and are composed of PMSAs. In actuality, the definitional rules are much more complex than this summary suggests. They represent detailed attempts to derive statistical units meaningful for public policy and economic decisions.

Figure 11.10

A hypothetical spatial arrangement of urban units within a Metropolitan Statistical Area. Sometimes official limits of the central city are very extensive and contain areas commonly considered suburban or even rural. Older eastern U.S. cities more often have restricted limits and contain only part of the high-density land uses associated with them. In the diagram, *A* may be thought of as suburban as well as urban.

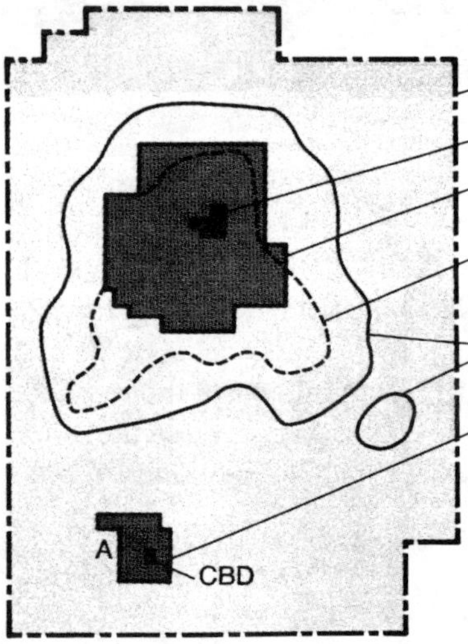

The Location of Urban Settlements

The American geographer Mark Jefferson (1863–1949) once observed, "Cities do not grow up of themselves. Countrysides set them up to do tasks that must be performed in central places." That astute observation reminds us of two essential facts. First, the urban settlement is part of a larger, partially nonurban economic whole, with which the city has close reciprocal relationships. These relationships—the functions of the city—are the reason for its existence. The city is a consumer of food, a processor of materials, and an accumulator and dispenser of goods and services. But it must depend on the "countryside" that established it for its essential supplies and as a market for its products and activities. Second, in order to perform adequately the tasks that support it and to add new functions as demanded by the larger economy, the city must be efficiently located. That efficiency may be marked by spatial or transportation centrality to the area served. It may derive from location at points whose physical characteristics permit the performance of assigned functions. Or placement may be related to the resources, productive regions, and transportation network of the nation, so that the effective performance of a wide array of activities is possible without reference to the requirements of the immediate local region.

In discussing urban settlement location, geographers usually mention the significance of site and situation, concepts already introduced in Chapter 1 (see p. 7). You will recall that *site* refers to the exact terrain features underlying the city, as well as—less usefully—to its absolute (globe grid) location. Classifications of cities according to site characteristics have been proposed, recognizing special placement circumstances: *break-of-bulk* locations, for example, such as river crossing points where cargos and people must interrupt a journey; *head-of-navigation* or *bay head* locations where the limits of water transportation are reached; or *railhead* locations where the railroad ended. In Europe, security and defense—island locations or elevations—were considerations in earlier settlement locations and figure in site classifications. Waterpower sites of earlier, and coal field sites of later, stages of the Industrial Revolution were noted in Chapter 9 and represent a union of environmental and cultural-economic considerations.

Situation implies location in relation to the physical and cultural characteristics of surrounding or external areas (see p. 7 and Figure 1.6). Although in many ways more important than site in understanding the functions and growth potentials of cities, situation is distinctive to each settlement and does not lend itself to easy generalization. References to site and situation are by their nature attempts at explanatory description. They express the conviction that the functional success or failure of a settlement is not a product of chance or accident but is rooted in geographical realities of position and relative location.

The site or situation that originally gave rise to an urban unit may not long remain the essential ingredient for its growth and development. Agglomerations, originally successful for whatever reason, may by their success attract people and activities totally unrelated to the initial localizing forces (Figure 11.11). By a form of "circular and cumulative causation" (see p. 313), a successful urban unit may acquire new populations and functions attracted by its already existing markets, labor force, and urban facilities.

The Functions of Cities

The key concept and reference of urban success is *function*—what cities actually do within the larger society and economy that established them. No city stands alone. Each is linked to other towns and cities in an interconnected city system; each provides services and products for its surrounding tributary region—its *hinterland* or trade area. Those linkages reflect complementarity and the processes of spatial interaction that we explored in Chapter 3; they are rooted in the different functions performed by different units within the urban system. However, not all of the activities carried on within a city are intended to connect that city with the outside world. Some are necessary simply to support the city itself. Together, these two levels of activity make up the **economic base** of an urban settlement.

Figure 11.11

Early urbanization in eastern Massachusetts. Maritime activities, situational advantages, port characteristics, and diversity of functions helped localize individual settlements and led to the early rise of urban agglomerations in eastern Massachusetts. Their development and success encouraged continued urbanization despite changing economic conditions.

Lester J. Cappon et al., ed; *Atlas of Early American History: The Revolutionary Era 1760–1790*. Copyright © 1976 by Princeton University Press. Map reprinted with permission of Princeton University Press.

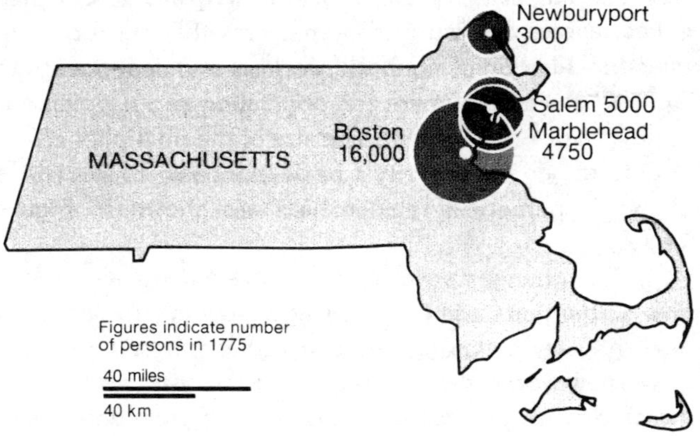

The Economic Base

Only part of a city's employed population is engaged in the production of goods or the performance of services for external markets. They are workers engaged in "export" activities, whose efforts result in money flowing into the community. Collectively, they constitute the **basic sector** of the city's total economic structure. Other workers support themselves by producing things for residents of the urban unit itself. Their efforts, necessary to the well-being and successful operation of the city, do not generate new money for it but comprise a **service** or **nonbasic sector** of its economy. These people are responsible for the internal functioning of the urban unit. They are crucial to the continued operation of its stores, professional offices, city government, local transit, and school systems.

The total economic structure of a city equals the sum of its basic and nonbasic activities. In actuality, it is the rare urbanite who can be classified as belonging entirely to one sector or another. Some part of most people's work involves financial interaction with residents of other areas. Doctors, for example, may have mainly local patients and thus are members of the nonbasic sector. But the moment they provide a service to someone from outside the community, they bring new money into the city and become part of the basic sector.

Base Ratios

Assuming it were possible to divide with complete accuracy the employed population of a city into totally separate basic and nonbasic components, a ratio between the two employment groups could be established. With some exceptions, this basic/nonbasic ratio is roughly similar for cities of similar size irrespective of their functional specializations. Further, as a city increases in size, the number of nonbasic personnel grows faster than the number of new basic workers. Thus, in cities with a population of 1 million, the ratio is about two nonbasic workers for every basic worker. The addition of 10 new basic employees implies the expansion of the labor force by 30 (10 basic, 20 nonbasic) and an increase in total population equal to the added workers plus their dependents. A **multiplier effect** thus exists, associated with economic growth. The term implies the addition of nonbasic workers and dependents to a city's total employment and population as a supplement to new basic employment. The size of the multiplier effect is determined by the city's basic/nonbasic ratio. These changing numerical relationships are shown in Figure 11.12.

The changes are understandable when we consider how settlements add functions and grow in population. A new industry selling services to other communities requires new workers who thus increase the basic work force. These new employees in turn demand certain goods and services, such as clothing, food, and medical assistance, which are provided locally. Those who perform such services must themselves have services available to them. For example, a grocery clerk must also buy groceries. The more nonbasic workers a city has, the more nonbasic workers are needed to support them, and the application of the multiplier effect becomes obvious.

We have also seen that the growth of cities may be self-generating—"circular and cumulative" in a way related not to the development of basic industry but to the attraction of what would be classified as *service* in an economic base study. Banking and legal services, a sizable market, a diversified labor force, extensive public services, and the like may attract or generate additions to the labor force not classically basic by definition. The multiplier effect in larger metropolitan centers is discerned in many guises.

In much the same way as settlements grow in size and complexity, so do they decline. When the demand for the goods and services of an urban unit falls, obviously there is a need for fewer workers. Thus both the basic and the services components of a settlement system are affected. There is, however, a resistance to decline that impedes the process and delays its impact. That is, cities can grow rapidly as migrants respond quickly to the need for more workers, but under conditions of decline those who have developed roots in the community are hesitant to leave or may be financially unable to move to another locale.

The Classification of Functions

Variations in basic employment structure among urban units characterize the specific functional role individual cities play within the national economy and provide a means for a classification and comparison of cities. Some settlements, particularly smaller ones, may be termed *unifunctional*; their basic sector is composed entirely of a single activity type. They are the small farm-market towns, the coal-mining or lumbering towns, the county seats (or even large national capitals such as Washington, D.C.), or university towns without other basic activities. Most cities, however, perform many export functions, and the larger the urban unit, the more *multifunctional* it becomes. Nonetheless, even in cities with a diversified economic base, one or a very small number of export activities tends to dominate the structure of the community and to identify its operational purpose within a system of cities.

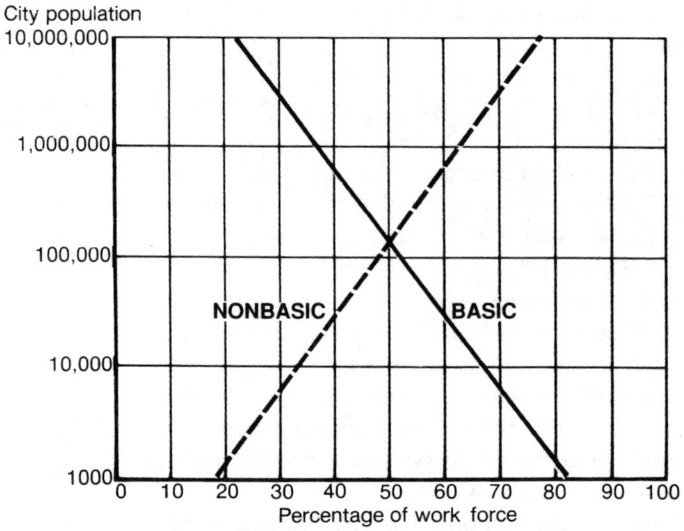

Figure 11.12

A generalized representation of the proportion of the work force engaged in basic and nonbasic activities by city size. As settlements become larger, a greater proportion of the work force is employed in nonbasic activities. Larger centers are therefore more self-contained.

Such functional specialization permits the classification of cities into categories: manufacturing, retailing, wholesaling, transportation, government, and so on. Such specialization may also evoke images when the city is named: Detroit, Michigan, and Tokyo, Japan, as manufacturing centers; Tulsa, Oklahoma, for oil production; Nice, France, as a resort; Ottawa, Canada, in public administration, and so on. Certain large regional, national, or world capitals—as befits major multifunctional concentrations—call up a whole series of mental associations, such as New York with banking, the stock exchange, entertainment, the fashion industry, port activities, and so forth.

While the functions of many cities are obvious to even the casual observer, the economic base of others is less evident. A variety of *qualitative* and *quantitative* functional classifications have been devised to categorize cities, most of them based on the employment structure of the urban labor force. We saw in Chapter 10 how national economies are dynamic and constantly evolving. Individual cities share in those changes and undergo alterations in their employment structure and functional roles. For most cities such changes are gradual, and persistent spatial patterns of economic specialization may be recognized. Those defined and mapped by Chauncy Harris (b. 1914) many years ago are still among the best known in the geographic literature (Figure 11.13). As we would expect, they show manufacturing cities to be concentrated in the industrial belt of the northeastern and midwestern states. Wholesale and retail trade are evident functional specializations of cities in a broad north-south midcontinental band. The mining towns shown were those on then productive and exploited ore bodies; resort and retirement centers are based on more enduring climatic and scenic resources.

Systems of Urban Settlements

The functional structure of a city affects its current size and growth prospects. At the same time, that structure is a reflection of both that city's location and its relationships with other units in the larger system of which all are a part. A simple but revealing threefold functional classification of urban settlements recognizes them as either transportation centers, special-function cities, or central places. Each class has its own characteristic spatial arrangement; together, the three classes help explain the distributional pattern and the size and functional hierarchies of the entire city system.

The spatial pattern of *transportation centers* is that of alignment—along seacoasts, major and minor rivers, canals, or railways. Routes of communication form the orienting axes along which cities developed and upon which at least their initial functional success depended (Figure 11.14). *Special function cities* are those engaged in mining, manufacturing, or other activities the localization of which is related to raw material occurrence, agglomeration economies, or the circular and cumulative attractions of constantly growing market and labor concentrations. Special function cities show a pattern of urban clustering—as the mining and manufacturing cities of the Ruhr district, the Midlands of England, or the Donets Basin, for example. More familiarly, they appear in the form of the multifunctional metropolitan concentrations recognized by the Metropolitan Statistical Areas of the United States, the Census Metropolitan Areas of Canada, or in such massive urbanized complexes as metropolitan Tokyo, Moscow, Paris, London, Buenos Aires, and others worldwide.

A common property of all settlements is centrality, no matter what their recognized functional specializations. Every urban unit provides goods and services for a surrounding area tributary to it. Some settlements, however, have that rural service and trade function as their dominant role, and these make up the third simplified category of cities: *central places.*

Central Place Theory

A **central place** is a nodal point for the distribution of goods and services to surrounding nonurban populations and an essential link in a system of interdependent urban settlements. Central places are universal components of urban systems in all parts of the globe under all economic and cultural systems, and everywhere tend to show size and spacing patterns peculiar to the tertiary (distributional) activity function they perform. Their special case, called *the marketing principle,* suggests spatial regularities in the urban system not seen where special function or transportation cities predominate. The examination of those regularities, first undertaken by the German geographer Walter Christaller (1893–1969) in 1933, comprises **central place theory.**

Christaller believed that his analysis could best be visualized in rather idealized circumstances. He and other later spatial modelers who developed his ideas proposed a set of simplifying assumptions. First is the assumption that his system of central places would develop on a uniform plain with no topographic barriers, channelization of traffic, or variations in agricultural productivity. Next, there is the supposition that the farm population would be evenly distributed across that idealized surface and that the characteristics of the people would be uniform; that is, they would possess similar tastes, demands, and incomes. Finally, it is stipulated that any good or service provided within a town would have a constant *range*—the maximum distance over which its sale could extend—extending uniformly in all directions from the town. Such assumptions are hardly realistic, but by employing them,

Figure 11.13

A functional classification of U.S. cities.

Source: Chauncy D. Harris, "A Functional Classification of Cities in the United States" in *Geographical Review*, 33, p. 88, 1943. Copyright © 1943 The American Geographical Society, New York, NY. Reprinted by permission.

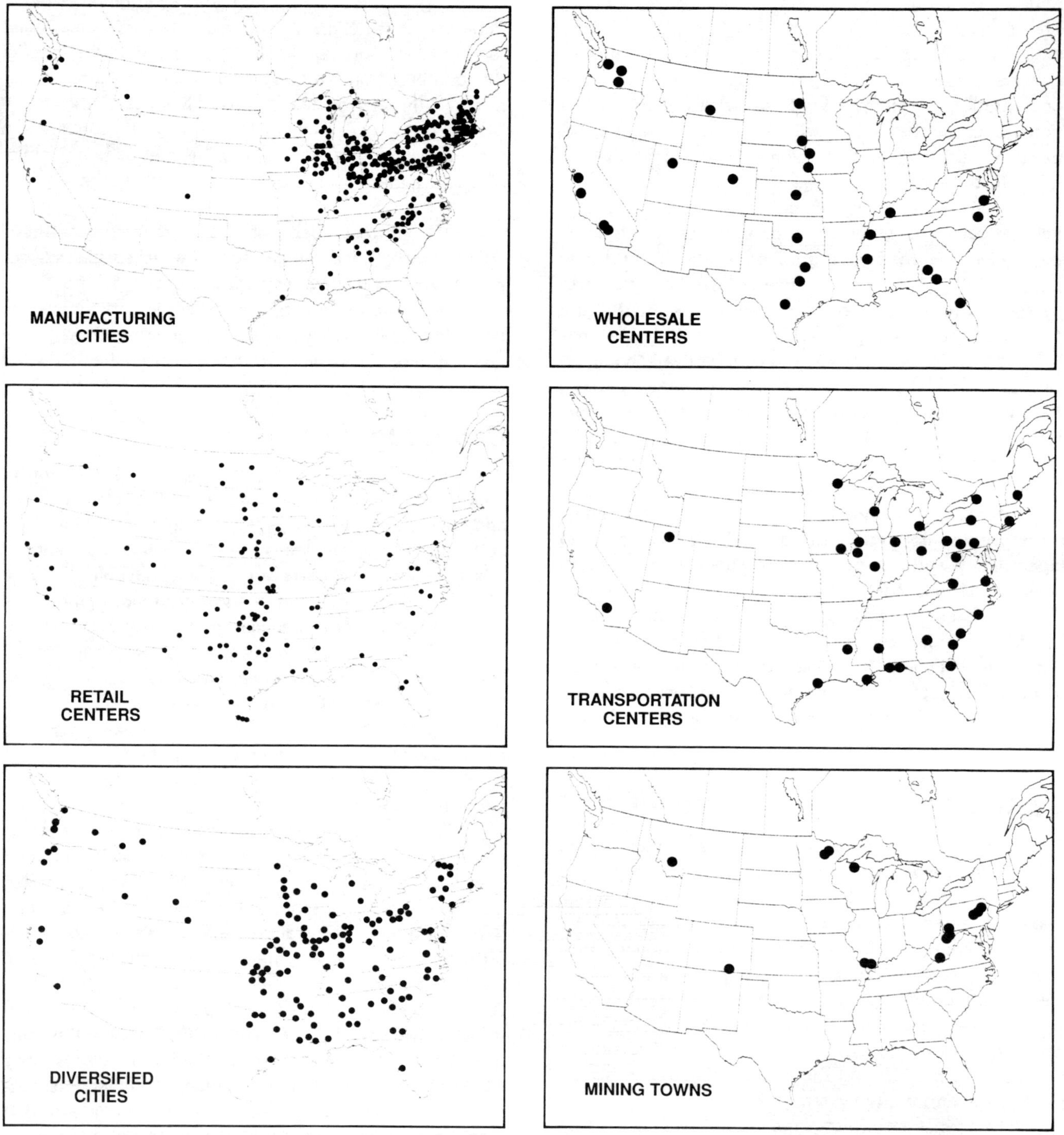

central place modelers were able to develop theoretical laws of central place patterning that could then be tested against observable regional realities.

Christaller and his followers added other economic and behavioral assumptions. They supposed that each kind of product or service supplied to the dispersed population has its own *threshold,* which is the minimum market (number of consumers or volume of purchasing power) needed to support the supply of the product. For example, a shoe store can prosper in an area only if there is sufficient demand for the shoes and money to buy them. There

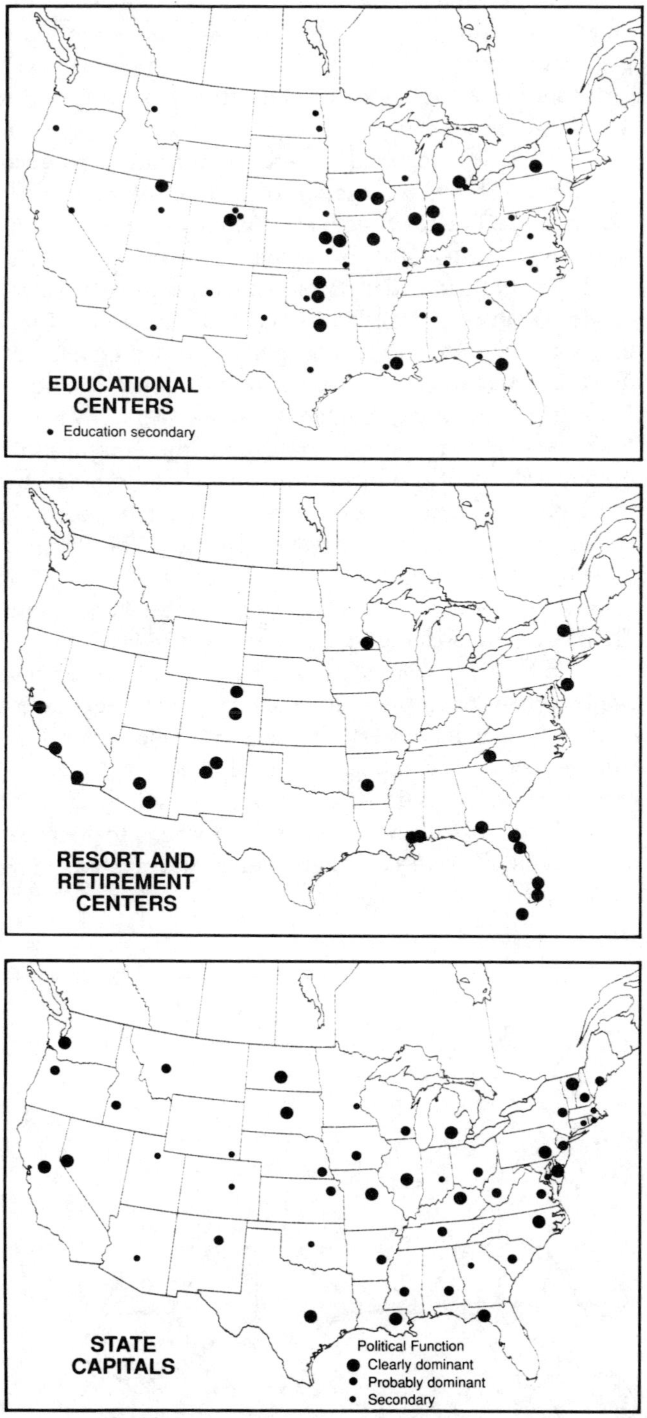

Figure 11.14

Urban alignments in Illinois. Railroads preceded settlement in much of the North American continental interior, and urban centers were developed—frequently by the railroad companies themselves—as collecting and distributing points expected to grow as the farm populations increased. Located at constant 5- to 6-mile (8- to 10-km) intervals in Illinois, the rail towns were the focal points of an expanding commercial agriculture. The linearity of the town pattern in 1940, at the peak of railroad influence, unmistakably marks the rail routes.

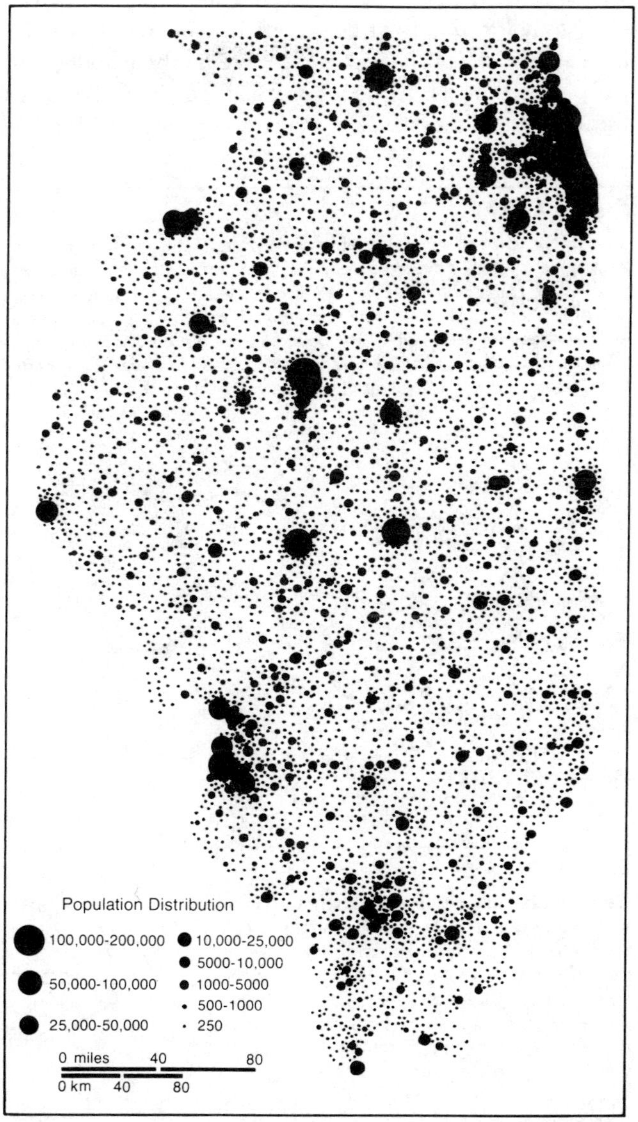

will be stores offering products with high thresholds (requiring high purchasing power) only if there is a large market area, that is, a large number of potential customers. High-threshold (or high-order) goods, such as grand pianos and sable coats, are either expensive or not in great demand. On the other hand, bread, a low-threshold (low-order) product, can be marketed from many locations because demand is high and the product is relatively inexpensive. Where multiple central places offer the same good or service for sale (always at the same store price), prospective customers will rationally limit the range of that good to a distance equal to half the separation between the competing centers. That is, people will invariably patronize the nearest central place supplying the item they need, for a longer journey would increase the acquisition cost of the purchase. That behavioral regularity assures that each central place is surrounded by a *complementary region* over which the town (and its merchants) has a sales monopoly based on travel range and price advantage.

Since the hypothetical plain must be completely subdivided into separate complementary regions, no area can be unserved and none can have equal service from two competing centers. That means that complementary regions cannot be circular (Figure 11.15). Rather, there exists a series of hexagonal market areas that covers the entire plain, for the hexagon is the most compact geometric form into which the entire hypothetical surface can be subdivided, leaving no unserved hinterland and no areas unassigned to a closest central place (Figure 11.16). The largest hexagonal market areas have as their nodes central places that supply goods and services from the lowest through the highest orders, that is, all goods and services supportable by the demand and the purchasing power of the entire area. Contained within or at the edge of the largest market areas are central places serving a smaller population and offering fewer goods and services to their respectively smaller hinterlands. As the diagram indicates, the central place pattern shows a *nesting* of complementary regions in which part or all of multiple lower-order service areas are contained within the hinterland of a higher-order center.

Figure 11.15

The derivation of complementary regions. (*a*) If the hypothetical region were totally covered by circular complementary regions, areas of overlap would occur. Since Christaller's assumption was that people will only shop at the nearest center, areas of overlap must be divided so that those on each side of the boundary are directed to their nearest service point. (*b*) Circular areas too small to cover the region completely result in impermissible unserved populations.

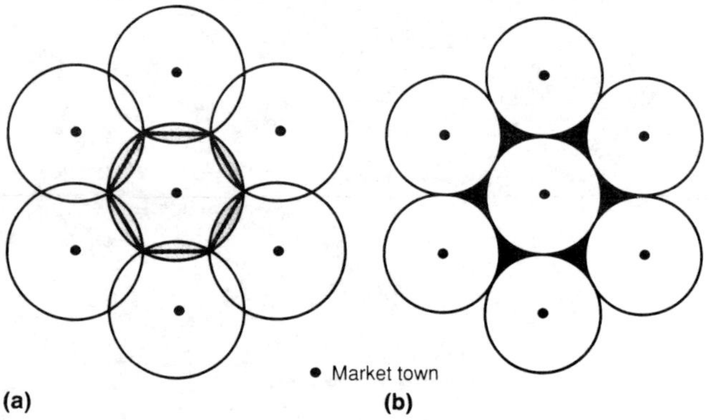

Christaller reached several important theoretical conclusions. First, the system of central places is interdependent. If one town were eliminated, the entire system would have to readjust in its spatial pattern, its offered goods, or both in order to provide consumers with the needed variety and availability of central place items. There exists a *hierarchy of central places* whose equilibrium must be maintained or reestablished when disturbed. Second, a necessary characteristic of that hierarchy is that towns at the same functional level in the central place system will be of the same size (since they provide the same goods and services) and will be evenly spaced. Of course, higher-order places will be larger and farther apart than towns at lower-order levels in the hierarchy. This means that there will be many more small than large towns. Third, size and spacing regularities will develop.

Under Christaller's marketing principle there is a single highest-order place and the number of places below it (after the second order) increases in a ratio—called the *k-value*—of 3. The sequence runs as follows: 1 highest-order center, 2 second-order places, 6 third-order, 18 fourth-order, 54 fifth-order, and so on through the number of hierarchical levels present within the system. The

Figure 11.16

Complementary regions and the pattern of central places. The two *A* central places are the largest on this diagram of Christaller's model. The *B* central places offer fewer goods and services for sale and serve only the areas of the intermediate-sized hexagons. The many *C* central places, which are considerably smaller and more closely spaced, serve still smaller hinterlands. The goods offered in the *C* places are also offered in the *B* and *A* places, but the latter offer considerably more and higher-order goods. Notice that places of the same size are equally spaced.

Source: From Arthur Getis and Judith Getis, "Christaller's Central Place Theory" in *Journal of Geography*, 1966. Copyright © 1966 National Council for Geographic Education, Macomb, IL. Reprinted by permission.

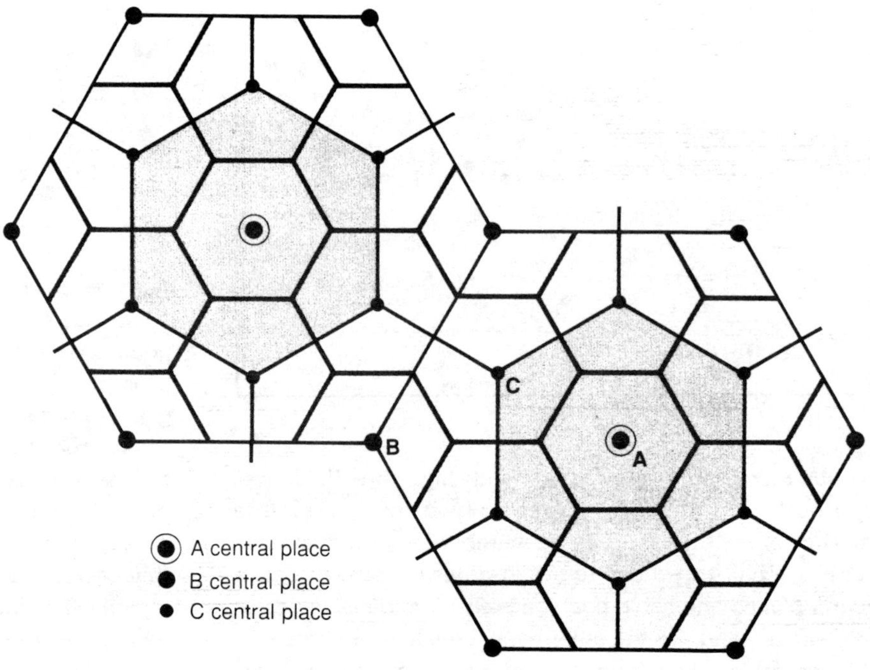

spacing regularity requires that the distance between centers at a given hierarchical level be equal to the square root of 3 times the distance between centers at the next lower level in the system.

Christaller also suggested two variants of the marketing ($k = 3$) principle. The *traffic* or *transport principle* involves a k-value of 4 and yields a hierarchy of places connected by the most efficient road network. The *administrative principle* proposes a $k = 7$ structure that allows each lower-order hinterland to nest within the complementary region of a higher-order center, producing the most efficient set of administrative regions.

Central place hierarchies have been recognized in studies of rural service centers in all parts of the world. Historical and archaeological records indicate that they existed in past urban landscapes as well. The theory, then, appears to be a valid statement about at least a limited arena of human spatial interaction. It is particularly applicable, of course, to agricultural areas. One has to stretch things a bit to see the model operating in highly industrialized districts where cities are more than just retailing centers. However, if we combine a Christaller-type approach with the ideas that help us understand the locational patterns of special-function and transportation-based cities, we begin to appreciate the spatial structure of the system of cities in advanced economies.

The Urban Hierarchy

Perhaps the most effective way to recognize how systems of cities are organized is to consider the **urban hierarchy.** Cities can be divided into size classes on the basis of their functional complexity and importance. One can measure the numbers and kinds of functions each city or metropolitan area performs. The hierarchy is then like a pyramid; the few large and complex cities are at the top, and the many smaller simpler ones are at the bottom. There are always more smaller cities than larger ones. One can envisage, say, a seven-level hierarchy where the complexity of cities increases as one rises in the pyramid.

When a spatial dimension is added to the hierarchy, as in Figure 11.17, it becomes clear that a spatial system of metropolitan centers, large cities, small cities, and towns exists. Goods, services, communications, and people flow up and down the hierarchy according to the location of the urban areas. The few high-level metropolitan complexes provide specialized functions for large regions, while the smaller cities serve smaller districts. Note that the cities interact with the area around them, but since cities of the same level provide roughly the same services, they tend not to serve each other unless each provides some very specialized activity, such as housing the political capital of a region or a major university. Thus, the cities of a given level in the hierarchy are not independent but interrelated with cities of other levels in the hierarchy. Together, all cities at all levels in the hierarchy constitute an urban system.

Rank-Size and Primacy

The observation that there are many more small than large cities within an urban system ("the larger the fewer") is a statement about hierarchy. For many large countries of

Figure 11.17

Hierarchical relationships among southern U.S. cities.

Source: From "Metropolitan Dominance and Integration," by Rupert B. Vance and Sara Smith in *The Urban South*, edited by Rupert B. Vance and Nicholas J. Demerath. © 1954 The University of North Carolina Press.

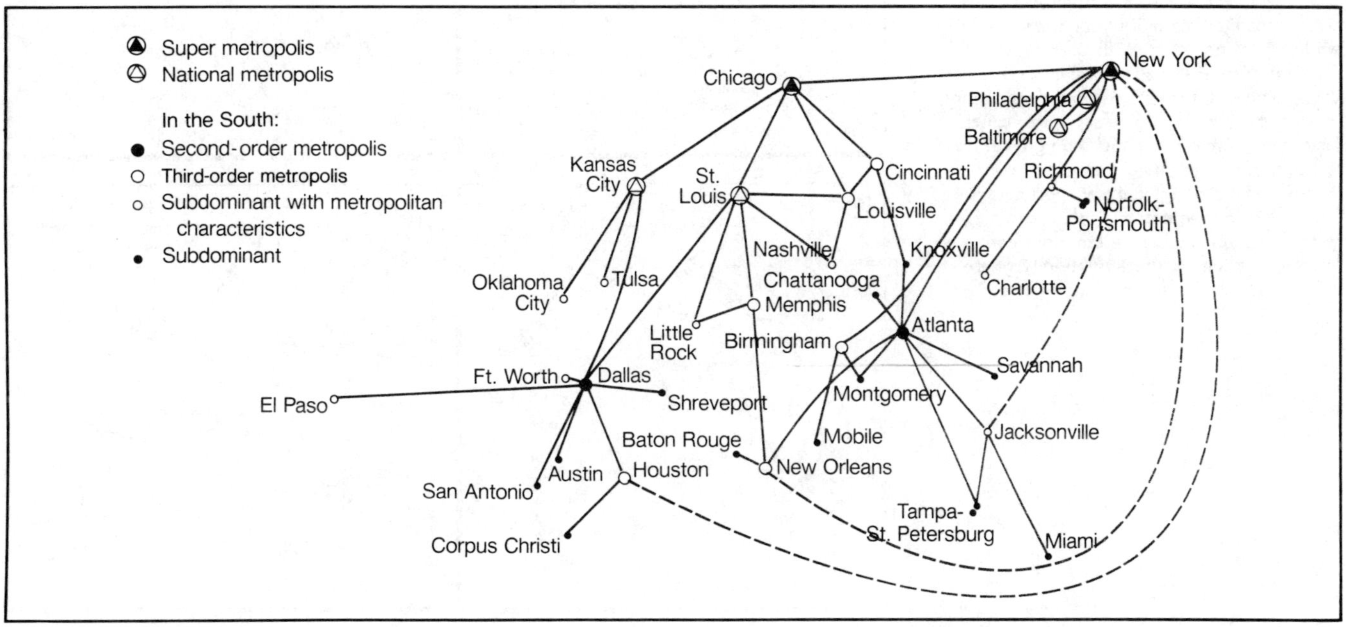

great regional diversity and advanced economy, the city size hierarchy is summarized by the **rank-size rule.** It tells us that the *n*th largest city of a national system of cities will be $1/n$ the size of the largest city. That is, the second-largest settlement will be 1/2 the size of the largest, the tenth-biggest city will be 1/10 the size of the first-ranked city.

The rank-size ordering may describe the size patterning of cities in complex economies where urban history is long and urbanizing forces are many and widely distributed. Although no national city system exactly meets the requirements of the rank-size rule, that of the Soviet Union or the United States closely approximates it. The rule is less applicable to countries with developing economies or where the urban size hierarchy has been distorted through concentration of functions in a single, paramount center.

In some countries the urban system is dominated by a **primate city,** one that is far more than twice the size (and therefore functional complexity) of the second-ranked city. In fact, there may be no clearly recognized "second city" at all, for a characteristic of a primate city hierarchy is one very large city, few or no intermediate-sized cities, and many subordinate smaller settlements. The capital cities of many developing countries display that kind of overwhelming primacy. In part, their primate city pattern is a heritage of their colonial past when economic development, colonial administration, and trade activities were concentrated at a single point (Figure 11.18). Kenya, Mozambique, Senegal, and several other African nations are examples. In other instances—Egypt and Mexico, for example—development and population growth has tended to concentrate disproportionately in a capital city whose very size attracts further development and growth. Many European countries also show a primate structure, often ascribed to the former concentration of economic and political power around the royal court in a capital city that was, perhaps, also the administrative and trade center of a larger colonial empire. Paris is the most frequently cited European primate city.

Where central places dominate, of course, the urban size hierarchy will be one of a relatively few classes of cities with distinct size differences separating the classes.

Inside the City

The structures, patterns, and spatial interactions that comprise and describe the system of cities make up only half of the story of urban settlements. The other half involves cities themselves, viewed as distinctive cultural landscapes spatially structured to perform their functions and organized to accommodate more or less efficiently the

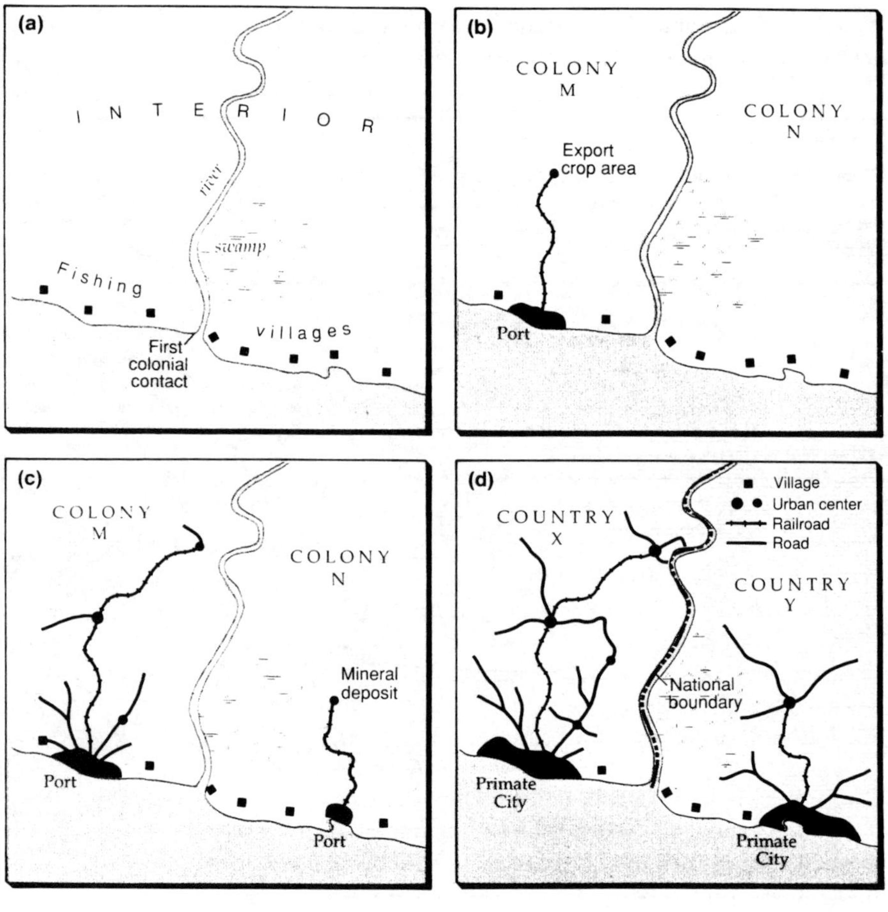

Figure 11.18

Primate city evolution. At first colonial contact (*a*), settlements are coastal and unconnected with each other. Joining a newly productive hinterland by European-built railroad to a new colonial port (*b*) begins to create a pattern of core-periphery relations and to focus European administration, trade, and settlement at the port. Mineral discoveries and another rail line in a neighboring colony across the river (*c*) mark the beginnings of a new set of core-periphery relationships and of a new multifunctional colonial capital nearby but unconnected by land with its neighbor. With the passage of time and further transport and economic development, two newly independent nations (*d*) display *primate city* structures in which further economic and population growth flows to the single dominating centers of countries lacking balanced regional transport networks, resource development, and urban structures. Both populations and new functions continue to seek locations in the primate city where their prospects for success are greatest.

land uses and populations they house. An understanding of the nature of cities is incomplete without a knowledge of their internal characteristics. In those characteristics, however, the world's cities are marked more by regional variation than by global uniformity. Like other built environments, settlements reflect not only their own histories but also the cultures and technologies of the societies that create them. Later in this chapter we will review world regional contrasts in urban land use patterns and social geographies. But first, let us examine in some detail the North American city to form a preliminary appreciation of how land uses are distributed, how social areas are formed, and how institutional controls affect the urban structure in a cultural context familiar to most of us. The generalizations we derive from that review can then become the benchmarks for our cross-cultural and world regional comparisons.

Generalizations valid particularly for older eastern United States central cities are possible, for it is a common observation that a recurring pattern of land use arrangements and population densities exists within them and their suburban fringes. It is deduced that those regularities have been created by the circumstances of differential accessibility, the impersonal operations of a free market in land, and the collective consequences of individual residential, commercial, and industrial locational decisions. The whole urban structure is, of course, also being modified by land use controls, obsolescence, transportation alterations, and governmental redevelopment programs.

The Competitive Bidding for Land

For its effective operation, the city requires close spatial association of the functions that it houses and of the populations carrying out those functions. As long as those functions were few and the population small, pedestrian movement and pack-animal or cart haulage were sufficient for the effective integration of the urban community. With the addition of large-scale manufacturing and the accelerated urbanization of the economy during the 19th century, however, functions and populations—and therefore city areas to contain them—grew beyond the interaction capabilities of pedestrian movement alone. Increasingly efficient and costly mass transit systems were installed. Even with their introduction, however, only land within walking distance of the mass transit routes or terminals could successfully be incorporated into the expanding urban structure.

Usable—because accessible—land, therefore, was a scarce commodity, and by its scarcity it assumed high market value and demanded intensive, high-density development. The industrial city of the mass transit era—the late 19th and early 20th centuries—because of its limited supply of developable land, was compact, was characterized by high residential and structural densities (Figure 11.19), and showed a sharp break on its margins between urban and nonurban uses. The older central cities of, particularly, the northeastern United States and southeastern Canada were of that vintage and pattern although, as we shall see, their resulting land use and social structures were not identical. In contrast, many cities of the western United States are more the product of the 20th century than the 19th, more influenced in their density and land use structures by the automobile than by mass transit systems. They spread more readily, evolved at lower densities, and display less tightly structured and standardized land use patterns.

Within the city, parcels of land were allocated among alternate potential users on the basis of the relative ability of those users to outbid their competitors for a chosen site. There was, in gross generalization, a continuous open auction in land in which users would locate, relocate, or be displaced in accordance with "rent-paying ability." The attractiveness of a parcel, and therefore the price that it could command at auction, was a function of its accessibility. Ideally, the most desirable and efficient location for all the functions and people of a city would be at the single point at which the maximum possible interchange could be achieved. Such total convergence of activity is obviously impossible.

Because uses must arrange themselves spatially, the attractiveness of a parcel was rated by its relative accessibility to all other land uses of the city. Store owners wished to locate where potential customers could easily reach them; factories needed a convenient assembling of their workers; residents desired easy connection with jobs, stores, and schools. Within the older central city, the radiating mass transit lines established the elements of the urban land use structure by freezing in the landscape a clear-cut pattern of differential accessibility. The convergence of that system on the city core gave that location the highest accessibility, the highest desirability, and

Figure 11.19

Duplexes, apartment buildings, and row houses like these in the Crown Heights district of Brooklyn were characteristic 19th-century residential responses to the price and scarcity of developable urban land. Where detached single-family dwellings were built, they were usually placed on far smaller lots than became the rule during the middle 20th century.

hence, the highest land values of the entire built-up area. Similarly, transit junction points were accessible to larger segments of the city than locations along single traffic routes. The latter were more desirable than parcels lying between the radiating lines (Figure 11.20).

Society deems certain functions desirable without regard to their economic competitiveness. Schools, parks, and public buildings were assigned space without being participants in the auction for land. Other uses, through the process of that auction, were assigned spaces by market forces. The merchants with the widest variety and highest order of goods and the largest threshold requirements bid most for and occupied parcels within the **central business district (CBD),** which became localized at the convergence of mass transit lines. The developers of tall office buildings, of principal hotels, and similar land uses were the successful bidders for slightly less accessible CBD parcels.

Comparable, but lower-order, commercial aggregations developed at the outlying intersections—transfer points—of the mass transit system. With time a distinctive retailing hierarchy emerged within the urban settlement, a hierarchy based on the purchasing power thresholds and complementary regions of city populations themselves. The direct counterpart of the rural central place hierarchy, it was capped by the central business district as the highest-order center, with major outlying business districts, neighborhood store clusters, and isolated low-order convenience stores completing the lower levels of the hierarchy.

Figure 11.20

In older cities, the central business district, localized by converging mass transit lines, had a superior accessibility that many functions sought. Competition for space and the resulting high land values demanded high-density use of a restricted area. That localization was not necessarily permanent, however. Business districts may "migrate" as that of St. Louis did—away from a no longer useful or sustainable focus on an originally important but later blighted waterfront. Urban redevelopment may convert an original core to other uses, but the central business district is re-established at a new focal point to serve the continuing need for a commercial, financial, and administrative heart of the metropolitan area. See also Figure 11.34.

Industry took control of parcels adjacent to essential cargo routes: rail lines, waterfronts, rivers, or canals. Strings of stores, light industries, and high-density apartment structures could afford and benefit from location along high-volume transit routes. The least accessible locations within the city were left for the least competitive uses: lower-density residences. A diagrammatic summary of this repetitive allocation of space among competitors for urban sites in American mass transit cities is shown in Figure 11.21.

Land Values and Population Density

Theoretically, the continuing land auction should yield two separate although interconnected distance decay patterns, one related to land values and the other to population density (as distance increases away from the CBD, the intensity of the distribution decreases). If one views the land value surface of the older central city as a topographic map (Figure 11.22), with hills representing high valuations and depressions showing low prices, a series of peaks, ridges, and valleys would reflect the differentials in

Figure 11.21

Generalized urban land use pattern. The location of various land uses in an idealized city where the highest bidder gets the most accessible land.

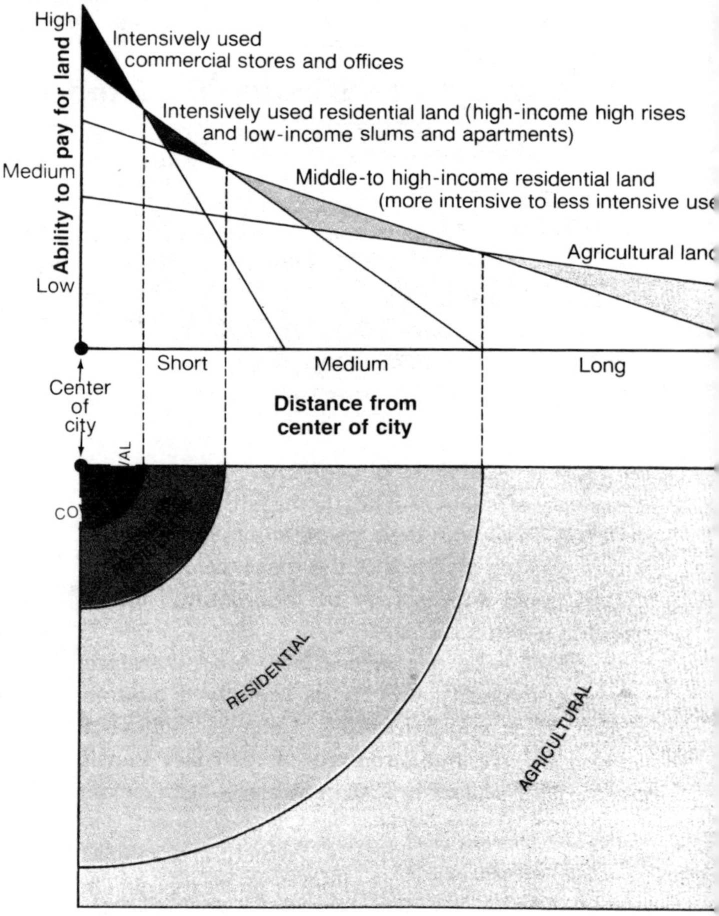

accessibility marked by the pattern of mass transit lines, their intersections, and the unserved interstitial areas. Dominating these local variations, however, is an overall decline of valuations with increasing distance away from the peak value intersection, the most accessible (by mass transit) and costly location of the central business district. As we would expect in a distance decay pattern, the drop in valuation is precipitous within a short linear distance from that point, and then the valuation declines at a lesser rate to the margins of the built-up area.

With one important variation, the population density pattern of the central city shows a comparable distance decay arrangement (Figure 11.23). The exception is the tendency to form a hollow *at the center,* the CBD, which represents the inability of all but the most costly apartments to compete for space against alternative uses desirous of supremely accessible parcels. Yet penalty of high land prices, accessibility is attr number of residential users. The result is the hig residential occupancy of parcels *near the cente* city—by those who are too poor to afford a long-di journey to work; who are consigned by their pover.y to overcrowding in obsolescent slum tenements near the heart of the inner city; or who are self-selected occupants of apartments whose high rents are made necessary by the price of land. Other urbanites, if financially able, may opt to trade off higher commuting costs for lower-priced land and may reside on larger parcels away from high-accessibility, high-congestion locations. Residential density declines with increasing distance from the city center as one exercises this option.

As a city grows in population, the peak densities no longer increase, and the pattern of population distribution becomes more uniform. Secondary centers begin to compete with the CBD for customers and industry, and the residential areas become less associated with the city center and more dependent on high-speed transportation arteries. Peak densities in the inner city decline, and peripheral areas increase in population concentration.

The validity of these generalizations may be seen on a time series graph of population density patterns for Des Moines, Iowa, over a 30-year period (Figure 11.24). The peak density was 2 miles from the CBD in 1960, but by 1980 it was at 3 miles. As the city expanded, density decreased close to the center, but beyond 3 miles from the center, population density increased.

Figure 11.22

Generalized pattern of land values. The "topography" represented in this diagram shows the major land-value peak of the CBD. The ridges radiating from that point mark mass transit lines and major business thoroughfares. Regularly occurring minor peaks indicate commercial agglomerations at intersections of mass transit lines or main traffic routes—points of accessibility superior to those of the interstitial troughs.

Source: Reproduced from B. J. L. Berry, *Commercial Structure and Commercial Blight.* Research Paper 85, Department of Geography *Research Series,* the University of Chicago, 1963.

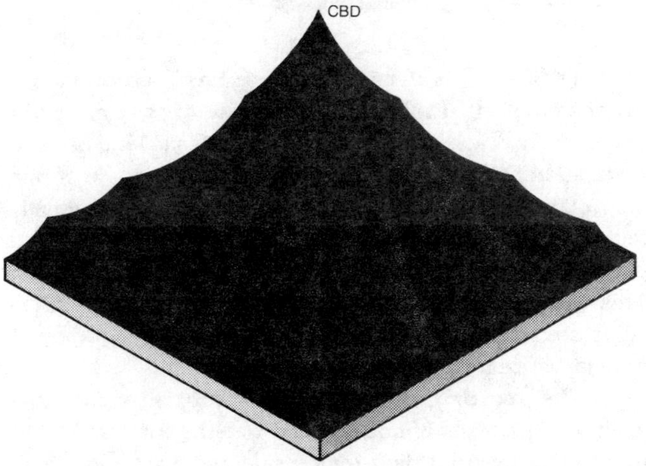

Models of Urban Land Use Structure

Generalized models of urban growth and land use patterns have been proposed to summarize the observable results of the organizing forces and controls of urban structure. Their starting point is the distinctive central business district that every older central city has. The core of this area is characterized by intensive land development: tall buildings, many stores and offices, and crowded streets. Just outside the core is the "frame" or fringe of the CBD; this

Figure 11.23

A theoretical population density curve. As distance from the area of multistory apartment buildings increases, the population density declines.

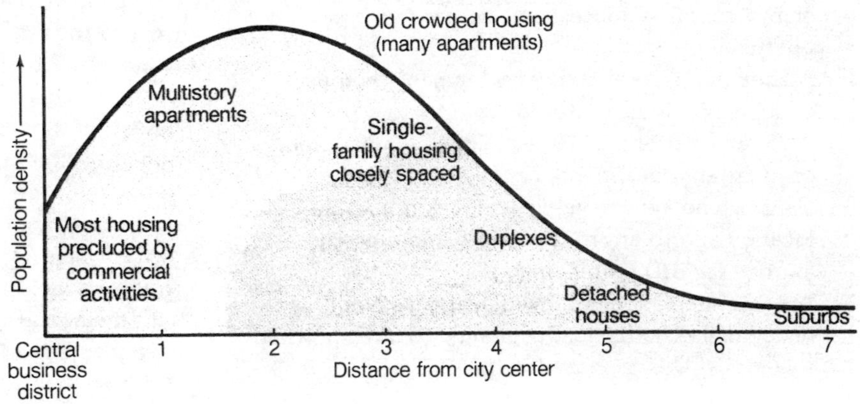

ure 11.24

Population density of Des Moines, Iowa, 1960–1980. The scale on the left is in natural logs (ln). For example, the natural log of 8100 people per square mile is equal to about 9, ln 3000 = 8, ln 1100 = 7, ln 400 = 6, and ln 150 = 5.

Source: From Arthur Getis, "Urban Population Spacing Analysis" in *Urban Geography*, Vol. 6, 1985. Copyright © 1985 V. H. Winston and Son, Inc., Silver Springs, Maryland. Reprinted by permission.

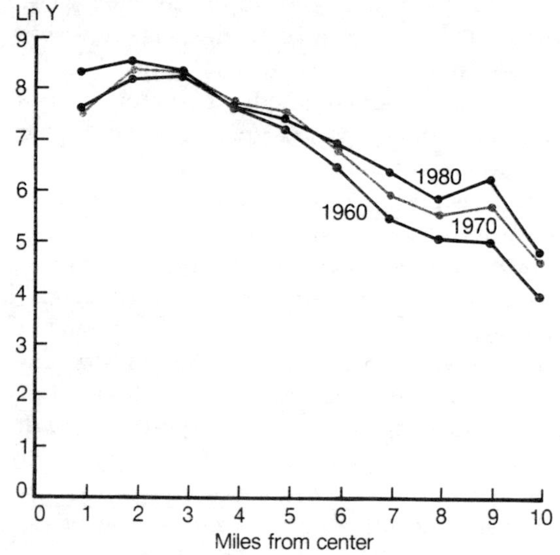

Figure 11.25

The core-frame concept. Commercial land is used intensively at the core of the central business district; the frame contains less intensive commercial uses. Some of the activities of the frame are related to each other and to the core.

From E. Horwood and R. Boyce, *Studies of the Central Business District and Urban Freeway Development*. Copyright © 1959 The University of Washington Press, Seattle, Washington. Reprinted by permission.

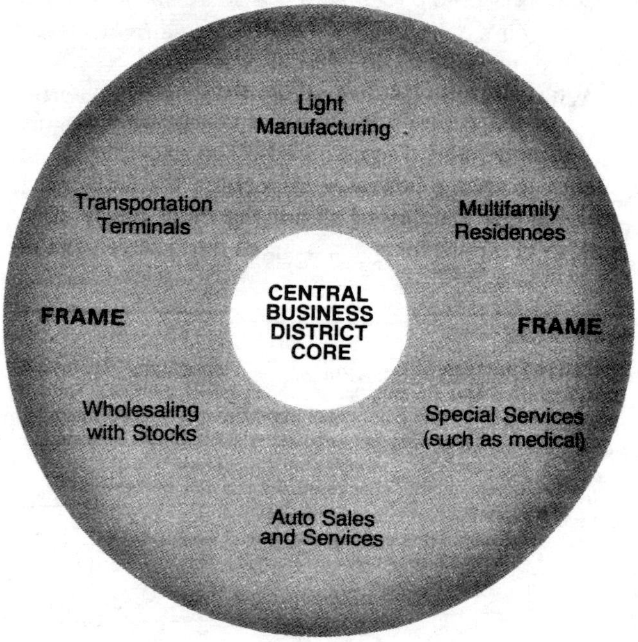

is an area of wholesaling activities, transportation terminals, warehouses, new-car dealers, furniture stores, and even light industries (Figure 11.25). Just beyond the central business district is the beginning of residential land uses.

The land use models depicted in Figure 11.26 diverge in their summaries of patterns outside the CBD. The **concentric zone model** (Figure 11.26*a*) was developed to explain the observed sociological patterning of American cities. The model recognizes five concentric circles of mostly residential diversity at increasing distance in all directions from the core.

1. The high-density CBD (core) with wholesaling, warehousing, light industry, and transport depots at its margins (frame).
2. A zone in transition marked by the deterioration of old residential structures abandoned, as the city expanded, by the former more wealthy occupants and now containing high-density, low-income slums, rooming houses, and perhaps ethnic ghettos.
3. A zone of "independent workingmen's homes" occupied by industrial workers.
4. A zone of better residences, single-family homes, or high-rent apartments occupied by those wealthy enough to exercise choice in housing location and to afford the longer, more costly journey to CBD employment.
5. A commuters' zone of low-density, isolated residential suburbs, just beginning to emerge when this model was proposed in the 1920s.

The model is dynamic; it imagines the continuous expansion of inner rings at the expense of the next outer developed circles. It suggests that changes in land use and population patterns occur as functions and people of one district begin to move into an adjacent zone, gradually taking over already occupied areas and properties, and eventually totally displacing former residents or activities. This sequence of *invasion* and *succession* yields a restructured land use pattern and a new social geography of population segregation by income level.

The **sector model** (Figure 11.26*b*) also concerns itself with patterns of housing and wealth, but it arrives at the conclusion that high-rent residential areas dominate expansion patterns and grow outward from the center of the city along major arterials, new housing for the wealthy being added in an outward extension of existing high-rent axes as the city grows. Middle-income housing sectors lie adjacent to the high-rent areas, and low-income residents occupy the remaining sectors of growth. There tends to be a "filtering-down" process as older areas are abandoned by the outward movement of their original inhabitants, with the lowest-income populations (closest to the center of the city and farthest from the current location of the wealthy) the dubious beneficiaries of the least desirable vacated areas. The accordance of the sector model with the actual pattern that developed in Calgary, Canada, is suggested in Figure 11.27.

Figure 11.26

Three classic models of the internal structure of cities.

Reprinted from "The Nature of Cities" by Chauncy D. Harris and Edward L. Ullman in volume CCXLII of *The Annals of the American Academy of Political and Social Science*. Copyright © 1945 The American Academy of Political and Social Science, Philadelphia, PA. Reprinted by permission.

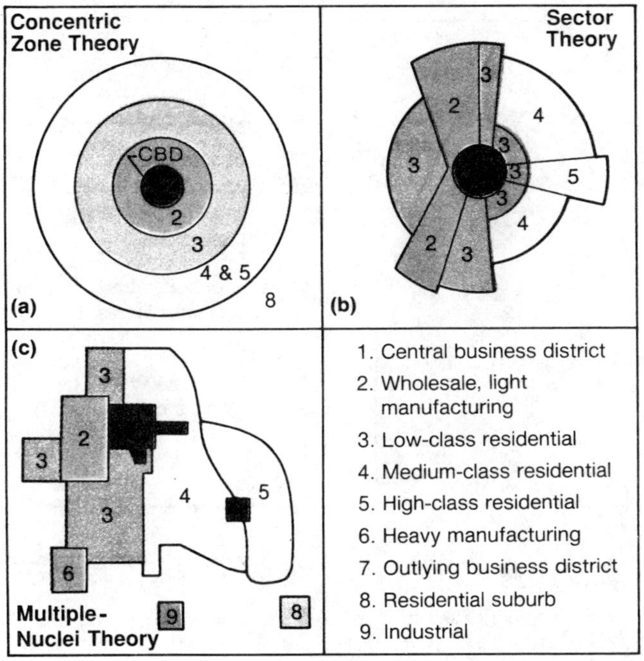

Figure 11.27

The land use pattern of Calgary, Alberta, 1961. The circular arrangement of uses suggested by the concentric zone theory (Figure 11.26*a*) might result if a city developed on a flat surface. In reality, hills, rivers, railroads, and highways affect land uses in uneven ways. Physical and cultural barriers and the evolution of cities over time tend to result in a sectoral pattern of similar land uses. Calgary's central business district is the focus for many of the sectors.

Source: From P. J. Smith, "Calgary: A Study in Urban Patterns," in *Economic Geography*, vol. 38, 1962. Copyright © 1962 *Economic Geography*, Worchester, MA. Reprinted by permission.

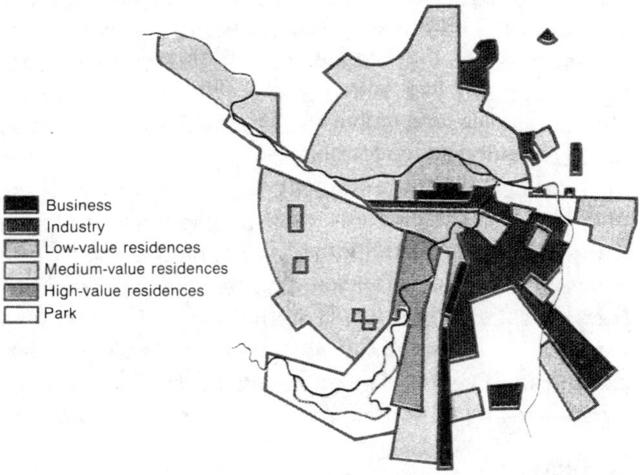

The concentric zone and sector models assume urban growth and development outward from a single central core, the site of original urban settlement that later developed into the central business district. These "single-node" models are countered by a **multiple-nuclei model** (Figure 11.26c), which postulates that large cities develop by peripheral spread from several nodes of growth, not just one. Individual nuclei of special function—commercial, industrial, port, residential—are originally developed in response to the benefits accruing from the spatial association of like activities. Peripheral expansion of the separate nuclei eventually leads to coalescence and the juxtaposition of incompatible land uses along the lines of juncture. Therefore, the urban land use pattern is not regularly structured from a single center in a sequence of circles or a series of sectors but based on separately expanding clusters of contrasting activities.

Social Areas of Cities

Although too simplistic to be fully satisfactory, these classical models of American city structure receive some confirmation from arrangements within urban areas of income levels, population age structures, ethnicity, housing stock characteristics, and the like. These indicate recurring patterns of a social geography of cities of North America.

The more complex cities are economically and socially, it appears, the stronger is the tendency for U.S. city residents to segregate themselves into groups based on *social status, family status,* and *ethnicity.* In a large metropolitan region with its more diversified population, this territorial behavior may be a defense against the unknown or the unwanted. Most people feel more secure when they are near those with whom they can easily identify. In traditional societies, these groups are the families and tribes. In modern societies, people cluster according to income or occupation (social status) and language or race (ethnic characteristics). Many of these groupings are fostered by the size and the value of the available housing. Land developers, especially in cities, produce homes of similar quality in specific areas. Of course, as time elapses, there is a change in the condition and quality of that housing, and new groups may replace old groups. In any case, neighborhoods of similar social characteristics evolve.

Social Status

The social status of an individual or a family is determined by income, education, occupation, and home value, though it may be measured differently in different cultures. In the United States, high income, a college education, a professional or managerial position, and high home value constitute high status. (High home value can mean an expensive renter apartment as well as a large house with extensive grounds.) There are many levels of status, and people tend to filter out into neighborhoods where most of the heads of households are of similar rank.

In most cities, people of similar social status are grouped in sectors whose points are in the innermost urban residential areas. The pattern in Chicago is illustrated in

Figure 11.28. If the number of people within a given social group increases, they tend to move away from the central city along an arterial connecting them with the old neighborhood. Major transport routes leading to the city center are the usual migration routes out from the center. Social status patterning agrees with sector theory.

Family Status

As the distance from the origin of each sector increases, the average age of the head of the household declines, or the size of the family increases, or both. Within a particular sector—say, that of high status—older people whose children do not live with them or young professionals without families tend to live close to the city center. Young families seeking space for child rearing may locate farther out, at or beyond the central city margin. However, where inner-city life is unpleasant, older people tend to migrate to the suburbs or to retirement communities. Within the lower-status sectors, the same pattern tends to emerge. Transients and single people are housed in the inner city, and families, if they find it possible or desirable, live farther from the center. The arrangement that emerges is a concentric-ring patterning according to family status.

Ethnicity

For some groups, ethnicity is a more important determinant of residential location than is social or family status. Areas of homogeneous ethnic identification appear in the social geography of cities as separate clusters or nuclei, reminiscent of the multiple-nuclei concept of urban structure. For some ethnic groups, cultural segregation is both sought and vigorously defended, even in the face of pressures for neighborhood change exerted by potential competitors for housing space, as we saw in Chapter 6. Certain ethnic or racial groups, however, including blacks, have had segregation in nuclear communities forced on them. Nearly every city has one or more black areas, which in many respects may be considered cities within a city, with their own self-contained social geographies of social status, income, and housing quality. The barriers to movement outside the assigned areas have always been high.

Figure 11.29 presents a summary model of these social geographic patterns. Of the three components, family status has undergone the most widespread change in recent years. Today the suburbs house large numbers of singles and childless couples, and areas near the central business districts have become popular for young professionals (see "Gentrification"). Much of this is a result of changes in family structure and the advent of large numbers of new jobs for professionals both in the suburbs and in the central business districts, but not in between.

Figure 11.28

A diagrammatic representation of the major social areas of the Chicago region. The central business district of Chicago is known as the Loop.

Source: Redrawn from Phillip Rees, "The Factorial Ecology of Metropolitan Chicago," M.A. Thesis, University of Chicago, 1968.

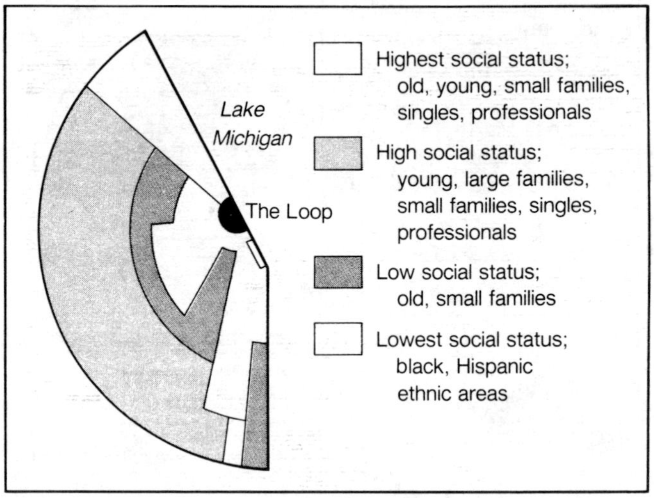

Figure 11.29

The social geography of North American cities.

Source: Redrawn from R. A. Murdie, *Factorial Ecology of Metropolitan Toronto.* Research Paper 116, Department of Geography *Research Series,* The University of Chicago, 1969.

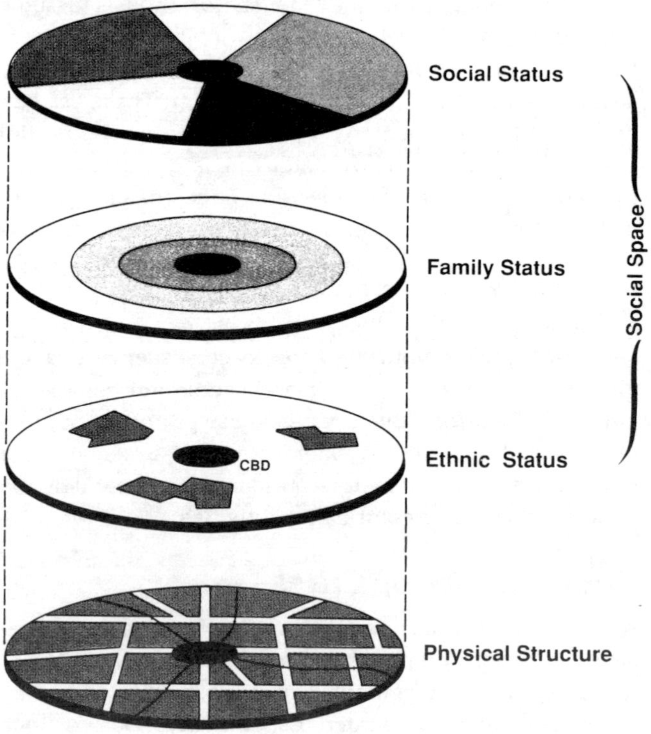

Gentrification

As urban governments and state and federal authorities attempt to find ways to revitalize ailing older American cities, a potentially significant private rehabilitation movement is occurring. **Gentrification** is the process by which certain deteriorated portions of the inner city—particularly areas of historical or architectural attraction—are taken over by higher-income groups and upgraded through renovation to desirable single-family or condominium residential areas. This movement, a reversal of the normal "filtering-down" process, does not yet offset the exodus to the suburbs that has taken place since the end of the World War II. It does, however, introduce a countercurrent to what had seemed to be the inevitable decline of older central cities. According to an estimate of the Urban Land Institute, 70% of all sizable American cities are experiencing a significant renewal of deteriorated areas, with gentrification especially noticeable in the major cities of the North and the East, from Montreal and Quebec down the Atlantic Coast to Charleston, South Carolina, and Savannah, Georgia.

The reasons for the upsurge of interest in central city housing reflect to some extent recent changes in North American family structure and in the employment structure of central business districts. Just a decade ago the suburbs, with their green spaces, were powerful attractions for young married couples who considered single-family houses with ample yards ideal places to raise children. Now that the proportion of single people (whether never or formerly married) and childless couples in the American population has increased, the attraction of nearby jobs and recreational activities has become a more important residential location factor. Many of the new jobs in the central business district are designed for professionals in the fields of banking, insurance, and financial services. These jobs are replacing the manufacturing employment of an earlier period and are the catalyst for the gentrification process in some of the depressed housing areas surrounding central business districts.

Institutional Controls

Within this century, particularly since World War II, institutional controls have strongly influenced the land use arrangements and growth patterns of most U.S. cities. Land use plans have been adopted, subdivision control and zoning ordinances have been enacted, and building codes have been designed to assure the nature and pattern of development. These nonmarket controls on private land use presuppose the possibility of achieving an ideal arrangement of uses, minimizing incompatibilities (residences adjacent to heavy industry, for example), maximizing all real estate values, and providing for the creation in appropriate locations of public and private uses needed for a balanced, orderly community. One evident objective has been to prevent the neighborhood deterioration so often resulting from undesirable adjacent uses or—through public planning and funding—to replace slum structures and areas.

Zoning ordinances and land use planning have sometimes been criticized as being unduly restrictive and unresponsive to changing land use needs and patterns of economic development. They have also, in some instances, served as devices to preserve established exclusive residential communities by setting lot or house size standards too high for lower-income populations. Bitter court battles have been waged with mixed results over "exclusionary" zoning practices that serve to separate rather than to unify the total urban structure. All institutional controls, of course, whatever their origin or objective, interfere with the open auction in urban land that initially gave rise to the city structure we have been reviewing.

Suburbanization in the United States

The North American mass transit city reached its climax by the late 1920s. The Great Depression of the 1930s and World War II (1939–1945) effectively halted urban expansion. However, the 20 years before the war saw the creation of a technological, physical, and institutional structure that resulted in the postwar period in a sudden and massive alteration of past urban forms. The improvement of the automobile increased its reliability, range, and convenience, freeing its owner from dependence on fixed-route public transit for access to home, work, and shopping. The new transport flexibility opened up vast acreages of nonurban land to urban development. The acceptance of a maximum 40-hour workweek guaranteed millions of Americans the time for a commuting journey not possible when workdays of 10 or more hours were common. Finally, to stimulate the economy by the ripple effect associated with a prospering construction industry, the Federal Housing Administration was established as part of the New Deal programs under President Franklin D. Roosevelt. It guaranteed creditors the security of their mortgage loans, thus reducing down-payment requirements and lengthening mortgage repayment periods. In addition, veterans of World War II were granted generous terms on new residences.

Demand for housing, pent up by years of economic depression and wartime restrictions, was loosed in a flood after 1945. As a result, a massive suburbanization of people and urban functions altered the existing pattern of urban

America. Between 1950 and 1970, the two most prominent patterns of population growth were the metropolitanization of people and, within metropolitan areas, their suburbanization. During the 1960s the interstate highway system was substantially completed and major metropolitan expressways put in place, allowing sites 20, 30, or more miles from workplaces to be within easy commuting distance. Growth patterns for the Chicago area, reflecting those developments, are shown in Figure 11.30. The high energy prices of the 1970s slowed the rush to the suburbs, but in the 1980s suburbanization again proceeded apace, although the tendency was as much for a "filling in" as it was for continued sprawl.

Residential land uses led the initial rush to the suburbs. Their emerging pattern showed individually uniform, spatially discontinuous housing developments beyond the boundaries of most older central cities. The new design was an unfocused sprawl because it was not tied to mass transit lines and the channelized pattern of nodes and links they imposed. Further, it represented a massive relocation of purchasing power to which retail merchants were quick to respond. The planned major regional shopping center became the suburban counterpart to higher-order central places and the central city's major outlying commercial districts (Figure 7.31). Smaller shopping plazas and a multiplying number of the strip shopping centers that had initially preceded the larger malls gradually completed the hierarchy. Faced with a newly suburbanized labor force, industry followed the outward move, drawn as well by the economies derived from modern single-story plants with plenty of parking space for employees and the new freedom from railroad access made possible by the motor truck and the expanding highway system (Figure 9.17). Service industries were also attracted by suburbanizing purchasing power and labor force, and along with the new shopping malls, office building complexes began to be developed; like the malls, these complexes localized at freeway intersections and along freeway frontage roads and major connecting highways.

In time, an established social and functional pattern of suburban land use emerged in the United States, giving evidence of a lower-density, more extensive repetition of the models developed to describe the structure of the central city. Multiple nuclei of specialized land uses appeared, expanded, and coalesced. Sectors of high-income residential use continued their outward extension beyond the central city limits, claiming the most scenic and most desirable suburban areas and segregating them by price and zoning restrictions. Middle-, lower-middle-, and lower-income groups found their own income-segregated portions of the fringe; and ethnic minorities, frequently relegated to the older industrial suburbs, found theirs. A generalized model of the metropolitan region is suggested in Figure 11.31.

Figure 11.30

A history of urban spread. In Chicago, as in most larger and older U.S. cities, the slow peripheral expansion recorded during the late 19th and early 20th centuries suddenly accelerated as the automobile suburbs began developing after 1945.

Source: Redrawn with permission from B. J. L. Berry, *Chicago: Transformation of an Urban System*. Copyright © 1976, Ballinger Publishing Company.

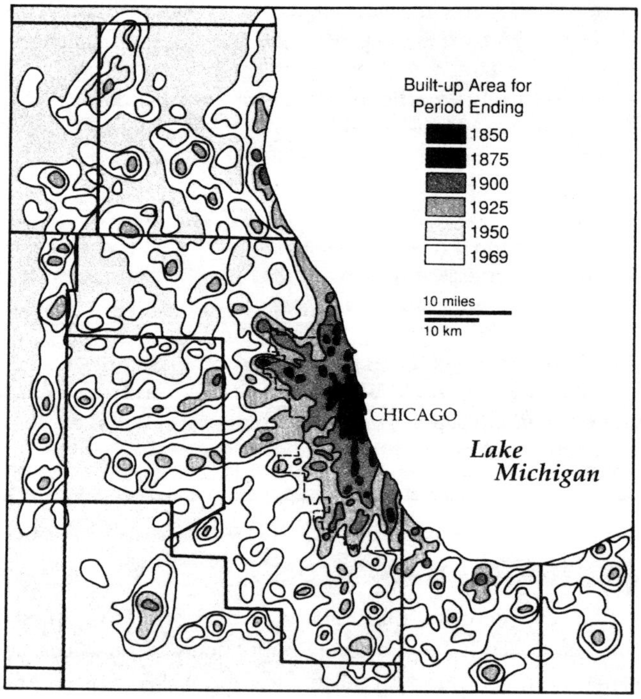

By the end of the 1980s, a new urban America had emerged on the perimeters of the major metropolitan areas. With increasing sprawl and the rising costs implicit in the ever-greater spatial separation of the functional segments of the fringe, peripheral expansion slowed, the supply of developable land was reduced (with corresponding increases in its price), and the intensity of land utilization grew. The suburbs, no longer dependent upon the central city, were reborn as vast, collectively self-sufficient outer cities, marked by landscapes of skyscraper office parks, massive retailing complexes, established industrial parks, and a proliferation of apartment and condominium districts (see "The Self-Sufficient Suburbs"). The new suburbia became more focused than formerly as its employment and service nodes began to rival older central business districts in size and complexity, though not necessarily in convenience of access. The journey to work for suburbanites was, on average, longer in the 1980s than it had been 10 years earlier.

Collectively, the new centers surpassed the central cities as generators of employment and income. Despite a boom in office building construction that occurred in many larger regional capitals, more office space in the 1980s was created in the suburbs than in the central cities of America. Tysons Corner, Virginia, for example, became the ninth-largest central business district in the United

Figure 11.31

A model of the United States metropolitan area. Note that aspects of the concentric-circle, sector, and multiple-nuclei patterns are evident and carried out into the suburban fringe. The "major regional shopping centers" of this earlier model are increasingly the cores of newly developing "outer cities."

Figure from *The North American City* by Maurice Yeates and Barry Garner. Copyright © 1980 by Maurice Yeates and Barry J. Garner. Reprinted by permission of Harper & Row, Publishers, Inc.

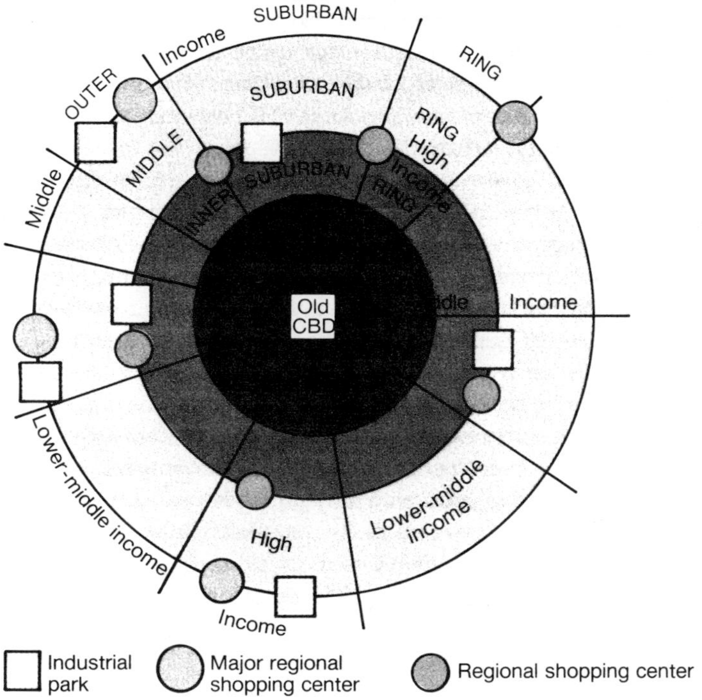

States (Figure 11.32). Regional and national headquarters of leading corporations, banking, professional services of all kinds, major hotel complexes and recreational centers—all formerly considered immovable keystones of central business districts—became parts of the new outer cities. And these outer cities themselves filled in and made more continuous the urban landscape of Megalopolis and other North American conurbations.

Figure 11.32

The low- and high-rise office buildings of Tysons Corner, Virginia, were a principal East Coast concentration of commercial land use in the late 1980s, when more than 70,000 office workers were employed in this major suburban "central business district."

The Self-Sufficient Suburbs

The suburbanization of the American population and of the commerce and industry dependent on its purchasing power and labor has led to more than the simple physical separation of the suburbanite from the central city. It has created a metropolitan area, many of whose inhabitants have no connection with the core city, feel no ties to it, and find satisfaction of all their needs within the peripheral zone.

A *New York Times* "suburban poll," conducted in the summer of 1978, revealed even then a surprising lack of interest in New York City on the part of those usually assumed to be intimately involved in the economic, cultural, and social life of the "functional city," the metropolitan area dominated by New York. The poll discovered that in 80% of suburban households, the principal wage earner did not work in New York City. The concept of the commuter zone obviously no longer has validity if *commuting* implies, as it formerly did, a daily journey to work in the central city.

Just as employment ties have been broken, now that the majority of suburbanites work in or near their outlying areas of residence, the cultural and service ties that were traditionally thought to bind the metropolitan areas into a single unit dominated by the core city have also been severed. The *Times* survey found that one-half of the suburbanites polled made fewer than five nonbusiness visits to New York City each year; one-quarter said that they never went there. Even the presumed concentration in the city of high-order goods and services—those with the largest service areas—did not constitute an attraction for most suburbanites. Only 20% of the respondents thought that they would journey to New York City to see a medical specialist, and only 7% or 8% to see a lawyer or an accountant or to make a major purchase. Even needs at the upper reaches of the central place hierarchy were satisfied within the fringe.

The suburbs, as the *Times* survey documented, have outgrown their former role as bedroom communities, are largely self-sufficient and divorced from the older inner city, and have little feeling of subordination to or dependence on it.

Those outer cities, the landscape realities of the postindustrial service and information economy, now exist in all regions of urbanized North America. The South Coast Metro Center in Orange County, California, the City Post Oak-Galleria center on Houston's west side, King of Prussia and the Route 202 corridor northwest of Philadelphia, the Meadowlands, New Jersey, west of New York City, and Oak Brook, Illinois, in the western Chicago suburbs are but a very few examples of the new urban forms.

Central City Change

The structure, the economic base, and the financial stability of the central city have been grievously damaged by the process of suburbanization and the creation of new outer cities. In earlier periods of urban growth, as new settlement areas or functional nodes developed beyond the existing city limits, annexation absorbed the new growth within the corporate boundaries of the expanding older city. The additional tax base and employment centers became part of the municipal whole. But in the states that recognized the right of separate incorporation for the new growth areas—states particularly in the eastern part of the United States—the ability of the city to continue to expand was restricted. Lower suburban residential densities made septic tanks and private wells adequate substitutes for expensive city-sponsored sewers and water mains; lower structural densities and lower crime rates meant less felt need for high-efficiency city fire and police protection. Where possible, suburbanites opted for a separation from the central city and for aloofness from the costs, the deterioration, and the adversities associated with it. Their homes, jobs, shopping, schools, and recreation all existed outside the confines of the city from which they had divorced themselves, and the close-knit fabric of urban life was torn.

The redistribution of population caused by suburbanization resulted not only in the spatial but also in the political segregation of the social groups of the metropolitan area. The upwardly mobile urbanite—younger, wealthier, and better educated—took advantage of the automobile and the freeway to leave the central city. The poorer, older, least-advantaged citizens were left behind. The central cities and the suburbs became increasingly differentiated. Large areas within the cities now contain only the poor and minority groups, a population little able to pay the rising costs of the social services that their numbers, neighborhoods, and condition require.

The services needed to support the poor include welfare payments, social workers, extra police and fire protection, health delivery systems, and subsidized housing. Central cities by themselves are hard pressed to support such an array and intensity of social services now that they have lost the tax bases represented by suburbanized commerce, industry, and upper-income residential uses. Lost, too, are the job opportunities formerly a part of the central city structure. Increasingly, the poor and the minorities are trapped in an inner city without the possibility of nearby employment and isolated by distance, immobility, and unfamiliarity from the few remaining low-skill jobs, which are now largely in the suburbs.

Except within central business districts of larger central cities, the traditional active, open auction of urban land has largely disappeared. No longer is there the vigorous economic competition for space that formerly led to the replacement of obsolescent uses and inefficient structures in a continuing process of urban modernization. In the vacuum left by the departure of private investors, the federal government, particularly since the landmark Housing Act of 1949, has periodically initiated urban renewal programs with or without provisions for a partnership with private capital. Under a wide array of programs instituted and funded since the late 1940s, slum areas have been cleared; public housing has been built; cleared land has been conveyed at subsidized costs to private developers for middle-income housing construction; cultural complexes and industrial parks have been created; and city centers have been reconstructed. With the continuing erosion of the urban economic base and the disadvantageous restructuring of the central city population base, the hard-fought governmental battle to maintain or revive the central city is frequently judged to be a losing one (Figure 11.33).

Figure 11.33

Some elaborate governmental projects for urban redevelopment, slum clearance, and rehousing of the poor have been failures. The demolition of the Wayne Miner public housing project in Kansas City, Missouri in 1987 was a recognition that public high-rise developments do not always meet the housing and social needs of their inhabitants.

World Urban Diversity

The city, Figure 11.4 reminds us, is a global phenomenon. It is also a regional and cultural variable. The descriptions and models that we have used to study the functions, land use arrangements, suburbanization trends, and other aspects of the U.S. city would not in all—or even many—instances help us understand the structures and patterns of cities in other parts of the world. Those cities have been created under different historical, cultural, and technological circumstances. They have developed different functional and structural patterns, some so radically different from our U.S. model that we would find them unfamiliar and uncharted landscapes indeed. The city is universal; its characteristics are cultural and regional.

The North American City

Even within the seemingly homogeneous North American culture realm the city shows subtle but significant differences—not only between older eastern and newer western U.S. cities, but between cities of Canada and those of the United States. Although the urban expression is similar in the two countries, it is not identical, and the truly "North American city" is more a myth than a reality. The Canadian city, for example, is more compact than its U.S. counterpart of equal population size, with a higher density of buildings and people and a lesser degree of suburbanization of populations and functions (Figure 11.34). Space-saving multiple-family housing units are more the rule in Canada, so a similar population size is housed on a smaller land area with much higher densities, on average, within the central area of cities. The Canadian city is better served by and more dependent on mass transportation than is the U.S. city. Since Canadian metropolitan areas have only one-quarter the number of miles of expressway lanes per capita as U.S. MSAs—and at least as much resistance to constructing more—suburbanization of peoples and functions is less extensive north of the border than south. It is likely to remain that way.

In social as well as physical structure, Canadian–United States contrasts are apparent. While cities in both countries are ethnically diverse—Canadian communities, in fact, have the higher proportion of foreign born—U.S. central cities exhibit far greater internal distinctions in race, income, and social status and more pronounced contrasts between central city and suburban residents. That is, there has been much less "flight to the suburbs" by middle-income Canadians. As a result, the Canadian city shows greater social stability, higher per capita average income, more retention of shopping facilities, and more employment opportunities and urban amenities than its U.S. central city counterpart. In particular, it does not have the rivalry from well-defined competitive "outer cities" of suburbia that so spread and fragment United States metropolitan complexes.

Figure 11.34

The central and outlying business districts of Toronto, easily visible in this photo, are still rooted firmly by mass transit convergence and mass transit usage. On a per capita basis, Canadian urbanites are two and a half times more dependent upon public transportation than are U.S. city dwellers. That reliance gives form, structure, and coherence to the Canadian central city, qualities now irretrievably lost in the sprawled and fragmented United States metropolis.

Photo © John A. Jakle.

The Western European City

If such significant urban differences are found even within the tightly knit North American region, we can only expect still greater divergences from the U.S. model at greater linear and cultural distance and in countries with long urban traditions and mature cities of their own. The political history of France, for example, has given to Paris an overwhelmingly *primate* position in its system of cities. Political, economic, and colonial history has done the same for London in the United Kingdom. On the other hand, Germany and Italy came late to nationhood, and no overwhelmingly dominant cities developed in their systems.

Nonetheless, a generally common heritage of medieval origins, Renaissance restructurings, and industrial period extensions has given to the cities of Western Europe features distinctly different from those of cities in other regions founded and settled by European immigrants. Despite wartime destructions and postwar redevelopments, many still bear the impress of past occupants and technologies, even back to Roman times in some cases. Although the European urban pattern we see today is primarily the product of the industrialization of the 19th and 20th centuries, frequently the location and shape of the settlement, the street pattern, and the layout of the older sections are more a reminder of the distant past than a reflection of modern requirements and responses. An irregular system of narrow streets may be retained from the random street pattern developed in medieval times of pedestrian and pack-animal movement. A system of main streets radiating from the city center and cut by circumferential "ring roads" tells us the location of high roads leading into town through the gates in city walls now gone and replaced by circular boulevards. Broad thoroughfares, public parks, and plazas mark Renaissance ideals of city beautification and the esthetic need felt for processional avenues and promenades.

Although each is unique historically and culturally, Western European cities as a group share certain common features that set them apart from the United States model, though they are less far removed from the Canadian norm. Cities of Western Europe have, for example, a much more compact form and occupy less total area than American cities of comparable population; most of their residents are apartment dwellers. Residential streets of the older sections tend to be narrow, and front, side, or rear yards or gardens are rare. European cities developed for pedestrians and still retain the compactness appropriate to walking distances. The sprawl of American peripheral and suburban zones is generally absent. At the same time, compactness and high density do not mean skyscraper skylines. Much of urban Europe predates the steel-frame building and the elevator. City skylines tend to be low, three to five stories in height, sometimes (as in central Paris) held down by building ordinance (Figure 11.35) or by prohibitions on private structures exceeding the height of a major public building, often the central cathedral.

Compactness, high densities, and apartment dwelling encouraged the development and continued importance of public transportation, including well-developed subway systems. The private automobile has become much more common of late, though most central city areas have not yet been significantly restructured with wider streets and parking facilities to accommodate it. The automobile is not the universal need in Europe that it has become in American cities. Home and work are generally more closely spaced in Europe—often within walking or bicycling distance—while most sections of towns have first-floor retail and business establishments (below upper-story apartments), bringing both places of employment and retail shops within convenient distance of residences.

A very generalized model of the social geography of the Western European city has been proposed (Figure 11.36). Its exact counterpart can be found nowhere, but many of its general features are part of the spatial social structure of most major European cities. In the historic core, now increasingly gentrified, residential units for the middle class, the self-employed, and the older generation of skilled artisans share limited space with preserved historic buildings, monuments, and tourist attractions. The old city fortifications may mark the boundary between the core and the surrounding transitional zone of substandard housing, 19th-century industry, and recent immigrants. The waterfront has similar older industry; newer plants are found on the periphery. Public housing and some immigrant concentrations may be near that newer industry, while other urban socioeconomic groups aggregate themselves in distinctive social areas within the body of the city. The European city does not characteristically feature the ethnic neighborhoods of U.S. cities although some, like London, do (see "The Caribbean Map in London," p. 185). Rather, its guest workers from Eastern Europe, the Near East, and North Africa tend to be relatively dispersed rather than concentrated. Nor is it characterized by inner-city deterioration and out-migration. Its core areas tend to be stable in population and to attract rather than repel the successful middle class and the upwardly mobile, conditions far different from comparable sections of older U.S. central cities.

Figure 11.35

Even in their central areas, many European cities show a low profile, like that of Paris seen here from the Eiffel Tower. Although taller buildings—20, 30, even 50 or more stories in height—have become more common in major cities since World War II, they are not the universal mark of central business districts that they have become in the United States, nor the generally welcomed symbols of city progress and pride.

The Socialist City

Cities of Eastern Europe and the Soviet Union make up a separate urban class—the socialist city. It is an urban form that shares many of the traditions and practices of Western European cities, but it differs from them in the centrally administered planning principles that have been designed to shape and control both new and older settlements. The particular concerns have been, first, limitation on size of cities to avoid supercity growth and metropolitan sprawl; second, assurance of an internal structure of neighborhood equality and self-sufficiency; and third, strict land use segregation. The socialist city has fully achieved none of these objectives, but by attempting them it has emerged as a distinctive urban form.

Figure 11.36

A model of the Western European city.

Source: Paul White, *The West European City: A Social Geography*. Copyright © 1984 Longman Group UK Limited, Essex, England. Reprinted by permission.

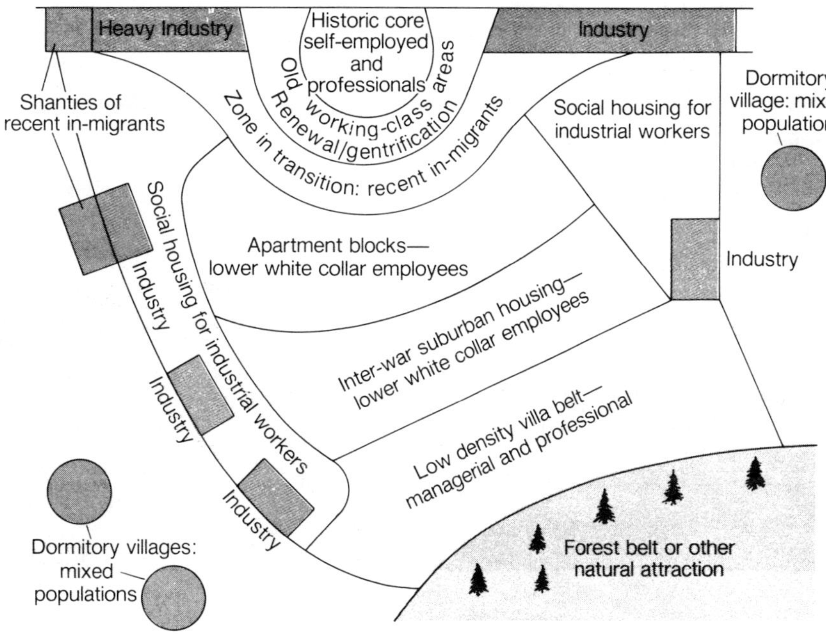

Within the vast domain of the Soviet Union, where the planning philosophies of the socialist city were formulated, urbanization has proceeded rapidly during the period of Communist rule. Urban population has increased from about 18% (1917) to 66% (1989) of total population, some 150 million new urban residents have been accommodated, and over 1500 new or largely rebuilt cities have been created in general accordance with the socialist city model. Although the Eastern European countries taken into the Soviet bloc following World War II were in general more urbanized than the USSR had been, they too have experienced an acceleration of urbanization on the Soviet model with recent industrialization. As a group, they add another 71 million urbanites to the 191 million Soviet city residents (1989).

Although city planning in the Soviet Union has not been rigorously or consistently applied—often thwarted because city governments responsible for planning are powerless to control or change site decisions made by economic ministries—the accepted planning guidelines have made themselves evident in the urban landscape. In general structural terms, the socialist (Soviet) city is compact, with relatively high building and population densities reflecting the nearly universal apartment dwelling, and with a sharp break between urban and rural land uses on its periphery. Like the older-generation Western European city, the socialist city depends nearly exclusively on public transportation.

It differs from its Western counterpart, of course, in its purely governmental rather than market control of land use and functional patterns. That control has dictated that the central area of cities should be reserved for public use, not occupied by retail establishments or office buildings on the Western, capitalist model. A large central square ringed by administrative and cultural buildings is the preferred pattern. Nearby, space is provided for a large recreational and commemorative park. Neither a central business district nor major outlying business districts are required or provided in the socialist city, for residential areas are expected to be largely self-contained in the provision of at least low-order goods and services, minimizing the need for a journey to centralized shopping locations. Those residential areas are made up of "microdistricts," assemblages of uniform apartment blocks housing perhaps 10,000 to 15,000 people, surrounded by broad boulevards and containing centrally sited nursery and grade schools, grocery and department stores, theaters, clinics, and similar neighborhood necessities and amenities (Figure 11.37*a*). Plans call for effective separation of residential quarters from industrial districts by landscaped buffer zones, but in practice many microdistricts are built by factories for their own workers and are located immediately adjacent to the workplace.

Since microdistricts are most easily and rapidly constructed on open land at the margins of expanding cities, high residential densities have been carried to the outskirts of town (Figure 11. 37*b*). The distance decay curve of declining population densities outlined on Figure 11.23 does not apply to the socialist city. Although egalitarian planning principles demand uniformity in the housing stock and a total prohibition of any form of residential segregation, in practice uniformity is not achieved and segregation does occur. To the extent that the housing stock is

Figure 11.37

(*a*) This scene from Bucharest, Romania, clearly shows important recurring characteristics of the socialist city: mass transit service to boulevard-bordered "superblocks" of self-contained apartment house microdistricts that contain their own shopping areas, schools, and other facilities. (*b*) High-density apartment houses bordered by wheat fields mark the urban margin of Poprad, Czechoslovakia; the Tatra Mountains are in the background.

(a)

(b)

built by industrial and institutional entities for their own workers (perhaps a third or more of all units), segregation of housing by place of employment is automatic. Further, some sections of most cities are seen as more prestigious and desirable than others, and those citizens able through income or influence to locate within them do so, establishing elite neighborhoods in a city of equals.

The Soviet ideal has long been for cities to achieve but not exceed a size in which (1) the provision of public services is efficient and at lowest possible per capita cost, (2) reasonable scale economies of agglomeration are possible, and (3) the diseconomies of peripheral spread, congestion, commuting costs, pollution, and the like are avoided. In particular, massive metropolitan sprawl and conurbations are to be avoided. Since domestic migration is controlled and Soviet citizens (and Eastern Europeans in general) must have permits to live and work in given locales, control of city size would appear simple. In reality, it has not proved so—again because economic ministries can make locational decisions without reference to individual city plans or goals. Thus, the Moscow metropolitan area approximates 10 million in an urban sprawl in many ways reminiscent of that of Western cities, and other major socialist cities show similar tendencies to spread.

Cities in the Developing World

Still farther removed from the United States urban model are the cities of Africa, Asia, and Latin America, where industrialization has come only recently or not at all, where modern technologies in transportation and public facilities are barely known or sparsely available, and where the structures of cities and the cultures of their inhabitants are far different from the urban world familiar to North Americans. The developing world is vast in extent and diverse in physical and social content; generalizations about it or its urban landscapes lack certainty and universality. Islamic cities of North Africa, for example, are entities sharply distinct from the black African, the Southeast Asian, or the Latin American city. Yet, by observation and consensus, some common features of Third World cities are recognizable. All, for example, have endured massive in-migrations from rural areas (see "Population Pressure in Cairo, Egypt"). As a result, most are ringed by vast squatter settlements high in density and low in public facilities and services. All, apparently, have populations greater than their formal functions and employment bases can support. In all, large numbers support themselves in the "informal" sector—as snack-food vendors, peddlers of cigarettes or trinkets, street-side barbers or tailors, errand runners or package carriers, and the like outside the usual forms of wage labor (Figure 10.5).

But the extent of acceptable generalization is limited, for the backgrounds, developmental histories, and current economies and administrations of Third World cities vary greatly. Some are still preindustrial, without a central commercial core, without industrial districts, without public transportation, without any meaningful degree of land use separation. Some are the product of Western colonialism, established as ports or outposts of administration and exploitation, built by Europeans on the Western model, though increasingly engulfed by later, indigenous urban forms. In some, Western-style skyscraper central areas and commercial cores have been newly constructed. In others, commerce is conducted in different forums and formats (Figure 11.38).

Population Pressure in Cairo, Egypt

Cairo is a vast, sprawling metropolis—hot, crowded, choked by automobiles, bothered by noxious fumes from traffic and by sand that blows from the Sahara. The city is representative of many of the principal cities of Third World countries, where population growth far outstrips economic development. A steady stream of migrants arrives in Cairo daily because it is the place where opportunities are perceived to be available, where people think that life will be better and brighter than in the crowded countryside. Cairo is the symbol of modern Egypt, a place where young people are willing to undergo deprivation for the chance to "make it." In addition, the population of the country was never more than 12 million people during its 7000-year history, but now better health care has led to a big drop in infant mortality, and the result is a country with an expected year 2000 population of 70 or 80 million.

Between 1960 and 1990, the population of Cairo increased sixfold. In 1960, the estimated population was 2.3 million; 30 years later some 14 million reside in the greater metropolitan area. Real opportunities continue to be low relative to the size of the metropolis. The poor, of whom there are millions, crowd into warrens of slums, often without tap water or sewage systems. Yet the city continues to grow onto valued farmland, thus decreasing the food available for the country's growing numbers.

The sewer and water systems, modernized in the 1930s, were intended to serve only 2 million people. Now it is not uncommon for burst drains to dump raw sewage into the city streets. When such conditions prevail in the hot summer, the chances of the spread of disease increase many times over. It is also not unusual to find people living on little boats on the river, under the arches of bridges, and in makeshift, poorly constructed buildings that, on occasion, cave in.

One's first impression when arriving in central Cairo is one of opulence and prosperity, a stark contrast to what lies outside the center. High-rise apartments, regional headquarters buildings of multinational corporations, and modern hotels stand amid clogged streets. This part of the city is clearly Western oriented, but the fancy restaurants, plush apartments, and high-priced cars stand only a short distance from the horrible slums that are home to masses of underemployed people.

The story of Cairo in Egypt can be repeated for city after city in the Third World. Bombay and Calcutta in India, São Paulo in Brazil, Kinshasa in Zaire, Lagos in Nigeria, Bangkok in Thailand, Jakarta in Indonesia, Karachi in Pakistan, and Mexico City are already housing millions more than their service structures are designed to accommodate, and one can foresee only worsening conditions in the future.

Wherever the automobile and modern transport systems are an integral part of the growth of Third World cities, the metropolis begins to take on Western characteristics. On the other hand, in places such as Bombay (India), Lagos (Nigeria), Jakarta (Indonesia), Kinshasa (Zaire), and Cairo (Egypt), where modern roads have not yet made a significant impact and the public transport system is limited, the result has been overcrowded cities centered on a single major business district in the old tradition. In such societies, the impact of urbanization and the responses to it differ from the patterns and problems observable in the cities of the United States.

The developing countries, emerging from earlier dominantly subsistence economies, have experienced disproportionate population concentrations, particularly in their national and regional capitals. Lacking or relatively undeveloped is the substructure of maturing, functionally complex smaller and medium-sized centers characteristic

Figure 11.38

The Great Bazaar of Istanbul, Turkey, with its miles of crowded streets, was built in the mid-17th century and now houses more than 3000 shops. It is a vibrant reminder that a successful, thriving commercial economy need not be housed in western-style business districts.

Figure 11.39

Modern office buildings in Lagos, Nigeria.

Figure 11.40

Millions of people of the developing world live in shantytown settlements on the fringes of large cities, without benefit of running water, electricity, sewage systems, or other public services. The residents of this *barrio* rising above Caracas, Venezuela, are part of their number.

of more advanced and diversified economies. The primate city dominates their urban systems (Figure 11.18). Greater Cairo contains nearly 30% of Egypt's total population; 30% of all Nicaraguans live in Managua; and Baghdad contains 24% of the Iraqi populace. Vast numbers of surplus, low-income rural populations have been attracted to these developed seats of wealth and political centrality in the hope of finding a job. Although attention may be lavished on creating urban cores on the skyscraper model of Western cities (Figure 11.39), most of the new urban multitudes have little choice but to pack themselves into squatter shanty communities on the fringes of the city, isolated from the sanitary facilities, the public utilities, and the job opportunities that are found only at the center. In the sprawling slum district of Nairobi, Kenya, called Mathare Valley, some 250,000 people are squeezed into 6 square miles (15.5 km²) and are increasing by 10,000 per year. Such impoverished squatter districts exist around most major cities in Africa, Asia, and Latin America (Figure 11.40). Proposed models of Third World urban structure demonstrate the limits of these generalizations. The same models (and others) help define some of the regional and cultural contrasts that distinguish those cities (Figure 11.41).

The large *Southeast Asian* city is shown in Figure 11.41*a* as a composite. The port and its associated areas were colonial creations, retained and strengthened in independence. Around them are found a Western-style central business district with European shops, hotels, and restaurants; one or more "alien commercial zones" where

Figure 11.41

Third World urban models.

Sources: (*a*) T. G. McGee, *The Southeast Asian City*, Figure 25, p. 128. Copyright © 1967 G. Bell and Sons, Ltd. (*b* & *c*) Figures from *Cities of the World: World Regional Development*, by Stanley D. Brunn and Jack F. Williams. Copyright © 1983 by Harper & Row, Publishers, Inc. Reprinted by permission.

(*d*) Ernst Griffin and Larry Ford, "A Model of Latin American City Structure" in *Geographical Review*, 70, p. 406, 1980. Copyright © 1980 The American Geographical Society, New York, NY. Reprinted by permission.

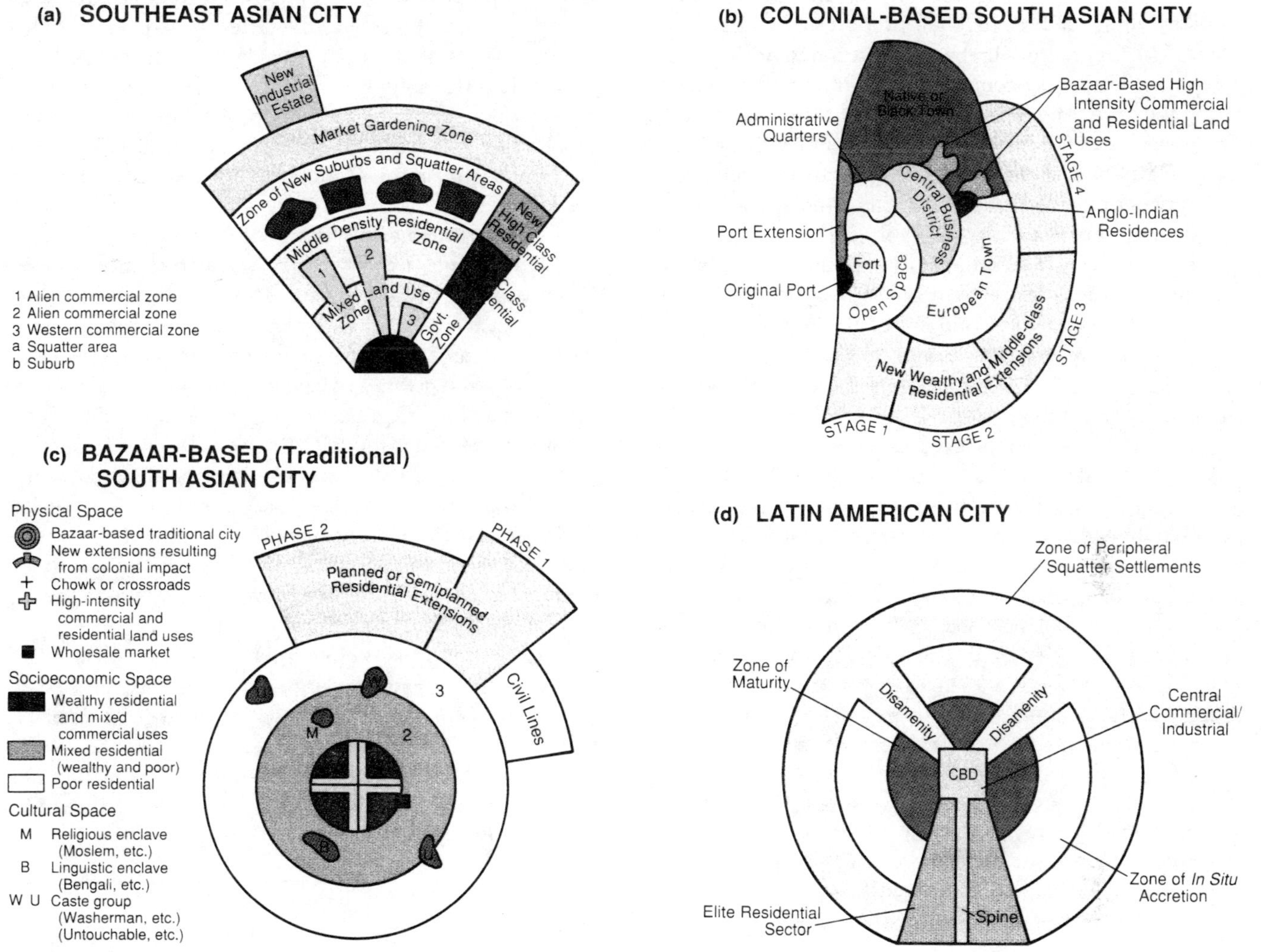

merchants of the Chinese and, perhaps, Indian communities have established themselves; and the more widespread zone of mixed residential, light industrial, and indigenous commercial uses. Central slums and peripheral squatter settlements house up to two-thirds of the total city population. Market gardening and recent industrial development mark the outer metropolitan limits.

The *South Asian* city appears in two forms. Figure 11.41*b* summarizes the internal structure of the colonial-based city, making clear the spatial separation of native and European residential areas, the mixed-race enclave between them, and the 20th-century new-growth areas housing the wealthier local elites. Figure 11.41*c* depicts the traditional bazaar city, its city center focused upon a crossroads around which the houses of the wealthier residents are situated. Merchants live above or behind their shops, and the entire city center is characterized by mixed residential, commercial, and manufacturing land uses. Beyond the inner core is, first, an upper-income residential area shared (but not in the same structures) with poorer servants. Still farther out are the slums and squatter communities, generally sharply segregated according to ethnic, religious, or caste status or native village of their inhabitants.

The *Latin American* city, modeled in Figure 11.41*d*, is more thoroughly "Westernized" than its Asian or African counterparts, but it shows many of the same land use arrangements. The focal core is a central business district on the North American and European model, served by a well-developed mass transit system and having the full range of urban services and amenities. It therefore retains its attraction for middle- and upper-income residents who contribute to its upgrading and modernization through new apartment construction. A commercial "spine," essentially an extension of the CBD, leads away from the core. Bordering it in an ever-widening elite sector are the upper-income housing districts, the best theaters and restaurants, modern office buildings, exclusive shops, major boulevards, and wealthiest suburbs. The area occupied by the urban elites far exceeds their proportion of total population. The far greater numbers of lower-income and destitute urban residents occupy the concentric zones spaced around the core. The inner "zone of maturity" houses the middle class in solid structures, well maintained. The "*in situ*" zone is one of more modest and mixed residential quality, under continuing reconstruction and upgrading as its residents improve their economic status. At its outer margins, however, it fades into the "peripheral squatter settlements" zone, the outer ring housing in the urban area's worst shantytowns and slums the most impoverished and most recent in-migrants to the city.

Each of these models presents a variant of the developing world's urban dilemma: an urban structure not fully capable of housing the peoples so rapidly thrust upon it. Populations of Third World cities in the late 1980s were growing by 3.5% annually, three times faster than urban growth in the Western world. The extreme case was Africa, where a 5% annual growth rate implied a doubling of city populations in 14 years. These numerical increases exceed urban support capabilities, and unemployment rates of 25% are common in cities of the developing world. There is little chance to reduce them as additional millions—through natural increase and in-migration—swell cities already overwhelmed by poverty, lack of sanitation facilities, inadequate water supplies, antiquated public transportation, and spreading slums. The problems, cultures, environments, and economies of Third World cities are tragically unique to them. The urban models that give us understanding of United States cities are of little assistance or guidance in such vastly different culture realms.

Summary

The city is the essential functional node in the chain of linkages and interdependencies that marks any society or economy advanced beyond the level of subsistence. The more complex and functionally integrated the society, the greater the degree of specialization and exchange that it demands for its maintenance, and the more urbanized it becomes. Although cities are among the oldest marks of civilization, only in this century have they become the home of the majority of the people in the industrialized countries and the inhospitable place of refuge for uncounted millions in modernizing Third World economies. At the global scale, four major world urban regions have emerged: Western Europe, South Asia, East Asia, and North America. Within those regions, metropolitan expansion has created massive conurbations.

All cities growing beyond their village origins take on functions uniting them to the countryside and to a larger system of cities; all, as they grow, become functionally complex. Their economic base may become diverse and is comprised of both *basic* and *service* activities. The former represent the functions performed for the larger economy and city system, the latter those present to satisfy the needs of the urban residents themselves. Functional classifications distinguish the economic roles of cities, while simple classification of them as transportation and special function cities or as central places helps define and explain their functional and size hierarchies and the spatial patterns they display within a system of cities.

As North American urban centers expanded in population size and diversity, they developed more or less structured land use patterns and social geographies based on market allocations of urban space, channelization of traffic, and largely voluntary socioeconomic aggregation. The observed regularity of land use arrangements has been summarized for U.S. cities by the concentric zone, sector, and multiple-nuclei models. Social area counterparts of land use specializations were based on social status, family status, and ethnicity. Since 1945, these older models of land uses and social areas have been modified by the suburbanization of people and functions that has led to the creation of new and complex outer cities and to the deterioration of the older central city itself.

Urbanization is a global phenomenon, and the U.S. models of city systems, land use structures, and social geographies are not necessarily or usually applicable to other cultural contexts. Models descriptive of Third World city structures do little to convey the nature of their universal dilemmas of intense overcrowding, poverty, and inadequate physical structure.

Key Words

basic sector 353
central business district (CBD) 362
central place 355
concentric zone model 364
economic base 353
gentrification 367
metropolitan area 351
multiple-nuclei model 365
multiplier effect 354
nonbasic (service) sector 353
primate city 360
rank-size rule 360
sector model 364
suburb 351

For Review

1. Consider the city or town in which you live or attend school or with which you are most familiar. In a brief paragraph, discuss that community's *site* and *situation*. Point out the connection, if any, between its site and situation and the basic functions that it earlier performed or now performs.
2. Describe the *multiplier effect* as it relates to the population growth of urban units.
3. What area does a *central place* serve, and what kinds of functions does it perform? If an urban system were composed solely of central places, what summary

statements could you make about the spatial distribution and the urban size hierarchy of that system?

4. Is there a hierarchy of retailing activities in the community with which you are most familiar? Of how many and what kinds of levels is that hierarchy composed? What localizing forces affect the distributional pattern of retailing within that community?
5. Briefly describe the urban land use patterns predicted by the *concentric zone,* the *sector,* and the *multiple-nuclei* models of urban development. Which one, if any, best corresponds to the growth and land use pattern of the community most familiar to you? How well does Figure 11.31 depict the land use patterns in the metropolitan area with which you are most familiar?
6. In what ways do *social status, family status,* and *ethnicity* affect the residential choices of households? What expected distributional patterns of urban *social areas* are associated with each? Does the social geography of your community conform to the predicted pattern?
7. How has suburbanization damaged the economic base and the financial stability of the United States central city?
8. In what ways does the Canadian city differ from the pattern of its U.S. counterpart? What design and social philosophies are incorporated in the planned socialist city that set it off from its Western European neighbor or the U.S. model?
9. Why are metropolitan areas in developing countries expected to grow larger than many Western metropolises by the year 2000?

Suggested Readings

Adams, John S. "Residential Structure of Midwestern Cities." *Annals of the Association of American Geographers* 60 (1970): 37–62.

Adams, John S., ed. *Contemporary Metropolitan America.* 4 vols. Cambridge, Mass.: Ballinger, 1976.

Agnew, John A., John Mercer, and David E. Sopher, eds. *The City in Cultural Context.* Winchester, Mass.: Allen & Unwin, 1984.

Bater, James H. *The Soviet City: Ideal and Reality.* London: Edward Arnold, 1980.

Berry, Brian J. L. *Comparative Urbanization: Divergent Paths in the Twentieth Century.* 2d rev. ed. New York: St. Martin's Press, 1981.

Borchert, John. "American Metropolitan Evolution." *Geographical Review* 57 (1967): 301–332.

Bourne, L. S., R. Sinclair, and K. Dziewnski, eds. *Urbanization and Settlement Systems: International Perspectives.* New York: Oxford University Press, 1984.

Brunn, Stanley D., and Jack F. Williams. *Cities of the World: World Regional Urban Development.* New York: Harper & Row, 1983.

Burgess, Ernest W. "The Growth of the City." In *The City,* edited by Robert E. Park, Ernest W. Burgess, and Roderick D. McKenzie, pp. 47–62. Chicago: University of Chicago Press, 1925.

Burtenshaw, D., M. Bateman, and G. J. Ashworth. *The City in Western Europe.* New York: John Wiley & Sons, 1981.

Christaller, Walter. *Central Places in Southern Germany.* Translated by C. W. Baskin. Englewood Cliffs, N.J.: Prentice-Hall, 1966.

Christian, Charles M., and Robert A. Harper, eds. *Modern Metropolitan Systems.* Columbus, Ohio: Charles E. Merrill, 1982.

Clark, David. *Post-Industrial America: A Geographical Perspective.* New York: Methuen, 1984.

Conzen, Michael P. "American Cities in Profound Transition: The New City Geography of the 1980s." *Journal of Geography* 82 (1983): 94–102.

Conzen, Michael P., ed. *World Patterns of Modern Urban Change.* Chicago: University of Chicago, Department of Geography, Research Paper 217–218. 1986.

Daniels, P. W. *Service Industries: A Geographical Appraisal.* London and New York: Methuen, 1985.

Dogan, Mattei, and John D. Kasarda, eds. *The Metropolis Era.* Vol. 1, *A World of Giant Cities*; Vol. 2, *Mega-Cities.* Newbury Park, Calif.: Sage, 1988.

Drakakis-Smith, David. *The Third World City.* London and New York: Methuen, 1987.

Erickson, Rodney A., and Marylynn Gentry. "Suburban Nucleations." *Geographical Review* 75 (1985): 19–31.

Fishman, Robert. *Bourgeois Utopias: The Rise and Fall of Suburbia.* New York: Basic Books, 1987.

Fox, Robert W. "The World's Urban Explosion." *National Geographic,* August 1984, pp. 179–185.

French, R. A., and F. E. Ian Hamilton, eds. *The Socialist City: Spatial Structure and Urban Policy.* New York: John Wiley & Sons, 1979.

Getis, Arthur and Judith Getis, "Christaller's Central Place Theory." *Journal of Geography* 65 (1966): 220–226.

Goldberg, Michael A., and John Mercer, *The Myth of the North American City.* Vancouver: University of British Columbia Press, 1986.

Griffin, Ernst, and Larry Ford. "A Model of Latin American City Structure." *Geographical Review* 70 (1980): 397–422.

Hakim, Belim Selim. *Arabic-Islamic Cities: Building and Planning Principles.* London: Routledge & Kegan Paul, 1986.

Hall, Peter. *The World Cities* 3d ed. New York: St. Martin's Press, 1984.

Harris, Chauncy D., and Edward L. Ullman, "The Nature of Cities." *Annals of the American Academy of Political and Social Science* 242 (1945): 7–17.

Hartshorn, Truman A. *Interpreting the City: An Urban Geography.* 2d rev. ed. New York: John Wiley & Sons, 1990.

Heinritz, Günter, and Elizabeth Lichtenberger, eds. *The Take-Off of Suburbia and the Crisis of the Central City.* Proceedings of an international symposium. *Erdkundliches Wissen* 76. Stuttgart: Franz Steiner Verlag, 1986.

Herbert, David T., and Colin J. Thomas. *Urban Geography: A First Approach.* New York: John Wiley & Sons, 1982.

Hoyt, Homer. *The Structure and Growth of Residential Neighborhoods in American Cities.* Washington, D.C.: U.S. Government Printing Office, 1939.

Jefferson, Mark. "The Law of the Primate City." *Geographical Review* 29 (1939): 226–232.

Karan, P. P., and W. A. Bladen. "Arabic Cities." *Focus* 33, no. 3 (1983).

Kelley, Allen C., and Jeffrey G. Williamson. "Population Growth, Industrial Revolutions, and the Urban Transition." *Population and Development Review* 10 (1984): 419–441.

King, Leslie J. *Central Place Theory*. Newbury Park, Calif.: Sage, 1984.

Knox, Paul. *Urban Social Geography* 2d ed. New York: Longman/John Wiley & Sons, 1987.

Kosambi, Meera, and John E. Brush, "Three Colonial Port Cities in India." *Geographical Review* 78 (1988): 32–47.

Lowder, Stella. *The Geography of Third World Cities*. Totowa, N.J.: Barnes & Noble, 1986.

McDowell, Bart. "Mexico City: An Alarming Giant." *National Geographic*, August 1984, pp. 138–178.

McGee, T. G. *The Southeast Asian City*. New York: Praeger, 1967.

Morton, Henry W., and Robert C. Stuart, eds. *The Contemporary Soviet City*. Armonk, N.Y.: M. E. Sharpe, 1984.

Muller, Peter O. *Contemporary Suburban America*. Englewood Cliffs, N.J.: Prentice-Hall, 1981.

Nader, George A. *Cities of Canada*. Vol. 1, *Theoretical, Historical and Planning Perspectives*; Vol. 2, *Profiles of Fifteen Metropolitan Centres*. Toronto: Macmillan of Canada, 1975.

Preston, Richard E. "The Structure of Central Place Systems." *Economic Geography* 47 (1971): 136–155.

Rose, Harold. *The Black Ghetto: A Spatial Behavioral Perspective*. New York: McGraw-Hill, 1971.

Shaw, Denis J. B. "Some Influences on Spatial Structure in the State Socialist City: The Case of the USSR." In *Soviet Geography Studies In Our Time*, edited by Lutz Holzner and Jeane M. Knapp, pp. 201–227. Milwaukee: College of Letters and Science, University of Wisconsin–Milwaukee and the American Geographical Society Collection of the Golda Meir Library, 1987.

Short, John R. *An Introduction to Urban Geography*. London: Routledge & Kegan Paul, 1984.

Smith, Neil, and Peter Williams, eds. *Gentrification of the City*. Winchester, Mass.: Allen & Unwin, 1986.

Stelter, Gilbert, and Alan F. J. Artibise, eds. *Power and Place: Canadian Urban Development in the North American Context*. Vancouver: University of British Columbia Press, 1986.

Vance, James E. *This Scene of Man: The Role and Structure of the City in the Geography of Western Civilization*. New York: Harper & Row, 1977.

White, Paul. *The West European City: A Social Geography*. London and New York: Longman, 1984.

Yeates, Maurice, and Barry Garner. *The North American City*. 3d ed. New York: Harper & Row, 1980.

C H A P T E R

Human Impact on the Environment

Pulp mill in LaTuque, Quebec.

TIMES BEACH, Missouri, used to be a tight-knit, working-class community along the Meramec River southwest of St. Louis. Now the people are gone. Empty houses and stores, abandoned cars, and rusting refrigerators stand silent. "Thanks for coming," says the sign on the Easy Living Laundromat to nonexistent customers. Wildflowers bloom on overgrown lawns. Flies buzz, squirrels chase each other up trees, and an occasional coyote prowls the streets. Were they allowed, visitors might think they were seeing an uncannily realistic Hollywood stage set. Instead, the ghost town is a symbol of what happens when hazardous wastes contaminate a community.

The trouble began in the summer of 1971, when Times Beach hired a man to spread oil on its 10 miles (16 km) of dirt roads to keep down the dust. Unfortunately, his chief occupation was removing used oil and other waste from a downstate chemical factory, and it was a mixture of these products that he sprayed all over town that summer and the next. The tens of thousands of gallons of purple sludge were contaminated with dioxin, the most toxic substance ever manufactured.

Effects of the dioxin exposure appeared almost immediately in two riding arenas that had been sprayed. Within days, hundreds of birds and small animals died, kittens were stillborn, and many horses became ill and died. Then residents began reporting a variety of medical disorders, including miscarriages, seizure disorders, liver impairment, and kidney cancer. Not until 1982 did the Environmental Protection Agency (EPA), alerted to high levels of dioxin in wastes stored at the chemical plant, make a thorough investigation of Times Beach. They found levels of the dioxin compound TCDD as high as 300 parts per billion in some of the soil samples; 1 part per billion is the maximum concentration deemed safe. The EPA purchased every piece of property in town and ordered its evacuation (Figure 5.1).

The story of Times Beach is one of many that could have been selected to illustrate how people can affect the quality of the water, air, and soil on which their existence depends.

Terrestrial features and ocean basins, elements of weather and characteristics of climate, flora, and fauna comprise the building blocks of that complex mosaic called the *environment*, or the totality of things that in any way may affect an organism. Humans exist within a natural environment—the sum of the physical world—that they have modified by their individual and collective actions. Forests have been cleared, grasslands plowed, dams built, and cities constructed. On the natural environment, then, has been erected a cultural environment, modifying, altering, or destroying the balance of nature that existed before human impact was expressed. This chapter is concerned with the interrelation between humans and the natural environment that they have so greatly altered.

Since the beginnings of agriculture, humans have changed the face of the earth, have distorted delicate balances and interplays of nature, and, in the process, have both enhanced and endangered the societies and the economies that they have erected. The essentials of the natural balance and the ways in which humans have altered it are not only our topics here but are also matters of social concern that rank among the principal domestic and international issues of our times. As we shall see, the fuels we consume, the raw materials we use, the products we create, and the wastes we discard all contribute to the harmful alteration of the **biosphere**, the thin film of air, water, and earth within which we live.

Ecosystems

The biosphere is composed of three interrelated parts: (1) the *troposphere*, some 6–7 miles (9.5–11.25 km) thick; (2) the *hydrosphere*, surface and subsurface waters in oceans, streams, lakes, glaciers, or groundwater —much of it locked in ice or earth and not immediately available for use; and (3) the upper reaches of the earth's *crust*, a few thousand feet at most, containing the soils that support plant life, the minerals that plants and animals require to exist, and the fossil fuels and ores that humans exploit. The biosphere is an intricately interlocked system, containing all that is needed for life, all that is available for living things to use, and, presumably, all that ever will be available. The ingredients of the biosphere must be, and are, constantly recycled and renewed in nature. Plants purify the air, the air helps to purify the water, the water and the minerals are used by plants and animals and are returned for reuse.

The biosphere, therefore, consists of two intertwined components: (1) a nonliving outside (solar) energy source and requisite chemicals and (2) a living world of

Figure 5.1
Times Beach, Missouri.

plants and animals. In turn, the biosphere may be subdivided into specific **ecosystems**, functional units that consist of all the organisms (plants and animals) and physical features (air, water, soil, and chemicals) existing together in a particular area. Figure 5.2 depicts one kind of ecosystem and in crude form suggests the most important principle concerning all ecosystems: everything is interconnected. Any intrusion or interruption in the balance that has been naturally achieved inevitably results in undesirable effects elsewhere in the system. Each organism occupies a specific *niche*, or place, within an ecosystem. In the energy exchange system, each organism plays a definite role; individual organisms survive because of other organisms that also live in that environment. The problem lies not in recognizing the niches but in anticipating the chain of causation and the readjustments of the system consequent on disturbing the occupants of a particular niche.

Food Chains and Renewal Cycles

Life depends on the energy and nutrients flowing through an ecosystem. The transfer of energy and materials from one organism to another is one link in a **food chain**, defined as a sequence of organisms, such as green plants, herbivores, and carnivores, through which energy and materials move within an ecosystem. Some food chains have only two links, as when human beings eat rice; most have only three or four. Because the ecosystem in nature is in a continuous cycle of integrated operation, there is no start or end to a food chain. There are, simply, nutritional transfer stages in which each lower level in the food chain transfers part of its contained energy to the next-higher-level consumer.

The *decomposers* pictured in Figure 5.2 are essential in maintaining food chains and the cycle of life. They cause the disintegration of organic matter—animal carcasses and droppings, dead vegetation, wastepaper, and so on. In the process of decomposition, the chemical nature of the material is changed, and the nutrients contained within it become available for reuse by plants or animals. *Nutrients*, the minerals and other elements that organisms need for growth, are never destroyed; they keep moving from living to nonliving things and back again. Our bodies contain nutrients that were once part of other organisms, perhaps a hare, a hawk, or an oak tree.

The idea of a cycle is important in furthering our understanding of the natural renewal of the elements essential to life, which include carbon, oxygen, hydrogen, and phosphorus. Because the supply of these materials is fixed, they must be continuously recycled through food chains. Figures 5.3 and 5.4 show two important cycles, the phosphorus cycle and the hydrologic cycle.

Phosphorus is an element essential to the survival of many species of plants and animals. It is an important component of DNA and of bones and teeth. Plants absorb phosphate from the soil, and other organisms acquire it by eating plants. It is returned to the soil through animal excretions or through the decay of plants and animals. When phosphate that has been used as an agricultural fertilizer is carried to rivers and eventually to the sea through runoff, only a small percentage is returned to the land through fish catches and the excretions (guano) of seabirds, which eat fish. Most of it remains in deep ocean waters until geologic processes cause uplifting of the ocean floor.

Ecosystems change constantly whether people are present or not, but humans have affected them more than has any other species. The impact of humans on ecosystems was small at first, with low population size, energy consumption, and technological levels. It has increased so rapidly and pervasively as to present us with widely recognized and varied ecological crises. Some of the effects of humans on the natural environment are the topic of the remainder of this chapter.

Impact on Water

The supply of water is constant. The system by which it continuously circulates through the biosphere is called the **hydrologic cycle** (Figure 5.4). In that cycle, water may change form and composition, but under natural environmental circumstances, it is purified in the recycling process and is again made available with appropriate properties to the ecosystems of the earth. *Evaporation* and

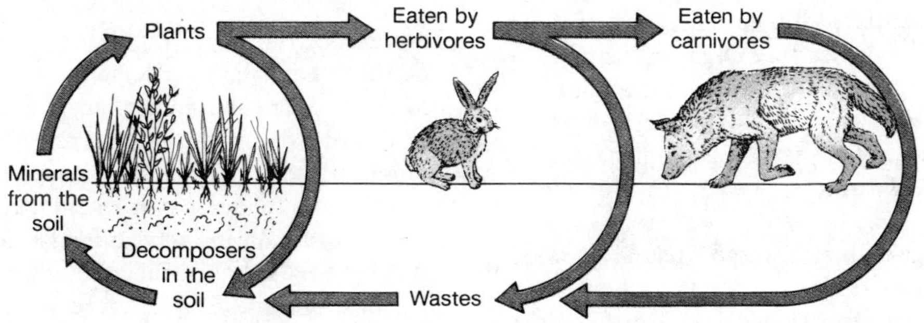

Figure 5.2
A simplified example of an ecosystem illustrating the interdependence of organisms and the physical environment.

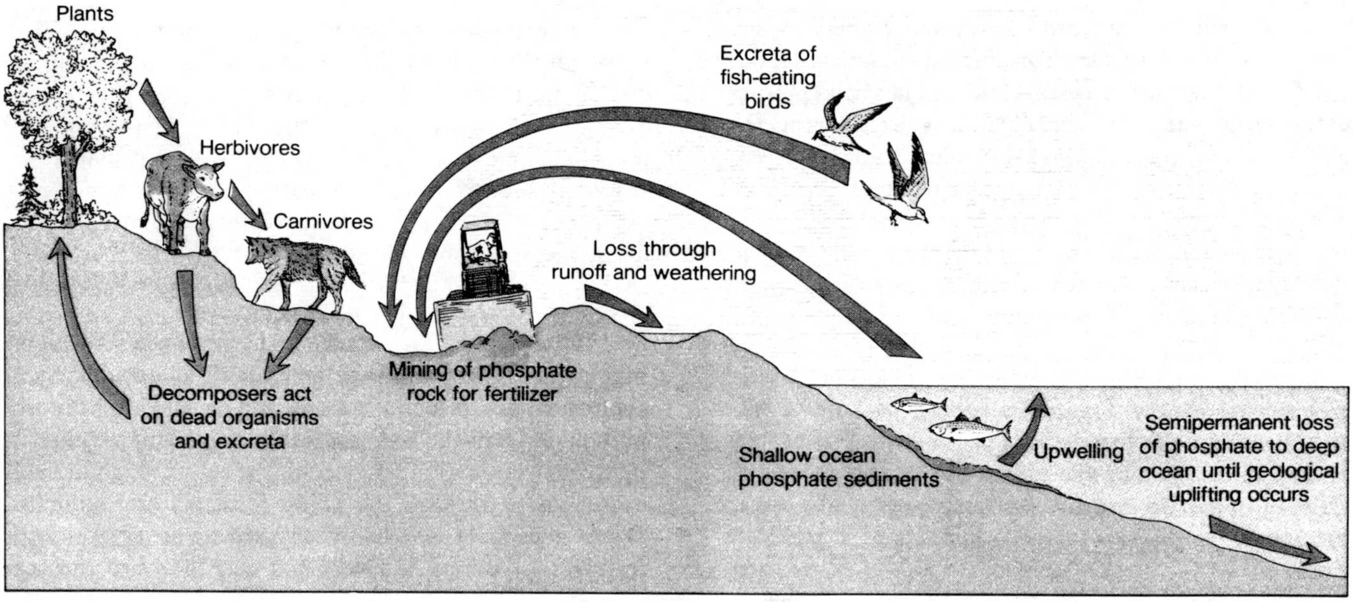

Figure 5.3

The phosphorus cycle. The natural phosphorus cycle is exceedingly slow and has been upset by human activity. Currently, phosphorus is being washed into the sea faster than it is being returned to the land, primarily because it is being mined for use as a fertilizer.

transpiration (the emission of water vapor from plants) are the mechanisms by which water is redistributed. Water vapor collects in clouds, condenses, and then falls again to earth. There it is reevaporated and retranspired, only to fall once more as precipitation.

Humans have a discernible impact on the hydrologic cycle by affecting the speed of water recirculation and by hastening the return of water to the sea. Massive withdrawals of water from surface supplies by thermal power plants or by industry for cooling purposes can abbreviate the normal cycle by prematurely converting water from a liquid to its vaporous state. Comparable accelerated vaporization occurs through evaporation from reservoirs, particularly in arid regions. The amount of usable water available at some point in the cycle is thereby reduced.

Because the need for water will continue to grow as the world population increases, many scientists fear a coming water crisis. Agriculture currently accounts for almost 80% of all the water used by people, most of it for irrigation. The Food and Agriculture Organization (FAO) of the United Nations has estimated that the amount of water used for irrigation will have to double to feed all those who will be alive in the year 2000. Regional water shortages will affect developed as well as less developed countries.

For example, agriculture in large parts of Kansas, Nebraska, Colorado, and Texas draws on the water stored in a vast underground formation called the Ogallala Aquifer (Figure 5.5). (An *aquifer* is water-bearing, porous rock lying between impermeable layers.) The water bed is less than 200 feet (60 m) thick in part of the area; yet, because large amounts are withdrawn for irrigated agriculture, the water table falls from 2 to 5 feet each year. Preliminary studies indicate that if present rates of pumping continue, as much as 40% of the irrigated acreage in portions of the Great or High Plains states will be lost by the year 2020.

Urbanization and industrialization also place growing demands on water supplies. Some 150 tons (40,000 gallons) of water are used to produce a ton of steel; 250 tons are needed to make a ton of paper. As more and more countries industrialize, pressure on the water supply will increase.

People's dependence on water has long led to efforts to control its supply. Such manipulation has altered the quantity and the quality of water in rivers and streams.

Modification of Streams

To prevent flooding, to regulate the water supply for agriculture and urban settlements, or to generate power, people have for thousands of years manipulated rivers by constructing dams, canals, and reservoirs. Although they generally have achieved their purposes, these structures can have unintended environmental consequences. These include reduction in the sediment load downstream, followed by a reduction in the amount of the nutrients available for crops and fish; an increase in the salinity of the soil; and subsidence (discussed later in this chapter).

Channelization, another method of modifying river flow, is the construction of embankments and dikes and

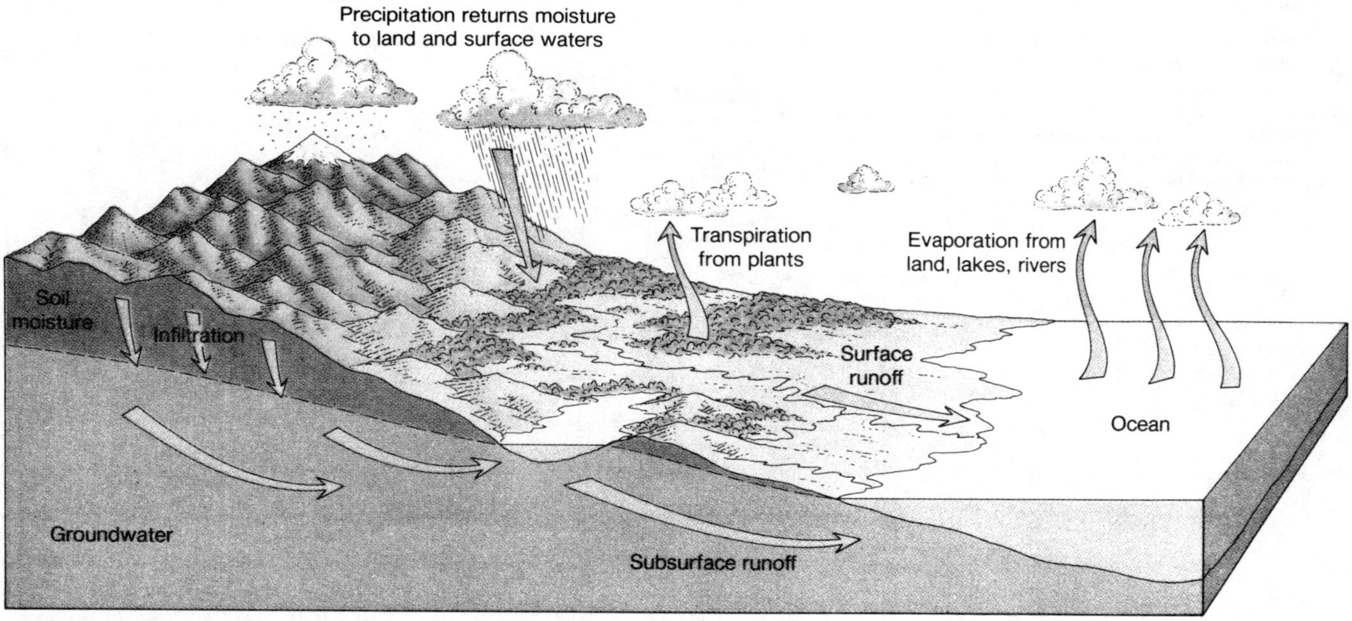

Figure 5.4

The hydrologic cycle. The sun provides energy for the evaporation of fresh and ocean water. The water is held as vapor until the air becomes supersaturated. Atmospheric moisture is returned to the earth's surface as solid or liquid precipitation to complete the cycle. Because precipitation is not uniformly distributed, moisture is not necessarily returned to areas in the same quantity as it has evaporated from them. The continents receive more water than they lose. The excess returns to the seas as surface water or groundwater. A global water balance, however, is always maintained.

the straightening, widening, and/or deepening of channels to control floodwaters or to improve navigation. Many of the great rivers of the world, including the Nile and the Huang He, are lined by embankment systems. Like dams, these systems can have unforeseen consequences. They reduce the natural storage of floodwaters, can aggravate flood peaks downstream, and can cause excessive erosion. Some of the possible ecological effects of channelization are shown in Figure 5.6.

Channelization and dam construction are deliberate attempts to modify river regimes, but other types of human action also affect river flow. Urbanization, for example, has significant hydrologic impacts, including a lowering of the water table, pollution, and increased flood runoff. Likewise, the removal of forest cover increases runoff, promotes flash floods, lowers the water table, and hastens erosion. Nevertheless, the primary adverse human impact on water is felt in the area of water quality. People withdraw water from lakes, rivers, or underground deposits to use for drinking, bathing, agriculture, industry, and many other purposes. Although the water that is withdrawn returns to the water cycle, it is not always returned in the same condition as at the time of withdrawal. Water, like other segments of the ecosystem, is subject to serious problems of pollution.

Figure 5.5

The giant Ogallala Aquifer, the country's largest underground water supply. Some 20 million acres (8 million hectares) of land are now watered by pumping from the aquifer, which supports about one-quarter of the country's cotton crop and a great deal of its corn and wheat. About 40% of the country's cattle are fed grain grown in the High Plains region. Approximately 300 gallons of water in the field are needed to grow enough wheat for one loaf of bread, 4200 gallons to produce one pound of beef—figures that help explain why the aquifer is being depleted faster than it can be replenished by nature.

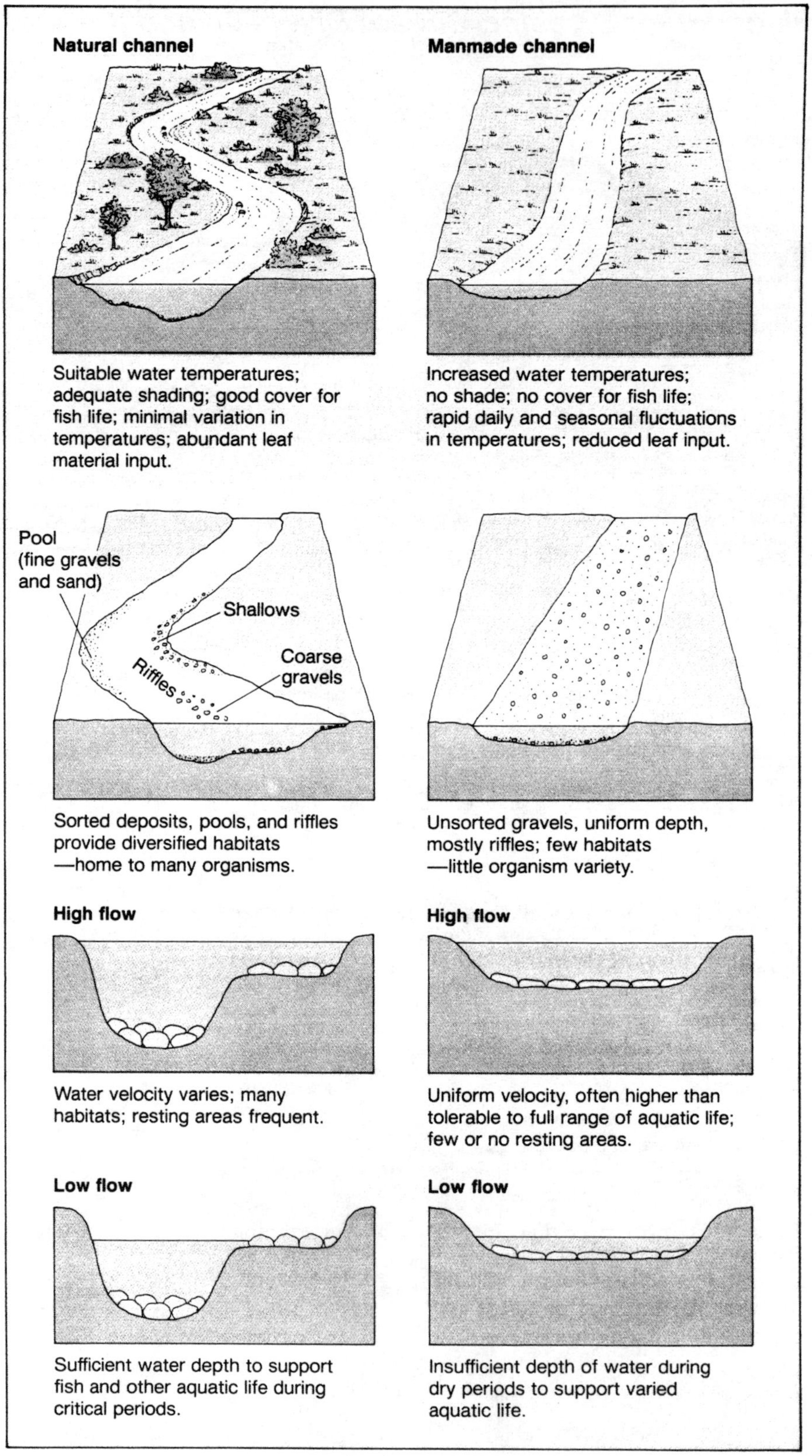

Figure 5.6

Comparison of a natural channel with a channelized stream. Some of the possible effects of channelization are depicted. Dredging and filling (the deposition of dredged materials to create new land) are other ways in which people modify the configuration of water bodies.

Blueprint for Disaster: Stream Diversion and the Aral Sea

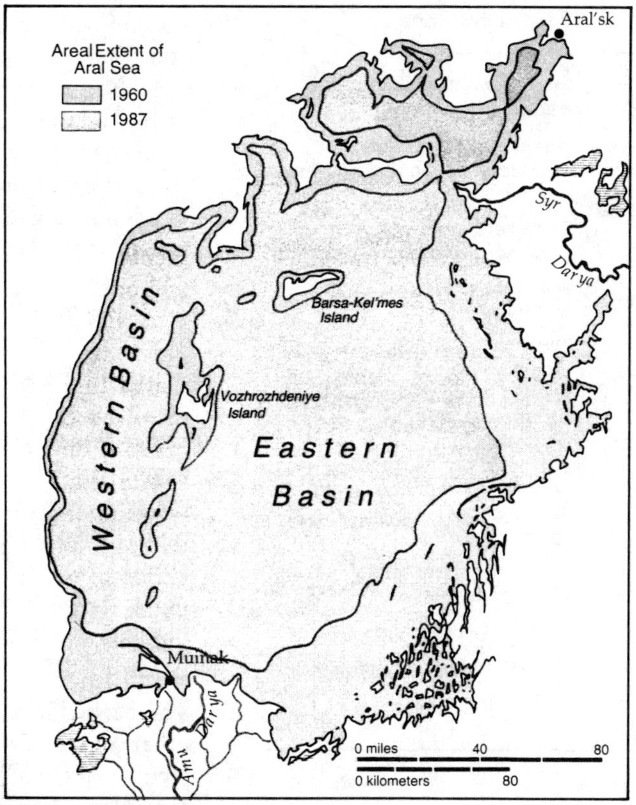

It used to be the fourth largest lake in the world, covering an area of 27,000 square miles, (69,900 km^2) larger than the state of West Virginia. Now the Soviet Union's Aral Sea is only the sixth largest lake, and it may cease to exist by the year 2010, having been converted instead into a number of lakes. The level of the sea has dropped 40 feet (12 m) since 1960 and the volume of the water is only one-third of what it used to be.

The shrinkage of the lake is just one consequence of diverting nearly all the water from its primary sources, the Amu Darya and the Syr Darya, to irrigate the agricultural fields of Turkmenia and Uzbekistan. These are some of the others:

- As the shoreline has receded, it has left behind 10,000 square miles (26,000 km^2) of salty desert waste. The high concentrations of salt, fertilizers, and pesticides are slowly turning the Aral into a dead sea. Because the water is too salty for most fish species to endure, the commercial fishing industry has already collapsed. Once a fishing port, Muinak is now 30 miles from the sea.
- In addition, wind storms whip up the salty grit from the dried-up seabed and deposit it on cropland hundreds of miles away. Thus the agricultural crops for which the Aral Sea was sacrificed are themselves at risk. This is not a minor concern: the irrigated land provides virtually all the cotton and at least one-fourth of the fruit, vegetables, and rice produced in the Soviet Union.
- Finally, the rivers that feed the Aral Sea have become sluggish sewers, contaminated by industrial and agricultural wastes and sewage. Although the rivers teem with viruses causing dysentery, typhoid, and other diseases, millions of people depend on them for drinking water. Over 65,000 cases of hepatitis have been reported in the region in the last 15 years, and child mortality rates are among the highest in the world.

The Meaning of Water Pollution

As a general definition, **environmental pollution** by humans means the introduction into the biosphere of wastes that, because of their volume or their composition or both, cannot be readily disposed of by natural recycling processes. In the case of water, the central idea is that pollution exists when water composition has been so modified by the presence of one or more substances that either it cannot be used for a specific purpose or it is less suitable for that use than it was in its natural state. Pollution is brought about by the discharge into water of substances that cause unfavorable changes in its chemical or physical nature or in the quantity and quality of the organisms living in the water. Pollution is a relative term. Water that is not suitable for drinking may be completely satisfactory for cleaning streets. Water that is too polluted for fish may provide an acceptable environment for certain water plants.

Human activity is not the only cause of water pollution. Leaves that fall from trees and decay, animal wastes, oil seepages, and other natural phenomena may affect water quality. There are natural processes, however, to take care of such pollution. Organisms in water are able to degrade, assimilate, and disperse such substances in the amounts in which they naturally occur. Only in rare instances do natural pollutants overwhelm the cleansing abilities of the recipient waters. What is happening now is that the quantities of wastes discharged by humans often exceed the ability of a given body of water to purify itself. In addition, humans are introducing pollutants, such as metals or inorganic substances, that cannot be broken down at all by natural mechanisms or that take a very long time to break down. Table 5.1 shows the kinds of water pollutants associated with different sources.

Table 5.1
Sources and Types of Water Pollutants

Source of waste	Kind of pollutant
Municipalities and residences	Human wastes, detergents, garbage, trash
Urban drainage	Suspended sediment, fertilizers, pesticides, road salt
Agriculture	
Crop production	Fertilizers, herbicides, pesticides, erosion sediment
Irrigation return flow	Mineral salts and erosion sediment
Industry	Wide range of pollutants, including biodegradable wastes in the paper and food-processing industries, heat discharge in a variety of activities, nondegradable wastes in the chemical and iron and steel industries, and radioactivity
Mining	Acids, sediment, metal wastes, culm (coal dust)
Electric power production	Heat and nuclear wastes
Recreation and navigation	Human wastes, garbage, fuel wastes

As long as there are people on earth, there will be pollution. Thus the problem is one not of eliminating pollution but of controlling it. This is particularly true in the technologically advanced countries. The more developed and affluent the country, the more resources it uses and the more it pollutes.

Agricultural Sources of Water Pollution

On a worldwide basis, agriculture probably contributes more to water pollution than does any other single activity. In the United States, agriculture is estimated to be responsible for about two-thirds of stream pollution. Agricultural runoff carries three main types of pollutants: fertilizers, biocides, and animal wastes.

Fertilizers

Agriculture is a chief contributor of *excess nutrients* to water bodies. Pollution occurs when nitrates and phosphates that have been used in fertilizers and are present in animal manure drain into streams and rivers, eventually accumulating in ponds, lakes, and estuaries. The nutrients hasten the process of **eutrophication**, or the enrichment of waters by nutrients. Eutrophication occurs naturally when nutrients in the surrounding area are washed into the water, but when the sources of enrichment are artificial, as is true of commercial fertilizers, the body of water may become overloaded with nutrients. The end result may be an oxygen deficiency in the water.

Scientists have estimated that as many as one-third of the medium- and large-size lakes in the United States have been affected by accelerated eutrophication. Symptoms of a eutrophic lake are prolific weed growth, large masses of algae, fish kills, rapid accumulation of sediments on the lake bottom, and water that has a foul taste and odor.

Figure 5.7 illustrates one form that overfertilization of a water body can take. Algae and other plants are stimulated to grow abundantly. When they die, the level of dissolved oxygen in the water decreases, primarily because of the bacteria acting on the dead and decomposing vegetation. Fish and plants that cannot tolerate the poorly oxygenated water are eliminated. In addition to being potentially lethal for fish, eutrophication affects the suitability of water for drinking and bathing, because excess nutrients have been shown to pose a health hazard to humans.

Biocides

The herbicides and pesticides used in agriculture are another source of chemical pollution of water bodies. Runoff from farms where such *biocides* have been applied contaminates both ground and surface waters. One of the problems connected with the use of biocides is that the long-term effects of such usage are not always immediately known. DDT, for example, was used for many years before people discovered its effect on birds, fish, and water plant life. One of the dilemmas that we are facing is that we have not yet discovered how to balance our need for increased agricultural yields in the short run—and the manipulation of the environment that it entails—with what is in our best interest in the long run.

Animal Wastes

A final agricultural source of chemical pollution is animal wastes, especially in countries where animals are raised intensively. This is a problem particularly in feedlots, where animals are crowded together at maximum densities to be fattened before slaughter. Large feedlots, such as the one pictured in Figure 5.8, may produce as much waste as would a large city. It is estimated that animal wastes in the United States total about 1.5 billion tons per year, with feedlots generating about half of the total. If not treated properly, the manure pollutes both soil and water with infectious agents and excess nutrients.

Other Sources of Water Pollution

As is evident from Table 5.1, agriculture is only one of the human activities that contributes to water pollution. Other sources are industry, mining, urban drainage, and sewage.

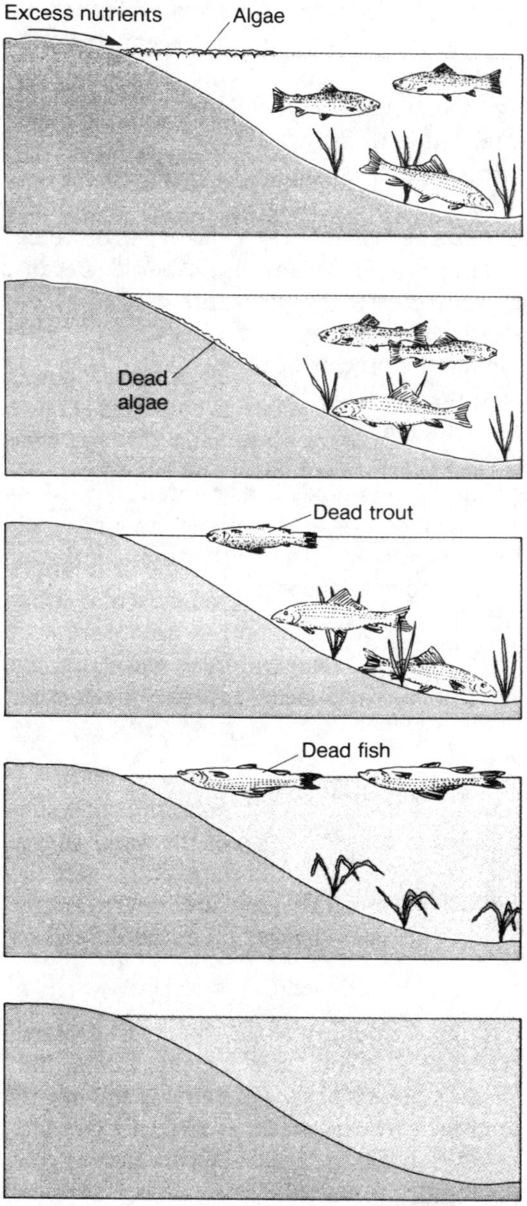

Nitrates and phosphates enrich the lake, stimulating the growth of blue-green algae.

When the algae die and are decomposed by oxygen-consuming bacteria, the level of dissolved oxygen in the lake decreases.

Fish that require less oxygen to survive (such as carp) increase in numbers while fish requiring more oxygen die.

Fish kills lower the oxygen content even more, until no fish can survive.

Diversity of plant and animal life continues to decrease until the lake is suitable only for bacteria that do not require oxygen.

Figure 5.7

Eutrophication is hastened by artificial sources of nutrients. Although eutrophication is primarily a result of agricultural activities, additional sources of nitrates in both surface and underground water supplies are urban drainage, industrial wastewater, and septic tanks.

Figure 5.8

A feedlot in Kansas. The sanitary disposal of organic wastes generated by such concentrations of animals is a problem only recently addressed by environmental protection agencies.

Industry

In developed countries, industry probably contributes as much to contamination of the water supply as does agriculture. In the United States, about half the water used daily is used by industry. Many industries dump organic and inorganic wastes into bodies of water. These may be acids, highly toxic minerals, such as mercury or arsenic, or, in the case of petroleum refineries, toxic organic chemicals. The nuclear power industry has caused some water pollution when radioactive material has seeped from tanks in which wastes have been buried, either at sea or underground.

Such pollution can have a variety of effects. Organisms not adapted to living in contaminated water may die; the water may become unsuitable for domestic use or irrigation; or the wastes may reenter the food chain, with deleterious effects on humans. A particularly unfortunate example occurred in Japan in 1953. A chemical plant in Minamata Bay that used mercury chloride in its manufacturing process discharged the waste mercury into the bay, where it settled with the mud. Fish that fed on organisms in the mud absorbed the mercury and concentrated it; the fish were in turn eaten by humans. Over 700 people died, and at least 9000 others suffered deformity or other permanent disability.

A similar contamination of the water supply with mercury is occurring today in the Amazon River and its tributaries. Because mercury attaches itself to gold, an estimated half-million Brazilian prospectors use the toxic liquid to separate gold from the accompanying mud. Each year, hundreds of tons of mercury are poured in or right next to the rivers, poisoning the water and the fish that swim in it.

Among the pollutants that have been discharged into the water supply in the United States are **polychlorinated biphenyls (PCBs)**, a family of related chemicals used as lubricants in pipelines and in a wide variety of electrical devices, paints, and plastics. During the manufacturing process, companies have dumped PCBs into rivers, from which they have entered the food chain. Several states have banned commercial fishing in lakes and rivers where fish have higher levels of PCBs than are considered safe. Although not all of the effects of PCBs on human health are known, they have been linked to birth defects, damage to the immune system, liver disease, and cancer. In 1977, the Environmental Protection Agency banned the direct discharge of PCBs into U.S. waters, but immense quantities remain in water bodies.

The petroleum industry is a significant contributor to the chemical pollution of water. Oceans are becoming increasingly contaminated by oil. Although massive oil spills like that from the tanker *Exxon Valdez* in March 1989 command public attention, the greatest single source of petroleum pollution in open ocean regions is the discharge of tank flushings and ballast from tanker holds. Other sources include tanker collisions and explosions, and seepage from offshore oil installations. Recent studies indicate that the Gulf of Mexico, the site of extensive offshore drilling, is among the most seriously polluted major bodies of water in the world.

Acid precipitation (sometimes called acid rain), a by-product of emissions from factories, power plants, and automobiles, has affected the water quality and ecology of thousands of lakes and streams in the world. Because the precipitation is caused by pollutants in the air, it is discussed later in this chapter.

Many industrial processes, as well as electric power production, require the use of water as a coolant (Figure 5.9). **Thermal pollution** occurs when water that has been heated is returned to the environment and has adverse effects on the plants and animals in the water body. Many plants and fish cannot survive changes of even a few degrees in water temperature. They either die or migrate. The species that depend on them for food must also either die or migrate. Thus the food chain has been disrupted. In addition, the higher the temperature of the water, the less oxygen it contains, which means that only lower-order plants and animals can survive.

Mining

Surface mining for coal, iron, copper, gold, and other substances contributes to contamination of the water supply through the wastes it generates. Rainwater reacts with the wastes, and dissolved minerals seep into nearby water bodies. The exact chemical changes produced depend on the composition of the coal or ore slag heaps and the reaction of the minerals with sediments or river water. In addition to altering the quality of the water, the contaminants have secondary effects on plant and animal life. Each year, for example, thousands of animals and migratory birds die in such western states as Arizona, Nevada, and California after drinking cyanide-laced waters at gold mines. The toxic water is poured over mounds of crushed rock to leach the gold and then settles into ponds and lakes that attract wildlife.

Urban Drainage

A host of pollutants derives from the activities associated with urbanization. The use of detergents has increased the phosphorus content of rivers, and salt (used for deicing roads) increases the chloride content of runoff. Water runoff from urban areas contains contaminants from garbage, animal droppings, litter, vehicle drippings, and the like. Because the sources of pollution are so varied, the water supply in any single area is often affected by diverse contaminants. This diversity complicates the problem of controlling water quality.

Contaminated drinking-water wells have been found in more than half of the states. Hundreds of wells in the New York metropolitan area have been closed in recent years because of chemical contamination, and thousands

Figure 5.9

Cooling towers at Three Mile Island nuclear plant near Harrisburg, Pennsylvania. Heated wastewaters are often significantly warmer than the waters into which they are discharged, disrupting the growth, reproduction, and migration of fish populations.

more may be closed in coming years. Chemicals have reached groundwater by seeping into aquifers from landfills, from ruptured gasoline and fuel-oil storage tanks, from septic tanks, and from fields sprayed with pesticides and herbicides. The pollution of aquifers is particularly troublesome because, unlike surface waters, groundwater lacks natural cleansing properties; it can remain contaminated for centuries.

Sewage

Sewage can also be a major water pollutant, depending on how well it is treated before being discharged. Raw, untreated human waste contains viruses responsible for dysentery, polio, hepatitis, spinal meningitis, and other diseases. In general, sewage causes less water pollution in countries such as the United States than in countries where sewage treatment is either not practiced or is not thorough.

North America is not without such sources of water pollution, however. Only half of the U.S. population lives in communities with sewage-treatment plants that meet the minimum goals set by the federal Clean Water Act. The aged sewer system that serves Boston and 42 nearby communities (soon to be replaced) discharges millions of gallons of raw sewage from some 2 million people into Boston Harbor annually. A number of cities in New Jersey and New York, including New York City, dump millions of gallons of treated sewage, called *sludge*, into the Atlantic Ocean each year, a practice that is supposed to end by 1992. And because pollution does not observe international borders, the waters and beaches of San Diego are regularly contaminated by the 15 million gallons of raw sewage that daily flow down the Tijuana River to the Pacific Ocean, where currents carry it northward.

Controlling Water Pollution

In recent years, concern over increased levels of pollution has brought about major improvements in the quality of some surface waters, both in the United States and abroad. The federal government in 1972 took the lead in regulating water pollution with the enactment of the Clean Water Act. Its objective was "to restore and maintain the chemical, physical, and biological integrity of the nation's waters." Congress established uniform nationwide controls for each category of major polluting industry and directed the government to pay most of the cost of new sewage-treatment plants. Since 1972 such plants have been built to serve over 80 million Americans, and industries have spent billions of dollars to comply with the Clean Water Act by reducing organic waste discharges.

The gains have been impressive. Many rivers and lakes that were ecologically dead or dying are now thriving. Once dumping grounds for all kinds of human and industrial waste, the Hudson, Potomac, Cuyahoga, and

Trinity rivers are cleaner, more inviting, and more productive than before, and they now support fishing, swimming, and recreational boating. Similarly, Seattle's Lake Washington and the Great Lakes are healthier than they were two decades ago. Recently, authorities have announced ambitious plans to clean up the waters of Chesapeake Bay, the country's largest estuary, and to undo much of the damage that has been inflicted on Florida's Everglades by improving the water quality of the Kissimmee River and Lake Okeechobee.

Environmental awareness in other countries has also prompted legislation and action. For example, the river Thames in southern England, which had become seriously contaminated by sewage and industrial wastes, is now cleaner than it has been in centuries. The enforcement of stringent pollution-control standards has halted the downward trend in quality. Algae and seaweed, fish and wildfowl have returned to the river in abundance.

Even the Mediterranean Sea is on its way to gradual recovery. When the 18 countries that border the sea signed the Convention for the Protection of the Mediterranean Sea against Pollution in 1976, all coastal cities dumped their untreated sewage into the sea, tankers spewed oily wastes into it, and tons of phosphorus, detergents, lead, and other substances contaminated the waters. Now many cities have built or are building sewage-treatment plants, ships are prohibited from indiscriminate dumping, and national governments are beginning to enforce control of pollution from land-based sources.

Such gains should not mislead us. While some of the most severe problems have been attacked, serious pollution still plagues about one-fourth of our country's rivers and streams, lakes and reservoirs. The solution to water pollution lies in the effective treatment of municipal and industrial wastes; the regulation of chemical runoff from agriculture, mining, and forestry; and the development of less polluting technologies. Although pollution-control projects are expensive, the long-term costs of pollution are even higher.

Impact on Air and Climate

The **troposphere**, the thin layer of air just above the earth's surface, contains all the air that we breathe. Every day thousands of tons of pollutants are discharged into the air by cars and incinerators, factories and airplanes. Air is polluted when it contains substances in sufficient concentrations to have a harmful effect on living things.

Air Pollutants

Truly clean air has probably never existed. Just as there are natural sources of water pollution, so there are substances that pollute the air without the aid of humans. Ash from volcanic eruptions, marsh gases, smoke from forest fires, and wind-blown dust are natural sources of air pollution.

Normally these pollutants are of low volume and they are widely dispersed throughout the atmosphere. On occasion, a major volcanic eruption may produce so much dust that the atmosphere is temporarily altered. The eruption in 1883 of a volcano on the island of Krakatoa, west of Java, produced dust that lingered in the area for over a year. In general, however, the natural sources of air pollution do not have a significant, long-term effect on air, which, like water, is able to cleanse itself.

Far more important than naturally occurring pollutants are the substances that people discharge into the air. These pollutants result primarily from burning fossil fuels (coal, gas, and oil) and other materials. Fossil fuels are burned in power plants that generate electricity, in many industrial plants, in home furnaces, and in cars, trucks, buses, and airplanes. Scientists estimate that about three-quarters of all air pollutants come from burning fossil fuels. The remaining pollutants largely result from industrial processes other than fuel burning, incinerating solid wastes, forest and agricultural fires, and the evaporation of solvents. Figure 5.10 depicts the major sources and types of air pollutants.

Air pollution is a global problem. A 1988 study by the World Health Organization concluded that most of the world's 1.8 billion city dwellers breathe bad air. Sulfur dioxide levels are considered unacceptable for some 625 million people in developing countries alone, and levels of smoke, dust, and other particulates are unacceptable for 1.25 billion. Also in 1988, the EPA released a report

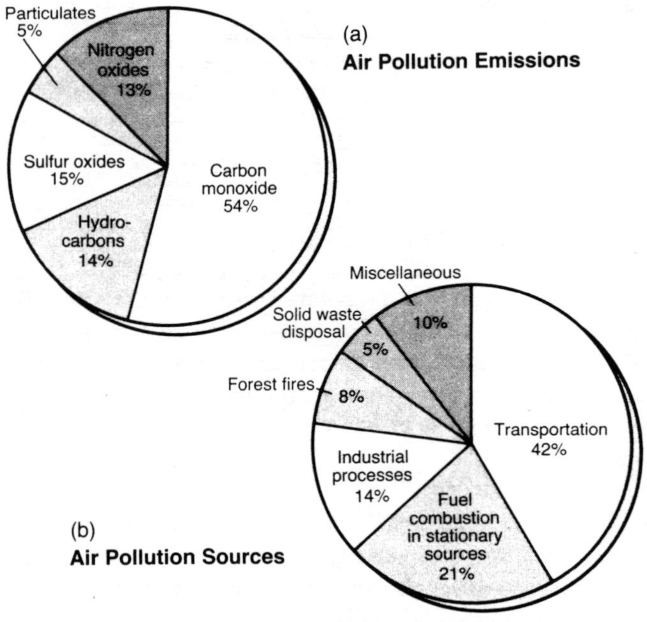

Figure 5.10

(a) Emissions of the principal air pollutants in the United States. Carbon monoxide, one of the most lethal gases known, comes primarily from vehicle exhausts. **(b) Sources of air pollution.** If pollution from vehicles were eliminated, emissions would be reduced by 42%.

showing that in this country over 2 billion pounds of toxic chemicals are emitted into the air annually by industry alone (Figure 5.11).

Factors Affecting Air Pollution

Many factors affect the type and the degree of air pollution found at a given place. Those over which people have relatively little control are climate, weather, wind patterns, and topography. These determine whether pollutants will be blown away or are likely to accumulate. Thus a city on a plain is less likely to experience a buildup than is a city in a valley.

Unusual weather can alter the normal patterns of pollutant dispersal. A *temperature inversion* magnifies the effects of air pollution. Under normal circumstances, air temperature decreases away from the earth's surface. A stationary layer of warm, dry air over a region, however, will prevent the normal rising and cooling of air from below. As described in Chapter 4, the air becomes stagnant during an inversion. Pollutants accumulate in the lowest layer instead of being blown away, so that the air becomes more and more contaminated. Normally inversions last for only a few hours, although certain areas experience them much of the time. Temperature inversions occur often in Los Angeles in the fall and Denver in the winter (Figure 5.12). If an inversion lingers long enough, say, several days, it can contribute to the accumulation of air pollutants to levels that seriously affect human health.

The air pollutants generated in one place may have their most serious effect in areas hundreds of miles away. Thus the worst effects of the air pollution that originates in New York City are felt in Connecticut and parts of Massachusetts. The chemical reaction that produces smog takes a few hours, and by that time air currents have carried the pollutants away from New York. In a similar fashion, New York is the recipient of pollutants produced in other places. Much of the acid rain that affects New England and eastern Canada originates in the coal-fired power plants along the lower Great Lakes and in the Ohio Valley that use extremely high smokestacks to disperse sulfurous emissions. And the coal-based industries in the Soviet Union and Europe produce sulfate, carbon, and other pollutants that are transported by air currents to the land north of the Arctic Circle, where they result in a contamination known as Arctic haze.

Other factors that affect the type and degree of air pollution at a given place are the levels of urbanization and industrialization. Population densities, traffic densities, the type and density of industries, and home heating practices all help to determine the kinds of substances discharged into the air at a single point. In general, the more urbanized and industrialized a place is, the more responsible it is for pollution. The United States may contribute as much as one-third of the world's air pollution, a figure roughly equivalent to the proportion of the world's fossil fuel and mineral resources consumed in this country.

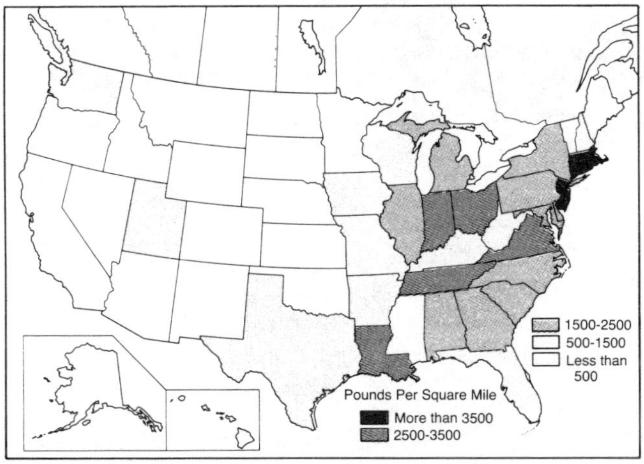

Figure 5.11

Millions of pounds of toxic industrial pollutants released into the air in each state in 1988. The EPA survey on which this map is based did not include pollution from automobiles, releases from toxic waste dumps, or pollution from companies that emit less than 75,000 pounds of toxic substances per year.

Figure 5.12

The brown cloud that hovers over Denver when temperature inversions keep air pollutants from dispersing. Denver frequently has the worst carbon monoxide level in the country from mid-November to mid-January, when cold winter air over the city is trapped by warm, still air at higher altitudes. The city has recently embarked on a campaign to reduce air pollution.

The sources of pollution are so many and varied that we cannot begin to discuss them all in this chapter. Instead, we will discuss three types of air pollution and their associated effects.

Acid Rain

Although acid *precipitation* is a more precise description, **acid rain** is the term generally used for pollutants, chiefly oxides of sulfur and nitrogen, that are created by burning fossil fuels and that change chemically as they are transported through the atmosphere and fall back to earth as rain, snow, fog, or dust. When sulfur dioxide is absorbed into water vapor in the atmosphere, it becomes sulfuric acid. Sulfur dioxide contributes about two-thirds of the acid in the rain. About one-third comes from nitrogen oxides, transformed into nitric acid in the atmosphere.

When washed out of the air by rain, snow, or fog, the acids change the *pH factor* (the measure of acidity/alkalinity on a scale of 0 to 14) of soil and water, setting off a chain of chemical and biological reactions (Figure 5.13). The average pH of normal rainfall is 5.6, slightly acidic, but acid rainfalls with a pH of 2.4—approximately the acidity of vinegar and lemon juice—have been recorded. It is important to note that the pH scale is logarithmic, so that 4.0 is ten times more acidic than 5.0, and *100* times more acidic than 6.0. Acid rain also coats the ground with particles of aluminum and toxic heavy metals, such as cadmium and lead.

Once the pollutants are airborne, winds can carry them hundreds of miles, depositing them far from their source. Approximately half of the acid rain that falls on eastern Canada originates in the United States, chiefly from coal-burning power plants in the Midwest (Figure 5.14). This issue has become a sore point in U.S.-Canadian relations because the effects of acid precipitation on water bodies, forests, and wildlife are deadly. The problem is that the acidity of a lake or stream need not increase much before it begins to disrupt the food chain.

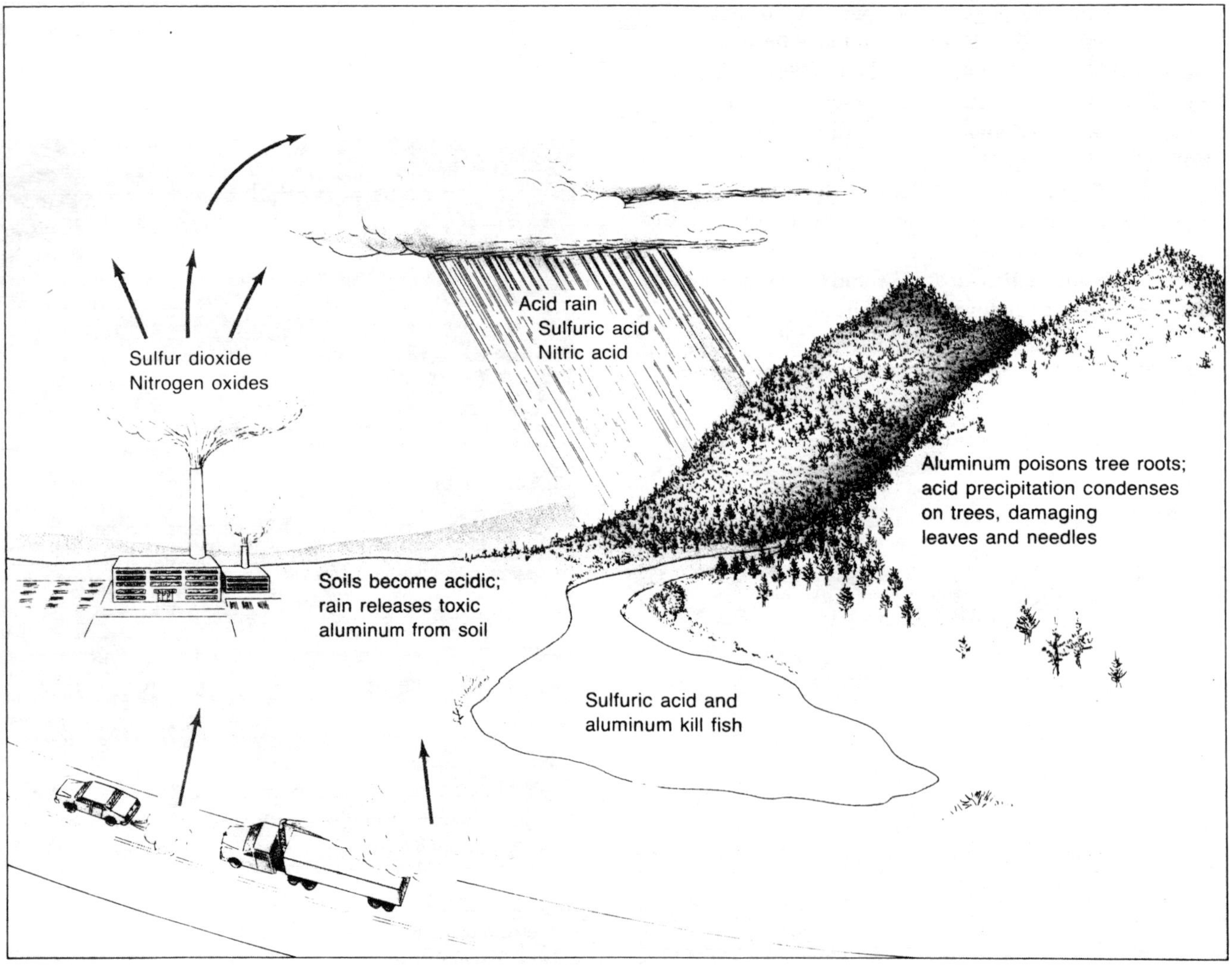

Figure 5.13

The formation and effects of acid rain. The acids in the precipitation damage soil, vegetation, and water.

Canada's Department of the Environment reports that acid rain has already damaged some 14,000 lakes in that country, rendering them almost fishless, and another 150,000 are in peril. Acid rains have also been linked to the disappearance or decline of fish populations in Scandinavia. Some 20% of Sweden's lakes are said to have been damaged by the pollution, and much of Norway's fish population has been exterminated.

While acid rain has for some years been recognized as a threat to freshwater organisms, recent evidence suggests that it may also damage marine life in coastal salt water. In this case the damage stems not from acidity but from the nitrogen in the rain, which causes eutrophication (discussed earlier in this chapter). Nitrogen is a nutrient that stimulates the excessive growth of algae. The algae in turn use up the supply of dissolved oxygen and reduce the amount of sunlight that penetrates the surface of the water.

As Figure 5.15 indicates, acid rain also affects soils and vegetation. It leaches toxic constituents like aluminum salts from the soil and kills microorganisms in the soil that break down organic matter and recycle nutrients through the ecosystem. Extensive forest damage has occurred in parts of North America, northern and western Europe, the USSR, and China. One can also see the corrosive effects of atmospheric acid on marble and limestone sculptures and buildings and on metals, such as iron and bronze.

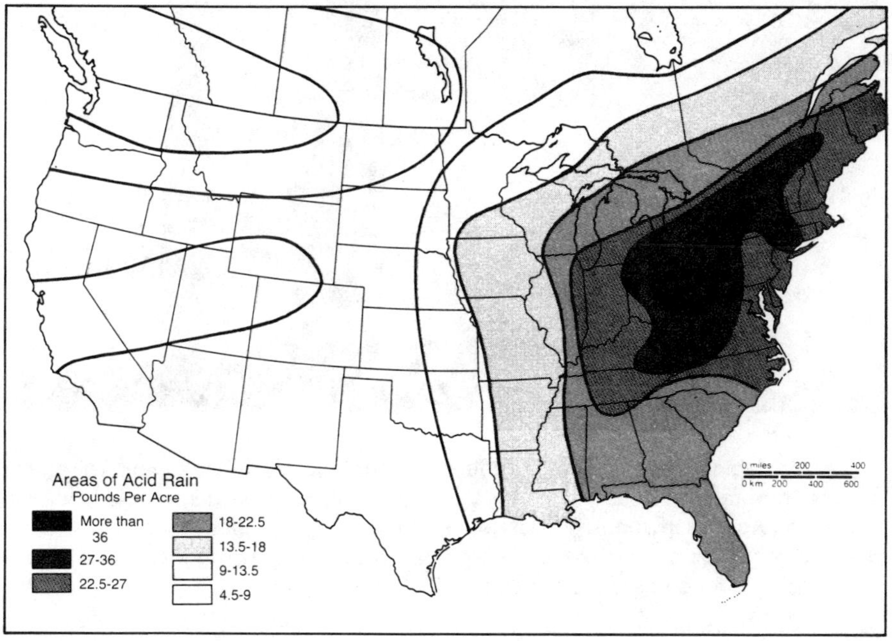

Figure 5.14

Where acid rain falls. In general, the areas that receive the most acid rain in the eastern United States and Canada are those that are least able to tolerate it. Their surface waters tend to be acidic rather than alkaline and are unable to neutralize the acids deposited by rain or snow.

Figure 5.15

Dead trees on Mount Mitchell, North Carolina. A combination of acid rain and ozone pollution has caused extensive damage to forests along the crest of the Appalachians from Maine to Georgia. The pollution weakens trees to the point where they cannot survive such natural stresses as temperature extremes, high winds, drought, or insects. In addition the high levels of lead and other heavy metals in the soil make it difficult for the forests to regenerate.

The Tall Smokestack Paradox

The dramatic increase in acid precipitation in recent decades is partly the result of an effort to curb air pollution. Written before concern about acid rain became widespread, the U.S. Clean Air Act of 1970 restricts the deposit of specific pollutants over the surrounding countryside and sets standards only for ground-level air quality.

Most of the millions of tons of sulfur and nitrogen oxides that are released into the atmosphere each year come from "stationary source" fuel combustion, primarily coal- and oil-burning power plants. In order to keep air in local communities clean enough to meet the air-quality standards, industries have since the early 1970s been building ever-taller smokestacks that discharge sulfur dioxide and other pollutants into the upper atmosphere. Stacks 1000 feet (305 m) high are now a common sight at utility plants and factories; previously, stacks 200–300 feet (60–90 m) high were the norm. The situation is not unlike disposing of garbage by throwing it over your backyard fence. It still comes down, but not in your yard. Ironically, the farther and higher the noxious emissions go, the longer they have to combine with other atmospheric components and moisture and form acids; thus, the taller stacks have directly aggravated the acid rain problem.

Recognizing the problem that the stacks created, the Environmental Protection Agency in June 1985 issued rules discouraging the use of tall smokestacks to disperse emissions. Plants now must seek ways to reduce rather than simply disperse emissions, for example, installing scrubbers on their stacks or using low-sulfur coal.

Ozone Pollution

While sulfur oxides are the chief cause of acid rain, oxides of nitrogen are responsible for the formation of *photochemical smog.* It is created when nitrogen oxides react with the oxygen present in water vapor in the air to form nitrogen dioxide. In the presence of sunlight, nitrogen dioxide reacts with hydrocarbons from automobile exhausts and industry to form new compounds, such as **ozone**. Warm, dry weather and poor air circulation promote ozone formation. The hotter and sunnier the weather, the more ozone and smog are created. In general, therefore, more ozone is produced during the summer months than during the rest of the year.

As many as 76 million Americans, or about one-third of the population, live in areas where ozone pollution exceeds the limit established by the federal Clean Air Act, 0.12 parts of ozone per million parts of air. The climate and topography of California are particularly conducive to ozone pollution. Its valleys are encircled by mountains that help hold air pollutants in the basins. When temperature inversions occur, the pollutants are effectively trapped, unable to escape to the stratosphere. Ozone levels in Los Angeles exceed the acceptable level nearly half the days in the year, and sometimes reach three times the acceptable level. Other metropolitan areas subject to ozone pollution are shown in Figure 5.16.

Ozone pollution is associated with respiratory problems. It has long been known to aggravate coughing and breathing problems for asthmatics, but recent studies indicate that exposure to ozone over a period of 6–8 hours also causes respiratory problems in healthy people. Chronic exposure to smog causes permanent damage to lungs, aging them prematurely, and is believed to increase the incidence of such respiratory ailments as pneumonia and emphysema. Because children have smaller breathing passages and less developed immune systems, they are especially susceptible to damage from the polluted air. In

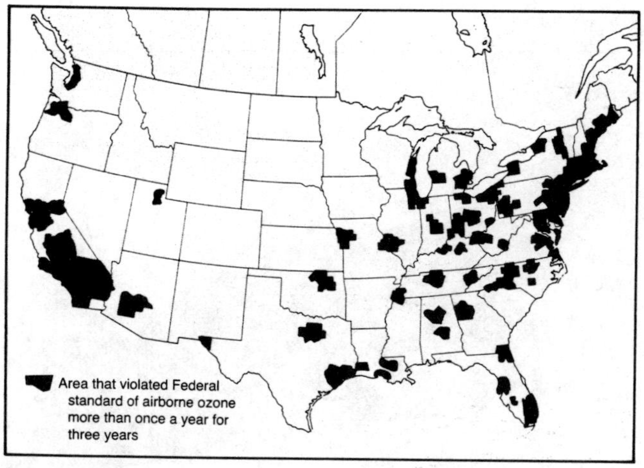

Figure 5.16
Urban areas where ozone levels violated the federal standard more than once a year for three years. By far the worst ozone pollution in the country occurs in the large Los Angeles basin. In 1988, it exceeded the federal ozone standard nearly half the year, 172 days.

addition to its effects on humans, ozone harms forests and agricultural crops (refer to Figure 5.15). Plants grown under smoggy conditions yield less and mature more slowly than normal.

Depletion of the Ozone Layer

Ozone, the same chemical that is a noxious pollutant near the ground, is essential in the stratosphere. There, some 6–15 miles above the ground, ozone forms a protective blanket, the **ozone layer**, which shields all forms of life on earth from overexposure to lethal ultraviolet (UV) radiation from the sun. Mounting evidence indicates that emissions from a variety of chemicals are destroying the ozone layer. Most important are a family of synthetic chemicals developed in 1931 known as **chlorofluorocarbons (CFCs)**. Commercially valuable, CFCs are used as coolants for refrigerators and air conditioners, as aerosol spray propellants, and as a component in foam packaging materials. Also implicated are *halons*, used in fire extinguishers.

After the gases are released, they rise through the lower atmosphere and, after a period of 7–15 years, reach the stratosphere (Figure 5.17). There UV light breaks their chemical bonds and releases chlorine and bromine atoms. Over time, a single one of these atoms can destroy tens of thousands (if not a potentially infinite number) of ozone molecules.

Every Southern Hemisphere spring, beginning around September, the atmosphere over the Antarctic loses more and more ozone. In 1987, researchers discovered what is popularly termed a "hole" as big as the continental United States in the ozone layer over Antarctica, extending northward as far as populated areas of South America (Figure 5.18). It is not truly a hole, but depletion in the ozone layer is as much as 60%. The hole disappears in the summer, when the winds change and the ozone-deficient air mixes with the surrounding atmosphere. A less dramatic but still serious depletion of the ozone shield occurs over the North Pole.

A depleted ozone layer allows more UV radiation to reach the earth's surface. Although the exact consequences of that increase won't be known for years, it is almost certain to cause a dramatic rise in the incidence of skin cancers. Some fear it may also damage the immune system of humans and other animals by impairing the cells that fight viral infections and parasitic disease. Because UV radiation also causes cell and tissue damage in plants, it is likely to have an adverse effect on crops. The most serious damage may occur in oceans. Increased amounts of UV affect the photosynthesis and metabolism of the microscopic plants called phytoplankton that flourish just below the surface of the Antarctic Ocean and are the base of the marine food chain, as well as playing a central role in the earth's CO_2 cycle.

Although 80 countries have indicated intent to ban production of CFCs by the end of the century, the destructive effects will continue for decades. The two most widely used forms of CFCs stay in the stratosphere breaking down ozone molecules for up to 100 years. Thus, even if use of the chemicals were to stop tomorrow, it would take the planet a century to replenish the ozone already lost.

In addition to their effect on the ozone layer, CFCs may have an impact on climate by contributing to the "greenhouse effect," the subject we explore next.

The Greenhouse Effect

In recent years, researchers have begun to realize that air pollution has an effect on the major factors that control the temperature of the earth's surface. Scientists have issued numerous warnings of a greenhouse effect that could cause earth temperatures to rise with significant impacts on earth's ecosystems. The theory of the **greenhouse effect** is that gases released by combustion concentrate in the atmosphere, where they function as an insulating barrier, absorbing infrared radiation that would otherwise be reflected back into the upper atmosphere. In other words, like glass in a greenhouse, the gases admit incoming solar radiation but retard its reradiation back into space (Figure 5.19). This causes a gradual warming of both the earth's surface and the lower atmosphere.

Carbon dioxide (CO_2), created mostly by burning fossil fuels, is the most plentiful of the gases and is thought to be responsible for about half of the warming. Equally important, taken together, is the accumulation of three other types of gases: (1) methane, from agriculture, swamps, and waste dumps; (2) nitrous oxides, from motor vehicles, industry, and chemical fertilizers; and (3) chlorofluorocarbons and halons, widely used industrial chemicals. Although these gases may be present in small amounts, some of them trap heat thousands of times more

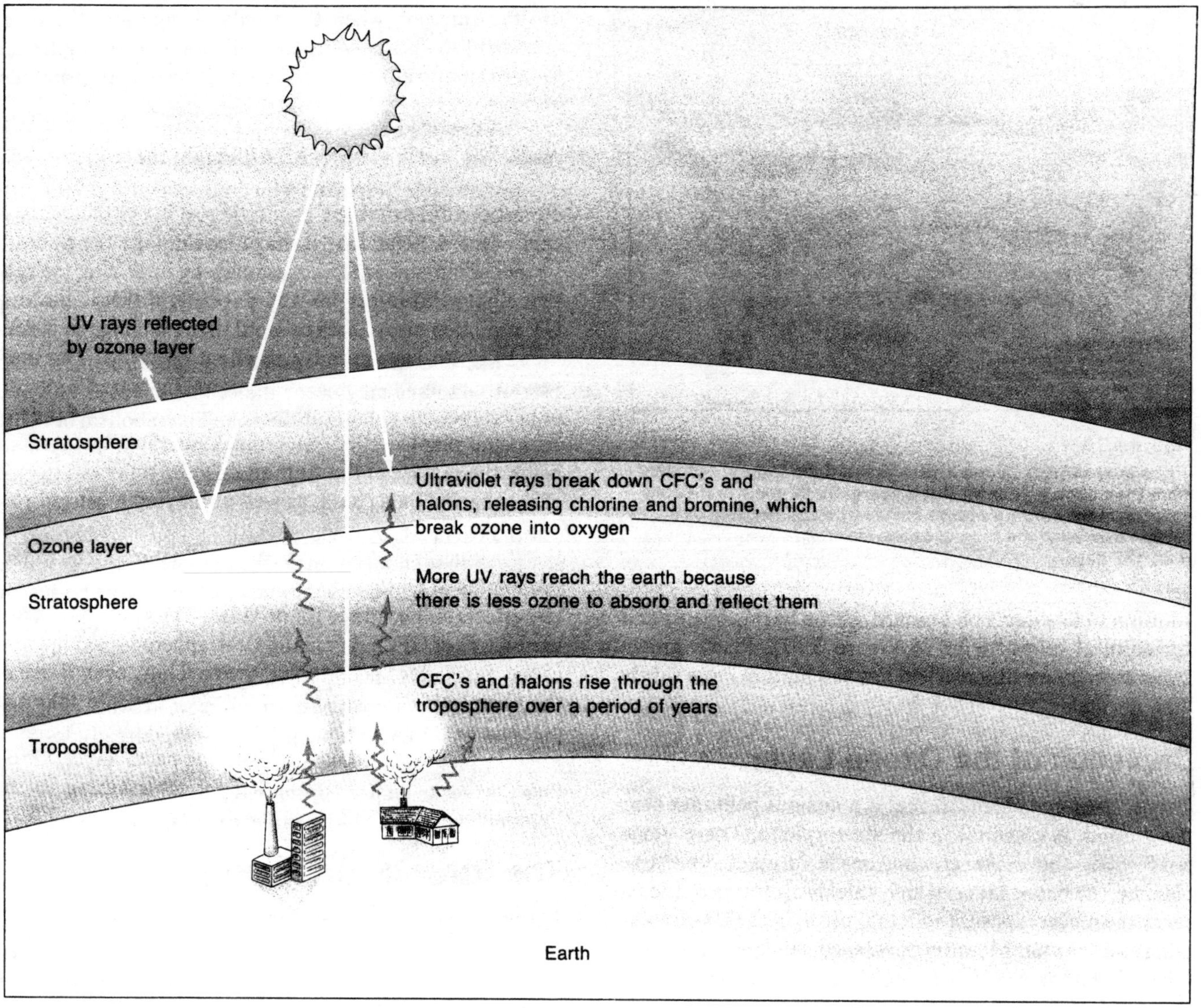

Figure 5.17

How ozone is lost. CFCs and halons released into the air rise through the troposphere without breaking down (as most pollutants do) and eventually enter the stratosphere. Once they reach the ozone layer, ultraviolet rays break them down, releasing chlorine (from CFCs) and bromine (from halons). These elements in turn disrupt ozone molecules, breaking them up into molecular oxygen and thus depleting the ozone layer.

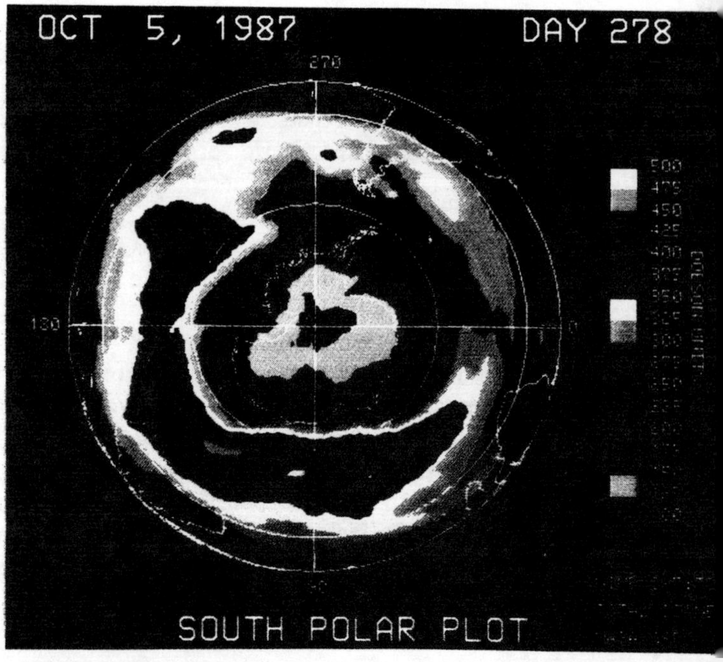

Figure 5.18

Ozone levels over the Southern Hemisphere in October 1987. The black and lavender colors over Antarctica show the areas of greatest depletion.

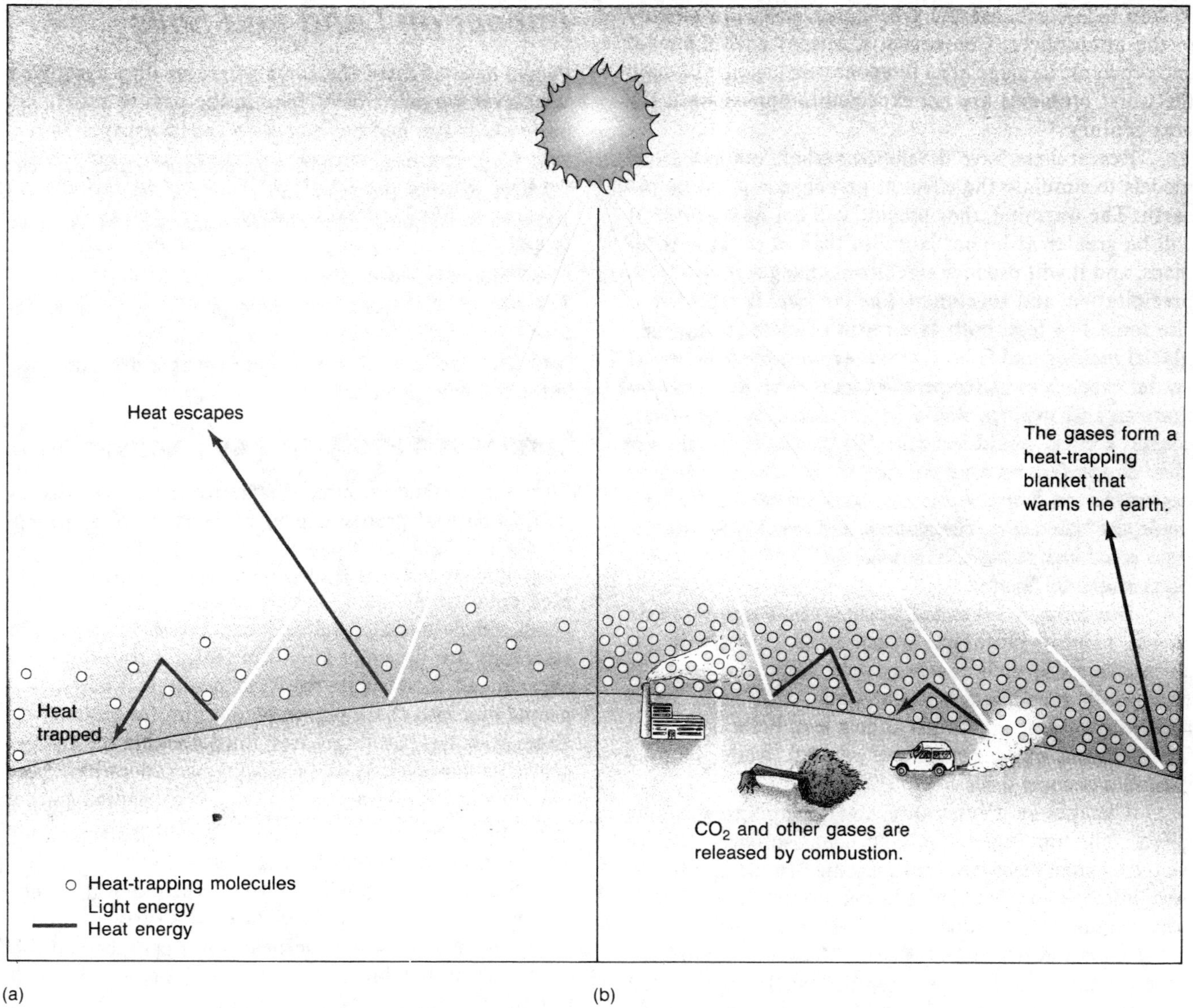

Figure 5.19

How the greenhouse effect works. When there is a low level of CO_2 and other gases in the air, as in (a), incoming solar radiation strikes the earth's surface, and the earth radiates the energy back into space as infrared light (heat). The greenhouse effect, depicted in (b), is the result of the billions of tons of CO_2 and other gases that are released into the air each year. They form a blanket that deflects the energy downward, preventing it from escaping into the atmosphere.

effectively than does CO_2. Fluorocarbon 12, for example, has 20,000 times the capacity of CO_2 to trap heat, and fluorocarbon 11 has 17,500 times the capacity of CO_2.

As the Industrial Revolution gained momentum in Europe and North America during the 19th century, the concentration of CO_2 in the atmosphere rose from its preindustrial level of about 274 parts per million (ppm) to over 340 ppm. (Just since 1958, concentrations of CO_2 have increased from 315 ppm.) Each year, we send about 5.5 billion tons of CO_2 into the atmosphere, only half of which is absorbed by the oceans and forests. Burning of the tropical rain forests not only adds to the emissions but also means the loss of trees that naturally absorb CO_2. The methane concentration in the lower atmosphere has already more than doubled from its preindustrial level and is currently increasing by just over 1% per year. The carbon monoxide concentration also seems to be increasing at a rate of slightly over 1% per year.

When CO_2 concentrations reach about 550 ppm (double pre-Industrial Revolution levels), average annual global temperatures are expected to rise by 4–9°F (2–5°C). Predictions vary as to when this doubling will occur. The year 2050 is commonly cited, but because gases other than CO_2 are contributing to the greenhouse effect, the warming may occur as early as 2030.

Proponents of the greenhouse effect theory believe that human activity has already put enough of the various gases into the atmosphere that there is no way to avoid a significant rise in temperature in the next century. That is, some warming is inevitable even if all emissions were to stop today, because the greenhouse gases are already in the atmosphere. Consequently, serious environmental and economic damage are a foregone conclusion, although the worst problems are not expected to appear until the next century.

Researchers have developed various mathematical models to simulate the effect of greenhouse gases on the earth. The warming, they predict, will not be uniform. It will be greater at higher latitudes than in equatorial regions, and it will produce significant changes in sea level, precipitation, and vegetation. The sea level is expected to rise some 1–4 feet, both as a result of some ice-cap and glacial melting and from *thermal expansion* of the water (water expands as its temperature increases). Most coastal marshes and swamps would be inundated by salt water; coastal erosion would increase. Water quality would decline as aquifers became polluted by salt. Such low-lying regions as the North American Gulf Coast, the Netherlands, the Nile Delta, Bangladesh, and much of Southeast Asia could lose substantial amounts of land. Many major ports might be flooded.

Warming of lakes and oceans would speed evaporation, causing more active convection currents in the atmosphere and thus fiercer storms. Important regional changes in precipitation would occur, with some areas receiving more precipitation, others less. Polar and equatorial regions might get heavier rainfall, while the mid-latitudes become drier.

Changes in temperature and precipitation would affect soils and vegetation. The composition of forests would change, as some areas became less favorable for certain species of plants, more hospitable to others. Hotter, drier weather would reduce crop yields in some areas, such as the corn and wheat belts of the Midwest. Conversely, more northerly agricultural regions, such as parts of Canada and the USSR, might become more productive.

It should be noted that not all scientists agree with these predictions. Some argue that global temperatures might stabilize or even decrease as the concentration of greenhouse gases increases. A hotter atmosphere, they say, would increase evaporation, sending up more water vapor that could condense into clouds. The increased cloud cover might reflect so much sunlight that it would slow the rate at which the earth would be heated. Others contend that the increased evaporation would produce more rainfall. As it falls, the rain would cool the land and subsequently cool the air over the land.

Finally, some researchers believe the geological record shows that large fluctuations in global temperature have always occurred independently of human activity, never as a result of it. These fluctuations are caused by such unpredictable events as variations in solar radiation, shifts in the earth's orbit and in ocean currents, meteoric activity, and volcanic eruptions.

Impact on Land and Soils

People have affected the earth wherever they have lived. Whatever we do or have done in the past to satisfy our basic needs has had an impact on the landscape. To provide food, clothing, shelter, transportation, and defense, we have cleared the land and replanted it, rechanneled waterways, and built roads, fortresses, and cities. We have mined the earth's resources, logged entire forests, terraced mountainsides, even reclaimed land from the sea. The nature of the changes made in any single area depends on what was there to begin with and how people have used the land. Some of these changes are examined in the following sections.

Landforms Produced by Excavation

Although we tend to think of landforms as "givens," created by natural processes over millions of years, people have played and continue to play a significant role in shaping local physical landscapes. Some features are created deliberately, others unknowingly or indirectly. Pits, ponds, ridges, trenches, subsidence depressions, canals, and reservoirs are the chief landform features resulting from excavations. Some date back to neolithic times, when people dug into chalk pits to obtain flint for toolmaking. Excavation has had its greatest impact within the last two centuries, however, as earth-moving operations have been undertaken for mining; for building construction and for agriculture; and for the construction of transport facilities such as railways, ship canals, and highways.

Surface mining, which involves the removal of vegetation, topsoil, and rocks from the earth's surface in order to get at the resources underneath, has perhaps had the greatest environmental impact. Open-pit mining and strip mining are the methods most commonly used.

Open-pit mining is used primarily to obtain iron and copper, sand, gravel, and stone. As Figure 5.20 indicates, an enormous pit remains after the mining has been completed because most of the material has been removed for processing. *Strip* mining is being increasingly employed in this country as a source of coal; more coal per year now comes from strip mines than from underground mines. Phosphate is also mined in this way. A trench is dug, the material is excavated, and another trench is dug, the soil and waste rock being deposited in the first trench, and so on. Unless reclamation is practiced, the result is a ridged landscape.

Landscapes marred by vast open pits or unevenly filled trenches are one of the most visible results of surface mining. Thousands of square miles of land have been affected, with the prospect of thousands more to come as the amount of surface mining increases. Damage to the aesthetic value of an area is not the only liability of surface mining. If the area is large, wildlife habitats are disrupted, and surface and subsurface drainage patterns are

(a)

(b)

Figure 5.20

(a) Aerial view of the Bingham Canyon open-pit copper mine in Utah. The pit is over 1500 feet (450 m) deep. The operations cover more than 1000 acres (405 hectares). (b) About 150 square miles (390 km^2) of land surface in the United States are lost each year to the strip mining of coal and other resources; far more is disrupted worldwide. Besides altering the topography, strip mining interrupts surface and subsurface drainage patterns, destroys vegetation, and places sterile and frequently highly acidic subsoil and rock on top of the new ground surface.

disturbed. In the United States in recent years, concern over the effect of strip mining has prompted federal and state legislation to increase regulation and to stop the worst abuses. Strip-mining companies are now expected to restore mined land to its original contours and to replant vegetation.

Landforms Produced by Dumping

Excavation in one area often leads to the creation, via dumping, of landforms nearby. Both surface and subsurface mining produce tons of waste and enormous spoil piles. In fact, in terms of tonnage, mining is the single greatest contributor to solid wastes, with about 2 billion tons per year left to be disposed of in this country alone. The normal custom is to dump waste rocks and mill tailings in huge heaps near the mine sites. Unfortunately this practice has secondary effects on the environment.

Carried by wind and water, dust from wastes pollutes the air, and dissolved minerals pollute nearby water sources. Occasionally the wastes cause greater damage, as happened in Wales in 1966, when slag heaps from the coal mines slid onto the village of Aberfan, burying over 140 schoolchildren. Such tragedies call attention to the need for less potentially destructive ways of disposing of mine wastes.

Another example of the combined effect of excavating and filling on the landscape is the agricultural terrace characteristic of parts of Asia. In order to retain water and increase the amount of arable land, terraces are cut into the slopes of hills and mountains. Low walls protect the patches of level land. The *tells* of the Middle East, pictured in Figure 5.21, are another type of landform produced by the accumulation of waste material.

Human impact on land has been particularly strong in areas where land and water meet. Dredging and filling operations undertaken for purposes of water control create landscape features like embankments and dikes. In many places, the actual shape of the shoreline has been altered, as builders in need of additional land have dumped solid wastes into landfills. In the Netherlands, millions of acres of land have been reclaimed from the sea by the building of dikes to enclose polders and canals to drain them. Farming practices in river valleys have had significant effects on deltas. For example, increased sedimentation has often extended the area of land into the sea.

Formation of Surface Depressions

The extraction of material from beneath the ground can lead to **subsidence**, the settling or sinking of a portion of the land surface. Many of the world's great cities are sinking because of the removal of *fluids* (groundwater, oil, and gas) from beneath them. Cities threatened by such subsidence are located on unconsolidated sediments (New Orleans, Bangkok), coastal marshes (Venice, Tokyo), or lake beds (Mexico City). When the fluids are removed, the sediments compact and the land surface sinks. Because many of the cities are on coasts or estuaries and are often only a few feet above sea level, subsidence makes them more vulnerable to flooding from the sea.

Groundwater abstraction has created serious subsidence in many places. Recent evidence indicates that the withdrawal of trillions of gallons of water from a 4500-square-mile (11,650-km^2) area of Arizona has resulted in widespread ground subsidence and the formation of more than a hundred earth fissures, jagged ground cracks that can be as much as 9 miles (14.5 km) long and 400 feet (120 m) deep.

The removal of *solids* (such as coal, salt, and gold) by underground mining may result in the collapse of land over the mine. *Sinkholes* or *pits* (circular, steep-walled depressions) and *sags* (larger and shallower depressions) are two types of landscape features produced by such collapse. If surface drainage patterns are disrupted, subsidence *lakes* may form in the depressions. Subsidence has become a more serious problem as towns and cities have expanded over mined-out areas.

Figure 5.21
Tell Hesi, northeast of Gaza, Israel. Here and elsewhere in the Middle East, the debris of millennia of human settlement gradually raised the level of the land surface, producing *tells*, or occupation mounds. The city literally was constantly rebuilt at higher elevations upon the accumulation of refuse of earlier occupants. In some cases, the striking landforms may rise hundreds of feet above the surrounding plains.

As one might expect, subsidence damages structures built on the land, including buildings, roads, and sewage lines. A dramatic example occurred in Los Angeles in 1963, when subsidence caused the dam at the Baldwin Hills Reservoir to crack. In less than 2 hours, the water emptied into the city, resulting in millions of dollars worth of property damage (Figure 5.22). The withdrawal of groundwater from beneath Mexico City has led to severe though differential subsidence. One of the reasons the 1985 earthquake in that city was so damaging was that subsidence had weakened building structures.

Soils

By design or by accident, people have brought about many changes in the physical, chemical, and biochemical nature of the soil and altered its structure, fertility, and drainage characteristics. The exact nature of the changes in any area depends on past practices as well as on the original nature of the land.

Over much of the earth's surface, the thin layer of topsoil upon which life depends is only a few inches deep, usually less than a foot (30 cm). Below it, the lithosphere is a complex mixture of rock particles, inorganic mineral matter, organic material, living organisms, air, and water. Under natural conditions, soil is constantly being formed by the physical and chemical decomposition of rock material and by the decay of organic matter. It is simultaneously being eroded, for **soil erosion**—the removal of soil particles, usually by wind or running water—is as natural a process as soil formation and occurs even when land is totally covered by forests or grass. Under most natural conditions, however, the rate of soil formation equals or exceeds the rate of soil erosion, so that soil depth and fertility tend to increase with time.

Figure 5.22
Subsidence caused cracking and emptying of the Baldwin Hills Reservoir of Los Angeles in December 1963.

When land is cleared and planted to crops, or when the vegetative cover is broken by overgrazing or other disturbances, the process of erosion accelerates. When its rate exceeds that of soil formation, the topsoil becomes thinner and eventually disappears, leaving behind only sterile subsoil or barren rock. At that point, the renewable soil resource has been converted through human impact into a nonrenewable and dissipated asset. Carried to the extreme of bare rock hillsides or wind-denuded plains, erosion spells the total end of agricultural use of the land.

Such massive destruction of the soil resource could endanger the survival of the civilization it has supported. For the most part, however, farmers devise ingenious ways to preserve and even improve the soil resource upon which their lives and livelihoods depend. Farming skills have not declined in recent years, but pressures upon farmlands have increased with population growth. Farming has been forced higher up onto steeper slopes, more forest land has been converted to cultivation, grazing and crops have been pushed farther and more intensively into semiarid areas, and existing fields have had to be worked more intensively and less carefully. Many traditional agricultural systems and areas that were ecologically stable and secure as recently as 1950, when world population stood at 2.5 billion, are disintegrating under the pressures of more than 5 billion.

The pressure of growing population numbers is having an especially destructive effect on tropical rain forests. Expanded demand for fuel and commercial wood, and a midlatitude market for beef that can be satisfied profitably by replacing tropical forest with cleared grazing land are responsible for some of the loss, but the major cause of *deforestation* is clearing the land for crops. Extending across parts of Asia, Africa, and Latin America, the tropical rain forests are the most biologically diverse places on earth, but some 40,000 square miles (over 100,000 km^2) are being destroyed every year. About 45% of their original expanse has already been cleared or degraded. Deforestation is discussed in more detail in Chapter 11, but it is important to note here that accelerated soil erosion quickly removes tropical forest soils from deforested areas. Lands cleared for agriculture almost immediately become unsuitable for that use partially because of soil loss (Figure 5.23).

Figure 5.23
Wholesale destruction of tropical forests guarantees environmental degradation so severe that the forest can never naturally regenerate itself. Exposed soils quickly deteriorate in structure and fertility and are easily eroded.

Figure 5.24
A drought-stricken area in Burkina Faso, one of the countries in the Sahel region of Africa where *desertification* has been accelerated by both climate and human activity. The cultivation of marginal land, overgrazing by livestock, and climatic change during the 1970s and 1980s have led to the destruction of native vegetation, to erosion, and to the expansion of the deserts of the world.

The tropical rain forests can succumb to deliberate massive human assaults and be irretrievably lost. With much less effort, and with no intent to destroy or alter the environment, humans are similarly affecting the arid and semiarid regions of the world. The process is called **desertification**, the spread of desertlike landscapes into arid and semiarid environments. Although climatic change can in some instances be a contributing cause, desertification is usually charged to increasing human pressures exerted through overgrazing, deforestation for fuel wood, clearing of original vegetation for cultivation, and burning (Figure 5.24).

Whatever the reason, the process begins in the same way: the disruption or removal of the native cover of grasses and shrubs through farming or overgrazing. If the disruption is severe enough, the original vegetation cannot reestablish itself, and the exposed soil is made susceptible to erosion during the brief, heavy rains that dominate precipitation patterns in semiarid regions. Water runs off the land surface instead of seeping in, carrying soil particles with it. When the water is lost through surface flow rather than seepage downward, the water table is lowered. Eventually, even deep-rooted bushes are unable to reach groundwater, and all natural vegetation is lost. The process is accentuated when too many grazing animals pack the earth down with their hooves, blocking the passage of air and water through the soil. When both plant cover and soil moisture are lost, desertification has occurred.

It happens with increasing frequency in many areas of the earth as pressures upon the land continue (Figure 5.25). Africa is most at risk; the United Nations has estimated that 40% of that continent's nondesert land is in danger of desertification. But nearly a third of Asia and a fifth of Latin America's land are similarly endangered. In countries where desertification is particularly extensive and severe (Algeria, Ethiopia, Iraq, Jordan, Lebanon, Mali, and Niger), per capita food production declined by some 40% between 1950 and the mid-1980s. The resulting

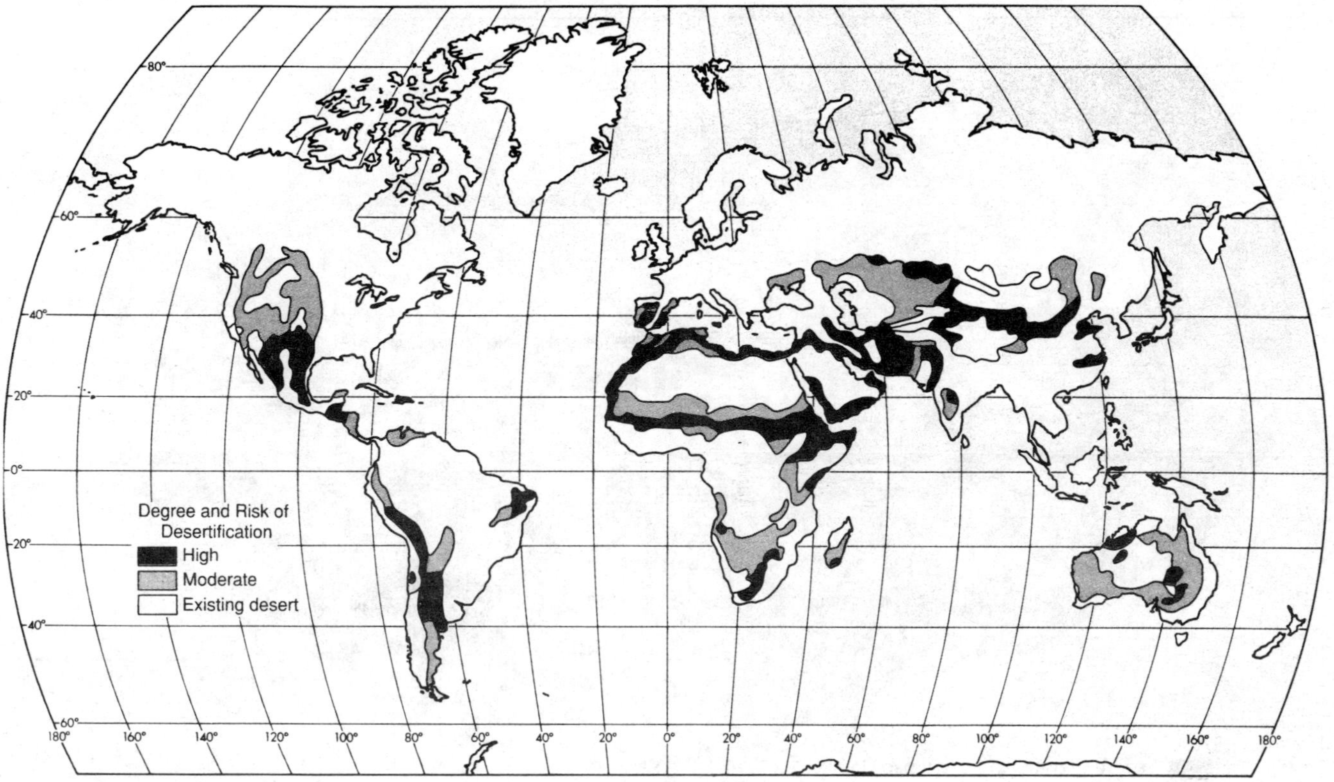

Figure 5.25

Desertification threatens about one-third of the world's land surface, some 18.5 million square miles (48 million km²). Once it occurs in its extreme form, large area desertification is probably irreversible.

threat of starvation spurs populations of the affected areas to increase their farming and livestock pressures on the denuded land, further contributing to their desertification.

Accelerated soil erosion is not limited to Africa and Asia. Indeed, in recent years, soil erosion in the United States has been at an all-time high (Figure 5.26). Wind and water are blowing and washing soil off pasturelands in the Great Plains, ranches in Texas, and farms in the Southeast. The country's croplands lose almost 2 billion tons of soil per year to erosion, an average annual loss of over 4 tons per acre. In some areas, the average is 15–20 tons per acre. Of the roughly 413 million acres (167 million hectares) of land that are intensively cropped in this country, over one-third are losing topsoil faster than it can be replaced naturally (it can take up to a century to replace an inch of topsoil). In parts of Illinois and Iowa where the topsoil was once a foot deep, less than half of it remains. Every hour about 40,000 tons of topsoil wash into the Mississippi River.

Like most processes, soil erosion has secondary effects. As the soil quality and quantity decline, croplands become less productive and yields drop. Streams and reservoirs experience accelerated siltation. In countries where the topsoil is heavily laden with agricultural chemicals, erosion-borne silt pollutes water supplies. The danger of floods increases as bottomlands fill with silt, and the costs of maintaining navigation channels grow.

Accelerated erosion is a primary cause of agricultural soil deterioration, but in arid and semiarid areas, salt accumulation is a contributing factor. In many parts of the world, irrigation has led to excessive salinity of the soil. These regions are naturally salty because evapotranspiration exceeds precipitation. Since irrigation water tends to move slowly, and thus to evaporate more rapidly, the salts carried in the water are absorbed into the ground. Over time, increased salinity makes the soil less productive. Thousands of once-fertile acres have been abandoned in Iran and Iraq; over 25% of the irrigated areas of Pakistan, Syria, and Egypt are affected by such salting up, or **salinization.** In the United States, the 250-mile-long (400 km) San Joaquin Valley of California is experiencing a similar buildup of harmful salty residues. Because the valley is a major supplier of fruits and vegetables to the rest of the country, the conversion of over 1 million acres (405,000 hectares) to a barren salt flat would have an impact well beyond California's borders.

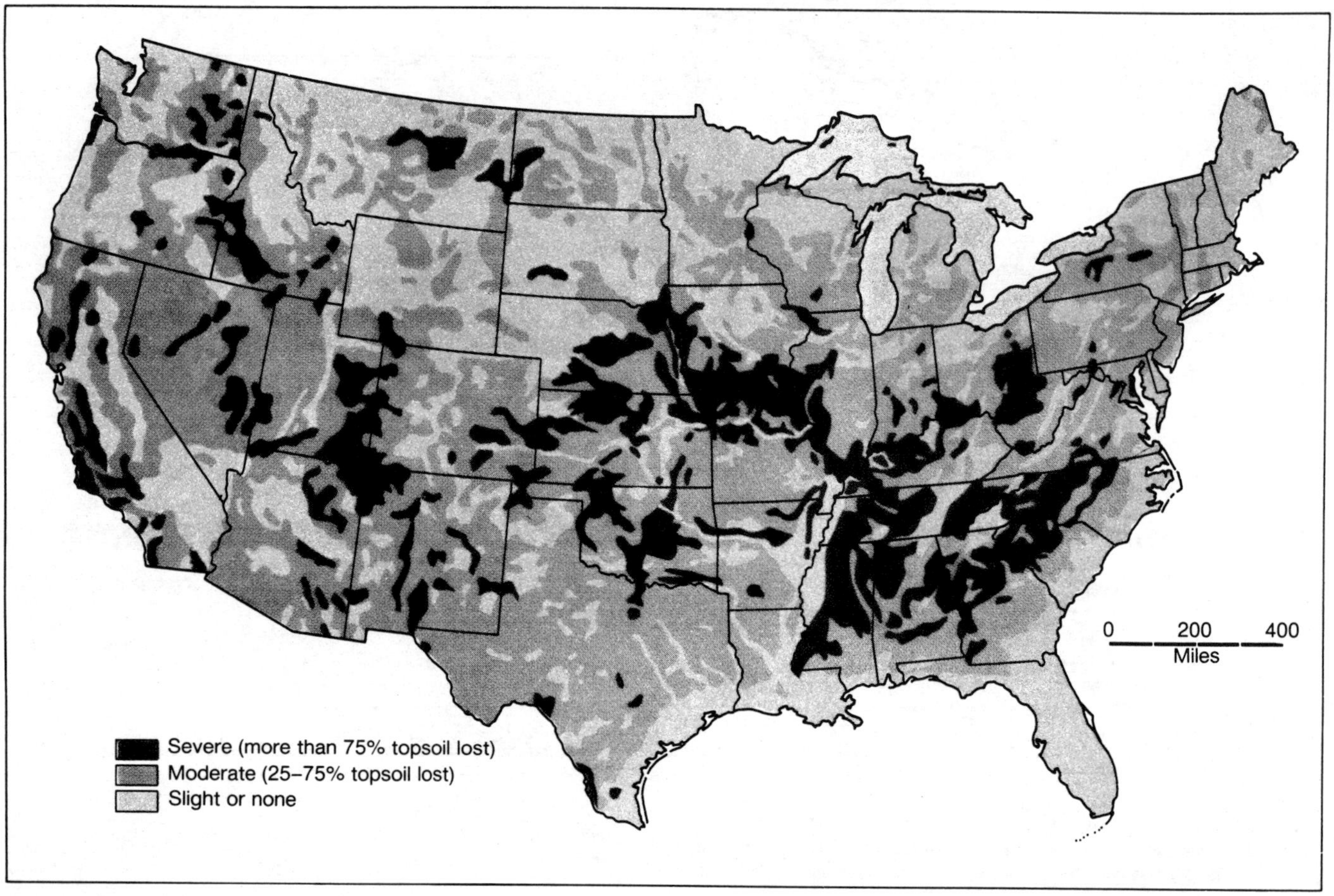

Figure 5.26

Soil erosion in the United States. Although many activities (mining, construction, and urbanization) contribute to erosion, agriculture and deforestation are particularly significant. In recent years, erosion has been most severe not in the southwestern Dust Bowl but in the moist, rolling-hill regions of western Mississippi, western Tennessee, and Missouri. Each of these areas loses about 10 tons of topsoil per acre of cropland annually. However, soil resource stress and depletion affect all parts of the country.

Impact on Plants and Animals

People have affected plant and animal life on the earth in several different ways. When human impact is severe enough, a species becomes *extinct*, that is, it no longer exists. Although fossil records show that extinction is a natural feature of life on earth, scientists estimate that in recent history the rate of extinction has increased a million times—an increase due to human activity. Species that are in danger of becoming extinct are called *endangered*, while a *threatened* species is one that is likely to become endangered within the foreseeable future. In the United States, about 500 species of plants and animals are on the list of endangered and threatened species that is maintained by the federal government and updated yearly; another 4000 species are candidates for the list.

In this section we examine some of the ways people have modified plant and animal life.

Introduction of New Species

The deliberate or inadvertent introduction of a plant or animal into an area where it did not previously exist can have damaging and unforeseen consequences. Introduced species have often left behind their natural enemies—predators and diseases—giving them an advantage over native species that are held in check by biological controls. The rabbit, for example, was purposely introduced into Australia in 1859. The original dozen pairs multiplied to a population in the thousands in only a few years and, despite programs of control, to an estimated 1 billion a hundred years later (Figure 5.27). Inasmuch as five rabbits eat about as much as one sheep, a national problem had been created. Rabbits had denuded much of the grassland on which sheep could graze.

Plants and animals found on islands are particularly prone to extinction. Because island species often occur on just one or a few islands, the loss of only a few individuals can be devastating to small populations. Goats, used as

Maintaining Soil Productivity

In much of the world, increasing population numbers are largely responsible for accelerated soil erosion. In the United States in the 1970s and 1980s, economic conditions contributed to a rate of soil erosion equal to that registered during the Dust Bowl era of the 1930s. Federal tax laws and the high farmland values of the 1970s encouraged farmers to plow virgin grasslands and to tear down windbreaks to increase their cultivable land and yields. The secretary of agriculture exhorted farmers to plant all of their land, "from fencerow to fencerow," to produce more grain for export. Land was converted from cattle grazing to corn and soybean production as livestock prices declined.

When prices of both land and agricultural products declined in the 1980s, farmers felt impelled to produce as much as they could in order to meet their debts and make any profit at all. To maintain or increase their productivity, many neglected conservation practices, plowing under marginal lands and using fields for the same crops every year.

Conservation techniques were not forgotten, of course. They were both practiced by many and persistently advocated by farm organizations and soil conservation groups. Techniques to reduce erosion by holding the soils in place are well-known. They include contour plowing, terracing, strip-cropping and crop rotation, erecting windbreaks and constructing water diversion channels, and practicing no-till farming (allowing crop residue such as cut corn stalks to remain on the soil surface throughout the winter). In addition, farmers can be paid to idle marginal, highly erodible land. Only by employing such practices can the country maintain the long-term productivity of soil, the resource base upon which all depend.

No-till farming

Strip-cropping

food for whaling crews, were imported to Guadalupe Island off the coast of Baja, California, in the 19th century. Without predators to keep their numbers low, the goats reproduced unchecked, with a devastating effect on the indigenous vegetation.

Introduced plants as well as animals can alter vegetative patterns. The Asiatic chestnut blight, for instance, has destroyed most of the native American chestnut trees in the United States, trees with significant commercial as well as aesthetic value. The cause was the importation of

Figure 5.27
Rabbits converging on a water hole in Australia during a drought. Deliberately imported to Australia, the rabbit became an economic burden and environmental menace, competing with sheep for grazing land and stimulating soil erosion.

chestnut trees from China to the United States. They carried a fungus fatal to the American chestnut tree but not to the Asiatic variety, which is largely immune to it.

The water hyacinth entered North America in 1884, when bulbs were given as souvenirs at an exposition in New Orleans. Before the end of the century, the spread of the plant had become a matter of concern. In one growing season, a single plant can produce over 60,000 offshoots. To the consternation of those who sail and fish, swamps, lakes, and canals in the southern United States have become clogged with hyacinths, which also affect fish and plankton.

These are just a few of the many examples that illustrate an often-ignored ecological truth: plant and animal life are so interrelated that when people introduce a new species to a region, whether by choice or by chance, there may be unforeseen and far-reaching consequences.

Habitat Disruption

One of the main causes of extinction has been the loss or alteration of habitats for wildlife. By clearing forest land, draining wetlands, extending farmland, and building cities, people modify or destroy the habitats in which plants and animals have lived. Tidal marshes have been subjected to dredging and filling for residential and industrial development. The loss of such areas reduces the essential habitat of fish, crustaceans, and mollusks. The whooping crane has been virtually eliminated in the United States because the marshes where it nested were drained, and roads and canals brought intruders into its habitat. Its comeback, sought by breeding programs in the United States and Canada, is still uncertain.

Many people fear that as countries in Africa and South America become more industrialized and more urbanized and expand their areas under cultivation, there will be an increasingly negative impact on wildlife. It is already known that in Africa, wild animals are vanishing fast, in part the victims of habitat destruction. In Botswana, for example, 250,000 antelope and zebra died in a decade, disoriented by fences erected to protect cattle. As selected animal species decline in numbers, balances among species are upset and entire ecosystems are disrupted.

The destruction of the world's rain forests, the most biologically diverse places on earth, is by some estimates causing the extinction of about 1000 plant and animal species every year. For every plant that becomes extinct, about 20 animals that depend on the plant also vanish.

Hunting and Commercial Exploitation

Another way in which people have affected plants and animals is through deliberate destruction. We have overhunted and overfished, for food, fur, hides, jewelry, and trophies. In the past, unregulated hunting harmed wildlife all over the world and was responsible for the destruction of many populations and species. Beaver, sea otter, alligator, and buffalo are among the species brought to the edge of extinction in the United States by thoughtless exploitation. Under protective legislation, their populations are now increasing, but hunting in developing countries still poses a threat to a number of species.

Three African animals whose existence is threatened by hunting, most of it illegal, are the elephant, rhinoceros, and mountain gorilla. Prized for its ivory tusks, the African elephant has been ruthlessly slaughtered. Ten million elephants are estimated to have been alive in the 1930s. By 1979 the population had dropped to 1.5 million, and a decade later, to only one-half million (Figure 5.28). The rhinoceros, killed for its horn, is now an endangered species. Worldwide, the population has declined from 70,000 in 1970 to about 10,000 today. Only 400 mountain gorillas are believed to exist, most of them in the Virunga Mountains of Uganda and Rwanda.

Figure 5.28

Elephant killed by poachers in Kenya. Although habitat alteration has contributed to the decline of the elephant population, the greatest threat comes from illegal poaching. By their voracious consumption of vegetation, elephants help shape ecosystems in both savanna woodlands and rain forests. As elephants disappear, so will many species that share its habitat, including zebras, gazelles, and giraffes.

Poisoning and Contamination

Humans have also affected plant and animal life by poisoning or contamination. In the last several years, we have become acutely conscious of the effect of insecticides, rodenticides, and herbicides, known collectively as *biocides*. The best known and most widely used has been DDT, although there are now thousands of compounds in use.

DDT was first used during World War II to kill insects that carried diseases, such as malaria and yellow fever. In the years following the war, tons of DDT and other biocides were used, sometimes to combat disease, sometimes to increase agricultural yields. Insects, after all, can destroy a significant percentage of a given crop, either when it is in the field or after it has been harvested.

In the last ten years, some of the side effects of these biocides have been well enough documented for us to question their indiscriminate use. Once used, a biocide settles into the soil, where it may remain or may be washed into a body of water. In either case, it is absorbed by organisms living in the soil or the mud. Through a process known as **biological magnification**, the biocide accumulates and is concentrated at progressively higher levels in the food chain (Figure 5.29). By remaining as residue in fatty tissues, a very small amount of a biocide produces unexpected effects, with predators accumulating larger amounts than their prey. The higher the level of an organism in the food chain, the greater the concentration of DDT will be, and that concentration may be lethal. Robins and other small birds die when they eat earthworms that have ingested DDT that has settled into the earth after being sprayed on trees.

DDT also causes a decrease in the thickness of the eggshells of some of the larger birds, causing a greater number of eggs to break than normally would. Peregrine falcons, bald eagles, and brown pelicans were among the birds nearly made extinct by this disruption of the reproductive process.

Although the use of DDT has declined as its effects have become apparent, other chlorinated hydrocarbon compounds have been developed and are in wide use. Indeed, the use of pesticides in the United States has more than doubled in the last 25 years. More than 2 billion pounds (900 million kg) containing more than 600 active ingredients are applied each year. The pesticides pollute water supplies, often contaminate the crops they are meant to protect, and sometimes sicken the farm workers who apply them. In addition, they too often are only temporarily effective.

Biocides may in fact exacerbate the problem their use is designed to eradicate. By altering the natural processes that determine which insects in a population will

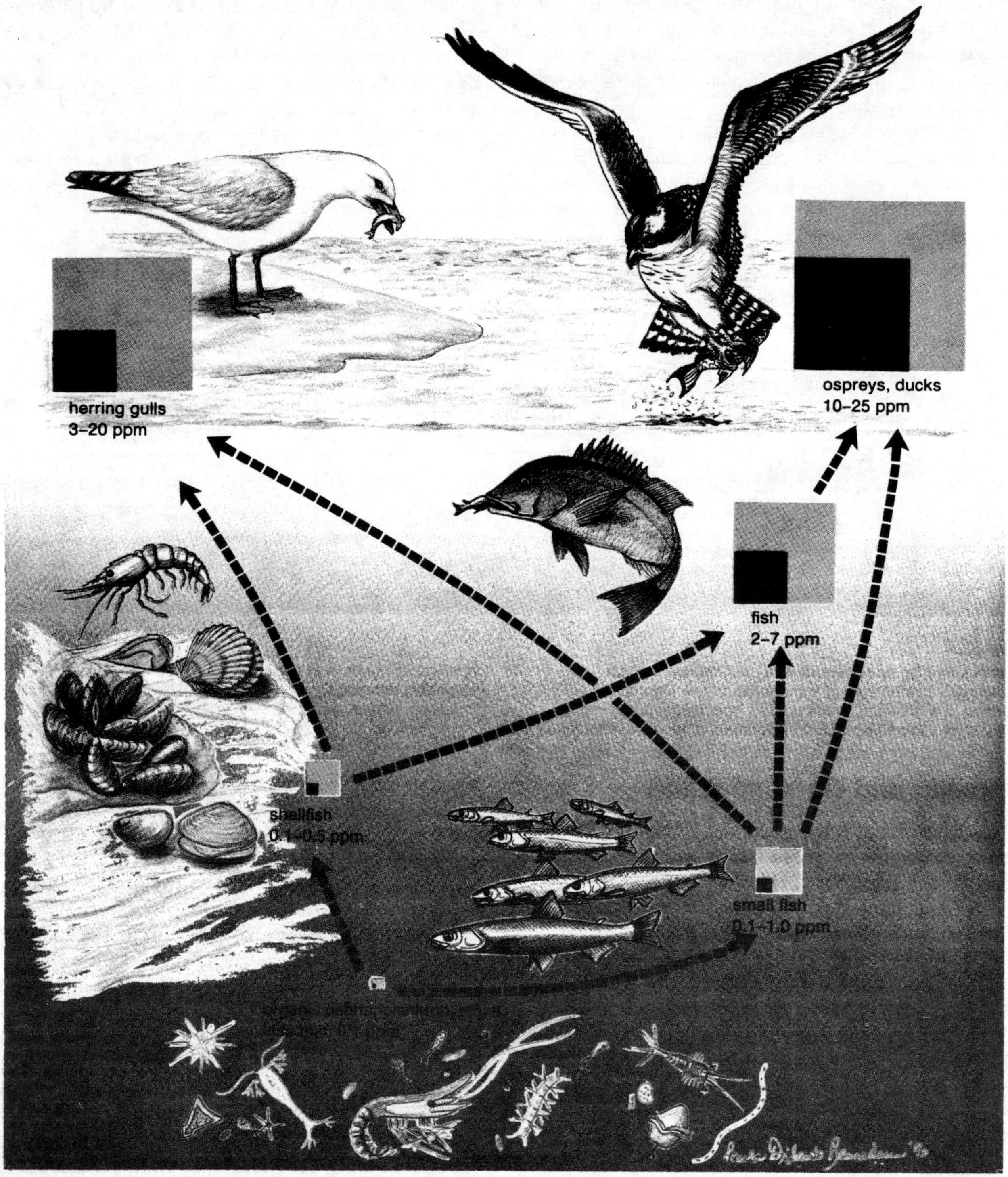

Figure 5.29

A simplified example of biological magnification. Although the level of DDT in the water and mud may be low, the impact on organisms at the top of the food chain can be significant. In this example, birds at the top of the chain have concentrations of residues as much as 250 times greater than the concentration in the water. PCBs and radioisotopes such as strontium-90 and cesium-137 undergo magnification in the food chain just as insecticides do.

survive, biocides spur the development of resistant species. If all but 5% of the mosquito population in an area are killed by an insecticide, the ones that survive are the most resistant individuals, and they are the ones that will produce the succeeding generations.

There are now insects whose total resistance to certain pesticides has led some scientists to conclude that the entire process of insecticide development may be self-defeating. Despite the enormous growth in the use of pesticides, crop loss to insect and weed pests has actually grown. According to Department of Agriculture figures, 32% of crops were lost to pests in 1945; 40 years later, such losses had increased to 37%.

Pesticides have also been known to increase problems by destroying the natural enemies of the intended target, leaving it to breed unchallenged (Figure 5.30). Examples include the tobacco budworm and the brown planthopper, which were relatively minor pests before intensive crop spraying destroyed rival pests. The budworm severely cut cotton production in Mexico and Texas in the 1960s. The brown planthopper continues to affect rice growing in Asia.

Solid-Waste Disposal

Modern technologies and the societies that have developed them produce enormous amounts of solid wastes that must be disposed of. The rubbish heaps of past cultures suggest that humankind has always been faced with the problem of ridding itself of materials it no longer needs. The problem for advanced societies, with their ever-greater variety, amount, and durability of refuse, is even more serious: how to dispose of the solid wastes produced by residential, commercial, and industrial processes. Although these account for much less tonnage than the wastes produced by mining or agriculture, they are everywhere, a problem with which each individual and each municipality must deal.

Municipal Waste

The wastes that communities must somehow dispose of include newspapers and beer cans, toothpaste tubes and old television sets, broken refrigerators and rusted cars. American communities are now facing twin crises in disposing of these wastes: the sheer volume of trash and the toxic nature of much of it. Solid-waste disposal is a greater problem in the United States than in any other country, for we throw away more trash per person than any country in the world. As Figure 5.31 indicates, the amount of household waste has doubled in the last 30 years, and currently stands at some 160 million tons per year, or about 1300 pounds (600 kg) per person. Solid-waste disposal costs are now the second largest expenditure of most local governments, ranking just behind education. Americans generate more than twice as much waste per person as do Japanese and Europeans, four times as much as Pakistanis or Indonesians.

Our volume of trash is the result of three factors—affluence, packaging, and open space. Craving convenience, Americans rely on disposable goods that they throw away after very limited use. Thus, although readily available substitutes are more economical, we annually throw out 16 billion baby diapers, 2 billion razors, 1.6 billion pens, and a million tons of paper towels and napkins. People in

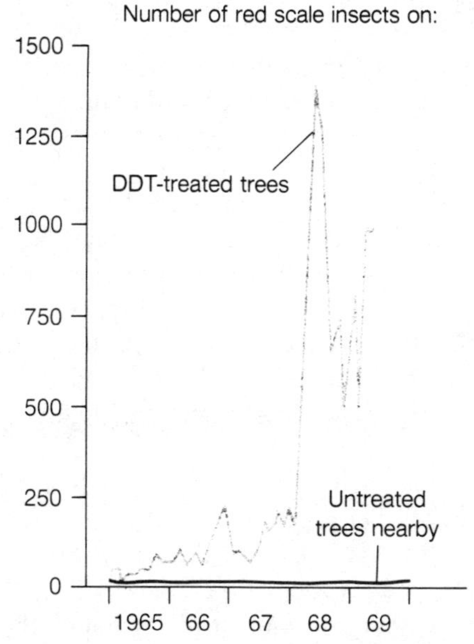

Figure 5.30
Using DDT to destroy sap-sucking red scale insects on lemon trees had the opposite effect, actually increasing the numbers of the pest. Spraying an infested crop kills perhaps 90% of the pests but also the insects that eat the pests. With their food abundant and their predators rare, the remaining pests recover faster than their enemies, whose prey is now scarce and harder to locate.

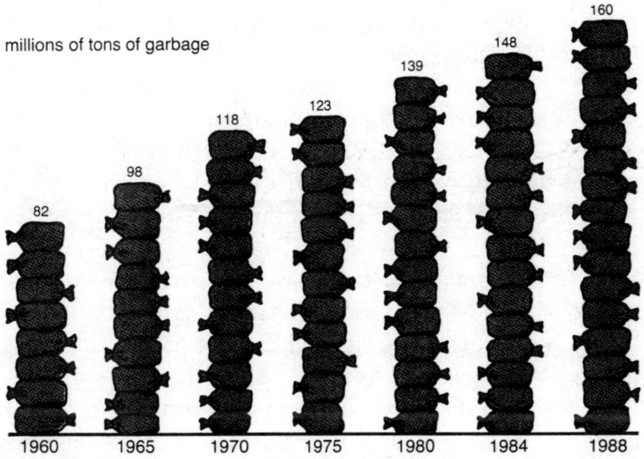

Figure 5.31
Municipal solid waste generated each year in the United States, in millions of tons. The figures do not include waste from industry or agriculture.

less affluent countries repair and recycle a far greater proportion of domestic products. In addition, nearly all consumer goods are encased in some sort of wrapping, whether it be paper, cardboard, plastic, or foam. An astounding one-third of the yearly volume of trash consists of these packaging materials. Finally, the United States has traditionally had ample space in which to dump unwanted materials. Countries that ran short of such space decades ago have made greater progress in reducing the volume of waste.

Although ordinary household trash does not meet the governmental designation of **hazardous waste**—defined as discarded material that may pose a substantial threat to human health or to the environment when improperly stored, transported, or disposed of—much of it is hazardous nonetheless. Products containing toxic chemicals include paint thinners and removers, furniture polishes, bleaches, oven and drain cleaners, used motor oil, and garden weed killers and pesticides.

Countries use various methods of disposing of solid wastes, and each has its own impact on the environment. Loading wastes onto barges and dumping them in the sea, long a practice for coastal communities, inevitably pollutes the ocean. Open dumps on land are a menace to public health, for they harbor disease-carrying rats and insects. Burning combustibles discharges chemicals and particulates into the air. In the United States, three methods of solid-waste disposal are employed: landfills, incineration, and recycling (Figure 5.32).

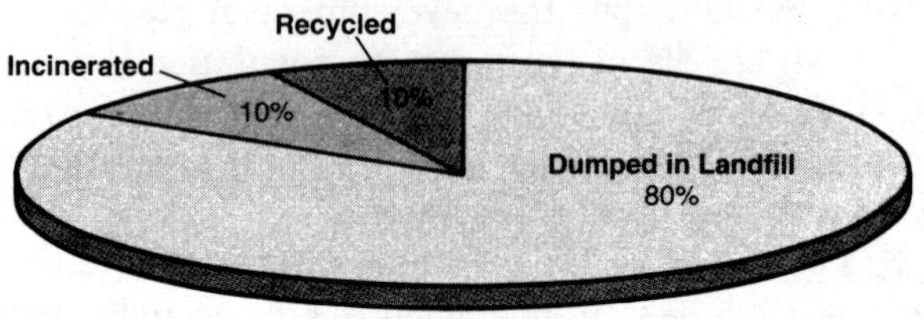

Figure 5.32
Methods of solid waste disposal in the United States.

Landfills

An estimated 80% of U.S. municipal solid waste is deposited in *sanitary landfills*, where each day's waste is compacted and covered by a layer of soil (Figure 5.33). "Sanitary" is a deceptive word. There are currently no federal standards to which local landfills must adhere, and while some communities and states regulate the environmental impact of dumps, many do not. Even if no commercial or industrial waste has been dumped at the site, most landfills eventually produce *leachate* liquids that contaminate the groundwater. The EPA estimates that only 15% of the country's landfills have liners to prevent contaminants from seeping into groundwater supplies. Indeed, over two-thirds of the dumps lack any type of system to monitor groundwater quality. Leachate occurs when precipitation entering the landfill interacts with the decomposing materials. Thus, heavy metals are leached from batteries and old electrical parts, while vinyl chlorides come from the plastic in household products. Typical leachate contains more than 40 organic chemicals, many of them poisonous.

Municipal landfill capacity is shrinking dramatically. The number of landfills in the United States fell from 18,000 in the late 1970s to 6000 in 1989, increasing pressure on those that remained open. A third of these will be closed within the next five years, either because they are full or because they pose a threat to the environment. Public opposition has made it increasingly difficult to site new landfills, forcing communities to look for alternative methods of waste disposal. A few in the northeast have resorted to long-distance, even interstate hauling, paying other communities to take in their garbage, but the chief effect has been to increase interest in incineration and recycling.

Incineration

The quickest way to reduce the volume of trash is to burn it, a practice that was common at this country's open dumps until it was halted by the Clean Air Act of 1970. Concern over air pollution also forced the closure of old, inefficient *incinerators* (facilities designed to burn waste), providing an impetus to designing a new generation of incinerators. By 1988, more than 100 municipal incinerators were operating in the United States, mostly in the northeast, burning about 10% of the national total of trash, and another 200 are expected to be operating soon. Connecticut burns about half of its household trash; and Massachusetts, New Jersey, and New York may soon do likewise. A Dade County facility burns 25% of the waste from 26 municipalities, including Miami. Most municipal incinerators are of the waste-to-energy type, which use extra-high (1800°F) temperatures to reduce trash to ash and simultaneously generate electricity or steam that is then sold to help pay operating costs.

A decade ago incinerators were hailed as the ideal solution to overflowing landfills, but it has become apparent that they pose environmental problems of their own by generating toxic pollutants in both air emissions and ash. Air emissions from incinerator stacks have been found to contain an alphabet soup of highly toxic elements, ranging from a (arsenic) to z (zinc), and including, among others, cadmium, dioxins, lead, and mercury, as well as significant amounts of such gases as carbon monoxide, sulfur dioxide, and nitrogen oxides. Emissions can be kept to acceptably low limits by installing electrostatic precipitators, filters, and scrubbers to capture pollutants before they are released into the outside air, although the devices add significantly to the cost of the plant. A greater problem is created by the concentration of toxins in the ash residue of burning. Incinerators typically reduce trash by only 75%. One-fourth remains as ash, which must then be buried in a landfill. Ideally the ash would be treated as hazardous waste, to lessen the danger of contaminating groundwater supplies.

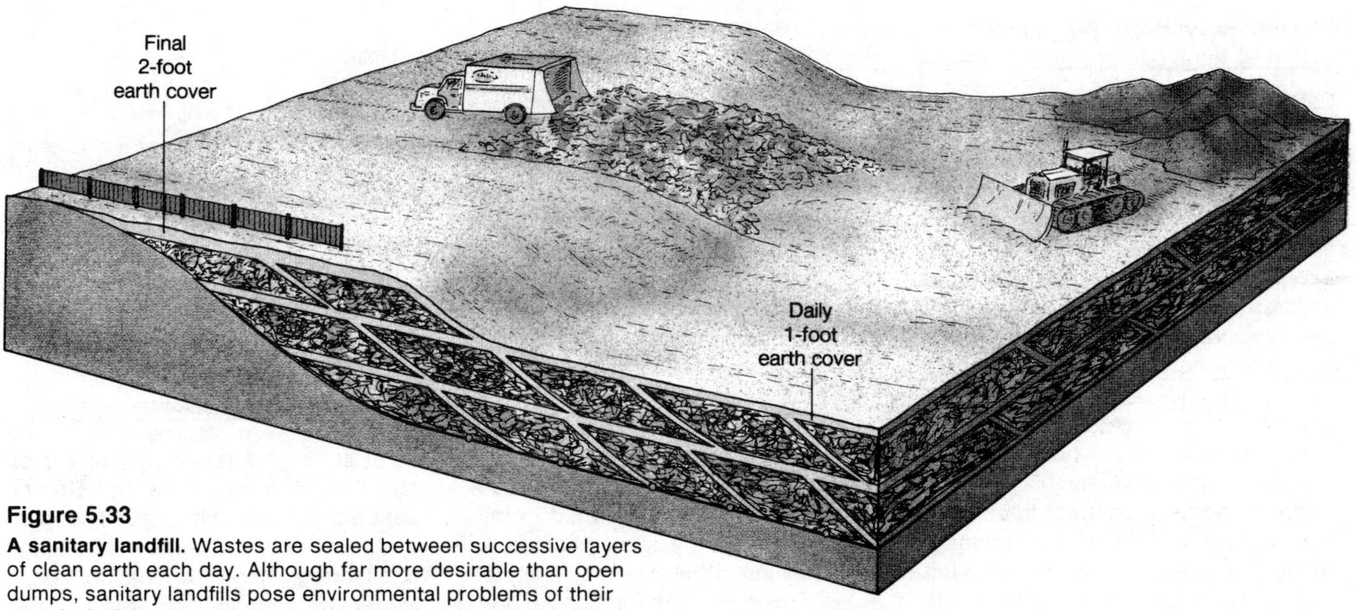

Figure 5.33
A sanitary landfill. Wastes are sealed between successive layers of clean earth each day. Although far more desirable than open dumps, sanitary landfills pose environmental problems of their own, including groundwater contamination and seepage of methane and hydrogen sulfide, gaseous products of decomposition.

Recycling

Awareness of the problems posed by landfills and incinerators has spurred interest in *recycling*, which reduces the amount of waste needing disposal by making a portion of it available for reuse. Thousands of businesses and at least 600 communities in the United States have initiated recycling programs for a variety of materials. Most programs collect paper, aluminum beverage cans, and glass. Some also compact leaves and other yard waste. Although recycling of plastics is more difficult, plastic milk and soda bottles are easily recycled.

At present about 10% of the country's waste is recycled, a figure that is likely to increase in the near future. By 1989, ten states had passed mandatory recycling laws, and the EPA has recommended a national goal of 25% waste recycling by 1992. The figure is not unattainable; Japan recycles more than half of its waste, and Western European countries about 30%.

Recycling has a beneficent impact on the environment. It saves natural resources by making it possible to cut fewer trees, burn less oil, mine less ore. Because it takes less energy to make things out of recycled material than out of virgin materials, recycling saves energy. It reduces the pollution of air, water, and land that stems from the manufacture of new materials and from other methods of waste disposal, and saves increasingly scarce landfill space for materials that cannot be recycled. These include hazardous wastes.

Hazardous Waste

The EPA has classified more than 400 substances as hazardous, posing a threat to human health or the environment. Industrial wastes have grown steadily more toxic. Currently about 10% of industrial waste materials are considered hazardous.

Every facility that either uses or produces radioactive materials generates *low-level waste*. Nuclear power plants produce about half the total low-level waste in the form of used resins, filter sludges, lubricating oils, and detergent wastes. Industries that manufacture radiopharmaceuticals, smoke alarms, radium watch dials, and other consumer goods produce low-level waste consisting of machinery parts, plastics, and organic solvents. Research establishments, including universities and hospitals, also produce radioactive waste materials. While some low-level waste will lose its contamination within months, some will remain radioactive for centuries.

High-level waste is nuclear waste with a relatively high level of radioactivity. It consists primarily of spent power reactor fuel assemblies—termed "civilian waste"—and waste generated as a by-product of the manufacture of nuclear weapons, or "military waste." The volume of high-level waste is not only great but increasing rapidly. Approximately one-third of a reactor's rods need to be disposed of every year.

By the end of the 1980s, nearly 70,000 spent-fuel assemblies were being stored in the containment pools of the country's commercial nuclear power reactors, awaiting more permanent disposition. Some 6000 more are added annually. "Spent fuel" is a misleading term: the assemblies are removed from commercial reactors not because their radiation is spent but because they have become too radioactive for further use. The assemblies will remain radioactively "hot" for thousands of years.

Unfortunately, no satisfactory method for disposing of any hazardous waste has yet been devised. Some wastes

Commerce in Poison

Early in March 1988, A.S. Bulk Handling Inc., a Norwegian shipping company, dumped 15,000 tons of a substance described as "raw material for bricks" in an abandoned quarry on the resort island of Kassa, just off Conakry, the capital of Guinea. Barefoot children played on the debris, but weekend vacationers from the mainland soon noticed that the island's vegetation had begun to shrivel. A government investigation revealed that the material was toxic incinerator ash from the city of Philadelphia, the first shipment under a contract to dispose of 85,000 tons of waste in the country. Guinea ordered the waste removed and its Norwegian broker jailed, and in July the ash was returned to Philadelphia.

That return shipment was the one bright spct in an otherwise dark picture. Between 1986 and 1988, over 3 million tons of hazardous waste were transported from Western Europe and the United States to countries in Africa and Eastern Europe, an irresponsible and reckless way to dispose of dangerous refuse. As the cost of toxic waste disposal in Western Europe and the United States soared as high as $2500 per ton, waste brokers sought cheaper sites in poorer countries. Nearly every West African country received offers from companies seeking to dispose of refuse. Waste brokers made enormous profits, for their offers to governments and private businesses ranged from $3 to $50 per ton.

It is evident why industrialized countries want to export waste, but why would a country like Guinea accept the shipments? Partly for economic reasons: poverty and a huge foreign debt. In addition, receiving countries have not been accurately informed about the nature of the wastes or the dangers they present. Incinerator ash, chlorinated solvents, PCBs, lead paint waste, and other hazardous substances have been described in disposal contracts by such phrases as "ordinary industrial waste," "complex organic matter," "non-explosive, non-radioactive, non-self-combusting materials," and, in the Guinea case, "raw material for bricks."

Exporting waste simply shifts toxic materials from one country to another, where they may do more damage to health or the environment. Waste that is deadly in the United States and Europe will be even more so in countries that lack the technology to dispose of it safely, and where the disposal sites have not been studied for their geological suitability. It is especially dangerous to dump wastes in areas with high rainfall, such as many of the West African countries, because it is likely to contaminate groundwater supplies.

Irate at this latest type of exploitation of poor countries, the Organization of African Unity in 1988 adopted a resolution condemning the dumping of all foreign wastes on that continent. Under the sponsorship of the United Nations, 117 countries in March of 1989 adopted a treaty aimed at regulating the trade in waste. The Basel Convention on the Control of Transboundary Movements of Hazardous Wastes and Their Disposal hopes to prevent the export of such wastes to unsafe or inadequate sites by requiring exporters to receive consent from receiving countries before shipping the waste. The treaty, which will become effective 90 days after 20 countries have formally adopted it, also requires both exporting and receiving countries to insure that the waste is disposed of in an environmentally sound manner. It remains to be seen how effectively the treaty is applied.

---

have been sealed in protective tanks and dumped at sea. Investigations have shown, however, that the tanks may be moved from the original dumping site by strong currents and may be crushed by water pressure, causing leakage of the wastes. Even without such physical damage, the life expectancy of the containers must be presumed to be far shorter than the half-life of their radioactive contents. Other radioactive wastes have been placed in tanks and buried in the earth. Millions of cubic feet of both high- and low-level waste are stored at various locations in the United States, one of which is shown in Figure 5.34. Several of these storage areas have experienced leakages, with seepage of waste into the surrounding soil and groundwater.

Another method of waste disposal, used for chemicals as well as for radioactive wastes, has been to inject them into deep steel- or concrete-lined wells. Because underground injection poses a threat of groundwater contamination and may contribute to earth tremors, the injection of wastes into or above strata that contain aquifers is being phased out.

The most common method of disposal of low-level waste is in landfills, often the local municipal dump, where the waste chemicals may leach through the soil and into the groundwater. By EPA estimates, the United States contains at least 25,000 legal and illegal dumps with hazardous waste. As many as 2000 are deemed potential ecological disasters. Some scientists contend that a safe dump exists only on paper, that even following the federal government's suggested guidelines for the best possible hazardous waste landfill will yield an insecure one (Figure 5.35).

Figure 5.34

Storage tanks under construction in Hanford, Washington. Built to contain high-level radioactive wastes, the tanks are shown before they were encased in concrete and buried underground.

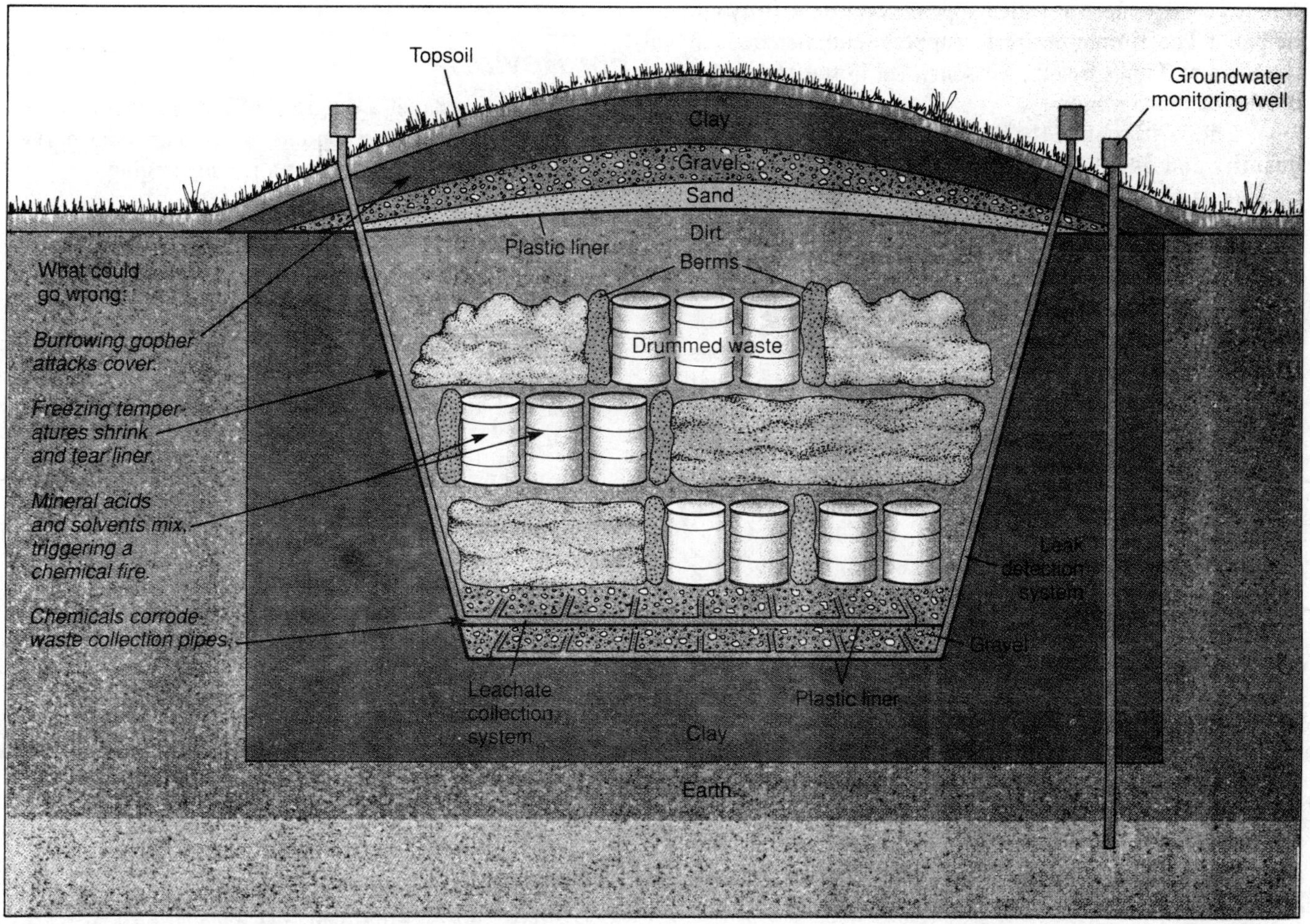

Figure 5.35

The closest thing to what experts consider a secure landfill for hazardous waste is a pit 60 feet (18 m) deep, sealed in a plastic liner to keep contaminated liquids from seeping into the soil. Toxic waste is placed in drums and covered with dirt. Leachate, the liquid mixture of wastes that seep to the bottom of the pit, passes into perforated collection pipes and is pumped to the surface for treatment. But landfills fail easily. Bulldozers can tear the plastic liner and leachate disintegrate it. The waste may crush collection pipes, and debris clog the perforations. If erosion breaches the landfill's protective cover, rain may mix with waste, overloading the collection system or causing the landfill to overflow.

Solid waste will never cease to be a problem, but its impact on the environment can be lessened by reducing the volume of waste that is generated, eliminating or reducing the production of toxic residues, halting irresponsible dumping, and finding ways to reuse the resources that waste contains. Until then, current methods of waste disposal will continue to pollute soil, air, and water.

Finally, all the common methods of disposing of the wastes that humans produce release contaminants into the surrounding environment, providing further evidence of the fact that people cannot manipulate, distort, pollute, or destroy any part of the ecosystem without diminishing its quality or disrupting its structure.

Summary

Humans are part of the natural environment and depend, literally, for their lives on the water, air, and other resources contained in the biosphere. But people have subjected the intricately interconnected systems of that biosphere—the troposphere, the hydrosphere, and the earth's crust—to profound and frequently unwittingly destructive alteration. All human activities have effects on the environment, effects that are complex and never isolated. An external action that impinges on any part of the web of nature inevitably triggers chain reactions, the ultimate consequences of which appear never to be fully anticipated.The simple desire to suppress summer dust in the town of Times Beach, Missouri, led in the end to the demise of the community.

Efforts to control the supply of water alter both the quantity and quality of water in streams, and structures such as dams and reservoirs often create unintended side effects. Pollutants associated with agriculture, industry, and other activities have degraded the quality of freshwater supplies, although regulatory efforts have brought about major improvements in some areas in recent years.

Combustion of fossil fuels has contributed to serious problems of air pollution. Some manifestations of that pollution, such as acid rain and depletion of the ozone layer, are matters of global concern. If the theory of the greenhouse effect is valid, air pollution will alter normal patterns of temperature and precipitation sometime during the next century.

Activities such as agriculture and mining have long helped to shape local landscapes, producing a variety of landforms, and have also contributed to degradation of air, water, and soil. In this century, pressure exerted by world population growth and economic expansion have accelerated deforestation, soil erosion, and desertification over much of the world.

People affect other living things—plants and animals—by importing them to areas where they did not previously exist, disrupting their habitats, hunting, and using biocides to eradicate them. The greatest human impact is occurring in the tropical rain forests, where agriculture, industrialization, and urbanization are contributing to the extinction of at least 1000 species a year.

Key Words

acid rain
biological magnification
biosphere
channelization
chlorofluorocarbons (CFCs)
desertification
ecosystem
environmental pollution
eutrophication
food chain
greenhouse effect
hazardous waste
hydrologic cycle
ozone
ozone layer
polychlorinated biphenyls (PCBs)
salinization
soil erosion
subsidence
thermal pollution
troposphere

For Review

1. Sketch and label a diagram of the *biosphere*. Briefly indicate the content of its component parts. Is that content permanent and unchanging? Explain.
2. How are the concepts of *ecosystem*, *niche*, and *food chain* related? How does each add to our understanding of the "web of nature"?
3. Draw a diagram of or briefly describe the *hydrologic cycle*. How do people have an impact on that cycle? What effect does urbanization have on it?
4. Is all environmental pollution the result of human action? When can we say that pollution of a part of the biosphere has occurred?
5. Describe the chief sources of water pollution. What steps have the United States and other countries taken to control water pollution?
6. What factors affect the type and degree of air pollution found at a place? What is *acid rain*, and where is it a problem? Describe the relationship of *ozone* to *photochemical smog*. Why has the ozone layer been depleted?
7. What causes the greenhouse effect? What impact might it have on the environment?
8. What kinds of landforms has excavation produced? Dumping? What are the chief causes and effects of subsidence?

9. How have people diminished the amount and productivity of soil? What types of areas are particularly subject to desertification? Why has soil erosion accelerated in recent decades?
10. Briefly describe the chief ways that humans affect plant and animal life. What is meant by *biological magnification*? Why may the use of biocides be self-defeating?
11. What methods do communities use to dispose of solid waste? What ecological problems does solid-waste disposal present? How does the government define *hazardous waste*, and how is it disposed of?

Selected References

Abrahamson, Dean E., ed. *The Challenge of Global Warming.* Covelo, Calif.: Island Press, 1989.

Bailes, Kendall E., ed. *Environmental History: Critical Issues in Comparative Perspective.* Lanham, Md.: University Press of America, 1985.

Bolin, Bert, et al., eds. *The Greenhouse Effect, Climatic Change, and Ecosystems.* SCOPE 29. New York: Wiley, 1986.

Brown, Lester R., et al. *State of the World.* Washington, D.C.: Worldwatch Institute. Annual.

Brown, Lester R. and Edward C. Wolf. *Soil Erosion: Quiet Crisis in the World Economy.* Worldwatch Paper 60. Washington, D.C.: Worldwatch Institute, 1984.

Clark, William C., and R. E. Munn, eds. *Sustainable Development of the Biosphere.* International Institute for Applied Systems Analysis. Cambridge: Cambridge University Press, 1986.

Dix, H. M. *Environmental Pollution: Atmosphere, Land, Water, and Noise.* New York: Wiley, 1981.

Dregne, Harold E. *Desertification of Arid Lands.* New York: Harwood Academic Publishers, 1983.

Drew, David. *Man-Environment Processes.* London: Allen & Unwin, 1983.

Ehrlich, Anne H., and Paul R. Ehrlich. *Earth.* London: Methuen, 1987.

Elsom, Derek. *Atmospheric Pollution: Causes, Effects and Control Policies.* New York: Basil Blackwell Inc., 1987.

French, Hilary F. *Clearing the Air: A Global Agenda.* Worldwatch Paper 94. Washington, D.C.: Worldwatch Institute, 1990.

Goldman, Benjamin A., et al. *Hazardous Waste Management: Reducing the Risk.* Council on Economic Priorities. Washington, D.C.: Island Press, 1986.

Goudie, Andrew. *The Human Impact on the Natural Environment.* 2d ed. Oxford, England: Basil Blackwell Ltd., 1986.

Longman, K. A., and J. Jenik. *Tropical Forest and Its Environment.* 2d ed. New York: Longman Scientific & Technical/Wiley, 1987.

Louma, Samuel N. *Introduction to Environmental Issues.* New York: Macmillan, 1984.

McCormick, John. *Acid Earth: The Global Threat of Acid Pollution.* London: Earthscan/International Institute for Environment and Development, 1985.

Miller, G. Tyler, Jr. *Living in the Environment.* 6th ed. Belmont, Calif.: Wadsworth, 1990.

Neal, Homer, and J. R. Schubel. *Solid Waste Management and the Environment.* Englewood Cliffs, N.J.: Prentice-Hall, 1987.

Postel, Sandra. *Air Pollution, Acid Rain, and the Future of Forests.* Worldwatch Paper 58. Washington, D.C.: Worldwatch Institute, 1984.

Pye, Veronica, Ruth Patrick, and John Quarles. *Groundwater Contamination in the United States.* Philadelphia: University of Pennsylvania Press, 1983.

ReVelle, Charles, and Penelope ReVelle. *The Environment.* 3d ed. Boston: Jones and Bartlett, 1988.

Scientific American, 261, no. 3 (September 1989). Special issue.

Shea, Cynthia P. *Protecting Life on Earth: Steps to Save the Ozone Layer.* Worldwatch Paper 87. Washington, D.C.: Worldwatch Institute, 1988.

Sheridan, David. *Desertification of the United States.* Council on Environmental Quality. Washington, D.C.: Government Printing Office, 1981.

Southwick, Charles H., ed. *Global Ecology.* Sunderland, Mass.: Sinauer Associates, Inc., 1985.

Thomas, William, ed. *Man's Role in Changing the Face of the Earth.* Chicago: University of Chicago Press, 1956.

Wellburn, Alan. *Air Pollution and Acid Rain.* New York: Wiley, 1988.

Worster, Donald, ed. *The Ends of the Earth: Perspectives on Modern Environmental History.* Cambridge: Cambridge University Press, 1988.

C H A P T E R

The Regional Concept

Hillside houses, Taxco, Mexico.

The questions geographers ask, we saw in Chapter 1 of this book, ultimately focus on matters of location and character of place. We asked how things are distributed over the surface of the earth, how physical and cultural features of areas are alike or different from place to place, how the varying content of different places came about, and what all these differences and similarities mean for people.

The Nature of Regions

In earlier chapters we examined some of the physical and cultural content of areas and some spatially important aspects of human behavior. We looked at the physical earth processes that lead to differences from place to place in the environment. We studied ways in which humans organize their actions in earth space—through political institutions, by economic systems and practices, and by cultural and social processes influencing their spatial behavior and interaction. Population and settlement patterns and areal differences in human use and misuse of earth resources all were considered as part of the mission of geography.

For each of the topics we studied, from landforms to cities, we found spatial regularities. We discovered that things are not irrationally distributed over the surface of the earth, but reflect an underlying spatial order based on understandable physical and cultural processes. We found, in short, that although no two places are *exactly* the same, it is possible and useful to recognize segments of the total world that are internally similar in some important characteristic and distinct in that feature from surrounding areas.

These regions of significant uniformity of content are the geographer's equivalent of the historian's "eras" or "ages": brief summary names for specific areas that are different in some important way from adjacent or distant territories. The **region**, then, is a device of areal generalization. It is an attempt to separate into recognizable component parts the otherwise overwhelming diversity and complexity of the earth's surface.

All of us have a general idea of the meaning of region, and all of us refer to regions in everyday speech and action. We visit "the old neighborhood" or "go downtown," we plan to vacation or retire in the "Sunbelt," or we speculate on the effects of weather conditions in the "Northeast" or the "Corn Belt" on grain surpluses or next year's food prices. In each instance we have mental images of the areas mentioned. Those images are based upon place characteristics and areal generalizations that seem useful to us and recognizable to our listeners. We have, in short, engaged in an informal place classification to pass along quite complex spatial, organizational, or content ideas. We have applied the **regional concept** to bring order to the immense diversity of the earth's surface.

What we do informally as individuals, geography attempts to do formally as a discipline—define and explain regions (Figure 13.1). The purpose is clear: to make the infinitely varying world around us understandable through spatial summaries. That world is only rarely subdivided into neat, unmistakable "packages" of uniformity. Neither the environment nor human areal actions present us with a compartmentalized order, any more than the sweep of human history has predetermined "eras," or all plant specimens come labelled in nature with species names. We all must classify to understand, and the geographer classifies in regional terms.

Regions are spatial expressions of ideas or summaries useful to the analysis of the problem at hand. Although as many possible regions exist as there are physical, cultural, or organizational attributes of area, the geographer selects for study those areal variables that contribute to the understanding of a specific topical or spatial

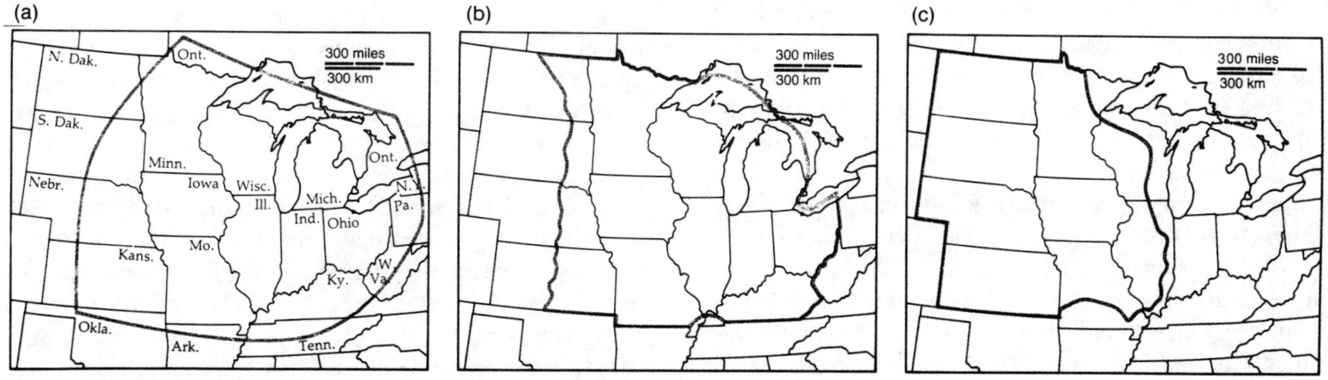

Figure 13.1
The Middle West as seen by different professional geographers. Agreement on the need to recognize spatial order and to define regional units does not imply unanimity in the selection of boundary criteria. Each of the sources concurs in the significance of the Middle West as a regional entity in the spatial structure of the United States and agrees on its core area. The sources differ, however, in their assessments of its limiting characteristics. ([a] John H. Garland, editor, *The North American Midwest.* Copyright © 1955 John Wiley & Sons, Inc., New York, NY; [b] John R. Borchert and Jane McGuigan, *Geography of the New World.* Copyright © 1961 Rand McNally, Chicago, IL; [c] Otis P. Starkey and J. Lewis Robinson, *The Anglo-American Realm.* Copyright © 1969 McGraw-Hill Publishing Company, New York, NY.)

problem. All other variables are disregarded as irrelevant. In our reference to the Corn Belt we delimit a portion of the United States showing common characteristics of farming economy and marketing practices. We dismiss—at that level of generalization—differences within the region based upon slope, soil type, state borders, or characteristics of population. The boundaries of the Corn Belt are assumed to be marked where the region's internal unifying characteristics change so materially that different agricultural economies become dominant and different regional summaries are required. It is the content of the region that suggests its definition and determines the basis of its delimitation.

Although regions may vary greatly, they all share certain common characteristics related to earth space.

1. **Regions have *location,* often expressed in the regional name selected, such as the Middle West, the Near East, North Africa, and the like. This form of regional name underscores the importance of *relative location* (see Chapter 1, p. 5).**
2. **Regions have *spatial extent.* They are recognitions of the fact that there are differences from place to place in the physical and cultural attributes of area, and that those differences are important.**
3. **Regions have *boundaries* based upon the areal spread of the features selected for study. Since regions are the recognition of the features defining them, their boundaries are drawn where those features no longer occur or dominate (Figure 13.2).**
4. **Regions, as we saw in Chapter 1 (p. 10), may be either *formal* or *functional.***

Formal regions are areas of essential uniformity throughout in one or a limited combination of physical or cultural features. In previous chapters we encountered such formal physical regions as the humid subtropical climate zone and the Sahel region of Africa, as well as formal (homogeneous) cultural regions in which standardized characteristics of language, religion, ethnicity, or livelihood existed. The endpaper maps of topographical regions and countries show other formal regional patterns. Whatever the basis of its definition, the formal region is the largest area over which a valid generalization of attribute uniformity may be made. Whatever is stated about one part of it holds true for its remainder.

The **functional region**, in contrast, is a spatial system defined by the interactions and connections that give it a dynamic, organizational basis. Its boundaries remain constant only as long as the interchanges establishing it remain unaltered. The commuting region shown in Figure 1.9 retained its shape and size only as long as the road pattern and residential neighborhoods on which it was based remained as shown.

5. **Regions are *hierarchically arranged.* Although regions vary in scale, type, and degree of generalization, none stands alone as the ultimate key to areal understanding. Each defines only a part of spatial reality.**

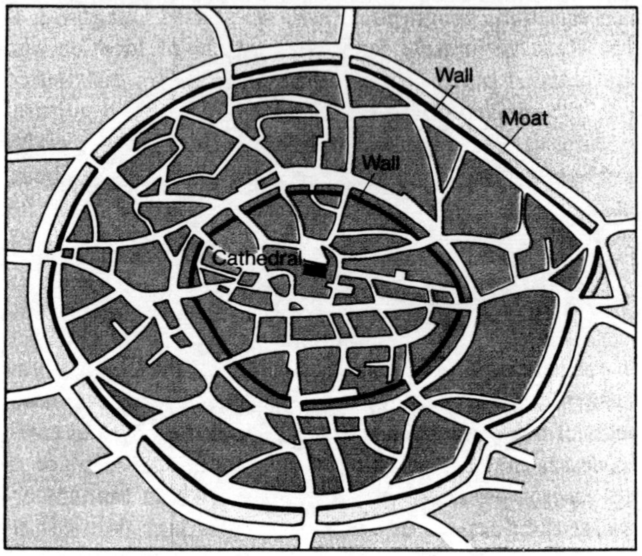

Figure 13.2
Aachen in 1649. The acceptance of regional extent implies the recognition of regional boundaries. At some defined point, *urban* is replaced by *nonurban,* the Middle West ends and the Plains begin, or the rain forest ceases and the savanna emerges. Regional boundaries are, of course, rarely as precisely and visibly marked as were the limits of the walled medieval city. Its sprawling modern counterpart may be more difficult to define, but the boundary significance of the concept of *urban* remains.

On a formal regional scale of size progression, the Delmarva Peninsula of the eastern United States may be seen as part of the Atlantic Coastal Plain, which is in turn a portion of the eastern North American humid continental climatic region, a hierarchy that changes the basis of regional recognition as the level and purpose of generalization alter (Figure 13.3). The central business district of Chicago is one land use complex in the functional regional hierarchy that describes the spatial influences of the city of Chicago and the metropolitan region of which it is the core. Each recognized regional entity in such progressions may stand alone and, at the same time, exist as a part of a larger, equally valid, territorial unit.

Such progressions may also reflect gradations in the intensity in spatial dominance of the phenomenon giving definition to the region. With particular attention to regions defined by the distribution of culture groups—but with application to regions delimited by other criteria—Donald Meinig has suggested the term *core* to mean a centralized zone of concentration and greatest homogeneity of regional character. Areas in which the culture (or other defining feature) is dominant, but with less intensity and totality of development, were labelled *domains.* Finally, Meinig proposed the term *sphere* to

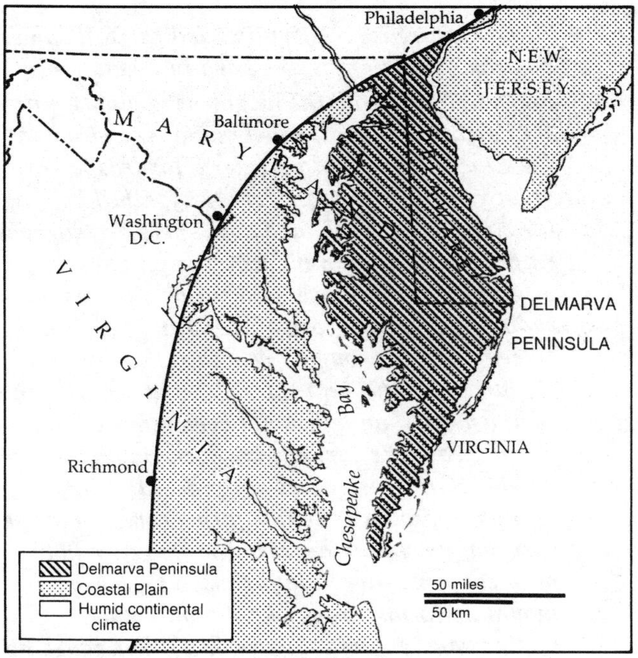

Figure 13.3

A hierarchy of regions. One possible nesting of regions within a *regional hierarchy* defined by differing criteria. Each regional unit has internal coherence. The recognition of its constituent parts aids in understanding the composite areal unit.

recognize the zone of broadest but least-intensive expression of the regional character, where some of its defining traits are encountered but where it is no longer spatially dominant.

These generalizations about the nature of regions and the regional concept are meant to instill firmly the understanding that regions are human intellectual creations designed to serve a purpose. Regions focus our attention upon spatial uniformities, and they bring clarity to the seeming confusion of the observable physical and cultural features of the world we inhabit. Regions provide the framework for the purposeful organization of spatial data.

The Structure of This Chapter

The remainder of the chapter contains examples of how geographers have organized regionally their observations about the physical and cultural world. Each study or vignette explores a different aspect of regional reality. Each organizes its data in ways appropriate to its subject matter and objective, but in each, some or all of the common characteristics of regional delimitation and structure may be recognized. Each of the regional examples is based upon the content of one of the earlier chapters of this book, and the page numbers preceding the different selections refer to the material within those chapters most closely associated with, or explained further by, the regional case study. As an aid to visualizing the application of the regional concept, the examples are introduced by reference to the traditions of geography that unite their subject matter and approach.

Part I. Regions in the Earth-Science Tradition

The simplest of all regions to define, and generally the easiest to recognize, is the formal region based upon a single readily apparent component or characteristic. The island is land, not water, and its unmistakable boundary is naturally given where the one element passes to the other. The terminal moraine may mark the transition from the rich, black soils of recent formation to the particolored clays of earlier generation. The dense forest may break dramatically upon the glade or the open prairies. The nature of change is singular and apparent.

The physical geographer, although concerned with all the earth sciences that explain the natural environment, deals at the outset with *single factor* formal regions. Many of the earth features of concern to physical geography, of course, do not exist in simple, clearly defined units. They must be arbitrarily regionalized by the application of boundary definitions. A stated amount of received precipitation, the presence of certain important soil characteristics, the dominance of particular plant associations—all must be decided upon as regional limits, and all such limits are subject to change through time or by purpose of the regional geographer.

Landforms as Regions *(See "Stream Landscapes in Humid Areas," page 68)*

The landform region exists in a more sharply defined fashion than such transitional physical features as soil, climate, or vegetation. For these latter areas, the boundaries depend upon definitional decisions made (and defended) by the researcher. The landform region, on the other hand, arises—visibly and apparently unarguably—from nature itself, independent of human influence and unaffected by time on the human scale. Landforms constitute basic, naturally defined regions of physical geographic concern. The existence of major landform regions—mountains, lowlands, plateaus—is unquestioned in popular recognition or scientific definition. Their influence on climates, vegetational patterns, even upon the primary economies of subsistence populations has been noted in earlier portions of this book. The following discussion of a distinct landform region, describing its constitution and its relationships to other physical and cultural features of the landscape, is adapted from a classic study by Wallace W. Atwood.

> ***The Black Hills Province.***[1] *The Black Hills rise abruptly from the surrounding plains (Figure 13.4). The break in topography at the margin of this province is obvious to the most casual*

1. From *The Physiographic Provinces of North America* by Wallace W. Atwood. © Copyright, 1940, by Ginn and Company (Xerox Corporation). From pp. 13–15.

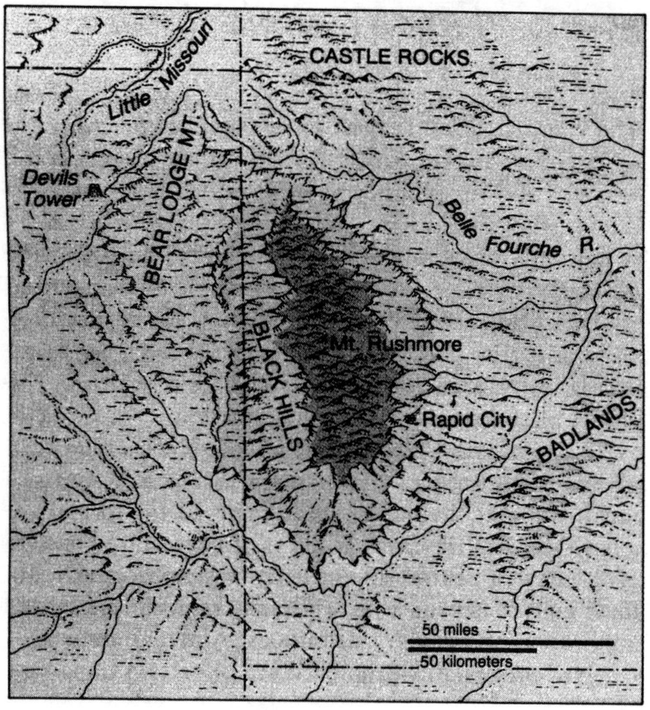

Figure 13.4
The Black Hills physiographic province.

observer who visits that part of our country. Thus the boundaries of this area, based on contrasts in topography, are readily determined.

One who looks more deeply into the study of the natural environment may recognize that in the neighboring plains the rock formations lie in a nearly horizontal position. They are sandstones, shales, conglomerates, and limestones. In the foothills those same sedimentary formations are bent upward and at places stand in a nearly vertical position. Precisely where the change in topography occurs, we find a notable change in the geologic structure and thus discover an explanation for the variation in relief.

The Black Hills are due to a distinct upwarping, or doming, of the crustal portion of the earth. Subsequent removal by stream erosion of the higher portions of that dome and the dissection of the core rocks have produced the present relief features. As erosion has proceeded, more and more of a complex series of ancient metamorphic rocks has been uncovered. Associated with the very old rocks of the core and, at places, with the sedimentary strata, there are a number of later intrusions which have cooled and formed solid rock. They have produced minor domes about the northern margin of the Black Hills.

With the elevation of this part of our country there came an increase in rainfall in the area, and with the increase in elevation and rainfall came contrasts in relief, in soils, and in vegetation.

As we pass from the neighboring plains, where the surface is monotonously level, and climb into the Black Hills area, we enter a landscape having great variety in the relief. In the foothill belt, at the southwest, south, and east, there are hogback ridges interrupted in places by water gaps, *or gateways, that have been cut by streams radiating from the core of the Hills. Between the ridges there are roughly concentric valley lowlands. On the west side of the range, where the sedimentary mantle has not been removed, there is a plateau-like surface; hogback ridges are absent. Here erosion has not proceeded far enough to produce the landforms common to the east margin. In the heart of the range we find deep canyons, rugged intercanyon ridges, bold mountain forms, craggy knobs, and other picturesque features (Figure 13.5). The range has passed through several periods of mountain growth and several stages, or cycles, of erosion.*

The rainfall of the Black Hills area is somewhat greater than that of the brown, seared, semiarid plains regions, and evergreen trees survive among the hills. We leave a land of sagebrush and grasses to enter one of forests. The dark-colored evergreen trees suggested to early settlers the name Black Hills. As we enter the area, we pass from a land of cattle ranches and some seminomadic shepherds to a land where forestry, mining, general farming, and recreational activities give character to the life of the people. In color and form, in topography, climate, vegetation, and economic opportunities, the Black Hills stand out conspicuously as a distinct geographic unit.

Dynamic Regions in Weather and Climate *(See "Air Masses," page 96)*

The unmistakable clarity and durability of the Black Hills landform region and the precision with which its boundaries may be drawn are rarely echoed in other types of formal physical regions. Most of the natural environment, despite its appearance of permanence and certainty, is dynamic in nature. Vegetations, soils, and climates change through time by natural process or by the action of humans. Boundaries shift, perhaps abruptly, as witness the recent migration southward of the Sahara. The core characteristics of whole provinces change as marshes are drained or forests are replaced by cultivated fields.

That complex of physical conditions that we recognize locally and briefly as weather and summarize as climate displays particularly clearly the temporary nature of much of the natural environment that surrounds us. Yet

Figure 13.5
The "Needles" of the Black Hills result from erosion along vertical cracks and crevices in granite.

even in the turbulent change of the atmosphere, distinct regional entities exist with definable boundaries and internally consistent horizontal and vertical properties. "Air masses" and the consequences of their encounters constitute a major portion of contemporary weather analysis and prediction. Air masses further fulfill all the criteria of *multifactor formal regions,* though their dynamic quality and their patterns of movement obviously mark them as being of a nature distinctly different from such stable physical entities as landform regions, as the following extract from *Climatology and the World's Climates* by George R. Rumney makes clear.

> ***Air Masses.***[2] *An* air mass *is a portion of the atmosphere having a uniform horizontal distribution of certain physical characteristics, especially of temperature and humidity. These qualities are acquired when a mass of air stagnates or moves very slowly over a large and relatively unvaried surface of land or sea. Under these circumstances surface air gradually takes on properties of temperature and moisture approaching those of the underlying surface, and there then follows a steady, progressive transmission of properties to greater heights, resulting finally in a fairly clearly marked vertical transition of characteristics. Those parts of the earth where air masses acquire their distinguishing qualities are called* source regions.
>
> *The height to which an air mass is modified depends upon the length of time it remains in its source region and also upon the difference between the initial properties of the air when it first arrived and those of the underlying surface. If, for example, an invading flow of air is cooler than the surface beneath as it comes to virtual rest over a source region, it is warmed from below, and convective currents are formed, rapidly bearing aloft new characteristics of temperature and moisture to considerable heights. If, on the other hand, it is warmer than the surface of the source region, cooling of its surface layers takes place, vertical thermal currents do not develop, and the air is modified only in its lower portions. The process of modification may be accomplished in just a few days of slow horizontal drift, although it often takes longer, sometimes several weeks. Radiation, convection, turbulence, and advection are the chief means of bringing it about.*
>
> *The prerequisite conditions for these developments are very slowly migrating, outward spreading, and diverging air and a very extensive surface beneath that is fairly uniform in nature. Light winds and relatively high barometric pressure characteristically prevail. Hence, most masses form within the great semipermanent anticyclonic regions of the general circulation, where calms, light variable winds, and overall subsidence of the atmosphere are typical.*

2. G. R. Rumney, *Climatology and the World's Climates,* Macmillan, 1968.

Four major types of source regions are recognized: continental polar, maritime polar, continental tropical, and maritime tropical. Polar air masses are continental when they develop over land or ice surfaces in high latitudes; these are cold and dry. They are maritime when they form over the oceans in high latitudes. An air mass from these sources is cold and moist. Similarly, tropical air is continental when it originates along the Tropics of Cancer and Capricorn over northern Africa and northern Australia and is therefore warm and dry. It is maritime when it forms along the Tropics over the oceans, where it develops as a mass of warm, moist air. A single air mass usually covers thousands of square miles of the earth's surface when fully formed.

An air mass is recognizable chiefly because of the uniformity of its primary properties—temperature and humidity—and the vertical distribution of these. Secondary qualities, such as cloud types, precipitation, and visibility, are also taken into account. These qualities are retained for a remarkably long time, often for several weeks, after an air mass has traveled far from its source region, and they are thus the means of distinguishing it from other masses of air.

The principal air masses of the Americas, their source regions, and seasonal movements are shown in Figure 13.6.

Ecosystems as Regions

(See "Ecosystems," page 122)

A traditional though oversimplified definition of geography as "the study of the areal variation of the surface of the earth" suggested that the discipline centered on the classification of areas and on the subdivision of the earth into its constituent regional parts. Considerations of organization and function were secondary and even unnecessary to the implied main purpose of regional study: the definition of cores and boundaries of areas uniform in physical or cultural properties.

Newer research approaches stress the need for the study of spatial relationships from the standpoint of *systems analysis,* which emphasizes the organization, structure, and functional dynamics within an area and provides for the quantification of the linkages between the things in space. The *ecosystem,* or *biome,* introduced in Chapter 5, provides a systems-analytic concept of great flexibility. It brings together in a single framework environment, humans, and the biological realm, permitting an analysis of the relationships between these components of area. Since that relationship is structured, structure rather than spatial uniformity attracts attention and leads to new understanding of the region. The ecosystem concept, particularly, provides a point of view for investigating the complex consequences of human impact upon the natural environment, the theme of Chapter 5.

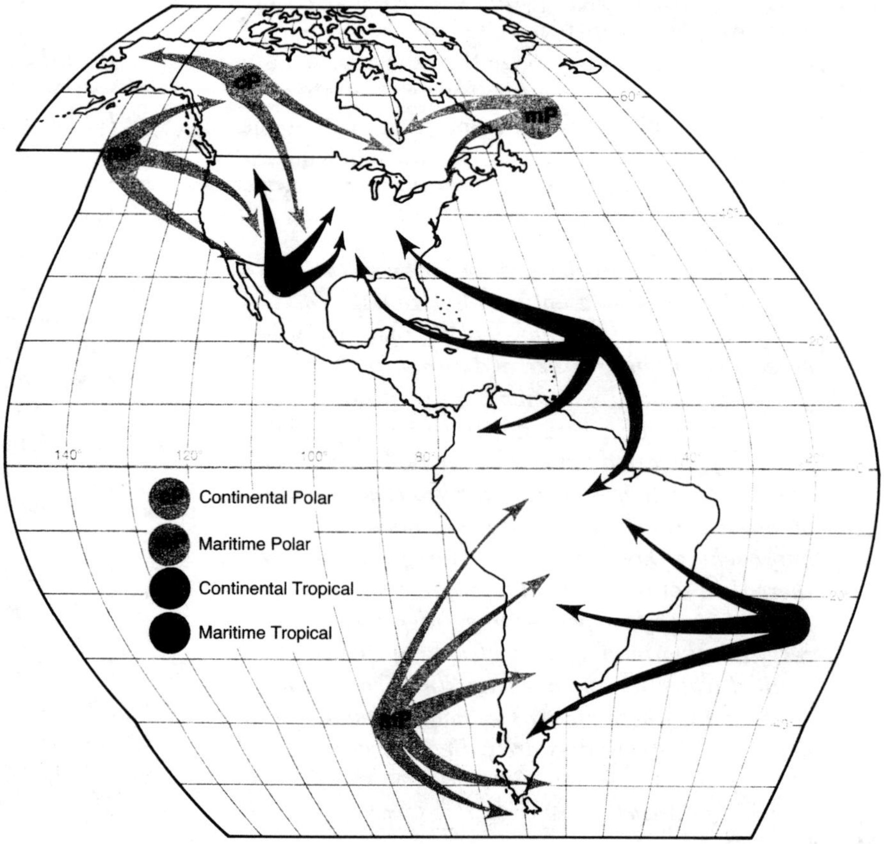

Figure 13.6
Air masses of North and South America, their source regions, and paths of movement.

Note how, in the following description drawn from an article by William J. Schneider, the ecosystem concept is employed in the recognition and analysis of regions and subregions of varying size, complexity, and nature. Introduced, too, is the concept of *ecotone,* or zone of ecological stress, in this case induced by human pressures upon the natural system.

> ***The Everglades.***[3] *The Everglades is a river. Like the Hudson or the Mississippi, it is a channel through which water drains from higher to lower ground as it moves to the sea. It extends in a broad, sweeping arc from the southern end of Lake Okeechobee in central Florida to the tidal estuaries of the Gulf Coast and Florida Bay. As much as 70 miles wide, but generally averaging 40, it is a large, shallow slough that weaves tortuously through acres of saw grass and past "islands" of trees, its waters, even in the wet season, rarely deeper than 2 feet. But, again like the Hudson or the Mississippi, the Everglades bears the significant imprint of civilization.*
>
> *Since the close of the Pleistocene, 10,000 years ago, the Everglades has been the natural drainage course for the periodically abundant overflow of Lake Okeechobee. As the lake filled during the wet summer months or as hurricane winds blew and literally scooped the water out of the lake basin, excess water spilled over the lake's southern rim. This overflow, together with rainfall collected en route, drained slowly southward between Big Cypress Swamp and the sandy flatlands to the west and the Atlantic coastal ridge to the east, sliding finally into the brackish water of the coastal marshes (Figure 13.7).*
>
> *Water has always been the key factor in the life of the Everglades. Three-fourths of the annual average 55 inches (140 cm) of rainfall occurs in the wet season, June through October, when water levels rise to cover 90% of the land area of the Everglades. In normal dry seasons in the past, water covered no more than 10% of the land surface. Throughout much of the Everglades, and prior to recent engineering activities, this seasonal rain cycle caused fluctuations in water levels that averaged 3 feet. Both occasional severe flooding and prolonged drought accompanied by fire imposed periodic stress upon the ecosystem. It may be that randomly occurring ecologic trauma is vital to the character of the Everglades.*
>
> *Three dominant biological communities—open water, saw grass, and woody vegetation—reflect small, but consistent, differences in the surface elevation of peat soils that cover the Everglades (Figure 13.8). The open-water areas occur at the lower soil elevations; inundated much of the year, they contain both sparse, scattered marsh grasses and a mat of algae. The saw grass communities develop on a soil base only a few inches higher than that in the surrounding open glades. The soil base is thickest under the tree islands. The few inches' difference in soil depth apparently governs the species composition of these three communities.*
>
> *Today, the Everglades is no longer precisely a natural river. Much of it has been altered by an extensive program of water management, including drainage, canalization, and the building of locks and dams. Large withdrawals of groundwater for municipal and industrial use have depleted the underlying aquifer and permitted the landward penetration of sea water through the aquifer and through the surface canals. Thousands of individual water-supply wells have been contaminated by encroaching saline water; large biotic changes have taken place in the former freshwater marshes south of Miami. Mangroves—indicators of salinity—have*

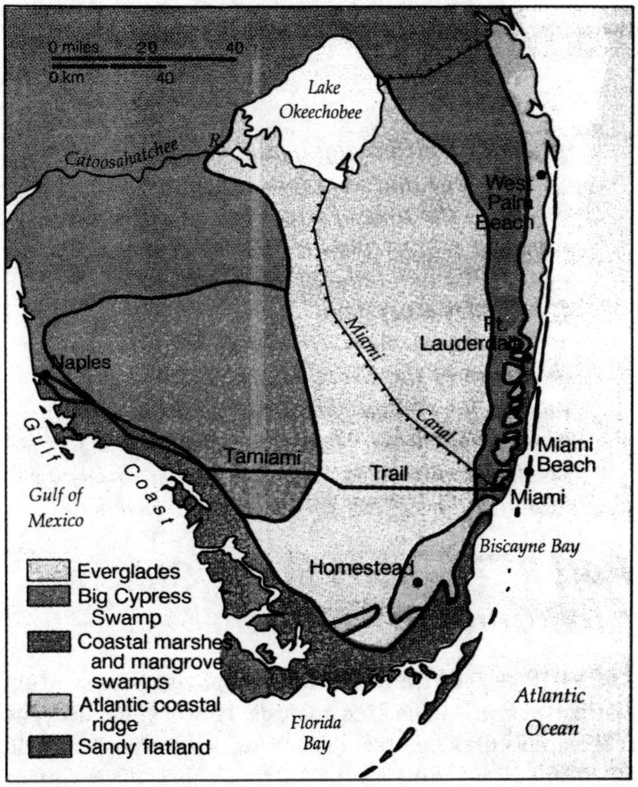

Figure 13.7

The Everglades is part of a complex of ecosystems stretching southward in Florida from Lake Okeechobee to the sea. Drainage and water-control systems have altered its natural condition.

3. From William J. Schneider, "Water and the Everglades." With permission from *Natural History*, November, 1966, pp. 32–40. Copyright American Museum of Natural History, 1966.

Figure 13.8
Open water, saw grasses, and tree islands, all visible in this scene, constitute separate *biomes* of the Everglades, which is home also of a teeming animal life, including the great egret shown in the foreground.

extended their habitat inland, and fires rage across areas that were formerly much wetter. The ecotone—*the zone of stress between dissimilar adjacent ecosystems—is altering as a consequence of these human-induced modifications of the Everglades ecosystem.*

The organization, structure, and functional dynamics of the Everglades ecosystems are thus undergoing change. The structured relationships of its components—in nature affected and formed by stress—are being subjected to distortions by humans in ways not yet fully comprehended.

Part II. Regions in the Culture–Environment Tradition

The earth-science tradition of geography imposes certain distinctive limits upon area analysis. However defined, the regions that may be drawn are based upon nature and do not result from human action. The culture–environment tradition, however, introduces to regional geography the infinite variations of human occupation and organization of space. There is a corresponding multiplication of recognized regional types and of regional boundary decisions.

Despite the differing interests of physical and cultural geographers, one element of study is common to their concerns: that of process. The "becoming" of an ecosystem, of a cultural landscape, or of the pattern of exchanges in an economic system is an important open or implied part of nearly all geographic study. Evidence of the past as an aid to understanding of the present is involved in much geographical investigation, for present-day distributional patterns or qualities of regions mark a merely temporary stage in a continuing process of change.

Population as Regional Focus

(See "World Population Distribution," page 178)

In no phase of geography are process and change more basic to regional understanding than in population studies. The human condition is dynamic and patterns of settlement are ever-changing. Although these spatial distributions are related to the ways that people utilize the physical environment in which they are located, they are also conditioned by the purposes, patterns, and solutions of those who went before. In the following extract taken from the work of Glenn Trewartha, a dean of American population geographers, notice how population regionalization—used as a focal theme—ties together a number of threads of regional description and understanding. The aspirations of colonist-conquerors, past and contemporary transportation patterns, physical geographic conditions, political separatism, and the history and practice of agriculture and rural landholdings are all introduced to give understanding to population from a regional perspective.

> ***Population Patterns of Latin America.***[4] *A distinctive feature of the spatial arrangement of population in Latin America is its strongly nucleated character; the pattern is one of striking clusters. Most of the population clusters remain distinct and are separated from other clusters by sparsely occupied territory. Such a pattern of isolated nodes of settlement is common in many pioneer regions; indeed, it was characteristic of early settlement in both Europe and eastern North America. In those regions, as population expanded, the scantily occupied areas between individual clusters gradually filled in with settlers, and the nodes merged. But in Latin America such an evolution generally did not occur, and so the nucleated pattern persists. Expectably, the individual population clusters show considerable variations in density.*
>
> *The origin of the nucleated pattern of settlement is partly to be sought in the gold and missionary fever that imbued the Spanish colonists. Their settlements were characteristically located with some care, since only areas with precious metals for exploitation and large Indian populations to be Christianized and to provide laborers could satisfy their dual hungers. A clustered pattern was also fostered by the isolation and localism that prevailed in the separate territories and settlement areas of Latin America.*
>
> *Almost invariably each of the distinct population clusters has a conspicuous urban nucleus. To an unusual degree the economic, political, and social life within a regional cluster centers on a single large primate city, which is also the focus of the local lines of transport.*
>
> *The prevailing nucleated pattern of population distribution also bears a relationship to political boundaries. In some countries . . . a single population cluster represents the core area of the nation. In more instances, however, a population cluster forms the core of a major political subdivision of a nation-state, so that a country may contain more than one cluster. A consequence of this simple population distribution pattern and its relation to administrative subdivisions is that political boundaries ordinarily fall within the sparsely occupied territory separating individual clusters. In Latin America few national or provincial boundaries pass through nodes of relatively dense settlement (Figure 13.9).*
>
> *Another feature arising out of the cluster pattern of population arrangement is that the* total national territory *of a country is often very different from the* effective national territory, *since the latter includes only those populated parts that contribute to the country's economic support.*
>
> *A further consequence of the nucleated pattern is found in the nature of the transport routes and systems. Overland routes between population clusters are usually poorly developed, while a more efficient network ordinarily exists* within *each cluster, with each such regional network joined by an overland route to the nearest port. Thus, the chief lines of transport connecting individual population clusters are often sea lanes*

4. From Glenn T. Trewartha, *The Less Developed Realm: A Geography of Its Population*. Copyright © 1972 by John Wiley and Sons, Inc. Reprinted from pp. 43–46.

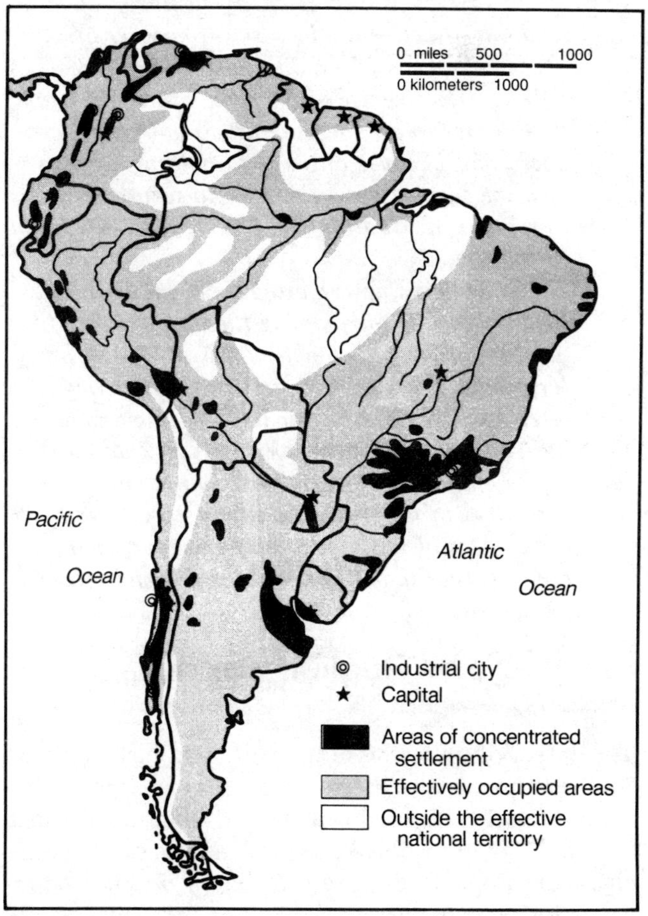

Figure 13.9

Basic settlement patterns of South America. Population clusters focused on urban cores and separated by sparsely populated rural areas were the traditional pattern in Latin American countries.

rather than land routes. Gradually, as highways are developed and improved, intercluster overland traffic tends to increase.

. . . The spectacular rates at which population numbers are currently soaring in Latin America are not matched by equivalent changes in their spatial redistributions. Any population map of Latin America reveals extensive areas of unused and underutilized land. Part of such land is highland and plagued by steep slopes, but by far the larger share of it is characterized by a moist tropical climate, either tropical wet or tropical wet and dry. Such a climatic environment, with its associated wild vegetation, soils, and drainage, admittedly presents many discouraging elements to the new settler of virgin lands. . . . [T]ropical climate alone is scarcely a sufficient explanation for the abundance of near-empty lands south of the Rio Grande. Cultural factors are involved as much as, if not more than, physical ones. One of the former is the unfortunate land-holding system that has been fastened on the continent, under which vast areas of potentially cultivable land are held out of active use by a small number of absentee landlords, who not only themselves make ineffective use of the land, but at the same time refuse to permit its cultivation by small operators. Because of the land-holding system, peasant proprietors are unable to secure their own lands, a situation that discourages new rural settlement. . . .

The changes now in progress in the spatial distributions accompanying the vast increase in numbers of people do not appear to involve any large-scale push of rural settlement into virgin territory. Only to a rather limited extent is new agricultural settlement taking place. Intercluster regions are not filling rapidly. The overwhelming tendency is for people to continue to pile up in the old centers of settlement in and around their cities, rather than to expand the frontier into new pioneer-settlement areas. . . .

Language as Region *(See "Language," page 211)*

The great culture realms of the world (outlined in Figure 7.39) are historically based composites of peoples. They are not closely identified with nation, language, religion, or technology, but with all these and more in varying combinations. Culture realms are therefore *multifactor regions* that obscure more than they clarify the distinctions between peoples that are so fundamental to the human mosaic of the earth. Basic to cultural geography is the recognition of small regions of single-factor homogeneity that give character to their areas of occurrence and that collectively provide a needed balance to the sweeping generalizations of the culture realm.

Language provides an example of such small-area variation, one partially explored in Chapter 7. The language families shown in Figure 7.18 conceal the identity of and the distinctions among the different official tongues of separate countries. These, in turn, ignore or submerge the language forms of minority populations, who may base their own sense of proud identity upon their regional linguistic separateness. In scale and recognition even below these ethnically identified regional languages are those local speech variants frequently denied status as identifiable languages and cited as proof of the ignorance and the cultural deprivation of their speakers. Yet such a limited-area, limited-population tongue contains all the elements of the classic culturally based region. Its area is defined; its boundaries are easily drawn; it represents homogeneity and majority behavior among its members; and it summarizes, by a single cultural trait, a collection of areally distinctive outlooks.

Gullah as Language.[5] *Isolation is a key element in the retention or the creation of distinctive and even externally unintelligible languages. The isolation of the ancestors of the quarter-million present-day speakers of Gullah—themselves called Gullahs—was nearly complete. Held by the hundreds as slaves on the offshore islands and in the nearly equally remote low country along the southeastern United States coast from South Carolina to the Florida border (Figure 13.10), the blacks retained both the speech patterns of the African languages—Ewe, Fanti, Bambara, Twi, Wolof, Ibo, Malinke, Yoruba, Efik—native to the slave groups and over 4000 words drawn from them. Folk tales written in the Gullah creole can be heard and understood by Krio-speaking audiences in Sierra Leone today.*

Forced to use English words for minimal communication with their white overseers, but modifying, distorting, and interjecting African-based substitute words into that unfamiliar language, the Gullahs kept intonations and word and idea order in their spoken common speech that made it unintelligible to white masters or to more completely integrated mainland blacks. Because the language was not understood, its speakers were considered ignorant, unable to master the niceties of English. Because ignorance was ascribed to them, the Gullahs learned to be ashamed of themselves, their culture, and their tongue, which even they themselves did not recognize as a highly structured and sophisticated separate language.

5. By Jerome Fellmann.

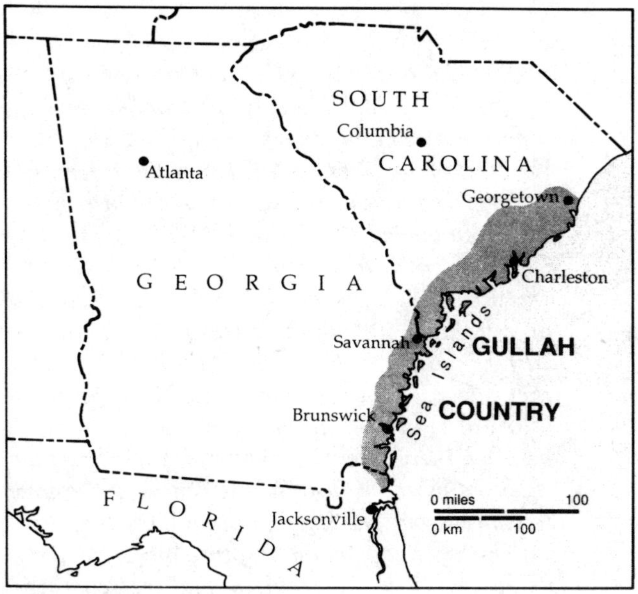

Figure 13.10
Gullah speakers are concentrated on the Sea Islands and the coastal mainland of South Carolina and Georgia. The isolation that promoted their linguistic distinction is now being eroded.

In common with many linguistic minorities, the Gullahs are losing their former sense of inferiority and gaining pride in their cultural heritage and in the distinctive tongue that represents it. Out of economic necessity, standard English is being taught to their schoolchildren, but an increasing scholarly and popular interest in the structure of their language and in the nature of their culture has caused Gullah to be rendered as a written language, studied as a second language, and translated into English.

In both the written and the spoken versions, Gullah betrays its African syntax patterns, particularly in its employment of terminal locator words: "Where you goin' at?" The same African origins are revealed by the absence in English translation of distinctive tenses: "I be tired" conveys the concept that "I have been tired for a period of time." Though tenses exist in the African root languages, they are noted more by inflection than by special words and structures.

"He en gut no morratater fer mak no pie wid" may be poor English, but it is good Gullah. Its translation—"He has no more sweet potatoes for making pie"—renders it intelligible to ears attuned to English but loses the musical lilt of original speech and, more importantly, obscures the cultural identity of the speaker, a member of a regionally compact group of distinctive Americans whose territorial extent is clearly defined by its linguistic dominance.

Political Regions *(See "Boundaries: The Limits of the State," page 248)*

The most rigorously defined formal cultural region is the national state. Its boundaries are presumably carefully surveyed and are, perhaps, marked by fences and guard posts. There is no question of an arbitrarily divided transition zone or of lessening toward the borders of the basic quality of the regional core. This rigidity of a country's boundaries, its unmistakable placement in space, and the trappings—flag, anthem, government, army—that are uniquely its own give to the state an appearance of permanence and immutability not common in other, more fluid, cultural regions. But its stability is often more imagined than real. Political boundaries are not necessarily permanent. They are subject to change, sometimes violent change, as a result of internal and external pressures. The Indian subcontinent illustrates the point.

Political Regions in the Indian Subcontinent.[6] *The history of the subcontinent since about 400 B.C. has been one of the alternating creation and dissolution of empires, of the extension of central control based upon the Ganges Basin, and of resistance to that centralization by the marginal territories of the peninsula. British India, created largely unintentionally by 1858, was only the last, though perhaps the most successful, attempt to bring under unified control the vast territory of incredibly complex and often implacably opposed racial, religious, and linguistic groupings.*

A common desire for independence and freedom from British rule united the subcontinent's disparate populations at the end of World War II. That common desire, however, was countered by the mutual religious antipathies felt by Muslims and Hindus, each dominant in separate regions of the colony and each unwilling to be affiliated with or subordinated to the other. When the British surrendered control of the subcontinent in 1947, they recognized these apparently irreconcilable religious differences and partitioned the subcontinent into the second and seventh most populous countries on earth. The independent state of India was created out of the largely Hindu areas constituting the bulk of the former colony. Separate sovereignty was granted to most of the Muslim-majority area under the name of Pakistan. Even so, the partition left boundaries, notably in the Vale of Kashmir, dangerously undefined or in dispute.

An estimated 1 million people died in the religious riots that accompanied the partition decision. In perhaps the largest short-term mass

6. By Jerome Fellmann.

migration in history, some 10 million Hindus moved from Pakistan to India, and 7.5 million Muslims left India for Pakistan, "The Land of the Pure."

Unfortunately, the purity resided only in common religious belief, not in spatial coherence or in shared language, ethnicity, customs, food, or economy. During its 23 years of existence as originally conceived, Pakistan was a sorely divided country. The partition decision created an eastern and a western component separated by more than 1000 miles (1600 km) of foreign territory and united only by a common belief in Mohammed (Figure 13.11). West Pakistan, as large as Texas and Oklahoma combined, held 55 million largely light-skinned Punjabis with Urdu language and strong Middle Eastern cultural ties. Some 70 million Bengali-speakers, making up East Pakistan, were crammed into an Iowa-size portion of the delta of the Ganges and Brahmaputra rivers. The western segment of the country was part of the semiarid world of western Asia; the eastern portion of Pakistan was joined to humid, rice-producing Southeast Asia.

Beyond the affinity of religion, little else united the awkwardly separated country. East Pakistan felt itself exploited by a dominating western minority that sought to impose its language and its economic development, administrative objectives, and military control. Rightly or wrongly, East Pakistanis saw themselves as aggrieved and abused. They complained of a per capita income level far below that of their western compatriots, claimed discrimination in the allocation of investment capital, found disparities in the pricing of imported foods, and asserted that their exports of raw materials, particularly jute, were supporting a national economy in which they did not share proportionally. They argued that their demands for regional autonomy, voiced since nationhood, had been denied.

When, in November of 1970, East Pakistan was struck by a cyclone and tidal wave that took an estimated 500,000 lives (see p. 2), the limit of eastern patience was reached. Resentful over what they saw as a totally inadequate West Pakistani effort of aid in the natural disaster that had befallen them, the East Pakistanis were further incensed by the refusal of the central government to convene on schedule a national assembly to which they had won an absolute majority of delegates. Civil war resulted, and the separate new state of Bangladesh was created. The sequence of political change in the subcontinent is traced in Figure 13.11.

The country and nation-state so ingrained in our consciousness and so firmly defined, as displayed on the map inside the front cover of this book, is both a recent and an ephemeral creation of the cultural regional landscape. It rests upon a claim, more or less effectively enforced, of a monopoly of power and allegiance resident in a government and superior to the communal, linguistic, ethnic, or religious affiliations that preceded it or that claim loyalties overriding it. As the violent recent history of the Indian subcontinent demonstrates, nationalism may be sought, but its maintenance is not assured by the initiating motivations.

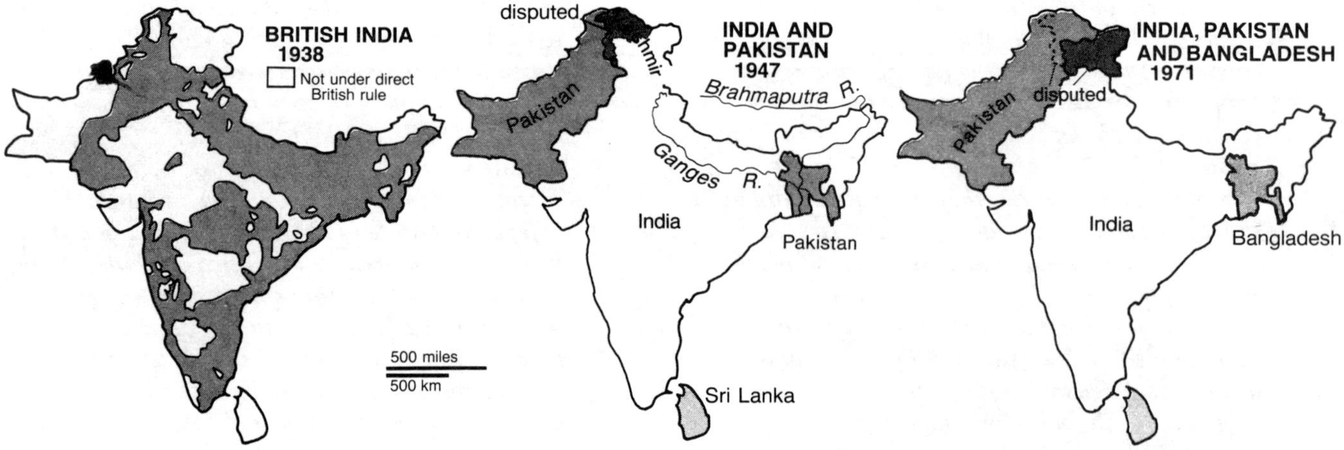

Figure 13.11

The sequence of political change on the Indian subcontinent. British India was transformed in 1947 to the countries of India and Pakistan, the latter a Muslim state with a western and an eastern component. In 1970 Pakistan was torn by civil war based on ethnic and political contrasts, and the eastern segment became the new country of Bangladesh.

Mental Regions *(See "Mental Maps," page 274)*

The regional units so far used as examples, and the methods of regionalization they demonstrate, have a concrete reality. They are formal or functional regions of specified, measurable content. They have boundaries drawn by some objective measures of change or alteration of content, and they have location upon an accurately measured global grid.

Individuals and whole cultures may operate, and operate successfully, with a much less formalized and less precise picture of the nature of the world and of the structure of its parts. The mental maps discussed in Chapter 9 represent personal views of regions and regionalization. The private world views they represent are, as we also saw, colored by the culture of which their holders are members.

Primitive societies, particularly, have distinctive world views by which they categorize what is familiar, and satisfactorily account for what is not. The Yurok Indians of the Klamath River area of northern California were no exception. Their geographic concepts were reported by T.T. Waterman, from whose paper, "Yurok Geography," the following summary is drawn.

> ***The Yurok World View.***[7] *The Yurok imagines himself to be living on a flat extent of landscape, which is roughly circular and surrounded by ocean. By going far enough up the river, it is believed, "you come to salt water again." In other words, the Klamath River is considered, in a sense, to bisect the world. This whole earth mass, with its forests and mountains, its rivers and sea cliffs, is regarded as slowly rising and falling, with a gigantic but imperceptible rhythm, on the heaving, primeval flood. The vast size of the "earth" causes you not to notice this quiet heaving and settling. This earth, therefore, is not merely surrounded by the ocean but floats upon it. At about the central point of the "world" lies a place which the Yurok call qe'nek, on the southern bank of the Klamath, a few miles below the point where the Trinity River comes in from the south. In the Indian concept, this point seems to be accepted as the center of the world.*
>
> *At this locality also the sky was made. Above the solid sky there is a sky-country, wo'noiyik, about the topography of which the Yurok's ideas are almost as definite as are his ideas of southern Mendocino County, for instance. Downstream from qe'nek, at a place called qe'nek-pul ("qe'nek-downstream"), is an invisible ladder leading up to the sky-country. The ladder is still thought to be there, though no one to my knowledge has been up it recently. The sky-vault is a very definite item in the Yurok's cosmic scheme. The structure consisting of the sky dome and the flat expanse of landscape and waters that it encloses is known to the Yurok as ki-we'sona (literally "that which exists"). This sky, then, together with its flooring of landscape, constitutes "our world." I used to be puzzled at the Yurok confusing earth and sky, telling me, for example, that a certain gigantic redwood tree "held up the world." Their ideas are of course perfectly logical, for the sky is as much a part of the "world" in their sense as the ground is.*
>
> *The Yurok believe that passing under the sky edge and voyaging still outward you come again to solid land. This is not our world, and mortals ordinarily do not go there; but it is good, solid land. What are breakers over here are just little ripples over there. Yonder lie several regions. To the north (in our sense) lies pu'lekūk, downstream at the north end of creation. South of pu'lekūk lies tsī'k-tsīk-ol ("money lives") where the dentalium-shell, medium of exchange, has its mythical abode. Again, to the south there is a place called kowe'tsik, the mythical home of the salmon, where also all have a "house." About due west of the mouth of the Klamath lies rkrgr', where lives the culture-hero wo'xpa-ku-mä ("across-the-ocean that widower").*
>
> *Still to the south of rkrgr' there lies a broad sea, kiolaaopa'a, which is half pitch—an Algonkian myth idea, by the way. All of these solid lands just mentioned lie on the margin, the absolute rim of things. Beyond them the Yurok does not go even in imagination. In the opposite direction, he names a place pe'tskuk ("up-river-at"), which is the upper "end" of the river but still in this world. He does not seem to concern himself much with the topography there.*
>
> *The Yuroks' conception of the world they live in may be summed up in the accompanying diagram (Figure 13.12).*

7. *University of California Publications in American Archaeology and Ethnology,* Vol. 16, No. 5 (1920). Selection from pp. 189–93.

Part III. Regions in the Locational Tradition

While location, as we have seen, is a primary attribute of all regions, regionalization in the locational tradition of geography implies far more than a named delimitation of earth space. The central concern is with the distribution of human activities and of the resources upon which those activities are based.

In this sense, world regionalization of agriculture and of the soils and climates with which it is related is within

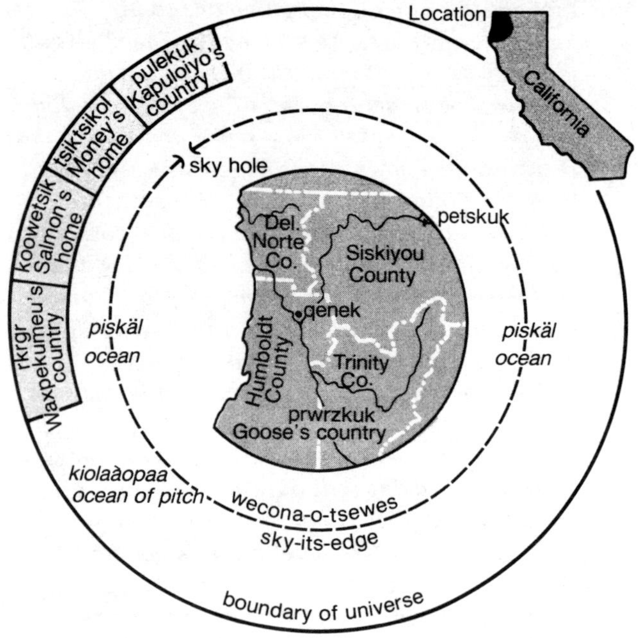

Figure 13.12
The world view of the Yurok as pieced together by T. T. Waterman during his anthropological study of the tribe. Qe'nek, in the center of the diagram, marks the center of the world in Indian belief.

the locational tradition. For practical and accepted reasons, however, such underlying physical patterns have been included under the earth-science tradition. But the point is made: the locational tradition emphasizes the "doing" in human affairs, and "doing" is not an abstract thing but an interrelation of life and the resources on which life depends.

The locational tradition, therefore, encourages the recognition and the definition of a far wider array of regional types than do the earth-science or culture–environment traditions. Any single pattern of economic activity or resource distribution invites the recognition of definable *formal regions.* The interchange of commodities, the control of urban market areas, the flows of capital, or the collection and distribution activities of ports are just a few examples of the infinite number of analytically useful *functional regions* that one may recognize.

Economic Regions *(See page 301)*

Economic regionalization is among the most frequent, familiar, and useful employments of the regional method. Through economic regions the geographer identifies activities and resources, maps the limits of their occurrence or use, and examines the interrelationships and flows that are part of the complexities of the contemporary world.

The economic region, examples of which were explored in Chapter 10, should be seen as potentially more than a device for recording what *is* in either a formal or a functional sense. It has increasingly become a device for examining what might or should be. The concept of the economic region as a tool for planning and a framework for the manipulation of the people, resources, and economic structure of a composite region first took root in the United States during the Great Depression years of the 1930s. The key element in the planning region is the public recognition of a major territorial unit in which economic change or decline is seen as the cause of a variety of interrelated problems including, for example, population out-migration, regional isolation, cultural deprivation, underdevelopment, and poverty.

Appalachia.[8] *Until the early 1960s, "Appalachia" was for most a loose reference to the complex physiographic province of the eastern United States associated with the Appalachian mountain chain. If thought of at all, it was apt to be visualized as rural, isolated, and tree-covered; as an area of coal mining, hillbillies, and folksongs (Figure 13.13).*

During the 1950s, however, the economic stagnation and the functional decline of the area became increasingly noticeable in the national context of economic growth, rising personal incomes, and growing concern with the elimination of the poverty and deprivation of every group of citizens. Less dramatically but just as decisively as the Dust Bowl or the Tennessee Valley of an earlier era, Appalachia became simultaneously a popularly recognized economic and cultural region and a governmentally determined planning region.

Evidences of poverty, underdevelopment, and social crisis were obvious to a country committed to recognize and eradicate such conditions within its own borders. By 1960, per capita income within Appalachia was $1400 when the national average was $1900. In the decade of the 1950s, mine employment fell 60% and farm jobs declined 52%; the rest of the country lost only 1% of mining jobs and 35% of agricultural employment. Rail employment fell with the drop in coal mining. Massive out-migration occurred among the young adults, with such cities as Chicago, Detroit, Dayton, Cleveland, and Gary the targets. Even with these departures, unemployment among those who remained in Appalachia averaged 50% higher than the national rate. Because of the departures, the remnant population—only 47% of whom lived, in 1960, in or near cities, as against 70% for the entire United States—was distorted in age structure. The very young and the old were disproportionally represented; the productive working-age groups were, at least temporarily, emigrants.

8. By Jerome Fellmann.

Figure 13.13

This deserted eastern Kentucky cabin is a mute reminder of the economic and social changes in Appalachia. Increases in per capita income, expansion of urban job opportunities, and improved highway networks have reduced the poverty and isolation so long associated with the region.

When these and other socioeconomic indicators were plotted by counties and by state economic areas, an elongated but regionally coherent and clearly bounded Appalachia as newly understood was revealed by maps (Figure 13.14). It extended through 13 states from Mississippi to New York, covered some 195,000 square miles (505,000 km²), and contained 18 million people, 93% of them white.

By 1963, awareness of the problems of the area at federal and state levels passed beyond recognition of a multifactor economic region to the establishment of a planning region. A joint federal-state Appalachian Regional Commission was created to develop a program designed to meet the perceived needs of the entire area. The approach chosen was one of limited investment in a restricted number of highly localized developments, with the expectation that these would spark economic growth supported by private funds.

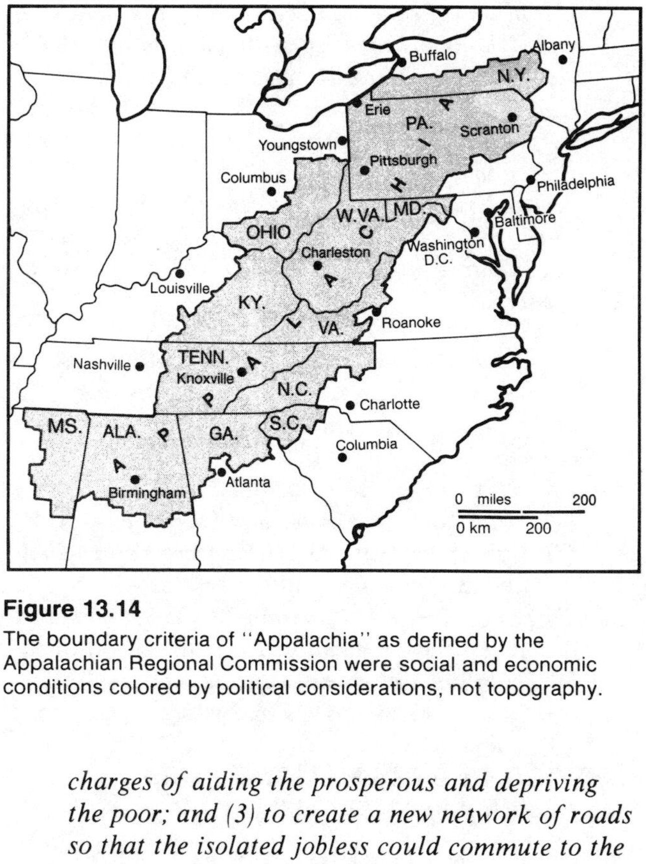

Figure 13.14

The boundary criteria of "Appalachia" as defined by the Appalachian Regional Commission were social and economic conditions colored by political considerations, not topography.

In outline, the plan was (1) to ignore those areas of poverty and unemployment that were in isolated, inaccessible "hollows" throughout the region; (2) to designate "growth centers" where development potential was greatest and concentrate all spending for economic expansion there; the regional growth potential in targeted expenditure was deemed sufficient to overcome charges of aiding the prosperous and depriving the poor; and (3) to create a new network of roads so that the isolated jobless could commute to the new jobs expected to form in and near the favored growth centers. Road construction would also, of course, open inaccessible areas to tourism and strengthen the economic base of the entire planning region.

In the years since the Appalachian Regional Commission was established, and in ways not anticipated, the economic prospects of Appalachia have altered. National energy crises at least temporarily revived mining and transportation employment; new industrial jobs multiplied as manufacturing has relocated to, or has been newly developed in, the Appalachia portion of the Sunbelt. New employment opportunities have exceeded local labor pools, and out-migrants have returned home from cities outside the region, bringing again a more balanced population pyramid.

By the late 1980s, the Appalachian Regional Commission had over the years committed more than $5 billion to the area, and over $10 billion more had come from other sources. The percentage of people living in poverty within the region declined from 31% in 1960 to 14% in 1980, and per capita income, which had been 79% of the national average in 1960, rose to 85% by 1986. Population flows stabilized, with the number of people coming into Appalachia about equaling those leaving. Some 2000 miles (3200 km) of road had been laid by the end of 1986, though their construction did not give the total regional access that had been hoped. Smaller towns and their hinterlands have been by-passed and remain as isolated as ever. Some basic social services have, for the first time, reached essentially everyone in the region. Each of the 397 counties of the commission territory, for example, has been supplied with a clinic or a doctor.

Despite these evidences of progress, Appalachia still remained economically distressed in the late 1980s. The mining and manufacturing recession of the earlier years of the decade was sorely felt within a region dependent upon coal mining and with an industrial base of primary and fabricated metals, wood products, textiles, and apparel—all eroded by imports replacing domestic products. Unemployment rates continued to be between 5% and 10% above the national average, and a third of the mining and manufacturing jobs disappeared between 1979 and 1987.

Regions of Natural Resources
(See "Coal," page 345)

The unevenly distributed resources upon which people depend for existence are logical topics of interest within the locational tradition of geography. Resource regions are mapped, and raw material qualities and quantities are discussed. Areal relationships to industrial concentrations and the impacts of material extraction upon alternate uses of areas are typical interests in resource geography and in the definition of resource regions.

Those regions, however, are usually treated as if they were expressions of observable surface phenomena, as if, somehow, an oil field were as exposed and two-dimensional as a soil region or a manufacturing district. What is ignored is that most mineral resources are three-dimensional regions beneath the ground. In addition to the characteristics of an area that may form the basis for regional delimitations and descriptions of surface phenomena, regions beneath the surface add their own particularities to the problem of regional definition. They have, for example, upper and lower boundaries in addition to the circumferential bounds of surface features. They may have an internal topography divorced from the visible landscape. Subsurface relationships—for example, mineral distribution and accessibility in relation to its enclosing rock or to groundwater amounts and movement—may be critical in understanding these specialized, but real, regions. An illustration drawn from the Schuylkill field of the anthracite region of eastern Pennsylvania helps illustrate the nature of regions beneath the surface.

The Schuylkill Anthracite Region.[9] *Nothing in the wild surface terrain of the anthracite country suggested the existence of an equally rugged subterranean topography of coal beds and interstratified rock, slate, and fire clays forming a total vertical depth of 3000 feet (900 m) at greatest development. Yet the creation of the surface landscape was an essential determinant of the areal extent of the Schuylkill district, of the contortions of its bedding, and of the nature of its coal content. A county history of the area reports, "The physical features of the anthracite country are wild. Its area exhibits an extraordinary series of parallel ridges and deep valleys, like long, rolling lines of surf which break upon a flat shore." Both the surface and the subsurface topographies reflect the strong folding of strata after the coal seams were deposited; the anthracite (hard) coal resulted from metamorphic carbonization of the original bituminous beds. Subsequent river and glacial erosion removed as much as 95% of the original anthracite deposits and gave to those that remained discontinuous existence in sharply bounded fields like the Schuylkill (Figure 13.15), a discrete areal entity of 181 square miles (470 km²).*

The irregular topography of the underground Schuylkill region means that the interbedded coal seams, the most steeply inclined of all the

9. By Jerome Fellmann.

anthracite regions (Figure 13.16), outcrop visibly at the surface on hillsides and along stream valleys. The outcrops made the presence of coal known as early as 1770, but not until 1795 did Schuylkill anthracite find its first use by local blacksmiths. Reviled as "stone coal" or "black stone" that would not ignite, anthracite found no ready commercial market, although it was used in wire and rolling mills located along the Schuylkill River before 1815 and to generate steam in the same area by 1830.

The resources of the subterranean Schuylkill region affected human patterns of surface regions only after the Schuylkill Navigation Canal was completed in 1825 (Figure 13.17), providing a passage to rapidly expanding external markets for the fuel and for the output of industry newly located atop the region. Growing demand induced a boom in coal exploitation, an exhaustion of the easily available outcrop coal, and the beginning of the more arduous and dangerous underground mining.

The early methods of mining were simple: merely quarrying the coal from exposed outcrops, usually driving on a slight incline to permit natural drainage. Deep shafts were unnecessary and, indeed, not thought of, since the presence of anthracite at depth was not suspected. Later, when it was no longer possible to secure coal from a given outcrop, a small pit was sunk to a depth of 30–40 feet (9–12 m); when the coal and the water that accumulated in the pit could no longer safely be brought to the surface by windlass, the pit was abandoned and a new one was started. Shaft mining, in which a vertical opening from the surface provides penetration to one or several coal beds, eventually became a necessity; with it came awareness of the complex interrelationships between seam thickness, the nature of interstratified rock and clays, the presence of gases, and the movement of subsurface water.

The subterranean Schuylkill region has a three-dimensional pattern of use. The configuration and the variable thickness of the seams demand concentration of mining activities. Minable coal is not uniformly available along any possible vertical or horizontal cross section because of the interstratification and the extreme folding of the beds. Mining is concentrated further by the location of shafts and the construction of passages, in their turn determined by both patterns of ownership and thickness of seam. In general, no seam less than 2 feet (.6 m)

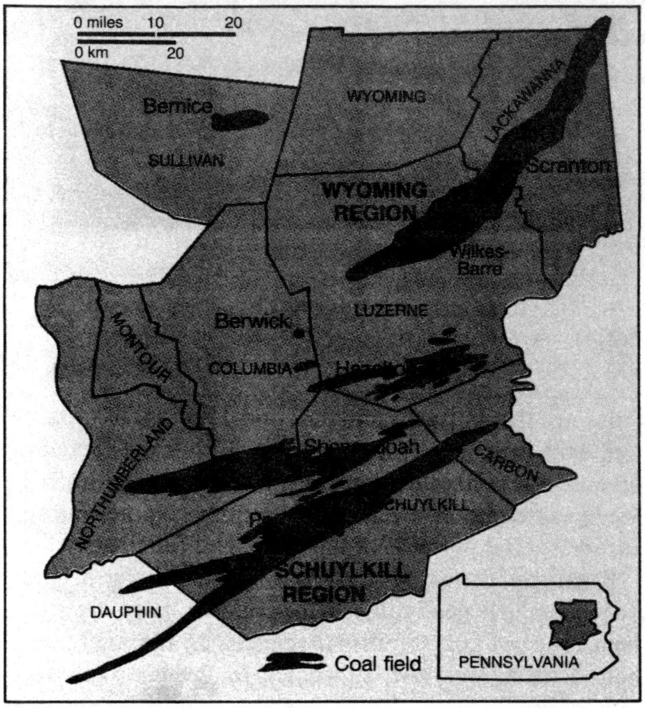

Figure 13.15

The anthracite regions of northeastern Pennsylvania are sharply defined by the geologic events that created them.

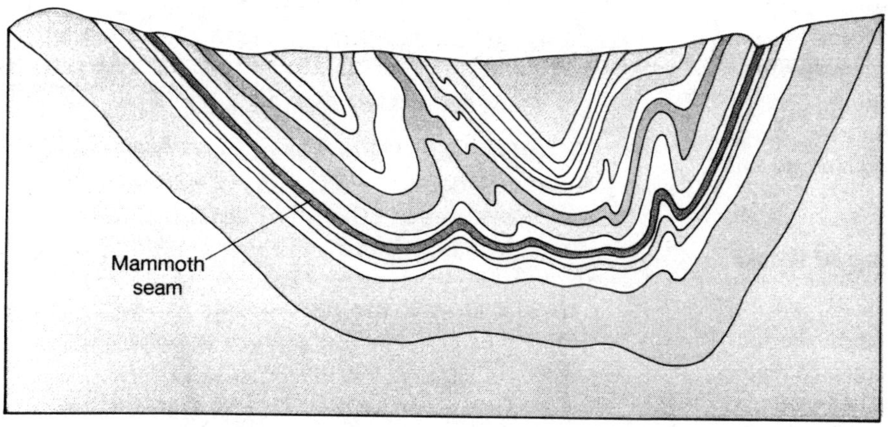

Figure 13.16

The deep folding of the Schuylkill coal seams made them costly to exploit. The Mammoth seam runs as deep as 1500 to 2000 feet (460-610 m) below the surface.

Figure 13.17
The Schuylkill Navigation Canal, like its counterpart Lehigh Canal shown here, provided an outlet to market for the anthracite region's coal resources after 1825.

is worked, and the absolute thickness—50 feet (15 m)—is found only in the Mammoth seam of the Schuylkill region.

Friable interstratified rock increases the danger of coal extraction and raises the costs of cave-in prevention. Although the Schuylkill mines are not gassy, the possibility of gas release from the collapse of coal pillars left as mine supports makes necessary systems of ventilation even more elaborate than those minimally required to provide adequate air to miners. Water is ever-present in the anthracite workings, and constant pumping or draining is necessary for mine operation. The collapse of strata underlying a river may result in sudden disastrous flooding.

The Schuylkill subterranean anthracite region presents a pattern of complexity in distribution of physical and cultural features and of interrelationships between phenomena every bit as great and inviting of geographical analysis as any purely surficial region.

Urban Regions *(See "The Making of World Metropolitan Regions," page 402)*

Urban geography represents a climax stage in the locational tradition of geography. Modern integrated, interdependent society on a world basis is urban-centered. Cities are the indispensable functional focuses of production, exchange, and administration. They exist individually as essential elements in interlocked hierarchical systems of cities. Internally, they display complex but repetitive spatial patterns of land uses and functions.

Because of the many-sided character of urbanism, cities are particularly good subjects for regional study. They are themselves, of course, formal regions. In the aggregate, their distributions give substance to formal regions of urban concentration. Cities are also the cores of functional regions of varying types and hierarchical orders. Their internal diversity of functional, land use, and socioeconomic patterns invite regional analysis. The employment of that approach in both its formal and functional modes is clearly displayed by Jean Gottmann, who, in examining the data and landscapes of the eastern United States at midcentury, recognized and analyzed Megalopolis. The following is taken from his study.

Megalopolis.[10] *The northeastern seaboard of the United States is today the site of a remarkable development—an almost continuous stretch of urban and suburban areas from southern New Hampshire to northern Virginia and from the Atlantic shore to the Appalachian foothills (Figure 13.18). The processes of urbanization,*

10. From Jean Gottman, *Megalopolis: The Urbanized Northeastern Seaboard of the United States,* Copyright 1961 by the Twentieth Century Fund, New York.

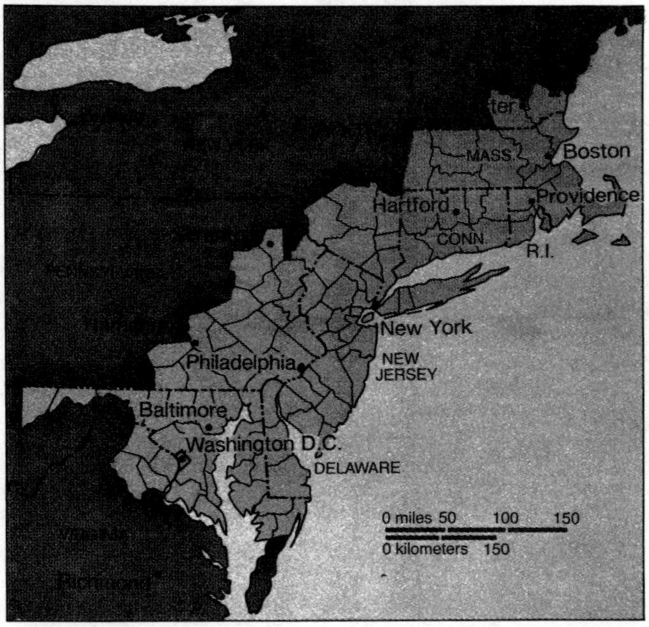

Figure 13.18

Megalopolis in 1960. The region was then composed of counties that, by U.S. census definition, were "urban" in population and economic characteristics. Much of the area is still distinctly "rural" in land use.

rooted deep in the American past, have worked steadily here, endowing the region with unique ways of life and of land use. No other section of the United States has such a large concentration of population, with such a high average density, spread over such a large area. And no other section has a comparable role within the country or a comparable importance in the world. Here has been developed a kind of supremacy, in politics, in economics, and possibly even in cultural activities, seldom before attained by an area of this size.

Great, then, is the importance of this section of the United States and of the processes now at work within it. And yet it is difficult to single this area out from the surrounding areas, for its limits cut across established historical divisions, such as New England and the Middle Atlantic states, and across political entities, since it includes some states entirely and others only partially. A special name is needed, therefore, to identify this special geographical area.

This particular type of region is new, but it is the result of age-old processes, such as the growth of cities, the division of labor within a civilized society, the development of world resources. The name applied to it should, therefore, be new as a place name but old as a symbol of the long tradition of human aspirations and endeavor underlying the situations and problems now found here. Hence the choice of the term Megalopolis, *used in this study.*

As one follows the main highways or railroads between Boston and Washington, D.C., one hardly loses sight of built-up areas, tightly woven residential communities, or powerful concentrations of manufacturing plants. Flying this same route one discovers, on the other hand, that behind the ribbons of densely occupied land along the principal arteries of traffic, and in between the clusters of suburbs around the old urban centers, there still remain large areas covered with woods and brush alternating with some carefully cultivated patches of farmland (Figure 13.19). These green spaces, however, when inspected at closer range, appear stuffed with a loose but immense scattering of buildings, most of them residential but some of industrial character. That is, many of these sections that look rural actually function largely as suburbs in the orbit of some city's downtown. Even the farms, which occupy the larger tilled patches, are seldom worked by people whose only occupation and income are properly agricultural.

Thus the old distinctions between rural and urban do not apply here anymore. Even a quick look at the vast area of Megalopolis reveals a revolution in land use. Most of the people living in the so-called rural areas, and still classified as "rural population" by recent censuses, have very little, if anything, to do with agriculture. In terms of their interests and work they are what used to be classified as "city folks," but their way of life and the landscapes around their residences do not fit the old meaning of urban.

In this area, then, we must abandon the idea of the city as a tightly settled and organized unit in which people, activities, and riches are crowded into a very small area clearly separated from its nonurban surroundings. Every city in this region spreads out far and wide around its original nucleus; it grows amidst an irregularly colloidal mixture of rural and suburban landscapes; it melds on broad fronts with other mixtures, of somewhat similar though different texture, belonging to the suburban neighborhoods of other cities.

Thus an almost continuous system of deeply interwoven urban and suburban areas, with a total population of about 37 million people in 1960, has been erected along the Northeastern Atlantic seaboard. It straddles state boundaries, stretches across wide estuaries and bays, and encompasses many regional differences. In fact,

Figure 13.19
The "Pine Barrens" of New Jersey is still a largely undisturbed natural enclave in the heart of Megalopolis. Its preservation is a subject of dispute between environmentalists and developers.

> *the landscapes of Megalopolis offer such variety that the average observer may well doubt the unity of the region. And it may seem to him that the main urban nuclei of the seaboard are little related to one another. Six of its great cities would be great individual metropolises in their own right if they were located elsewhere. This region indeed reminds one of Aristotle's saying that cities such as Babylon had "the compass of a nation rather than a city."*

The description of Megalopolis in 1960 represents, as does any regional study, a captured moment on the continuum of areal change. When that moment is lucidly described and the threads of areal character clearly delineated, the regional study serves as both a summary of the present and an augury of a future whose roots may be discerned in it. In the years since Gottmann described it, Megalopolis has continued to develop along the lines he summarized. Urbanization has continued, physically as well as functionally, to encroach upon the rural landscapes without regard for state boundaries or even the metropolitan cores that dominated in 1960. As then, new growth centers—becoming equivalents of central cities in their own right—are linked to the expanding transportation corridors, though now the lines of importance are increasingly expressways rather than railroads.

In the south, the corridors around Washington, D.C., are the Capital Beltway in Fairfax County and an extension from it westward to Loudoun County and Dulles Airport in Virginia and, to the north into Montgomery County, Maryland, the I-270 corridor. The Virginia suburbs, focusing on Tysons Corner (Figure 10.34), have specialized in defense-related industries, but vast office building complexes and commercial centers are rapidly converting rural land to general urban uses. The Maryland suburbs specialize in health, space, and communications interests in their own new office, industrial, and commercial "parks" and complexes. Still farther to the north, in the "Princeton Corridor," a 26-mile (42 km) stretch along Route 1 from New Brunswick to Trenton, New Jersey, huge corporate parks provide office and research space to companies relocating from the New York area and to new technology firms attracted by proximity to Princeton University. Morristown, New Jersey, and White Plains in Westchester County, New York, are other similar concentrations of commercial, industrial, and office developments north and west of New York City. To the east, Stamford, Connecticut, with 150,000 daily in-commuters has become the headquarters center for major national corporations and has emerged as a truly major central city new on the scene in its present form since Megalopolis was first described. And still farther to the north, the I-495 beltway around Boston has joined Route 128 as a magnet for electronics, computer, electrical, and other technology-related companies.

Summary

The region is a mental construct, a created entity whose sole function is the purposeful organization of spatial data. The scheme of that organization, the selection of data to be analyzed, and the region resulting from these decisions are reflections of the intellectual problem posed.

This chapter has not attempted to explore all aspects of regionalism and of the regional method. It has tried, by example, only to document its basic theme: the geographer's regions are arbitrary but deliberately conceived devices for the isolation of things, patterns, interrelations, and flows that invite geographic analysis. In this sense, all geographers are regional geographers, and the regional examples of this chapter may logically complete our survey of the four traditions of geography.

Key Words

formal region	region
functional region	regional concept

For Review

1. What do geographers seek to achieve when they recognize or define regions? On what basis are regional boundaries drawn? Are regions concrete entities whose dimensions and characteristics are agreed upon by all who study the same general segment of earth space? Ask three fellow students who are not participants in this course for their definition of the "South." If their answers differed, what implicit or explicit criteria of regional delimitation were they employing?
2. What are the spatial elements or the identifying qualities shared by all regions?
3. What is the identifying characteristic of a *formal region?* How are its boundaries determined? Name three different examples of formal regions drawn from any earlier chapters of this book. How was each defined and what was the purpose of its recognition?
4. How are *functional regions* defined? What is the nature of their bounding criteria? Give three or four examples of functional regions that were defined earlier in the text.
5. The ecosystem was suggested as a viable device for regional delimitation. What regional geographic concepts are suggested in ecosystem recognition? Is an ecosystem identical to a formal region? Why or why not?
6. National, linguistic, historical, planning, and other regions have been recognized in this chapter. With what other regional entities do you have acquaintance in your daily affairs? Are fire protection districts, police or voting precincts, or zoning districts regional units identifiable with geographers' regions as discussed in this chapter or elsewhere in this book? How influenced are you in your private life by your, or others', regional delimitations?

Selected References

Freeman, T. W. "The Regional Approach." *A Hundred Years of Geography.* Chicago: Aldine, 1961.

Gottmann, Jean. *Megalopolis Revisited: 25 Years Later.* College Park: Institute for Urban Studies, University of Maryland, 1987.

McDonald, James R. "The Region: Its Conception, Design and Limitations." *Annals of the Association of American Geographers* 56 (1966): 516–28.

Meinig, Donald. "The Mormon Culture Region: Strategies and Patterns in the Geography of the American West, 1847–1964." *Annals of the Association of American Geographers* 55 (1965): 191–220.

Minshull, Roger. *Regional Geography.* Chicago: Aldine, 1967.

Murphey, Rhoads. "The Region." *The Scope of Geography.* 3d ed. New York and London: Methuen, 1982.

Whittlesey, Derwent. "The Regional Concept and the Regional Method." In *American Geography: Inventory and Prospect.* Preston James and Clarence F. Jones, eds. Syracuse, N.Y.: Syracuse University Press (for the Association of American Geographers), 1954.

Chapter 3

Spatial Interaction and Spatial Behavior

Early in January of 1849 we first thought of migrating to California. It was a period of National hard times . . . and we longed to go to the new El Dorado and "pick up" gold enough with which to return and pay off our debts.

Our discontent and restlessness were enhanced by the fact that my health was not good. . . . The physician advised an entire change of climate thus to avoid the intense cold of Iowa, and recommended a sea voyage, but finally approved of our contemplated trip across the plains in a "prairie schooner."

Full of the energy and enthusiasm of youth, the prospects of so hazardous an undertaking had no terror for us, indeed, as we had been married but a few months, it appealed to us as a romantic wedding tour.[1]

So begins Catherine Haun's account of her and her husband's 9-month journey from Iowa to California. They were just two of the quarter-million people who traveled across the continent on the Overland Trail in one of the world's great migrations. The migrants faced months of grueling struggle over badly marked routes that crossed swollen rivers, deserts, and mountains. The weather was often foul, with hailstorms, drenching rains, and burning summer temperatures. Graves along the route were a silent testimony to the lives claimed by buffalo stampedes, Indian skirmishes, cholera epidemics, and other disasters.

What inducements were so great as to make emigrants leave behind all that was familiar and risk their lives on an uncertain venture? Catherine Haun alludes to economic hard times gripping the country and to their hope for riches to be found in California. Like other migrants, the Hauns were attracted by the climate in the West, which was said to be always sunny and free of disease. Finally, like most who undertook the perilous journey West, the Hauns were young, moved by restlessness, a sense of adventure, and a perception of greater opportunities in a new land. They, like their predecessors back to the beginnings of humankind, were acting in space and across space on the basis of acquired information and awareness of opportunity, prepared to pay the price in time, money, and hardship costs, of overcoming distance.

In human geography, a fundamental question is What considerations influence how individual human beings use space and act within it? Related analytical queries are these: Are there discernible controls on human spatial behavior? How does distance affect human interaction? How do our perceptions of areas influence our spatial activities? How do we overcome the consequences of distance in the exchange of commodities and information? How are movement and migration decisions (such as that of the Hauns) reached? These questions address geography's concern with understanding spatial interaction.

Spatial interaction means the movement of peoples, ideas, and commodities within and between areas. The Hauns were engaging in spatial interaction (Figure 3.1). International trade, the movement of semitrailers on the expressways, radio broadcasts, and business or personal telephone calls are more familiar examples. Such movements and exchanges are designed to achieve effective integration between different points of human activity. Movement of whatever nature satisfies some felt need or desire. It represents the attempt to smooth out the spatially differing availability of required resources, commodities, information, or opportunities. Because commodity movements represent simple demonstrations of the principles underlying all spatial interactions, let us turn to them first.

Figure 3.1

Cross country movement was slow, arduous, and dangerous early in the 19th century, and the price of long-distance spatial interaction was far higher in time and risks than a comparable journey today.

1. Source: From Catherine Haun, "A Woman's Trip Across the Plains in 1849," in Lillian Schlissel, *Women's Diaries of the Westward Journey*. Copyright © 1982 Schocken Books, New York, NY.

Bases for Interaction

Neither the world's resources nor the products of people's efforts are uniformly distributed. Commodity flows are responses to these differences, links between points of supply and locales of demand. Such response may not be immediate or even direct. Matters of awareness of supplies or markets, the presence or absence of transportation connections, costs of movement, ability to pay for things wanted and needed—all, and more, are factors in the structure of trade. Underlying even these, however, is a set of controlling principles governing spatial interaction.

A Summarizing Model

The conviction that spatial interaction reflects areal differences led the geographer Edward Ullman (1912–1976) to speculate on the essential conditions that affect such interactions and to propose an explanatory model. He observed that spatial interaction is effectively controlled by three flow-determining factors that he called *complementarity, transferability,* and *intervening opportunity.*

Complementarity

For two places to interact, one place must have a supply of an item for which there is an effective demand in the other, as evidenced by desire for the item, purchasing power to acquire it, and means to transport it. The word describing this circumstance is **complementarity.** *Effective* supply and demand are important considerations. Mere differences from place to place in commodity surplus or deficit are not enough to initiate exchange. Greenland and the Amazon basin are notably unlike in their natural resources and economies, but their degree and volume of interaction is minimal. Supply and market must come together, as they do in the flow of seasonal fruits and vegetables from California's Imperial Valley to the urban markets of the American Midwest and East or in the movement of manganese from the USSR to the steel mills of Western Europe. Declining domestic petroleum supplies and increasing consumption in the United States and Europe materially altered opportunity patterns for the oil producers of OPEC (Organization of Petroleum Exporting Countries) between 1956 and 1986 (Figure 3.2). More generalized patterns of complementarity underlie the exchanges of the raw materials and agricultural goods of less-developed countries for the industrial commodities of the developed nations.

Transferability

Even when complementarity exists, spatial interaction occurs only when conditions of **transferability**—acceptable costs of an exchange—are met. Spatial movement responds not just to availability and demand but to considerations of time and cost. Transferability is an expression of the mobility of a commodity; it is a function of the characteristics and value of the product, of the distance—measured in time and money penalties—over which it must be moved, and of the ability of the commodity to bear the costs of movement. If the time and money costs of traversing a distance are too great, exchange does not occur. That is, mobility is not just a physical matter but an economic one as well. If a given

Figure 3.2

Changes in seaborne trade in oil, 1956–1986. Complementarity is so basic in initiating interaction that even relatively low-value bulk commodities move in trade over long distances. Even before the price inflations of the 1970s, petroleum was the most important commodity in international trade. The line widths are proportional to average daily movements; the same scale is employed throughout.

Redrawn with permission from The BP International Limited, *BP Statistical Review of the World Oil Industry 1977; BP Statistical Review of World Energy, 1987.*

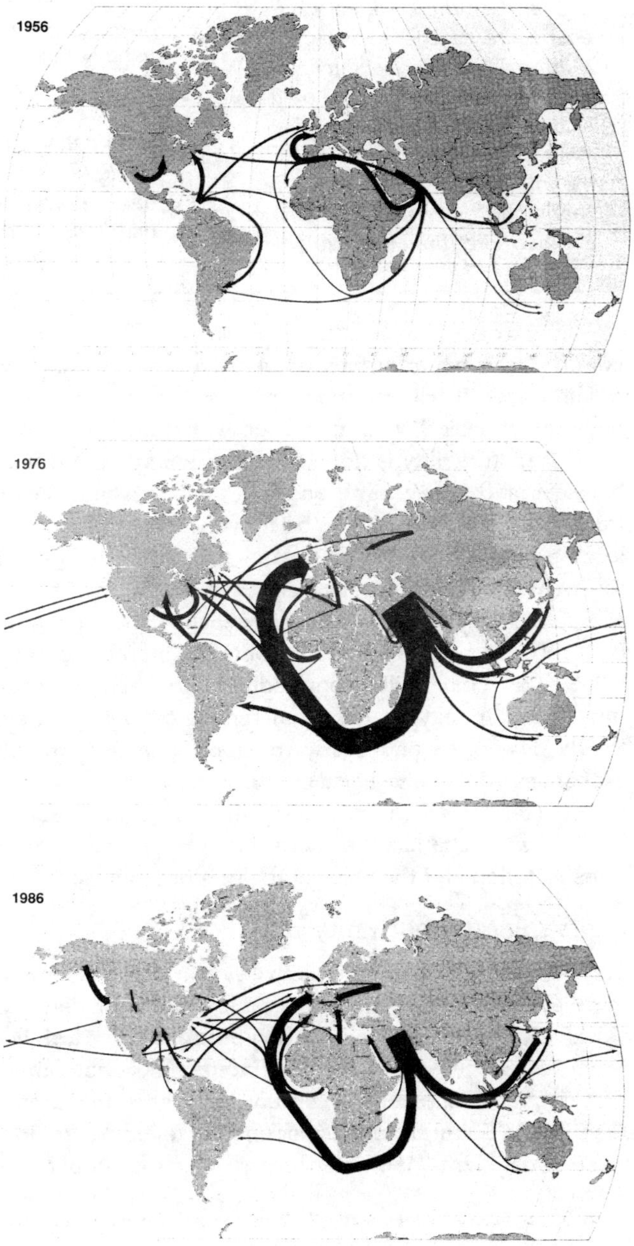

Roads as Barriers in Medieval Europe

Europe at the start of the 12th century was poised for rapid change and growth, sparked by population increase, spread of settlement, and the revival of commerce. That revival and the promise it held for economic specialization and spatial interaction were, however, hampered by the terrible condition of the continent's road network, as the following passage makes clear.

As much use as possible was made of navigable waterways for the movement of bulky and heavy goods, but most were transported overland on the backs of pack animals. Wherever possible the Roman roads were still followed, but these had never been built north of the Danube and east of the Rhine, and elsewhere bridges had in many places collapsed and the roads had fallen into disrepair. Furthermore, the cities on which the Roman roads had focussed were not necessarily those which the early medieval traveller wished to visit.

Most roads were merely tracks. Bridges were very few, and rivers were generally forded. Many roads, especially in central and eastern Europe, were not clearly defined, consisting of multiple paths through the forest and waste. The Alps aroused considerable fears, and the roads across their passes were avoided whenever possible. Crossing the passes was strenuous and dangerous, but the most hazardous part of the journey was usually the approach to the high mountains. Here, where the roads ran alternately through defiles and wide valleys, they could be more easily blocked and the traveller harassed by robbers and burdened with tolls.

It is convenient to divide Medieval trade into, first, local or short-distance trade and, second, long-distance commerce. The former consisted essentially of the products of agriculture and forestry, which were transported to the nearest monastery, market or town. In terms of volume the local trade was much the greater. Long-distance commerce consisted in general of luxury goods. Their price, when they reached their destination, included not only the production cost, but also the value added by transport by sea-going ship, by river boat and pack animal, together with the cost of tolls and doubtless a high insurance against all risks which the merchants ran.

Source: Norman J. G. Pounds, *An Historical Geography of Europe, 450 B.C.–A.D. 1330*. Copyright © 1973 Cambridge University Press, New York, NY.

commodity is not affordable upon delivery to an otherwise willing buyer, it will not move in trade, and the potential buyer must seek a substitute or go without.

Transferability is not a constant condition. It differs between places, over time, and in relation to what is being transferred and how it is to be moved. The opening of a logging road will connect a sawmill with stands of timber formerly inaccessible (nontransferable). An increasing scarcity of high-quality ores will enhance the transferability of lower-quality mine outputs by increasing their value. Low-cost bulk commodities not economically moved by air may be fully transferable by rail or water. Poorly developed and costly transportation may inhibit exchanges even at short distance between otherwise willing traders (see "Roads as Barriers"). In short, transferability expresses the changing relationships between the costs of transportation and the value of the product to be shipped.

Intervening Opportunity

Complementarity can be effective only in the absence of more attractive alternative sources of supply or demand closer at hand or cheaper. **Intervening opportunities** serve to reduce supply-demand interactions that otherwise might develop between distant complementary areas. A supply of Saharan sand is not enough to assure its flow to sand-deficient Manhattan Island because supplies of sand are more easily and cheaply available within the New York metropolitan region. A purchaser, for considerations of cost, convenience, and awareness, is unlikely to buy identical commodities at a distance when a suitable nearby supply is available. When it is, the intervening opportunity serves to demonstrate complementarity at a shorter distance.

Similarly, markets and destinations are sought, if possible, close at hand. Growing metropolitan demand in California reduces the importance of midwestern markets for western fruit growers. The intervening opportunities offered by Chicago and Philadelphia reduce the number of job seekers from Iowa searching for employment in New York City. More people from New England take winter vacations in Florida, which is relatively near and accessible, than in Southern California, which is not. That is, opportunities that are discerned closer at hand reduce the pull of opportunities offered by a distant destination (Figure 3.3). Patterns of spatial interaction are dynamic, reflecting the changeable structure of apparent opportunity.

Measuring Interaction

Complementarity, transferability, and intervening opportunity—the controlling conditions of commodity movement—are also a basis for understanding all forms of spatial interaction, including the placing of long-distance

Figure 3.3

Atlantic City gambling. When casino gambling began in Atlantic City, New Jersey, in 1978, the city became an "intervening opportunity," a closer alternative for northeastern gamblers than the Nevada resorts. Some 60 million adults reside within 400 miles (650 km) of Atlantic City. Some of its growing number of visitors might otherwise have journeyed to Reno or Las Vegas. For most, ho Atlantic City provided a convenient opportunity to satisfy a den for which the Nevada alternative was too distant and too costly.

Source: Graph data from New Jersey Casino Control Commission and the Atlantic City Public Relations Office.

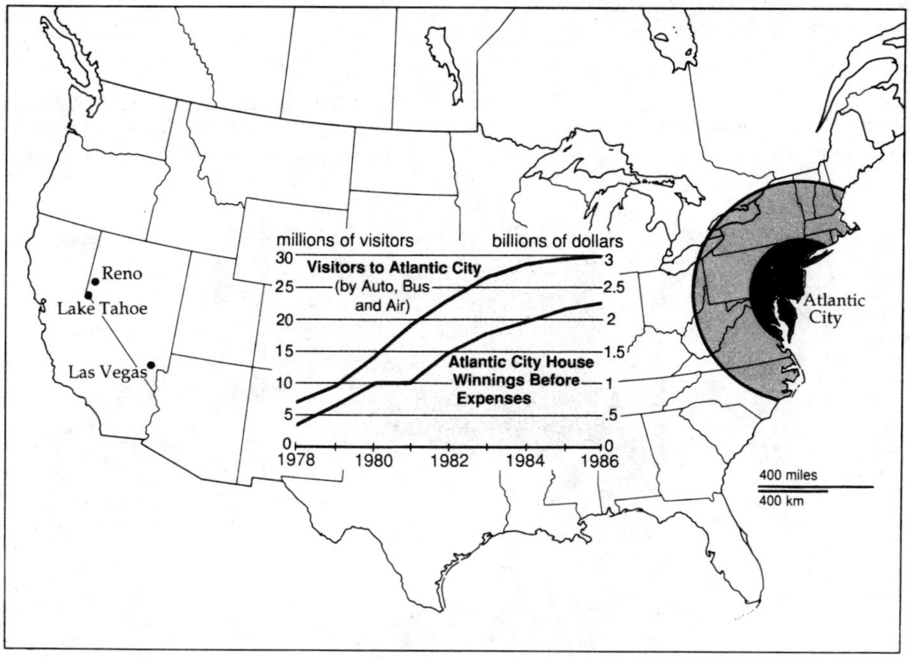

phone calls, the residential locational decisions of commuters, and the once-in-a-lifetime transcontinental adventure of the Hauns. Interaction of whatever nature between places is not, of course, meaningfully described by the movement of a single commodity, by the habits of an individual commuter, or the one-time decision of a migrant. The discovery of an Inuit (Eskimo) ivory carving in a Miami gift shop does not establish significant interaction between the Arctic coast and a Florida resort. The study of unique events is suggestive but not particularly informative. We seek general principles that govern the frequency and intensity of interaction both to validate the three preconditions of spatial exchange and to establish the probability that any given potential interaction will actually occur.

Our interest is similar to that of the physical scientist investigating, for example, the response of a gas to variations in temperature and pressure. The concern there is with *all* of the gas molecules and the probability of their collective reactions; the responses of any particular molecule are not the topic of interest. Similarly, we are concerned here with the probability of aggregate, not individual, behavior.

Distance Decay

In all manner of ways, our lives and activities are influenced by the **friction of distance.** That phrase reminds us that distance has a retarding effect on human interaction because increasing penalties in time and cost are associated with longer-distance, more expensive interchanges. We visit nearby friends more often than distant relatives; we go more frequently to the neighborhood convenience store than to the distant regional shopping center. Telephone calls or mail deliveries between nearby towns are greater in volume than those to more distant locations. Our common experience, clearly supported by maps and statistics recording all kinds of flows, is that most interactions occur over short distances. That is, interchange decreases as distance increases, a reflection of the fact that transferability costs increase with distance. More generally stated, **distance decay** describes the decline of an activity or function with increasing distance from its point of origin. As the examples in Figure 3.4 document, near destinations have a disproportionate pull over more distant points in commodity movements. However, it is also evident that the rate of distance decay varies with the type of activity.

Study of all manner of spatial interconnections has led to the very general conclusion that interaction between places is inversely related to the square of the distance separating them. That is, volume of flow between two points 50 miles (80 km) apart would probably be only one-quarter of that between centers at 25 miles (40 km) separation. Such a rigid *inverse-square* relationship is well documented in the physical sciences. For social, cultural, and economic relations, however, it is at best a useful approximation. In human interaction, linear distance is only

Figure 3.4

The shape of distance decay. The geographer W. Tobler summarized the concept of distance decay in proposing his "first law of geography: everything is related to everything else, but near things are more related than distant things." Distance decay curves vary with the type of flow.

Sources: (*a*) A generalized statement of distance decay; (*b*) United States data for 1963; (*c*) Chicago Area Transportation Study, *A Summary of Travel Characteristics, 1977.*

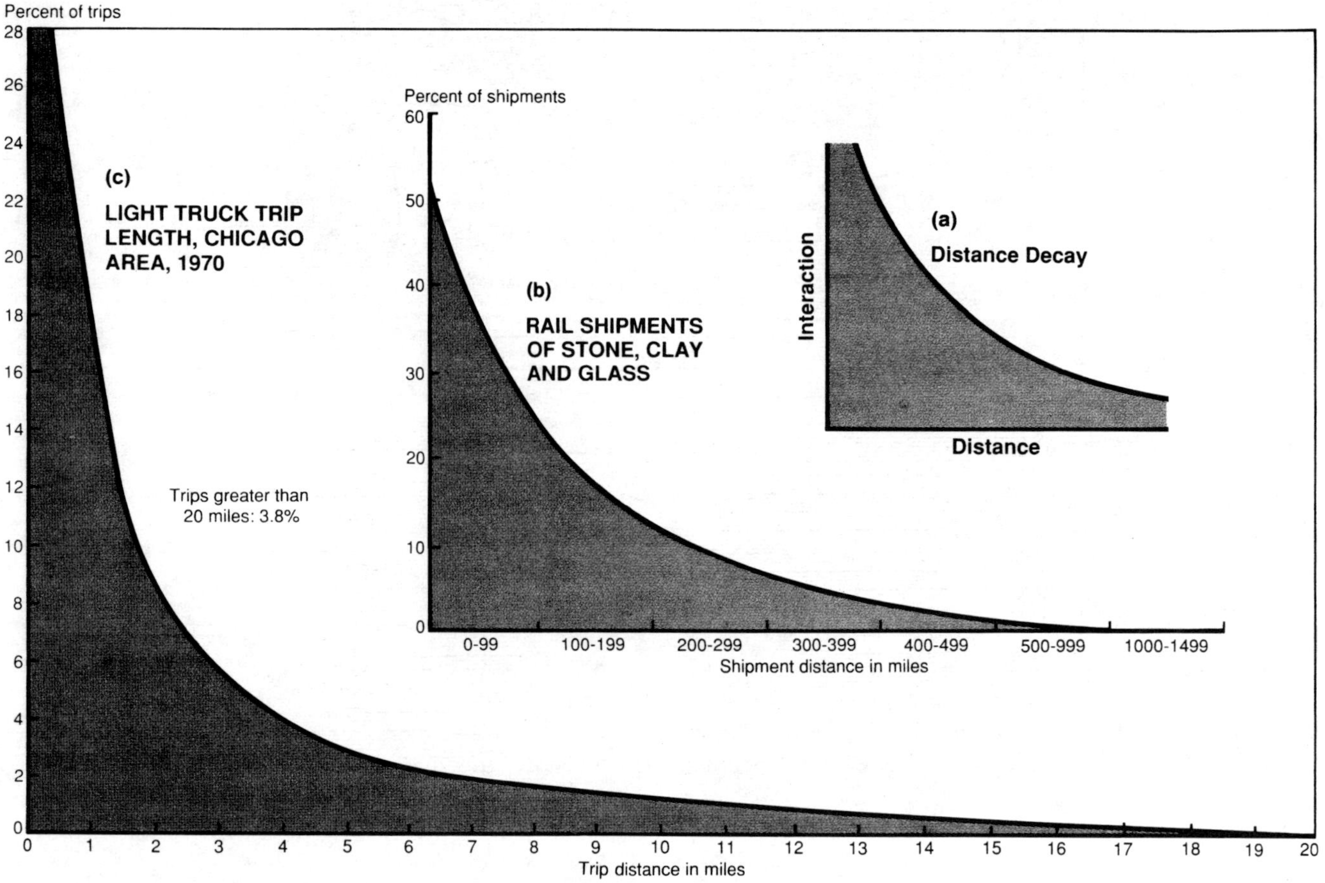

one aspect of transferability; cost and time are often more meaningful measures of separation. When the friction of distance is reduced by lowered costs or increased ease of flow, the slope of the distance decay curve is flattened and more total area is effectively united than when those costs are high. When telephone calls are charged by uniform area rates rather than strictly by distance, more calls are placed to the outer margins of the rate area than expected. Expressways extend commuting travel ranges to central cities and expand the total area conveniently accessible for weekend recreation. Figure 3.4 shows that shipping distances for high-cost truck transport are, on the average, shorter than for lower-cost rail hauls.

The Gravity Concept

Interaction decisions are not based on distance or distance-cost considerations alone. The large regional shopping center attracts customers from a wide radius because of the variety of shops and goods its very size promises. We go to distant big cities "to seek our fortune" rather than to the nearer small town. We are, that is, attracted by the expectation of opportunity that we associate with larger rather than smaller places. That expectation is summarized by another model of spatial interaction, the **gravity model,** drawn from the physical sciences.

In the 1850s, Henry C. Carey (1793–1879), in his *Principles of Social Science,* observed that the law of "molecular gravitation" was an essential condition in human existence and that the attractive force existing between areas was akin to the force of gravity. That is, the physical laws of gravity and motion developed by Sir Isaac Newton (1642–1727) have, according to Carey, applicability to the aggregate actions of humans.

Newton's *law of universal gravitation* states that any two objects attract each other with a force that is proportional to the product of their masses and inversely proportional to the square of the distance between them. Thus, the force of attraction, F, between two masses, M_i and M_j, separated by distance, d, is

$$F = g\frac{M_i M_j}{d_{ij}^2},$$

where g is the "gravitational constant."

Carey's interest was in the interaction between urban units and in the observation that a large city is more likely to attract an individual than is a small hamlet. His first interest could be quickly satisfied by simple analogy. The expected interaction (I) between two places, i and j, can be calculated by converting physical mass in the gravity model to population size (P), so that

$$I_{ij} = \frac{P_i P_j}{D_{ij}^2}.$$

Exchanges between any set of two cities, A and B, can therefore be quickly estimated:

$$I_{AB} = \frac{\text{population of } A \times \text{population of } B}{(\text{distance between } A \text{ and } B)^2}.$$

In social, rather than physical, science applications of the gravity model, distance may be calculated by travel time or travel cost modifications rather than by straight line separation. Whatever the unit of measure, however, the model assures us that although spatial interaction always tends to decrease with increasing distance between places, at a given distance it tends to expand with increases in their size.

Carey's second observation—that large cities have greater drawing power for individuals than small ones—was subsequently addressed by the *law of retail gravitation,* proposed by William J. Reilly (1899–1970) in 1931. Using the population and distance inputs of the gravity model, Reilly concluded that the *breaking point* (*BP*) or boundary marking the outer edge of either of the cities' trade area could be located by the expression

$$BP = \frac{d_{ij}}{1 + \sqrt{\frac{P_2}{P_1}}}$$

where

BP = distance from city 1 to the breaking point (or boundary)
d_{ij} = distance between city 1 and city 2
P_1 = population of city 1
P_2 = population of city 2

Any farm or small-town resident located between the two cities would be inclined to shop in one or the other of them according to that resident's position relative to the calculated breaking point. Since the breaking point between cities of unequal size will lie farther from the larger of the two, its spatially greater drawing power is assured (Figure 3.5).

Later studies in location theory, city systems, trade area analysis, and other social topics all suggest that the gravity model can be used to account for a wide variety of flow patterns in human geography, including population migration, commodity flows, journeys to work or to shop, telephone call volumes, and the like. Each such flow pattern suggests that size as well as distance influences spatial interaction. Carey's observation made nearly 150 years ago initiated a type of analysis that has continuing relevance. In modified form it is used today for a variety of practical studies that help us better understand the "friction of distance."

Figure 3.5

The *law of retail gravitation* provides a quick determination of the trade boundary (or breaking point) between two cities. In the diagram, cities 1 and 2 are 125 miles apart. Reilly's law tells us that the breaking point between them lies 50.7 miles distant from city 1. A potential customer located at *M*, midway (62.5 miles) between the cities, would lie well within the trade zone of city 2. A series of such calculations would define the "trade area" of any single city.

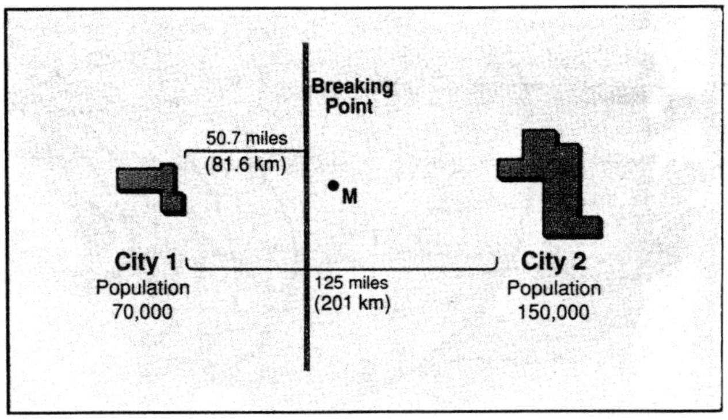

Interaction Potential

Spatial interaction models of distance decay and gravitational pull deal with only two places at a time. The world of reality is rather more complex. All cities, not just city pairs, within a regional system of cities have the possibility of interacting with each other. Indeed, the more specialized the goods produced in each separate center—that is, the greater their collective complementarity—the more likely it is that such multiple interactions will occur. A **potential model,** also based on Newtonian physics, provides an estimate of the interaction opportunities available to a center in such a multicentered network. It tells us the relative position of each point in relation to all other places within a region. It does so by summing the size and distance relationships between all points of potential interaction within an area.

The concept of potential is applicable whenever the measurement of the intensity of spatial interaction is of concern—as it is in studies of marketing, land values, broadcasting, commuting patterns, and the like. A frequent application is to *population potential,* a measurement of the aggregate nearness of people to each other or, more usefully, of the relative accessibility of a point to all people within a particular region (Figure 3.6).

Movement Biases

Distance decay and the gravity and potential models help us understand the bases for interaction in an idealized spatial structure in which there are no natural or cultural impediments to flow and no restrictions on routes followed

Figure 3.6

Population potentials of the United States, 1790–1950. The population potential model assesses the accessibility of any single point to the total population. High potentials indicate high degrees of accessibility. According to the formulation employed, "250" may indicate the line along which there is a potential for 250,000 interactions per year between the population per square mile and all the rest of the nation's population. In a very generalized way, the maps suggest the relative distribution of population for the year mapped.

Source: From William Warntz, "Macroscopic Analysis and Some Patterns of the Geographical Distribution of Population in the United States, 1790–1950," in *Northwestern University Studies in Geograpny*, No. 13, 1967. Reprinted by permission.

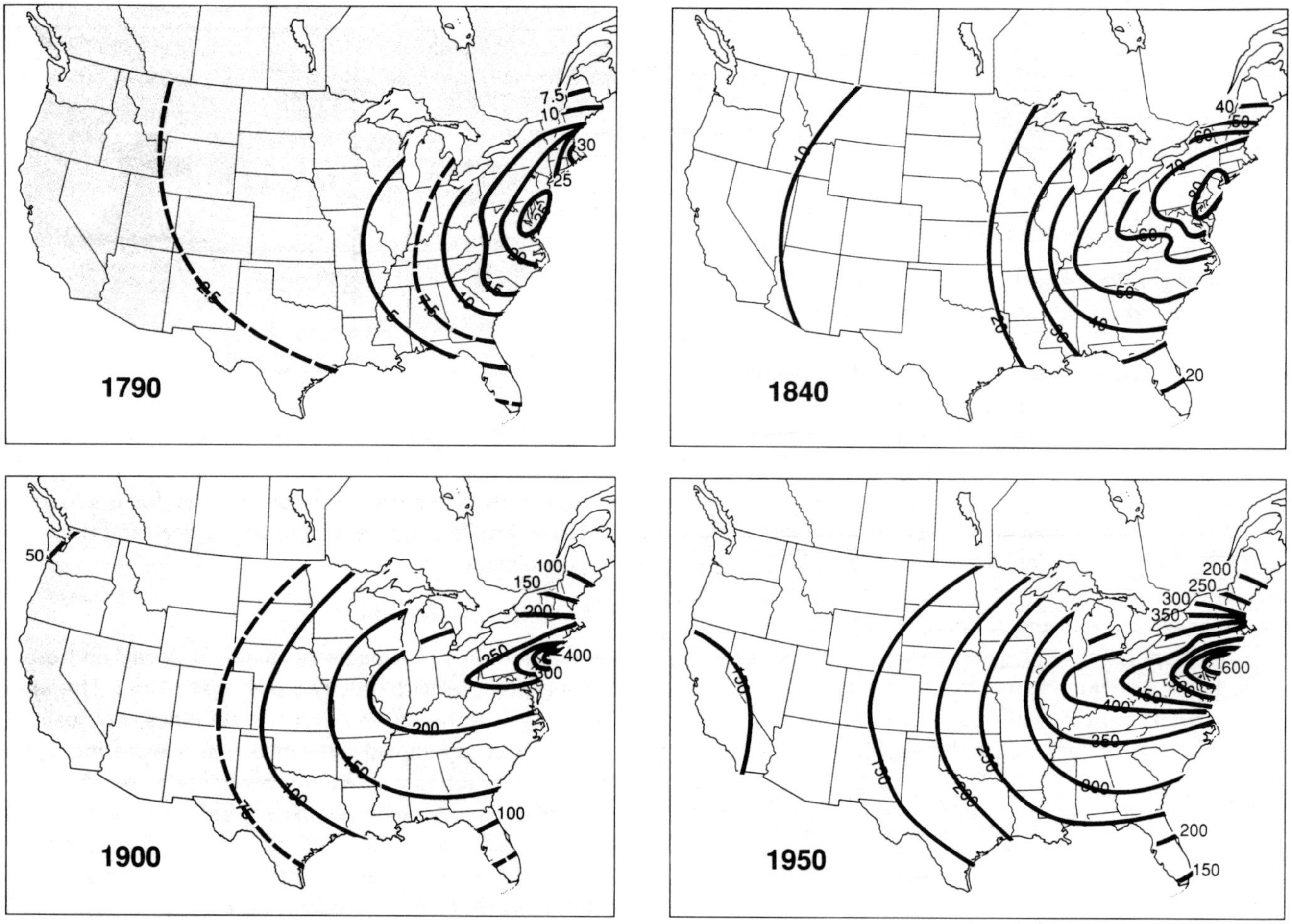

and in which only rational interaction decisions are made. Even under those idealized conditions, the pattern of spatial interaction that has been initiated helps shape the conditions under which future interactions will occur. An initial structure of centers and connecting flows will tend to freeze into the landscape a mutually reinforcing continuation of that same pattern. The predictable flows of shoppers to existing shopping centers make those centers attractive to other merchants. New store openings increase customer flow; increased flow strengthens the developed pattern of spatial interaction.

Such an aggregate regularity of flow is called a **movement bias.** We have already noted a *distance bias* favoring short movements over long. There is also *direction bias,* in which of all possible directions of movement, actual flows are restricted to only one or a few. Direction bias is simply a statement that from a given origin, flows are not random (Figure 3.7); rather, certain places have a greater attraction than do others. The movement patterns from an isolated farmstead are likely oriented to a favored shopping town. On a larger scale, in North America or Siberia long-distance freight movements are directionally biased in favor of east-west flows. Direction bias reflects the orientation and also the intensity of flow. Movements from a single point—from Novosibirsk in the USSR, for example, or from Winnipeg, Canada, or Kansas City in the United States—may occur in all directions; they are in reality more intense along the east-west axis.

Such directional biases are in part a reflection of *network bias,* a shorthand way of saying that the presence or absence of connecting channels strongly affects the likelihood that spatial interaction will occur. A set of routes

Figure 3.7

Direction bias. (a) When direction bias is absent, movements tend to be almost random, occurring in all possible directions, but less likely between points, such as *A* and *X*, not directly connected. (b) Direction bias indicating predominantly north-south movements. Direction bias implies greatest intensity of movement within a restricted number of directions.

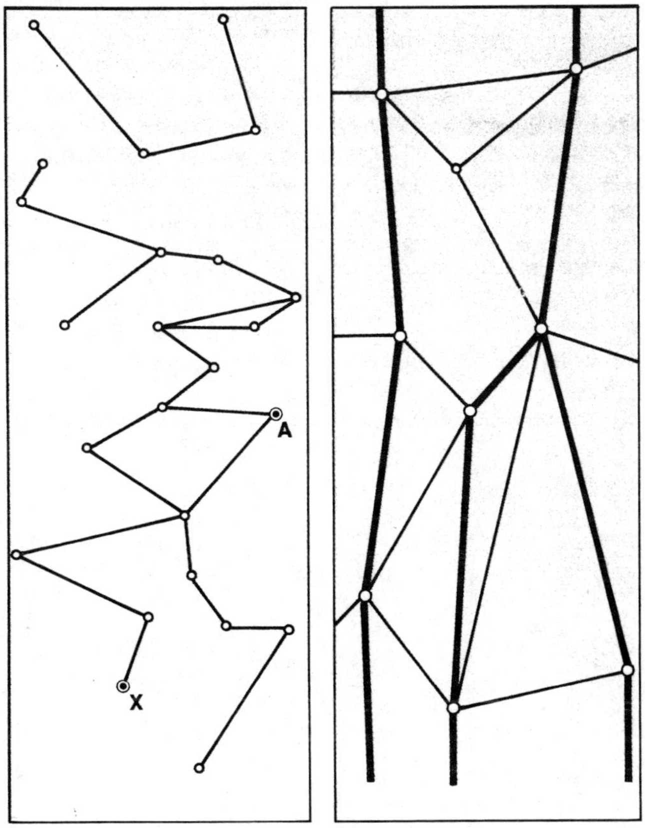

and the set of places that they connect are collectively called a **network** (see "The Structure of Networks"). Flows cannot occur between all points if not all points are linked. In Figure 3.7*a*, the interchange between *A* and *X*, though not necessarily impossible, is unlikely because no established roadway connects them. In information flows, a worker on the assembly line is less likely to know of company production plans than is a secretary in the executive offices; these two workers are tied into quite different information networks.

A recognition of movement biases helps to refine the coarser generalizations of spatial interaction based on complementarity, intervening opportunity, and transferability. Other modifying statements have been developed, but each further refinement moves us away from aggregate behavior toward less predictable individual movements and responses. The spatial interaction questions we ask and the degree of refinement of the answers we require determine the modifications we must introduce into the models we employ.

Human Spatial Behavior

Humans are not commodities and individually do not necessarily respond predictably to the impersonal dictates of spatial interaction constraints. Yet, to survive, people must be mobile and collectively do react to distance, time, and cost considerations of movement in space and to the implications of complementarity, transferability, and intervening opportunity. Indeed, one of the most current and exciting thrusts of geographic inquiry involves how individuals make spatial behavioral decisions and how those separate decisions may be summarized by models and generalizations to explain collective actions.

Two aspects of human spatial behavior concern us. The first is the daily or temporary use of space—the journeys to stores or to work or to school. The second is the longer-term commitment related to decisions to travel, to migrate, or to settle away from the home territory. Both aspects imply a time dimension. Humans' spatial actions are not instantaneous. They operate over time, frequently imparting a rhythm to individual and group activity patterns and imposing choices among time-consuming behaviors. Elements of both aspects are also embodied in how individuals perceive space and act within it and how they respond to information affecting their space-behavioral decisions.

Individual Activity Space

One of the realities of life is that groups and nations draw boundaries around themselves and divide space into territories that are, if necessary, defended. Some see the concept of **territoriality**—the emotional attachment to and the defense of home ground—as a root explanation of much of human action and response. It is true that some individual and collective activity appears to be governed by territorial defense responses: the conflict between street groups in claiming and protecting their "turf" (and their fear for their lives when venturing beyond it) and the sometimes violent rejection by ethnic urban neighborhoods of an advancing black, Hispanic, or other population group. On a more individualized basis, each of us claims as **personal space** the zone of privacy and separation from others our culture or our physical circumstances require or permit. North Americans demand greater face-to-face separation in conversations than do Latin Americans. Personal space on a crowded beach or in a department store is acceptably more limited than it is in our homes or in a library (Figure 3.8)

For most of us, our personal sense of territoriality is a tempered one. We regard our homes and property as defensible private domains but open them to innocent visitors, known and unknown, or to those on private or official business. Nor do we confine our activities so

The Structure of Networks

Movements of commodities, people, or information take place between points and along pathways that are organized into networks. The distribution of pathways creates a pattern of connectivity (or connectedness) between places, with pattern and flow mutually reinforcing. If places are close, we expect flows between them to be great. If flows are great, connecting routeways (roads, phone lines) are improved to speed interchange and reduce the friction of distance.

Networks may be viewed as structures composed of *nodes* (origins, destinations, or intersections) and *links* (connections or routes within the network). Network pattern helps shape spatial interaction. Points well connected, even at a distance, will tend to have more exchanges than nearby poorly connected or nonconnected points. The friction of distance is a statement not just of linear distance but of network distance. Although an infinite number of network patterns are theoretically possible, the five shown here suggest the impact of network design on human interaction patterns. Each network has five identically placed nodes; each differs in the number and layout of its links.

Networks are also described by their density, their connectivity, and their organization. *Density* of network is a calculation of length of route per unit area; it is a rough measure of accessibility. *Connectivity* is the relationship between the number of nodes and the number of links in a network. The greater the connectivity, usually, the more developed the national or regional economy served by the network. *Organization* of network refers to the hierarchy of nodes and links it may display. Not all nodes are of equal importance; not all links are of identical capacity.

Diagrams of simplified network patterns.

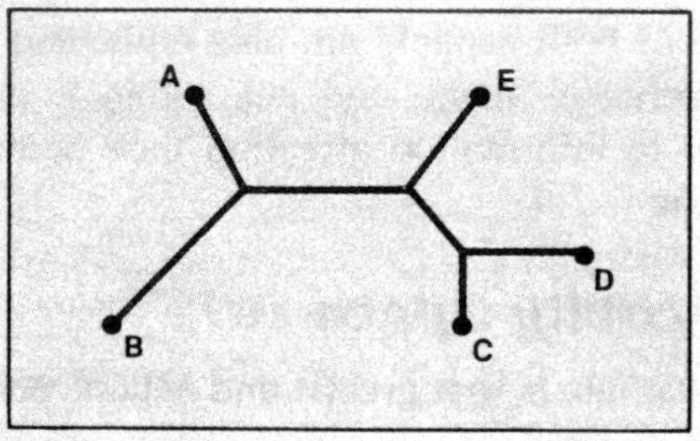

(a) BRANCHING NETWORK: shortest total path connections, lowest construction costs; poor nearby connections.

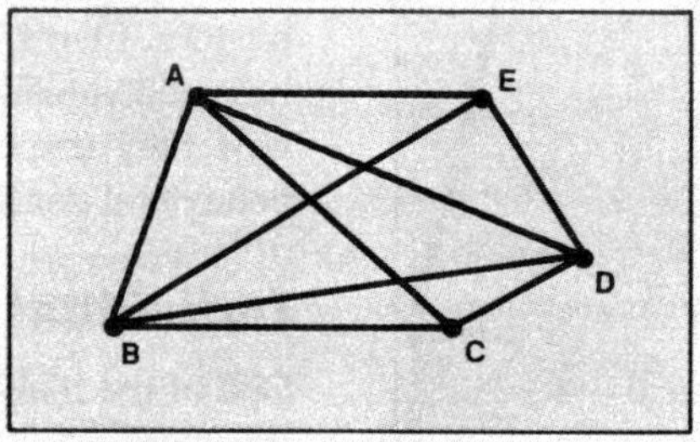

(b) CIRCUIT NETWORK: complete connection between points; lowest user costs.

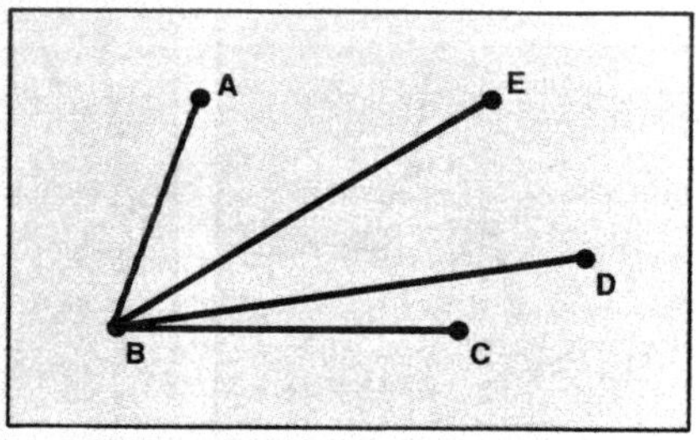

(c) CENTER-ORIENTED OR HIERARCHY NETWORK: the shortest set of connections between a central node and all other points; the hub-and-spoke system of airlines.

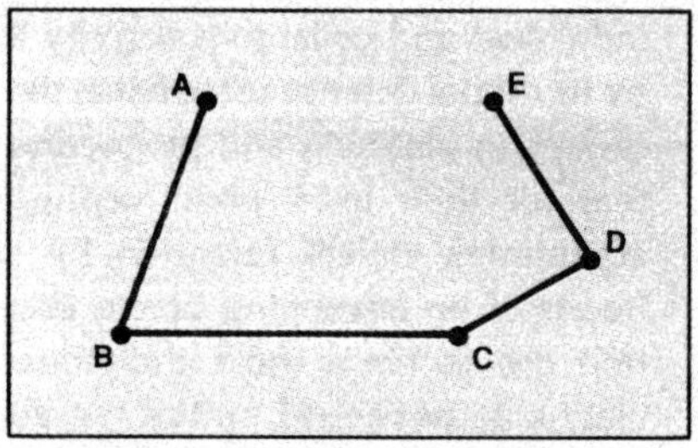

(d) PAUL REVERE'S RIDE: the shortest path between a given beginning point and all other points; a minimum distance solution to pipeline placement.

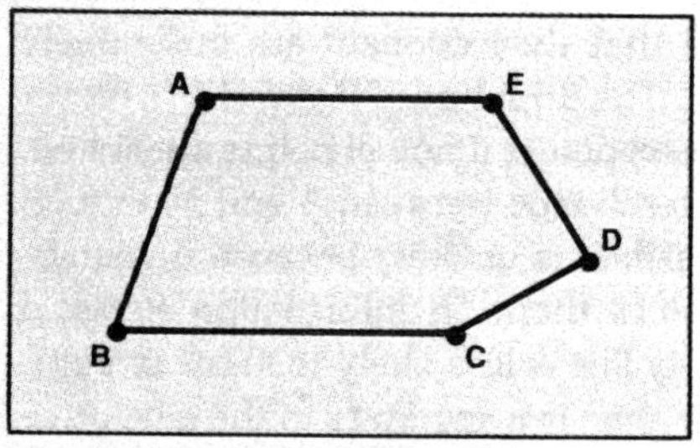

(e) TRAVELING SALESMAN NETWORK: the shortest cycle around a set of nodes; the most effective shopping trip pattern.

Figure 3.8

Our demanded *personal space* is not necessarily uniform in shape or constant in size. We tolerate strangers closer to our sides than directly in front of us; we accept more crowding in an elevator than in a store. We tend to distance ourselves from others in a library but accept the press of the crowd on a popular beach—as do these vacationers along the Costa Blanca in Spain.

exclusively within controlled home territories as street-gang members do within theirs. Rather, we have a more-or-less extended home range, an **activity space** or area within which we move freely on our rounds of regular activity, sharing that space with others who are also about their daily affairs. Figure 3.9 suggests probable activity spaces for a suburban family of five for a day. Note that the activity space is different and for the mapped day rather limited for each individual, even though two members of the family use automobiles. If a week's activity were shown, more paths would be added to the map.

The types of trips that individuals make and thus the extent of their activity space depend on at least three interrelated variables: their stage in the life cycle, the means of mobility at their command, and the demands or opportunities implicit in their daily activities. The first, *stage in the life cycle*, refers to membership in specific age groups. School-age children usually travel short distances to lower schools and longer distances to upper-level schools. After-school activities tend to be limited to walking or bicycling to nearby locations. Greater mobility is characteristic of high school students. Adults responsible for household duties make shopping trips and trips related to child care, as well as journeys away from home

Figure 3.9

Activity space for each member of a family of five for a typical weekday. Routes of regular movement and areas recurrently visited help to foster a sense of territoriality and to color one's perceptions of space.

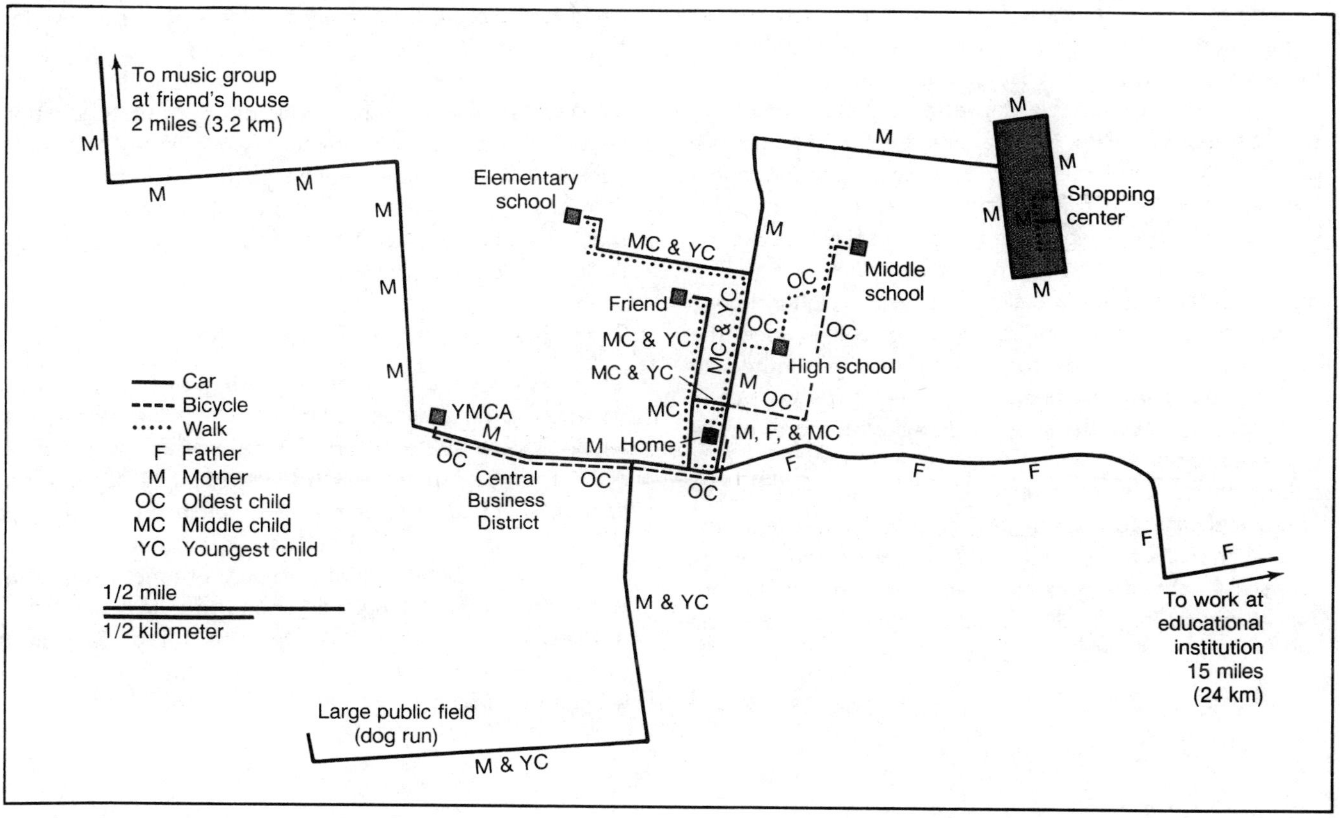

Figure 3.10

The space-time prism. An individual's daily prism has both geographical limits and totally surrounding space-time walls. The *time* (vertical axis) involved in movement affects the space that is accessible, along with the time and space available for other than travel purposes. (*a*) When collecting firewood for household use may take an entire day, as it does in some deforested Third World countries, no time or space is left for other activities, and the gatherer's space-time prism may be represented by a straight line. (*b*) Walking to and from work or school and spending the requisite number of hours there leave little time to broaden one's area of activity. (*c*) The automobile permits an extension of the geographical boundaries of the driver's space-time prism; the range of activity possibilities and locations is expanded for the highly mobile.

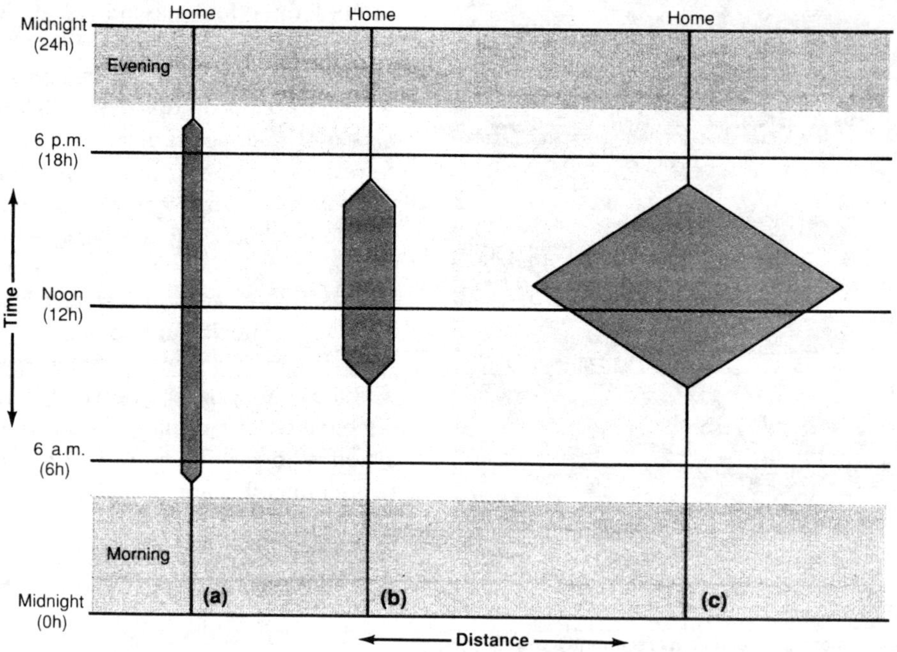

for social, cultural, and recreational purposes. Wage-earning adults usually travel farther from home than other family members. Elderly people may, through infirmity or interests, have less extensive activity spaces.

The second variable that affects the extent of activity space is *mobility,* or the ability to travel. An informal consideration of the cost and effort required to overcome the friction of distance is implicit. Where incomes are high, automobiles are available, and the cost of fuel is reckoned minor in the family budget, mobility may be great and individual activity space large. In societies or neighborhoods where cars are not a standard means of conveyance, the daily nonemergency activity space may be limited to walking, bicycling, or taking infrequent trips on public transportation. Wealthy suburbanites are far more mobile than are residents of inner-city slums, a circumstance that affects ability to seek or retain work and to have access to medical care, educational facilities, and social services.

A third factor limiting activity space is the individual assessment of the existence of possible activities or *opportunities.* In primitive societies, where the needs of daily life are satisfied at home, the impetus for journeys away from home is minimal. Without stores, schools, factories, or even roads, expectations and opportunities are limited. Not only are activities spatially restricted, but **awareness space**—knowledge of opportunity locations beyond normal activity space—is minimal, distorted, or absent. In low-income neighborhoods of modern cities in any nation, poverty and isolation restrict access to information about opportunities and therefore reinforce other limitations on activity space.

The Tyranny of Time

The daily activities of humans—eating, sleeping, traveling between home and destination (and back), working or attending classes—all consume time as well as involve space. An individual's spatial reach is constrained because one cannot be in two different places simultaneously or engage simultaneously in activities that are spatially separate. Further, since there is a finite amount of time within a day and each of us is biologically bound to a daily rhythm of day and night, sleeping and eating, time tyrannically limits the spatial choices we can make and the activity space we can command.

Our daily space-time constraints—our *time-geography*—may be represented by a **space-time prism,** the volume of space and length of time within which our activities must be confined. Its size and shape are determined by our mobility; its boundaries define what we can or cannot accomplish spatially or temporally (Figure 3.10). If our circumstances demand that we walk to work or school (3.10*b*), the sides of our prism are steep and the space available for our activities is narrow. We cannot use time spent in transit for other activities, and the area reasonably accessible to the pedestrian is limited. The space-time prism for the driver (3.10*c*) has flattened sides and the individual's spatial range is wide. The dimensions of the prism determine what spatially defined activities are possible, for no activity can exceed the bounds of the prism. Since most activities have their own time constraints, the choices of things you can do and the places you can do them are strictly limited. Defined class hours, travel time from residence to campus, and dining hall location and opening and closing hours, for example, may be the constraints on your *space-time path* (Figure 3.11). If you also need part-time work, your choice of jobs is constrained by their respective locations and work hours, for the job, too, must fit within your daily space-time prism.

Distance and Human Interaction

People make many more short-distance trips than long ones, a statement in human behavioral terms of the concept of *distance decay*. If we drew a boundary line around our activity space, it would be evident that we take trips to the boundary much less often than short-distance trips around the home. The tendency is for the frequency of trips to fall off very rapidly beyond an individual's **critical distance**—the distance beyond which cost, effort, and means strongly influence our willingness to travel. Figure 3.12 illustrates the point with regard to journeys from the homesite.

Regular movements defining our individual activity space are undertaken for different purposes and are differently influenced by time and distance considerations. The kind of activities individuals engage in can be classified according to type of trip: journeys to work, to school, to shop, for recreation, and so on. People in nearly all parts of the world make these same types of journeys, though the spatially variable requirements of culture, economy, and personal circumstance dictate their frequency, duration, and significance to an individual (Figure 3.13). A small child, for example, will make many trips up and down the block but is inhibited by parental admonitions from crossing the street. Different but equally effective distance constraints control adult behavior.

Figure 3.11

School-day space-time path for a hypothetical college student.

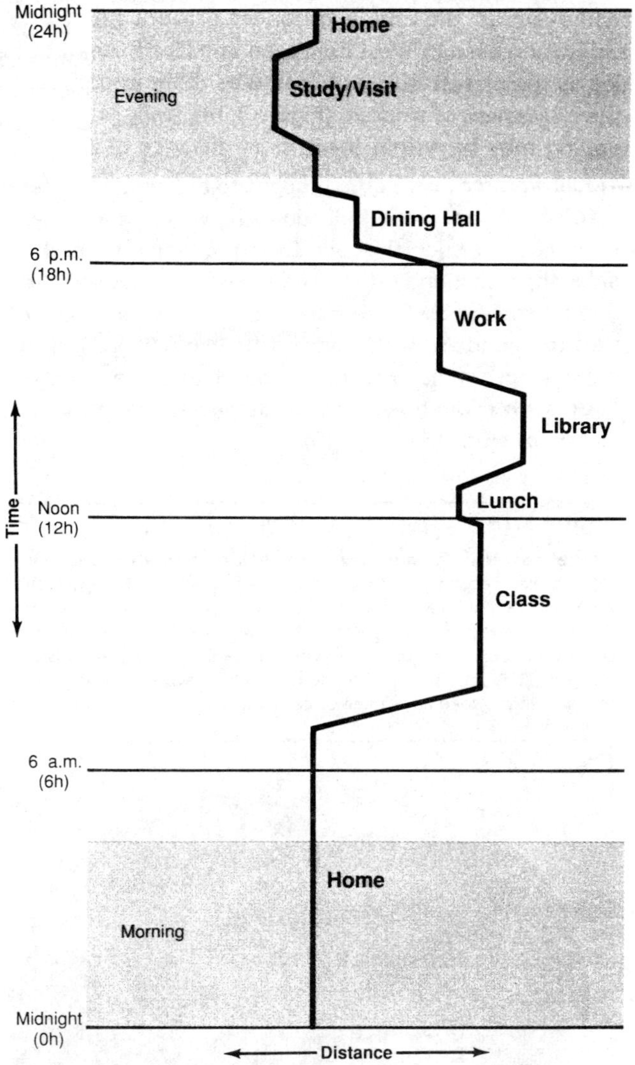

Figure 3.12

Critical distance. This general diagram indicates how most people observe distance. For each activity, there is a distance beyond which the intensity of contact declines. This is called the *critical distance* if distance alone is being considered or the *critical isochrone* (from Greek *isos*, "equal," and *chronos*, "time") if time is the measuring rod. The distance up to the critical distance is identified as a *frictionless zone*, in which time or distance considerations do not effectively figure in the trip decision.

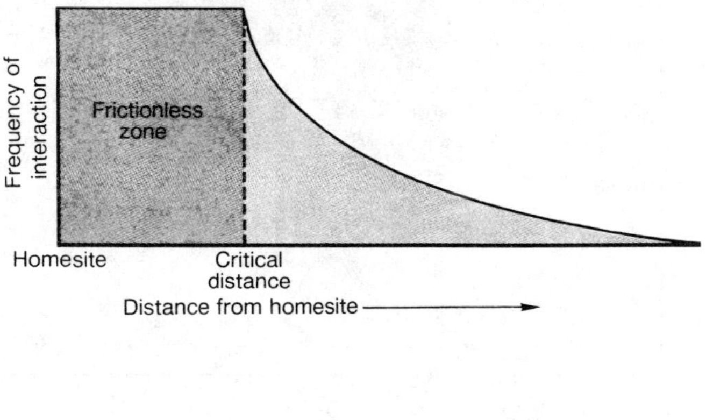

The journey to work plays a decisive role in defining the activity space of most adults. Formerly restricted by walking distance or by the routes and schedules of mass transit systems, the critical distances of work trips have steadily increased in West European and North American cities as the private automobile figures more importantly in the movement of workers (Figure 3.14). Daily or weekly shopping may be within the critical distance of an individual who may give little thought to the cost or the effort involved. That same individual, however, may relegate shopping for special goods to infrequent trips and consider their cost and effort. The majority of our social contacts tend to be at short distance within our own neighborhoods or with friends who live relatively close at hand; longer social trips to visit relatives are less frequent. In all such trips, however, the distance decay function is clearly at work (Figure 3.15).

Spatial Interaction and the Accumulation of Information

Critical distances, even for the same activity, are different for each person. The variables of life-cycle stage, mobility, and opportunity, together with an individual's interests and demands, help to define how often and how far a person will travel. On the basis of these variables, we can make inferences about the amount of information a person is likely to acquire about his or her activity space and the area beyond. The accumulation of information about the opportunities and rewards of spatial interaction helps increase and justify movement decisions.

For information flows, however, space has a different meaning than it does for the movement of commodities. Communication, for example, does not

Figure 3.13

Travel patterns for purchases of clothing and yard goods of (*a*) rural cash-economy Canadians and (*b*) Canadians of the Old-Order Mennonite sect. These strikingly different travel behaviors demonstrate the great differences that may exist in the action spaces of different culture groups occupying the same territory in midwestern Canada. They suggest that "modern" rural Canadians, owning cars and wishing to take advantage of the variety of goods offered in the more distant regional capital, are willing to travel longer distances than are people of a traditionalist culture, who have different mobility and whose different demands in clothing and other consumer goods are satisfied in nearby small settlements.

Source: From Robert A. Murdie, "Cultural Differences in Consumer Travel," in *Economic Geography*, 41, No. 3, July 1965. Copyright © 1965 Clark University, Worcester, MA. Reprinted by permission.

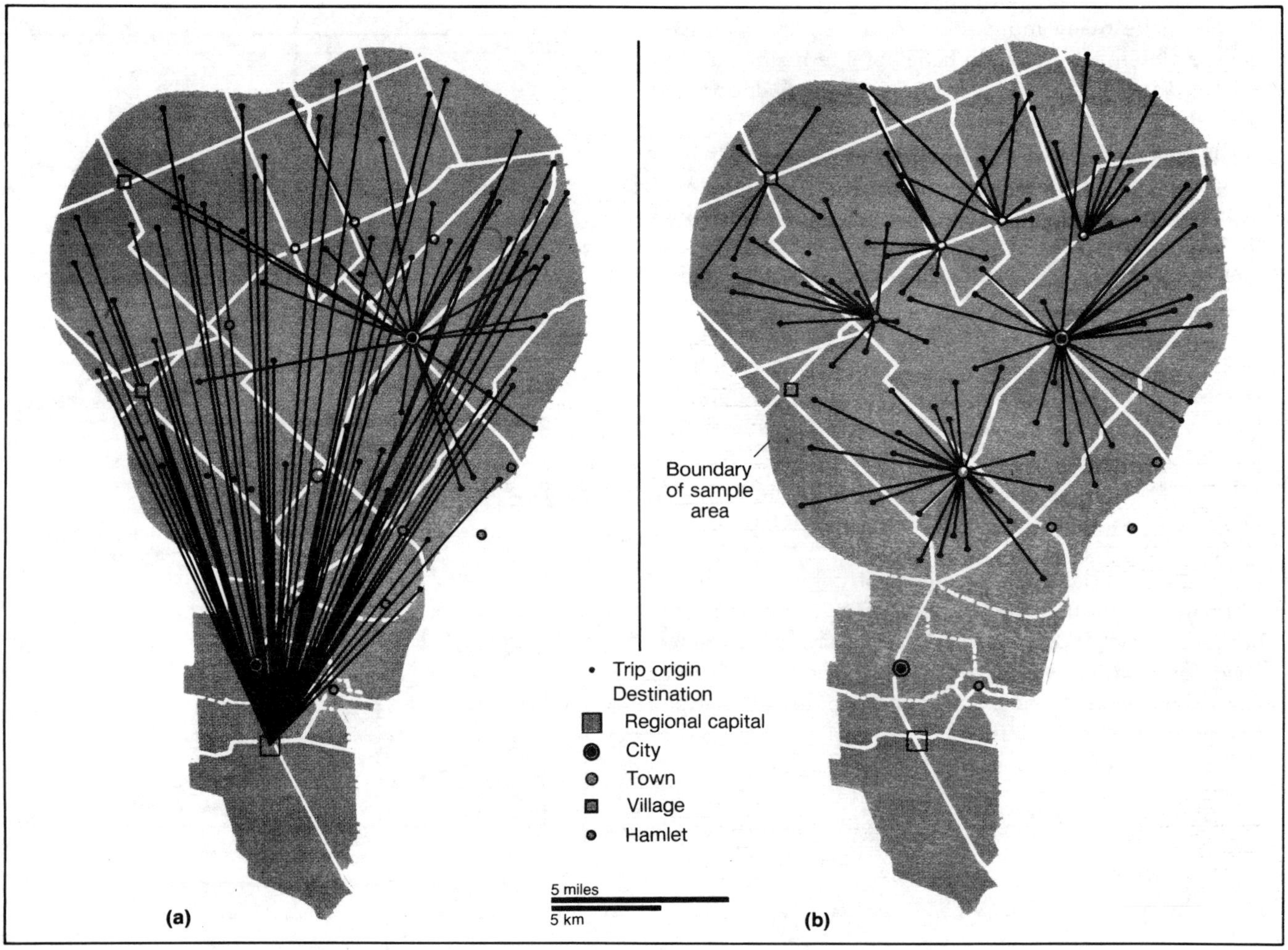

Figure 3.14

The frequency distribution of work and nonwork trip lengths in minutes in Toronto. Work trips are usually longer than other recurring journeys. In the United States, the average commuting trip in 1975 covered 9 miles (14.5 km) and took about 20 minutes. For North America, at least, 20 to 30 minutes seemed to be the upper limit of commuting time before people began to consider time and distance to work seriously in their choice of homesite. Most nonwork trips are relatively short.

Source: *Metropolitan Toronto and Region Transportation Study,* The Queen's Printer, Toronto, 1966, Fig. 42.

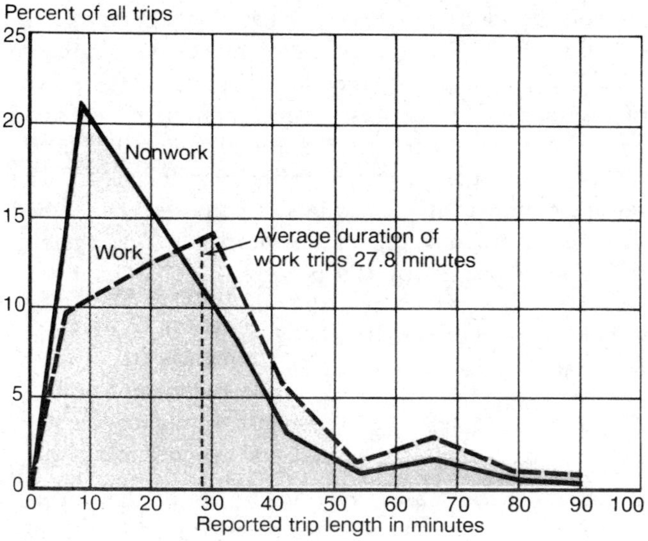

Figure 3.15

Social interaction as a function of distance. Visits with neighbors on the same street are frequent; they are less common with neighbors around the corner and diminish quickly to the vanishing point after a residential relocation. Friends exert a greater spatial pull, though the distance decay factor is clearly evident. Visits with relatives offer the greatest incentive for longer-distance (though relatively infrequent) journeys.

Source: From Frederick P. Stutz, "Distance and Network Effects on Urban Social Travel Fields," in *Economic Geography,* 49, No. 2, 1973. Copyright © 1973 Clark University, Worcester, MA. Reprinted by permission.

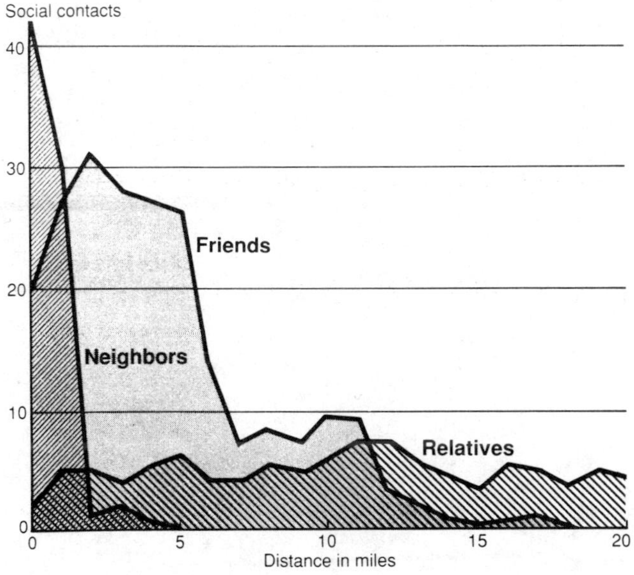

necessarily imply the time-consuming physical relocations of freight transportation (though in the case of letters and print media it usually does). Indeed, in modern telecommunications, the process of information flow may be instantaneous regardless of distance. The result is space-time convergence to the point of the obliteration of space. A Bell System report tells us that in 1920, putting through a transcontinental telephone call took 14 minutes and eight operators. By 1940, the call completion time was reduced to less than 1½ minutes. In the 1960s, direct distance dialing allowed a transcontinental connection in less than 30 seconds, and electronic switching has now reduced the completion time to that involved in dialing a number and answering a phone. Communication satellites have made worldwide personal and mass communication immediate and data transfers for economic integration instantaneous. The same technologies that have led to communications space-time convergence have tended toward a space-cost convergence (Figure 3.16). Domestic mail, which once charged a distance-based postage, is now carried nationwide or across town for the same price. In the modern world, transferability is no longer a consideration in information flows.

Information Flows

Spatially significant information flows are of two types: individual (person-to-person) exchanges and mass (source-to-area) communication. A further subdivision into formal and informal interchange recognizes, in the former, the need for an interposed channel (radio, press, postal service, or telephone, for example) to convey messages. Informal communication requires no such institutionalized message carrier.

Figure 3.16

Space-cost convergence in telephone tolls between New York and San Francisco, 1915–1988. The rates shown are a smooth-curve summary of frequently adjusted charges for station-to-station daytime 3-minute calls.

Sources: Data from *Historical Statistics of the United States;* National Association of Regulatory Utility Commissioners; and AT&T.

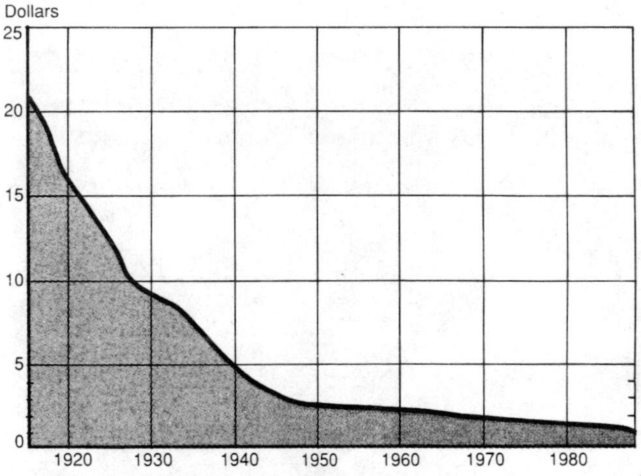

Short-range informal individual communication is as old as humankind itself (Figure 3.17). Contacts and exchanges between individuals and within small groups tend to increase as the complexity of social organization increases, as the size and importance of the population center grow, and as the range of interests and associations of the communicating person expands. Each individual develops a **personal communication field,** the informational counterpart of that person's activity space. Its size and shape are defined by the individual's contacts in work, recreation, shopping, school, and other regular activities. Those activities, as we have seen, are functions of the age, sex, education, employment, income, and so on of each person. An idealized personal communication field is suggested in Figure 3.18.

In common with other spatial interactions, the frequency of interpersonal exchange decreases with increasing distance. Most communication is with those closest at hand and most frequently seen. The letter "home" is less frequent than the jotted reminder to spouse or roommate; the long-distance call is less common than the phoned message to the office, to a fellow student, or to a team or club member. We experience more human interactions at short distances than at long distances. Where population densities are high, such as in cities (particularly central business districts during business hours), the spatial interaction between individuals can be at a very high level. This is one reason why these centers of commerce and entertainment, like the ancient culture hearths discussed in Chapter 2, are often also centers for the development and dissemination of new ideas.

Whether formal or informal in structure, each interpersonal exchange constitutes a link in the individual's personal communication field. Each person, in turn, is a node in the communication field of those with whom he or she makes or maintains contact. The total number of such separate informal networks equals the total count of people alive. Despite the number of such networks, all people, in theory, are interconnected by multiple shared nodes (Figure 3.19). Experimentation has demonstrated that through such interconnections no person in the United States is more than five links removed from any other person, no matter where located or how unlikely the association.

Each individual serving as a link in a communication network performs a *gatekeeper* function, acting as a receiver and selective transmitter of information. In Figure 3.20, the gatekeeper, *G*, receives information in the form of observed events, expressed ideas, or imparted data and retransmits some portion of that information flow in the form of messages conveyed to the next individual. The retransmission is influenced by *G*'s purposes, values, and assigned or assumed role in the information flow system. It is also influenced by *G*'s perceptions of what the receiver, *R*, is interested in, wishes to know, or can absorb. In turn, *R* receives data flow independent of *G*, modifying or enhancing the meaning of *G*'s transmission.

Although the gatekeeper role is inherent and frequently unrecognized in all personal communication, it becomes an important, recognized, and assigned role in mass communication. *Mass communication* is the formal, structured transmission of information in essentially a one-way flow between single points of origin and broad areas of reception. There are few transmitters and many receivers, and the gatekeeper function is central to the

Figure 3.17

Short-range individual and small-group exchanges take many forms. From them, we gain much of the information that influences our opinions and directs our actions.

Figure 3.18

A personal communication field is determined by individual spatial patterns of communication related to work, shopping, business trips, social visits, and so on.

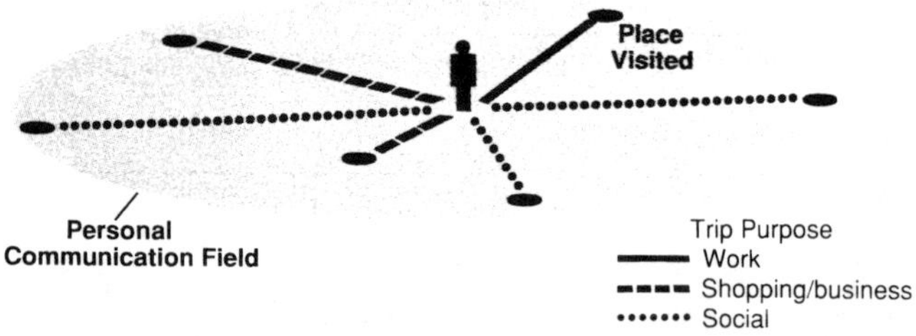

Figure 3.19

Separate population sets are interconnected by the links between individuals. If link *A–B* exists, everyone in the two sets is linked.

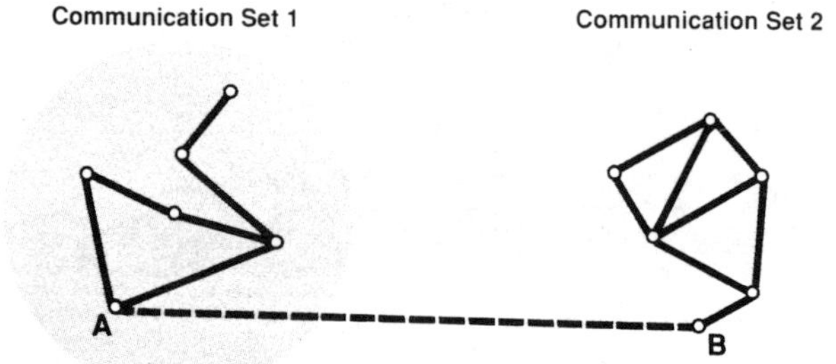

maintenance and control of the passage of information to mass audiences. News directors of radio and television stations and editors of newspapers and magazines select from the many news stories or feature items available from reporters, wire services, or contributors those items the directors and editors perceive to be of concern or value to their audiences (Figure 3.21).

The gatekeeper function may be performed in an open-information society by individuals and organizations responding to competitive and social pressures tempered by an independent assessment of accuracy, validity, and interest. In controlled-information societies, the gatekeeper function is ultimately exercised by government agencies through precensorship. However used, for whatever purpose, the gatekeeper position results in a constriction of information flows, a specialization of audience, and an isolation of region and society.

Spatial Implications

The mass media are by nature space filling. From single origin points they address their messages by print, radio, or television to potential receivers within a defined area. The number and location of disseminating points, therefore, are related to their spatial coverage characteristics, to the minimum size of area and population necessary for their support, and to the capability of the potential audiences to receive their message. The coverage area is determined both by the nature of the medium and by the corporate intent of the agency. There are no inherent spatial restrictions on the dissemination of printed materials. In the United States, much book and national magazine publishing has localized in New York City, as have the services supplying news and features for sale to the print media located there and elsewhere in the nation. Paris,

Figure 3.20

The gatekeeper model. Flows of information, *I*, to the gatekeeper, *G*, are sorted and selectively transmitted, *T*, to the receiver, *R*. The receiver also has independent sources of confirming or disputing information; his or her anticipated or actual feedback reaction to the gatekeeper's transmission may color the information selected for passage.

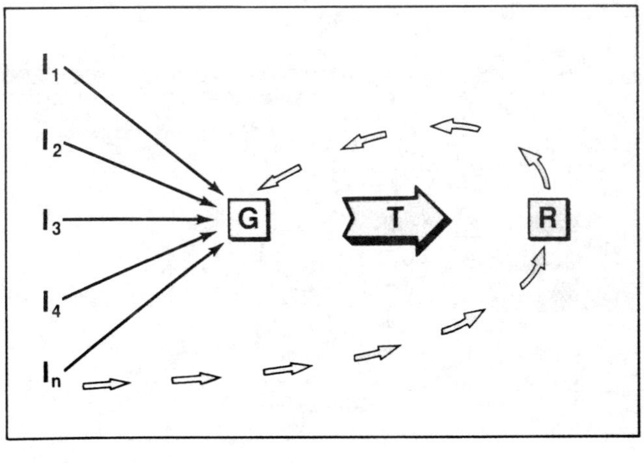

Figure 3.21

This now-dated view of the world from the British Broadcasting Company (BBC) and *Pravda* gives a timeless demonstration of the role of gatekeepers. Pick up a newspaper and see how much of it is devoted to local, national, and international news. You will discover that in general, as distance increases, the information about places decreases. The gatekeepers of national newspapers or network television news programs select items of interest to a broad audience. The editors of a local newscast or newspaper, on the other hand, have their judgment of newsworthiness colored by their recognition of the need to balance national and world news with strong infusions of information of purely local interest and concern. Perhaps the reason that there was a greater amount of London news on the BBC than there was Moscow news in *Pravda* is that London contains a greater proportion of the population of the British Isles than does Moscow of the USSR.

Source: From J. P. Cole and P. Whysall, "Places in the News: A Study of Geographical Information" in *Bulletin of Quantitative Data for Geographers,* no. 7, 1968. Copyright © 1968 Nottingham University, Nottingham, England. Reprinted by permission.

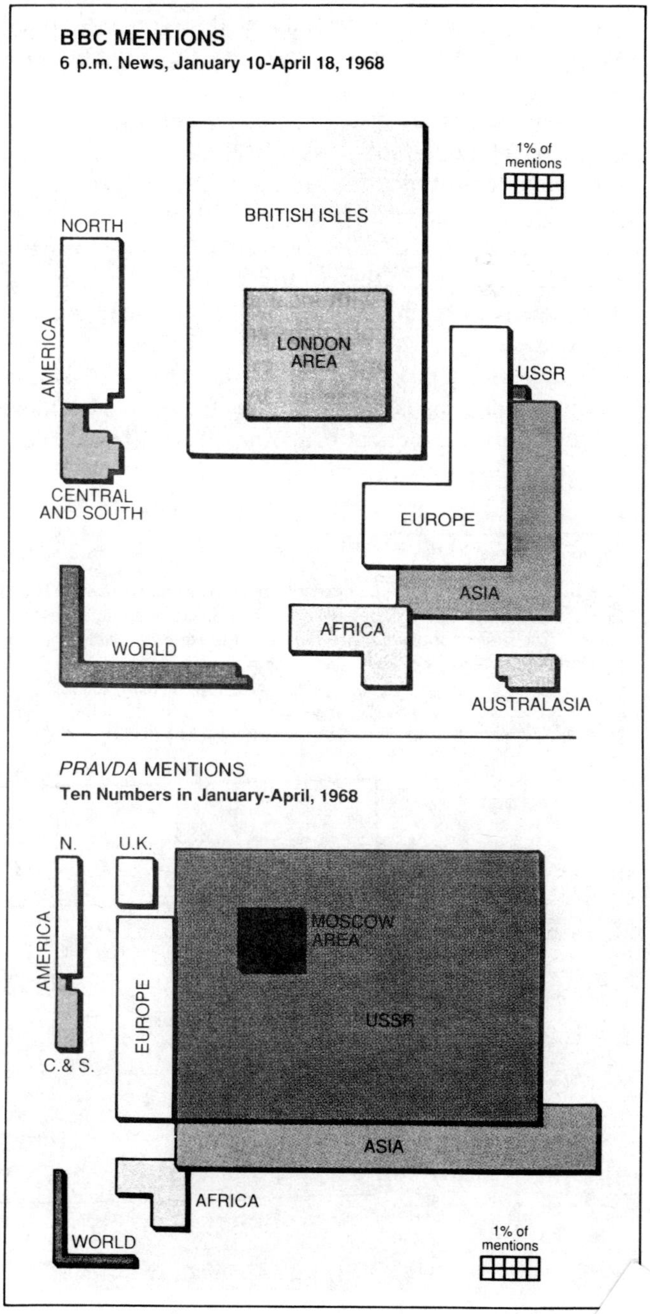

Buenos Aires, Moscow, London—indeed, the major metropolises and/or capital cities of other nations—show the same spatial concentration. Regional journals emanate from regional capitals; major metropolitan newspapers, though serving primarily their home market, are distributed over (or produce special editions for distribution within) tributary areas whose size and shape depend on the intensity of competition from other metropolises. A spatial information hierarchy has thus emerged.

When radio and television were introduced here and abroad, the tendency was for the larger cities to adopt these electronic media first and for the smaller cities to develop stations only later. The larger centers, through their size and sophistication, provided the awareness and acceptance of technology, the audience size, and the revenue generation necessary to introduce and support the new media. A process of hierarchical diffusion (see p. 50) spread the technology to smaller, localized outlets elsewhere in the nation.

Hierarchies are also reflected in the market-size requirements for different levels of media offerings. National and international organizations are required to expedite information flows (and, perhaps, to control their content), but market demand is heavily weighted in favor of regional and local coverage. In the electronic media, the result has been national networks with local affiliates acting as the gatekeepers of network offerings and adding to them locally originating programs and news content. A similar market subdivision is represented by the regional editions of national newspapers or magazines. At a different scale, the spatial distribution of newspapers in Kansas (Figure 3.22) shows their hierarchical pattern. Numerous small-town and rural weekly papers give partial coverage of the state. Evening papers, localized in cities of 10,000 or more, collectively serve all of Kansas. There are fewer larger-city morning papers; their circulation areas contain several weekly and evening papers. The centers with morning papers and those with Sunday editions of either morning or evening papers represent the regional capitals of the state.

The technological ability to fill space with messages from different mass media is unavailing if receiving audiences do not exist. In illiterate societies, publications cannot inform or influence. Unless the appropriate receivers are widely available, television and radio broadcasts are a waste of resources. Perhaps no invention in history has done more to weld isolated individuals and groups of purely person-to-person communicators into national societies exposed to centralized information flows than has the low-cost transistor radio. Its battery-powered transportability converts the remotest village and the most isolated individual into a receiving node of entertainment, information, and political messages. The direct satellite broadcast of television programs to community antennae or communal sets brings that mass medium to remote parts of Arctic Canada, India, Indonesia, and other world areas able to invest in the technology but as yet unserved by ground stations. In general, of course, the ability to participate in both formal interpersonal and mass media communication is a function of economic development and national wealth, as Figure 3.23 indicates.

Figure 3.22

The hierarchy of newspaper coverage in Kansas in the 1970s. The counties in sparsely populated western Kansas had only weekly papers. The more populous eastern part of the state had daily and Sunday papers, with wide-area distribution.

Source: From J. A. Jakle, S. Brunn, and C. C. Roseman, *Human Spatial Behavior: A Social Geography* (Copyright © 1976), reissued 1985 by Waveland Press, Inc., Prospect Heights, Illinois, p. 130. Reprinted with permission.

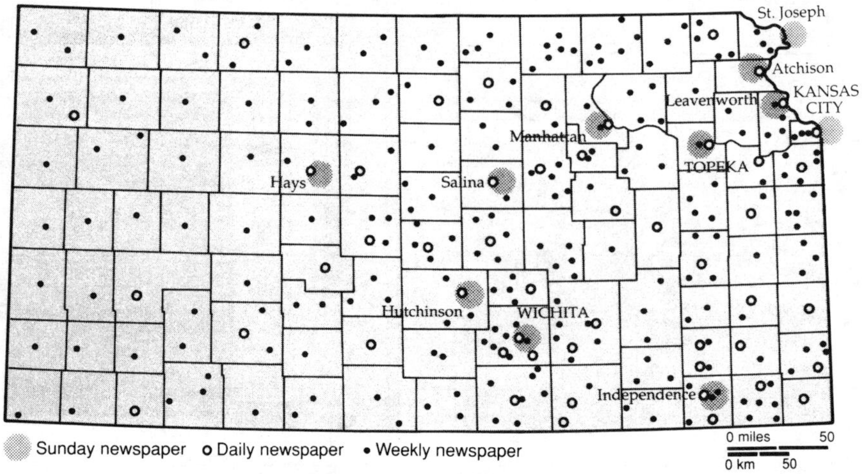

Information and Perception

Human spatial interaction, as we have seen, is conditioned by a number of factors. Complementarity, transferability, and intervening opportunity help pattern the movement of commodities and peoples. Flows between points and over area are influenced by distance decay and partially explained by gravity and potential models. Individuals in their daily affairs operate in activity spaces that are partly determined by stage in life cycle, mobility, and a variety of socioeconomic characteristics and that are roughly coextensive with their personal communication field. In every instance of spatial interaction, however, decisions are based on information about opportunity or feasibility of movement, exchange, or want satisfaction.

More precisely, actions and decisions are based on **place perception**—the awareness we have, as individuals, of home and distant places and the beliefs we hold about them. Place perception involves our feelings and understandings, reasoned or irrational, about the natural and cultural characteristics of an area and about its opportunity structure. Whether our view accords with that of others or truly reflects the "real" world seen in abstract descriptive terms is not the major concern. Our perceptions are the important thing, for the decisions people make about the use of their lives or about their actions in space are based not necessarily on reality but on their perceptions of reality.

Perception of Environment

Psychologists and geographers are interested in determining how we arrive at our perceptions of place and environment both within and beyond our normal activity space. The images we form firsthand of our home territory have been in part reviewed in the discussion of mental maps in Chapter 1. The perceptions we have of more distant places are less directly derived (Figure 3.24). In technologically advanced societies, television and radio, magazines and newspapers, books and lectures, travel brochures and hearsay all combine to help us develop a mental picture of unfamiliar places and of the interaction opportunities they may contain. Again, however, the most effectively transmitted information seems to come from

Figure 3.23

Potentials for participation in formal communication. When literacy levels are low, the mail and print media have little power to inform. To be influential, mass radio communication depends on set ownership.

Data from *UNESCO STATISTICAL YEARBOOK*. Copyright © Unesco 1986. Reproduced by permission of Unesco; Reproduced with permission from *World Military and Social Expenditures* 1987–88 by Ruth Leger Sivard. Copyright © 1987 by World Priorities, Box 25140, Washington, DC, USA.

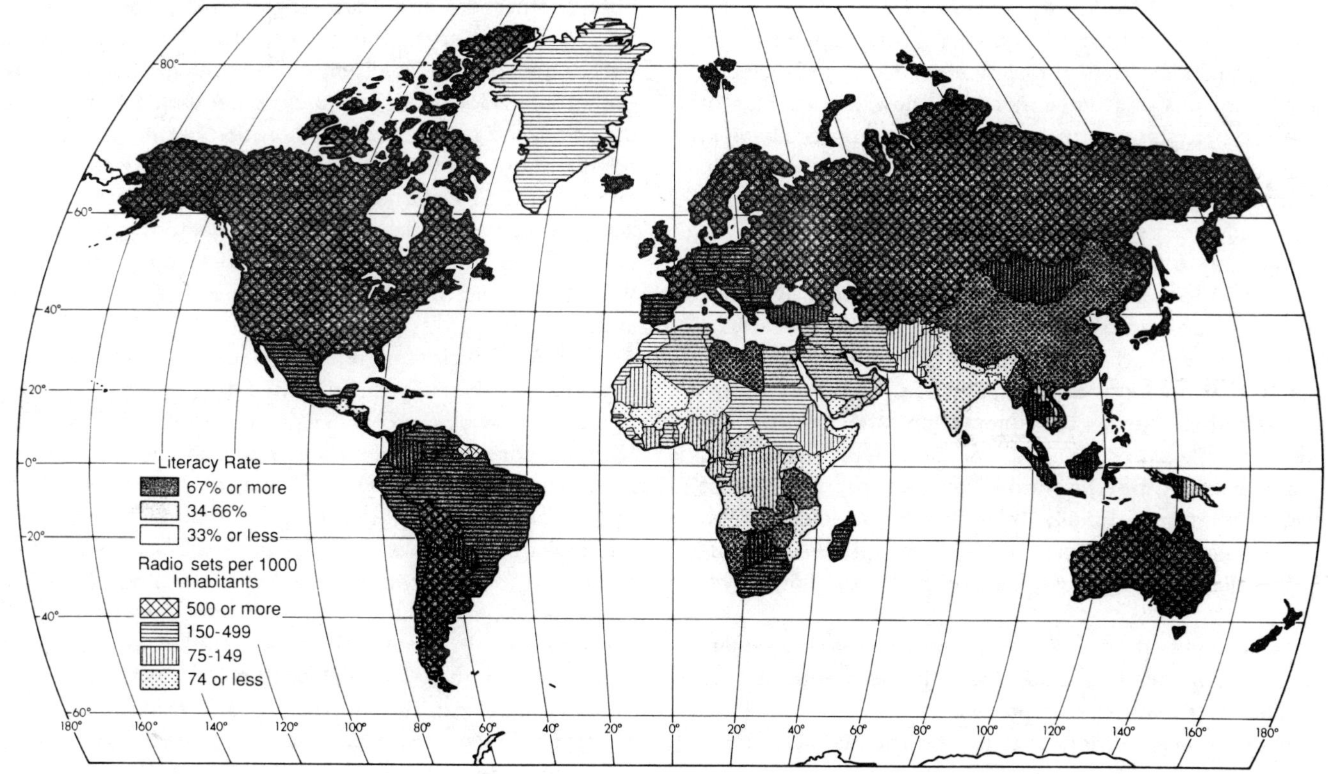

Figure 3.24

A view of the world as fashioned from newscasts heard by a Peace Corps volunteer working in the Central African Republic. The map is based on several months of listening to the Africa Service of the Voice of America (VOA) during 1987. Both areas and topics shown were those in the news and selected for emphasis by the broadcasters, whose choices helped form the mental maps and awareness patterns of their listeners. The "Russian Block" took on a three-dimensional reality not intended by the VOA.

Courtesy of Hilary Hope Getis.

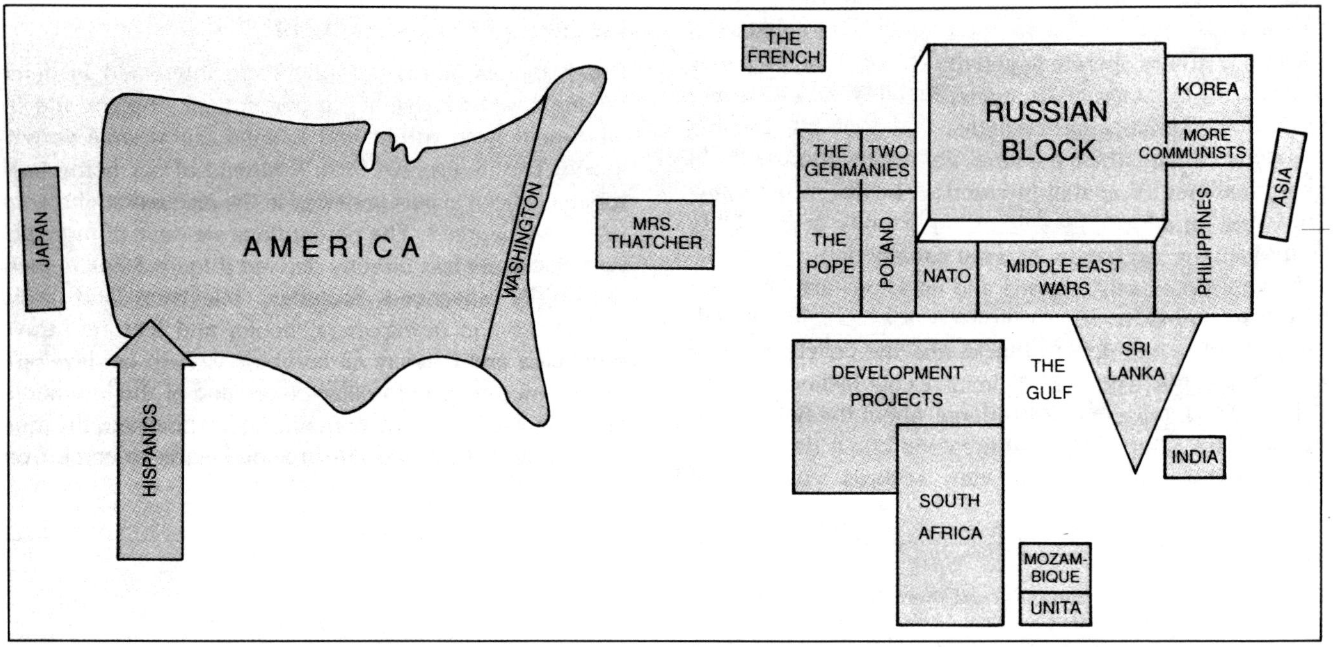

word-of-mouth reports. These may be in the form of letters or visits from relatives, friends, and associates who supply information that helps us develop lines of attachment to relatively unknown areas.

There are, of course, barriers to the flow of information, including that of distance decay. Our knowledge of close places is greater than our knowledge of distant points; our contacts with nearby people theoretically yield more information than we receive from afar. Yet, in crowded areas with maximum interaction potential, people commonly set psychological barriers around themselves so that only a limited number of those possible interactions and information exchanges actually occur. We raise barriers against information overload and to preserve a sense of privacy that permits the filtering out of information that does not directly affect us. Cost may represent another barrier to interaction and information exchange. Relatives, friends, and associates living long distances apart may find it difficult to afford the time or money to interact. Finally, there are the obvious barriers to interaction, such as mountains, oceans, rivers, and differing religions, languages, ideologies, and political systems.

Barriers to information flow give rise to what we earlier (page 64) called *direction bias*—in the present usage, a tendency to have greater knowledge of places in some directions than in others. Not having friends or relatives in one part of a country may represent a barrier to individuals, so that interest in and knowledge of the area beyond the "unknown" region are low. In the United States, both northerners and southerners tend to be less well-informed about each other's areas than about the western part of the country. Traditional communication lines in the United States follow an east-west rather than a north-south direction, the result of early migration patterns, business connections, and the pattern of the development of major cities. In the Soviet Union, directional bias favors a north-south information flow within the western part of the country and less familiarity with areas far to the east. Within the eastern USSR, however, east-west flows dominate.

When information about a place is sketchy, blurred pictures develop. These influence the impression—the perception—we have of places and cannot be discounted. Many important decisions are made on the basis of incomplete information or biased reports, such as decisions to visit or not, to migrate or not, to hate or not, even to make war or not. Awareness of places is usually accompanied by opinions about them, but there is no necessary relationship between the depth of knowledge and the perceptions held. In general, the more familiar we are with a locale, the more sound will be the factual basis of our mental image of it. But individuals form firm impressions of places totally unknown to them personally, and these may color interaction decisions.

Figure 3.25

Residential preferences of Canadians. Each of these maps shows the residential preference of a sampled group of Canadians from the provinces of British Columbia, Ontario, and Quebec, respectively. Note that each group of respondents prefers its own area but that all like the Canadian and United States west coasts.

Source: From Herbert A. Whitney, "Preferred Locations in North America: Canadians, Clues, and Conjectures," in *Journal of Geography*, 83, no. 5, 1984. Copyright © 1984 National Council for Geographic Education, Indiana, PA. Reprinted by permission.

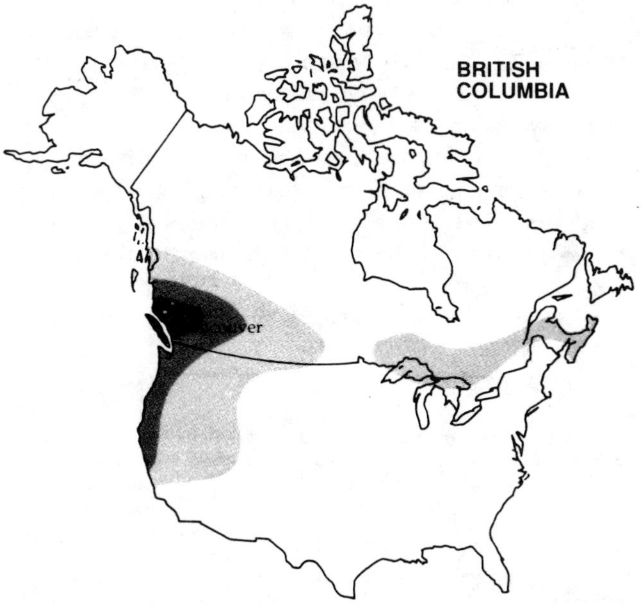

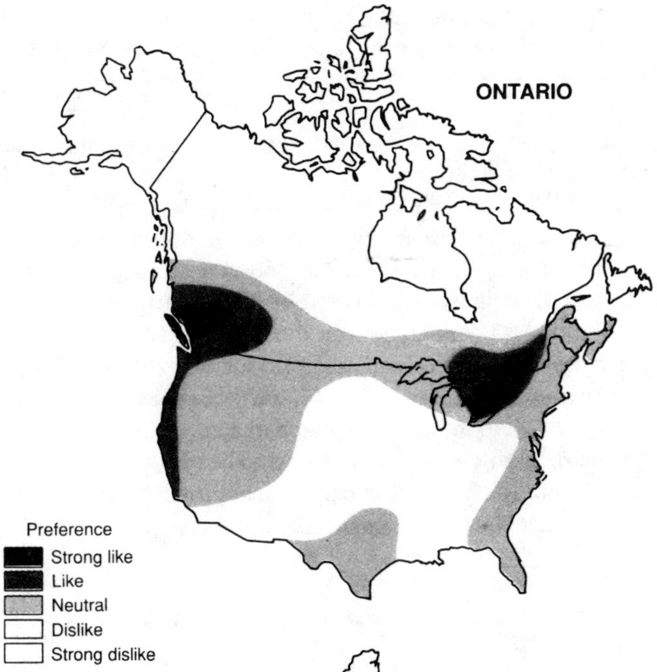

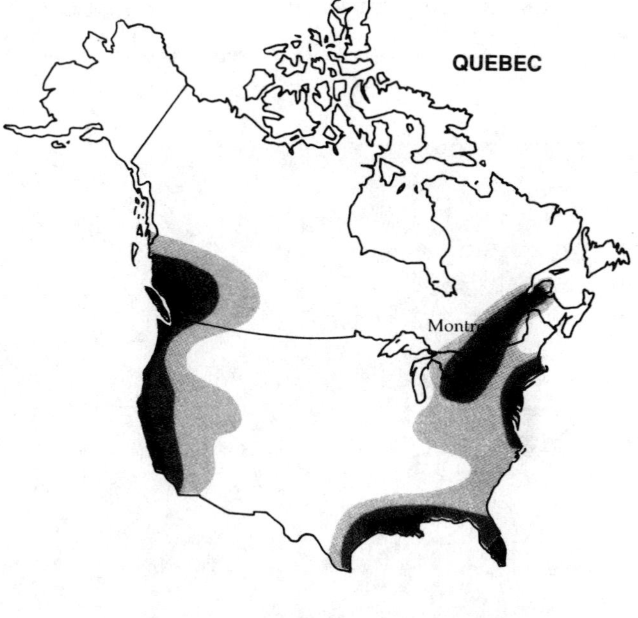

One way to determine how individuals envisage home or distant places is to ask them what they think of different locales. For instance, they may be asked to rate places according to desirability—perhaps residential desirability—or to make a list of the 10 best and the 10 worst cities in their country of residence. Certain regularities appear in such inquiries. Figure 3.25 presents some residential desirability data as elicited from college students in three provinces of Canada. These and comparable mental maps derived from studies conducted by researchers in many countries suggest that near places are preferred to far places unless much information is available about the far places. Places of similar culture are favored, as are places with high standards of living. Individuals tend to be indifferent to unfamiliar places and to dislike unfamiliar areas that have competing interests (such as disliked political and military activities or conflicting economic concerns) or a physical environment perceived to be unpleasant.

On the other hand, places perceived to have superior climates or landscape amenities are rated highly in mental map studies and favored in tourism and migration decisions. The southern and southwestern coast of England is attractive to citizens of generally wet and cloudy Britain, and holiday tours to Spain, the south of France, and the Mediterranean islands are heavily booked by the English. A U.S. Census Bureau study indicated that "climate" was, after work and family proximity, the most often reported reason for interstate moves by adult individuals of all ages during the mid-1970s. International studies reveal a similar migration motivation based not only on climate but also on concepts of natural beauty and amenities.

Perception of Natural Hazards

Less certain is the negative impact on spatial interaction or relocation decisions of assessments of *natural hazards,* processes or events in the physical environment that are not caused by humans but that have consequences harmful to them. Distinction is made between chronic, low-level hazards (health-affecting mineral content of drinking water, for example) and high-consequence/low-probability events such as hurricanes, earthquakes, landslides, and the like. Remedial low-level hazards do not appear to create negative space perceptions, though highly publicized chronic natural conditions, such as cancer-related radon emissions (Figure 3.26), may be an exception. Space perception studies do reveal, however, a small but measurable adverse assessment of locales deemed "dangerous," no matter what the statistical probability of the hazard's occurring.

Mental images of home areas do not generally include as an overriding concern an acknowledgment of potential natural dangers. The cyclone that struck the delta area of Bangladesh on November 12, 1970, left dead at least 500,000 people, yet after the disaster the movement of people into the area swelled population above precyclone levels. The July 28, 1976, earthquake in the Tangshan area of China devastated a major urban-industrial complex with casualties presumed to lie between 700,000 and 1 million. Rebuilding began almost immediately, as it usually does following earthquake damage (Figure 3.27). The human response to even such major and exceptional natural hazards is duplicated by a general tendency to discount dangers from more common hazard occurrences. Johnstown, Pennsylvania, has suffered recurrent floods, and yet its residents rebuild; violent storms strike the Gulf and East coasts of the United States, and people remain or return. Californians may be concerned about Kansas tornadoes if contemplating a move there but be unconcerned about earthquake dangers at home.

Why do people choose to settle in areas of high-consequence hazards in spite of the potential threat to their lives and property? Why do hundreds of thousands of people live along the San Andreas fault in California, build houses in Pacific coastal areas known to experience severe erosion during storms, return to flood-prone river valleys in Europe or Asia, or avalanche-threatened Andean valleys? What is it that makes the risk worth taking? Ignorance of natural hazard danger is not necessarily a consideration. People in seismically active regions of the

Figure 3.26

Areas with potentially high radon levels. The radon "scare" began in 1984 with the discovery that a Pennsylvania family was being exposed in its home to the equivalent of 455,000 chest X rays per year. With the estimate that as many as 20% of the nation's annual lung-cancer deaths may be attributable to radon, homeowners and seekers are becoming aware of a new but localized environmental hazard.

Source: "Areas with Potentially High Radon Levels," August 1987. U.S. Environmental Protection Agency, Office of Air and Radiation, Washington, DC.

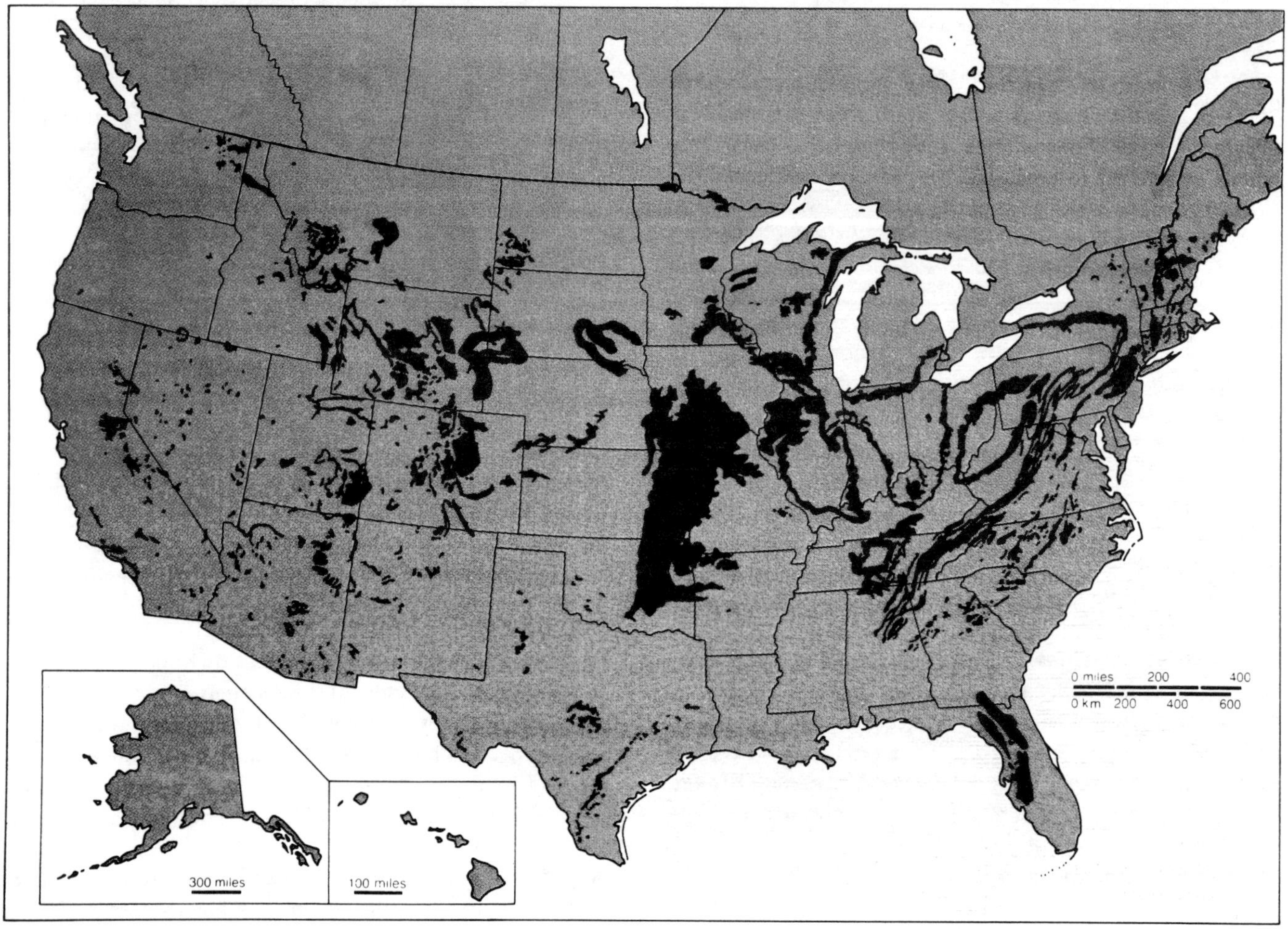

United States and Europe, at least, do believe that damaging earthquakes are a possibility in their districts but, research indicates, are reluctant to do anything about the risk. Similar awareness and reticence accompanies other low-incidence/high-consequence natural dangers. Less than one-tenth of 1% of respondents to a federal survey in the mid-1970s gave "natural disaster" as the reason for their interstate residential move.

There are many reasons why natural hazard risk does not deter settlement or adversely affect space-behavioral decisions. Of importance, of course, is the persistent belief that the likelihood of an earthquake or a flood or other natural calamity is sufficiently remote so that it is not reasonable or pressing to modify behavior because of it. Cost and political unpopularity appear to be principal considerations in unwillingness of community officials to take preventive or mitigating action against hazards. People are, as well, influenced by their innate optimism and by predictive uncertainty about timing or severity of a calamitous event. And, in fact, the prediction of hazards is not an exact science but rather is based on the calculation of the probability of occurrence of uncommon events. People are also influenced by their past experiences in high-hazard areas. If they have not suffered much damage in the past, they may be optimistic about the future. If, on the other hand, past damage has been great, they may think that the probability of repetition in the future is low (Table 3.1), though among Great Plains wheat farmers a heightened awareness of drought hazard seems to be a product of length of personal farming experience.

Perception of place as attractive or desirable may be quite divorced from any understanding of its hazard potential. Attachment to locale or region may be an expression of emotion and economic or cultural attraction, not just a rational assessment of risk. The culture hearths of antiquity discussed in Chapter 2 and shown on Figure 2.15 were for the most part sited in flood-prone river valleys; their enduring attraction was undiminished by that potential danger. The home area, whatever disadvantages an outside observer may discern, exerts a force not easily dismissed or ignored. Indeed, high-hazard areas are often sought out because they possess desirable topography or scenic views, as do, for instance, coastal areas subject to storm damage. Once people have purchased property in a known hazard area, they may be unable to sell it for a reasonable price even if they so desire. They think that they have no choice but to remain and protect their investment. The cultural hazard—loss of livelihood and investment—appears more serious than whatever natural hazards may exist.

Carried further, it has been observed that spatial adjustment to perceived natural hazards is a luxury not affordable to impoverished people in general or to the urban and rural poor of Third World countries in particular. Forced by population growth and economic necessity to exert ever-greater pressures upon fragile environments or to

Figure 3.27

Destruction from the San Francisco earthquake and fire. The first shock struck San Francisco early on the morning of April 18, 1906, damaging the city's water system. Fire broke out and raged for 3 days. It was finally stopped by dynamiting buildings in its path. When it was over, some 500 people were dead or missing, and 25,000 buildings had been destroyed. Locally, the event is usually referred to as the Great Fire of 1906, suggesting a denial of the natural hazard in favor of assigning blame to correctable human error. Postdestruction reconstruction began at once. Rebuilding following earthquake damage is the rule, though the immediate return of population to northern Italian areas after a major quake in 1976 was followed by an abrupt exodus after a subsequent, much weaker shock.

Table 3.1 Common Responses to the Uncertainty of Natural Hazards

Eliminate the Hazard	
Deny or Denigrate Its Existence	*Deny or Denigrate Its Recurrence*
"We have no floods here, only high water."	"Lightning never strikes twice in the same place."
"It can't happen here."	"It's a freak of nature."
Eliminate the Uncertainty	
Make It Determinate and Knowable	*Transfer Uncertainty to a Higher Power*
"Seven years of great plenty. . . . After them 7 years of famine."	"It's in the hands of God."
"Floods come every 5 years."	"The government is taking care of it."

Source: Burton and Kates, *The Perception of Natural Hazards in Resource Management*, 3 *NAT. RES. J.* 435 (1964). Reprinted by permission.

occupy at higher densities hazardous hillside and flood-plain slums, their margin of safety in the face of both chronic and low-probability hazards is minimal to non-existent (Figure 3.28).

Migration

When continental glaciers began their retreat some 11,000 years ago, the activity space and awareness space of Stone Age humans were limited. As a result of pressures of numbers, need for food, changes in climate, and other inducements, those spaces were collectively enlarged to encompass the world. **Migration**—the permanent relocation of residential place and activity space—has been one of the enduring themes of human history. It has contributed to the evolution of separate cultures, to the diffusion of those cultures and their components by interchange and communication, and to the frequently complex mix of peoples and cultures found in different areas of the world. Portions of the story of migration have been touched on already in Chapter 2; other elements of it are part of later discussions of population (Chapter 4) and ethnicity (Chapter 6). Our interest here is with migration as an unmistakable, recurring, and near-universal expression of human spatial behavior.

Migration embodies all of the principles of spatial interaction and space relations we have already reviewed. Complementarity, transferability, and intervening opportunities and barriers all play a role. Space information and perception are important, as are the sociocultural and economic characteristics of the migrants and the distance relationships between their original and prospective locations of settlement.

Although all spatial movement incorporates these same principles, not all long-distance interaction involves migration. The planned two-way trip, for example, does not include residential relocation. The exciting, perhaps historically decisive, journey of exploration about which we read in history was a temporary movement in space in which return home was an indisputable part of the plan. The more common contemporary vacation or business or social trip may enhance the mental maps or enlarge the awareness space of the participant—it may even be the prelude to later migration—but by definition it is a round-trip venture. Even Catherine Haun's cross-country trek, she reports, was begun with the thought of return.

Migration, for whatever reason undertaken, is presumed to be a permanent relocation, though seasonal and short-term moves are often imprecisely considered "migratory" (Figure 3.29). Naturally, the length of the move and its degree of disruption of established activity space patterns raise distinctions important in the study of migration. A change of residence from the central city to the

Figure 3.28

Many of the poor of Rio de Janeiro, Brazil, occupy steep hillside locations above the reach of sewer, water, and power lines that hold the more affluent at lower elevations. Heavy rains during late February 1988 caused mudflows from the saturated hillsides, wiping away the shacks and shelters that had insecurely clung to them and depositing the homes and hopes of the poor in richer neighborhoods below.

Figure 3.29

Fulani movement paths. Nomadic people have regular routes of "migration" that involve near-continuous movement in response to the needs of their herds or flocks. Their moves are always temporary and involve a circuit, not a one-way flow. The groups of Fulani families herd their cattle within the savanna of West Africa. On their southward trek during the dry season, the groups separate; when returning north during the wet season, they reunite. Migration circuits may be changed annually in response to the availability of pasture and the location of areas free of the tsetse fly, the transmitter of "sleeping sickness."

Source: Robin Baker, *The Mystery of Migration*, New York: The Viking Press, 1981. p. 230.

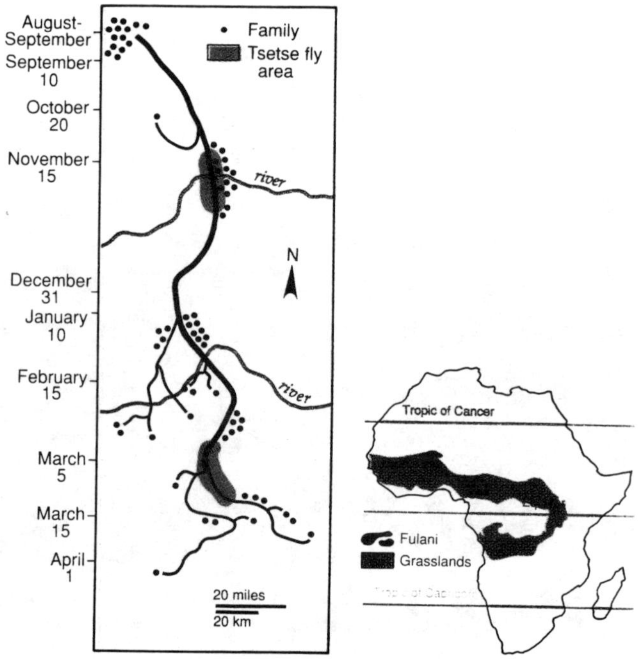

suburbs certainly changes both residence and activity space of schoolchildren and of adults in many of their non-working activities, but the working adults may still retain the city—indeed, the same place of employment there—as an action space. On the other hand, immigration from Europe to the United States and the massive farm-to-city movements of rural Americans late in the 19th and early in the 20th centuries meant a total change of all aspects of behavioral patterns.

Principal Migration Patterns

Migration flows may be discussed at different scale, from massive intercontinental torrents to individual decisions to move to a new house or apartment within the same metropolitan area. At each scale, the factors influencing the spatial interaction are different. The impact of the flow on the world's patterns of population and cultural landscapes also varies with scale.

At the broadest scale, *intercontinental* movements range from the earliest peopling of the habitable world to the most recent flight of Asian refugees to countries of Europe or the Western Hemisphere. The very structure of the United States, Canada, Australia and New Zealand, Argentina, Brazil, and other South American nations—as Chapter 4 suggests—is a reflection and result of massive flows of immigrants that began as a trickle during the 16th and 17th centuries and reached a flood during the 19th and early 20th (Figure 4.19). More recently, World War II and its immediate aftermath involved more than 25 million permanent population relocations, all of them international but not all intercontinental.

Intracontinental and *interregional* migrations involve movements between countries and within countries, most commonly in response to individual and group assessments of improved economic conditions or changes in life cycle, but often reflecting flight from disastrous environmental or political conditions. The some 15 million labor migrants ("guest workers") in West European countries in the early 1980s represented an influx of job seekers from southeastern Europe and the Mediterranean area including North Africa (Figure 3.30). North America has its counterparts in the hundreds of thousands of immigrants (many illegal) coming to the United States each year from Mexico, Central America, and the Caribbean region. The Hauns, whose westward trek opened this chapter, were part of a massive 19th-century regional shift of Americans that continues today (Figure 3.31). The USSR has experienced a similar, though eastward, flow of people in this century.

In the 20th century, nearly all countries have experienced a great movement of peoples from agricultural areas to the cities, continuing a pattern of *rural-to-urban migration* that first became prominent during the 18th- and 19th-century Industrial Revolution in advanced economies. While the rate of urban growth is decreasing in the more developed nations, Third World urbanization continues apace, as is discussed more fully in Chapter 11.

Figure 3.30

International labor flows. (*a*) Major international labor migration flows, 1960–1980. The industries of Western Europe, oil and construction work in Arabia, mining in South Africa, urban and agricultural opportunities in the United States, and similar economic attractions elsewhere sparked international labor movements from widely scattered source regions. (*b*) European "guest worker" flows. For prosperous but labor-short Western Europe (West Germany, France, Belgium, the Netherlands, and Switzerland), major source regions for labor immigrants in the early 1980s were southern and southeastern Europe, Turkey, and North Africa.

Sources: (*a*) Adrienne Kols and Dana Lewison. *Migration, Population Growth, and Development. Population Reports,* Series M, Number 7. Baltimore, Johns Hopkins University, Population Information Program, 1983. (*b*) From Günther Glebe and John O'Loughlin, editors, *Foreign Minorities in Continental European Cities,* Vol. 84 of *Erkundliches Wissen.* Copyright © 1987 Franz Steiner Verlag, Wiesbaden, West Germany. Reprinted by permission.

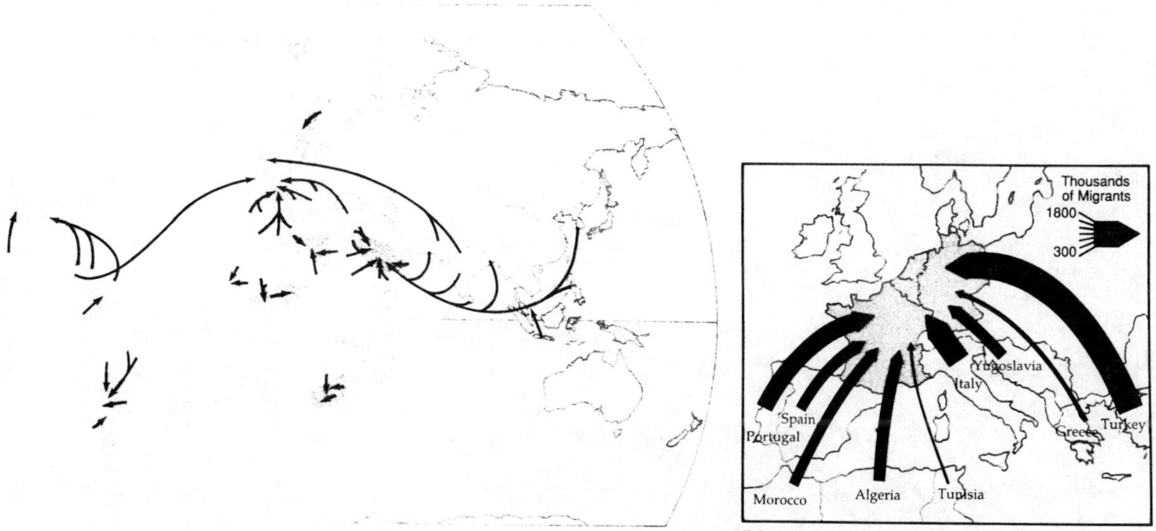

(b)

Within metropolitan areas, and at the narrowest scale, *localized residential shifts,* including suburbanization and neighborhood relocations, are particularly important in Western countries with free housing markets. Although not migrations in the sense of longer-distance relocations, they contribute to the frequent observation that one-sixth of the American population changes residence each year. Such movements reflect individual changes in life cycle, income level, job location, and other personal circumstances. Collectively, they contribute to distinctive patterns of urban social geography, also a topic reviewed in Chapter 11.

Types of Migration

Migrations may be forced or voluntary or, in many instances, reluctant relocations imposed on the migrants by circumstances.

In *forced migrations,* the relocation decision is made solely by people other than the migrants themselves (Figure 3.32) Perhaps 10 to 12 million Africans were forcibly transferred as slaves to the Western Hemisphere from the late 16th to early 19th centuries. Half or more were destined for the Caribbean and most of the remainder for Central and South America, though nearly a million arrived in the United States. Australia owed its earliest European settlement to convicts transported after the 1780s to the British penal colony established in southeastern Australia (New South Wales). More recent involuntary migrants include millions of Soviet citizens coercively relocated from countryside to cities and from the western areas to labor camps in Siberia and the Soviet Far East beginning in the late 1920s. Nigeria forcibly expelled 2 million foreign workers in 1983 and another 700,000 in 1985 to reduce unemployment among its own citizens. Other examples abound.

Less than fully voluntary migration—*reluctant relocation*—of some 6 million Indonesians has taken place under an aggressive governmental campaign begun in 1969 to move people from densely settled Jawa (Java, roughly 1500 people per square mile, or 580 per km²) to other islands and territories of the country in what has been called the "biggest colonization program in history." Refugees from war and political turmoil or repression (see "The Search for Asylum") numbered 12 million in 1988, according to United Nations figures. These were the officially designated refugees; unofficially, the total might have reached 20 million or more. In the past, refugees sought asylum mainly in Europe and other developed areas. More recently, the flight of people is primarily from developing countries to other developing regions. About 80% of the officially recorded refugees have found asylum in Southeast Asia, southern and eastern Africa, the Near East and South Asia, and South and Central America. Africa alone housed over 4 million refugees in 1988, including more than 1 million Ethiopians (Figure 3.33).

The great majority of migratory movements, however, are *voluntary,* representing individual response to the factors influencing all spatial interaction decisions. At root, migrations take place because the migrants believe that their opportunities and life circumstances will be better at their destination than they are at their present location.

Figure 3.31

Westward shift of population, 1790–1990. Two hundred years of western migration and population growth are recorded by the changing U.S. center of population. (The "center of population" is that point at which a rigid map of the United States would balance, reflecting the identical weights of all residents in their location on the census date.) The westward movement was rapid for the first 100 years of census history and slowed between 1890 and 1950. Some of the post-1950 acceleration reflects population growth in the Sunbelt. However, the two different locations for the population center in 1950 and the symbol change indicate the geographical pull on the center of population exerted by the admission of Alaska and Hawaii to statehood.

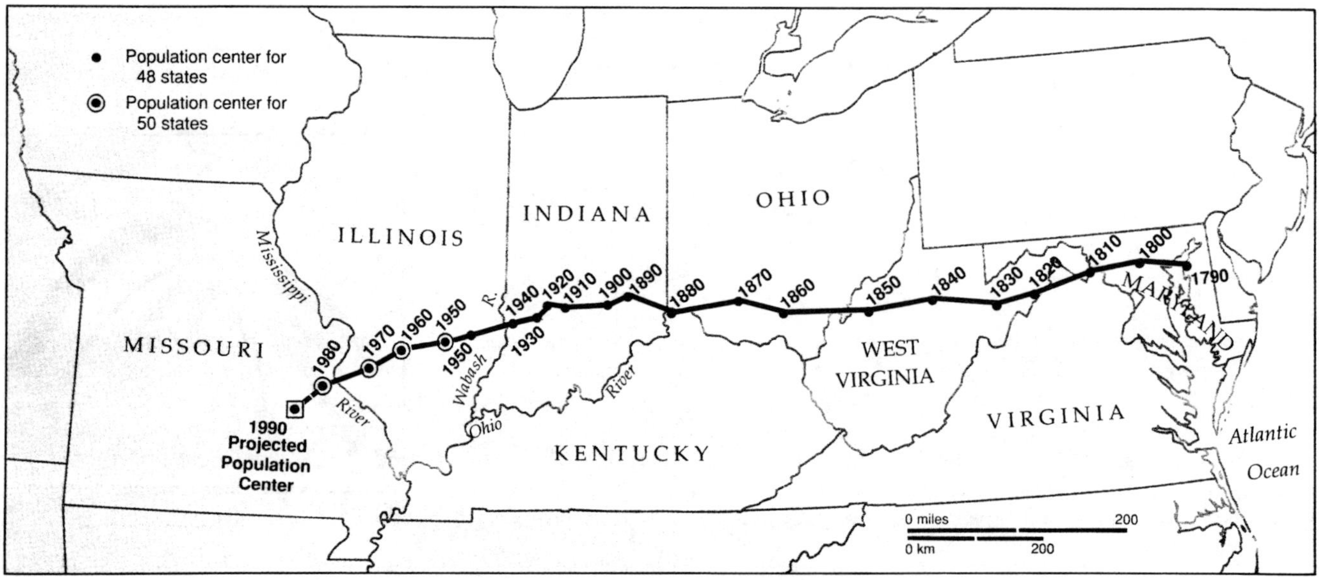

Figure 3.32

Forced migrations: the Five Civilized Tribes. Between 1825 and 1840, some 100,000 southeastern Amerindians were removed from their homelands and transferred by the army across the Mississippi River to "Indian Territory" in present-day Oklahoma. By far the largest number were members of the Five Civilized Tribes of the South: Cherokees, Choctaws, Chickasaws, Creeks, and Seminoles. Settled, Christianized, literate small farmers, their forced eviction and arduous journey—particularly along what the Cherokees named their "Trail of Tears" in the harsh winter of 1837–1838—resulted in much suffering and death.

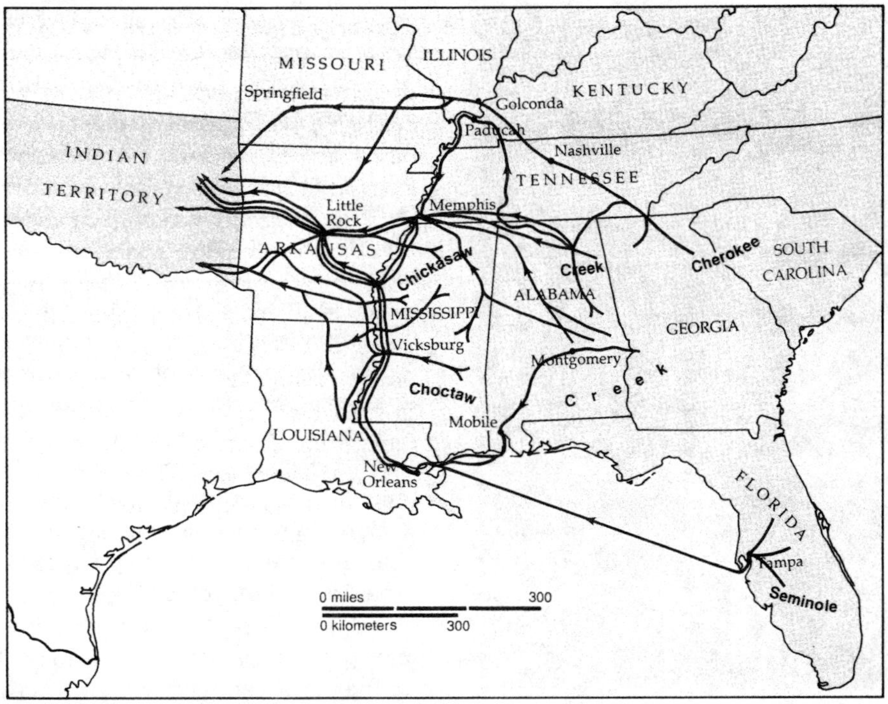

The Search for Asylum

In the decade after World War II, some 45 million people relocated under pressure: Finnish Karelians and Poles from areas annexed by the Soviet Union; Indian Muslims and Pakistani Hindus joining their co-religionists following partition of the subcontinent; Palestinian Arabs and Jews from Arab countries at the founding of Israel; Dutch settlers from Indonesia and French residents of North Africa; Koreans and Vietnamese; Africans caught up in countless colonial and civil wars. And the flow continues unchecked. Between 1977 and 1987, 750,000 people sought asylum in Western Europe—200,000 in 1986 alone. To the political and cultural refugees in Europe must be added the economic immigrants—North Africans in France, Asians in Britain, Turks in West Germany. These amounted to 70% of all immigration into Europe during the 1960s and 1970s. And to *them* add subsequent immigrants representing the reunion of families. Opportunity and flight are intermixed in migration flows.

According to the Geneva conventions on the subject, *refugees* are people who leave their own country, unable to return for fear of persecution on racial, religious, ethnic, or political grounds. *Asylum seekers* are refugees asking a different country to take them. The petitioned nation is under no obligation to agree. Among European nations, the policy in the late 1980s was to discourage applicants. West Germany, formerly one of the most generous of havens, began to forbid non-European asylum seekers from taking a job for 5 years, to restrict their freedom of movement, and to force their residence in designated camps even if they had relatives within the country. Border controls were tightened across Western Europe. Formerly accessible Sweden began to refuse entry to applicants with unsubstantiated claims for asylum; Belgian police turned back petitioners of doubtful merit or improper papers.

A basic question concerns an individual's right to free movement. In Western culture that right has traditionally been considered absolute within a nation's borders and for departing emigrants, though international free movement into nations is everywhere controlled. In other cultures a different view may hold—one in which the emigrant is considered a traitor. This is the Russian view, both Czarist and Marxist, and the view of contemporary China and East European countries. For these societies, and others, the state—not the individual—has the human right to determine place of residence, including residence within the nation itself. The geographical analysis of migration flows must take account of spatially variant viewpoints.

Figure 3.33

Fleeing war, repression, and famine, millions of people in developing nations, such as these Ethiopians, have become reluctant migrants from their homelands.

Controls on Migration

Economic considerations crystallize most migration decisions, though nomads fleeing the famine and spreading deserts of the Sahel obviously are impelled by different economic imperatives than is the executive considering a job transfer to Montreal or the resident of Appalachia seeking factory employment in the city. Among the aging, affluent populations of highly developed countries, retirement amenities figure importantly in perceptions of residential attractiveness of areas. Educational opportunities, changes in life cycle, and environmental attractions or repulsions are but a few other possible migration motivations.

Negative home conditions that impel the decision to migrate are called **push factors.** They might include loss of job, lack of professional opportunity, overcrowding or slum clearance, or a variety of other influences. The presumed positive attractions of the migration destination are known as **pull factors.** They include all the positive attributes perceived to exist at the new location—job opportunities, better climate, lower taxes, more room, and so forth. Very often migration is a result of both perceived push and pull factors. It is *perception* of the areal pattern of opportunities and want satisfaction that is important here, whether or not perceptions are supported by objective reality.

The concept of **place utility**—an individual's existing or anticipated degree of satisfaction or dissatisfaction with a place—helps us to understand the decision-making process that potential migrants undergo. The decision to migrate is a reflection of the appraisal—the perception—by a potential migrant of the current homesite as opposed to other sites of which something is known. In that appraisal, the decision maker considers not only present relative place utility but also expected place utility in a highly personal assessment that likely differs from the place utilities assigned by other potential migrants. The individual has a felt certainty about the home area; for other areas, place utility can only be estimated on the basis of information available to her or him. Information may be derived from past travel, from impressions gained from books or magazines, or importantly, from letters or conversations with friends or relatives familiar with the proposed destination. The decision maker can do no more than sample information about the new sites; and, of course, there may be sampling errors.

Individual sampling error may be reduced when the migrant progresses through a series of transitional relocation stages or when the migrant follows the example of known predecessors. **Step migration** involves the place transition from, for example, rural to central city residence through a series of less extreme locational changes—from farm to small town to suburb and, finally, to the major central city itself. **Chain migration** assures that the mover is part of an established migrant flow from a common origin to a prepared destination. An advance group of migrants, having established itself in a new home area, is followed by second and subsequent migrations originating in the same home district and frequently united by kinship or friendship ties. Ethnic and foreign-born enclaves in major cities and rural areas in a number of countries are the immediate result, as we shall see more fully in Chapter 6.

Sometimes the chain migration is specific to occupational groups. For example, nearly all newspaper vendors in New Delhi, in the north of India, come from one small district in Tamil Nadu, in the south of India. Most construction workers in New Delhi come either from Orissa, in the east of India, or Rajasthan, in the northwest. The diamond trade of Bombay, India, is dominated by a network of about 250 related families who come from a small town several hundred miles to the north. A corollary of all out-migration flows is **counter** (or **return**) **migration**—the likelihood that as many as 25% of all migrants will return to their place of origin.

Once established, origin and destination pairs of places tend to persist. Areas that dominate a locale's in- and out-migration patterns make up the **migration field** of the place in question. As we would expect, areas near the point of origin make up the largest part of the migration field, though larger cities more distantly located may also be prominent as the ultimate destination of hierarchical step migration. Some migration fields reveal a distinctly *channelized* pattern of flow. The channels link areas that are in some way tied to one another by past migrations, by economic trade considerations, or some other affinity. The flow along them is greater than otherwise would be the case (Figure 3.34), but it does not necessarily involve individuals with personal or family ties. The former streams of southern blacks and whites to northern cities, of Scandinavians to Minnesota and Wisconsin, and of U.S. retirees to Florida and Arizona, or their European counterparts to Iberia or the Mediterranean coast are all examples of **channelized migration.** In the case of "guest workers" in European cities, channelized flows may be the result of past or present labor recruitment by destination countries.

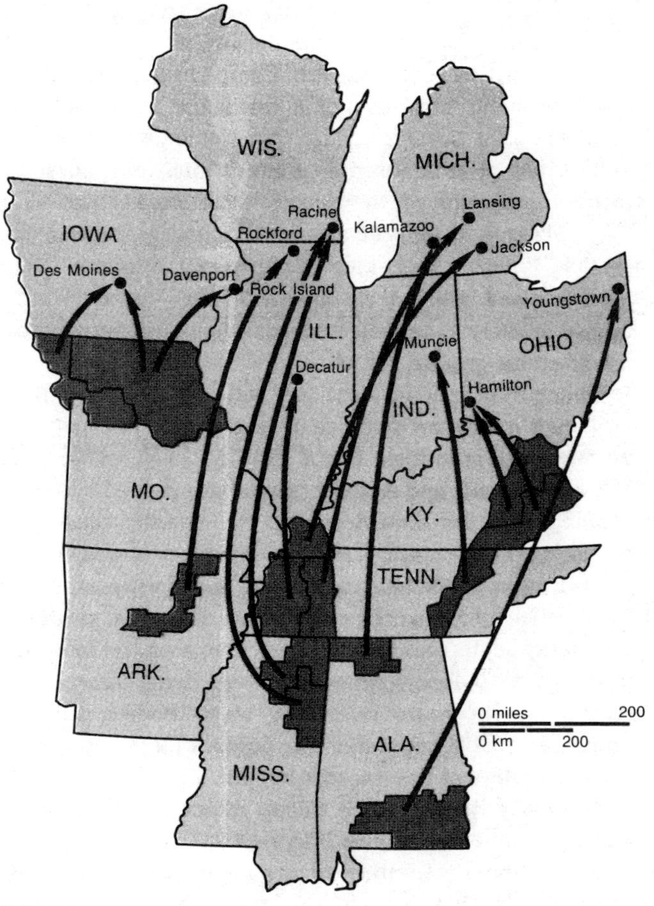

(a)

Figure 3.34

Channelized migration flows. (*a*) Place-specific flows from the rural South to midwestern cities of medium size. Distance is not necessarily the main determinant of flow direction. Perhaps through family and friendship links, the rural southern areas are tied to particular midwestern destinations. (*b*) International migration streams into Senegal are less place specific, as these immigration paths of resident immigrants suggest. Distance from home country is a prime consideration, but some flows ignore proximity and follow predecessor migrants to more distant areas.

Sources: (a) Reproduced by permission from *Proceedings of the Association of American Geographers*, C. C. Roseman, Volume 3, page 197, American Association of Geographers, 1971. (*b*) K. C. Zachariah and Julien Condé, *Migration in West Africa: Demographic Aspects.* Map 12. New York: Published for the World Bank by Oxford University Press, 1981.

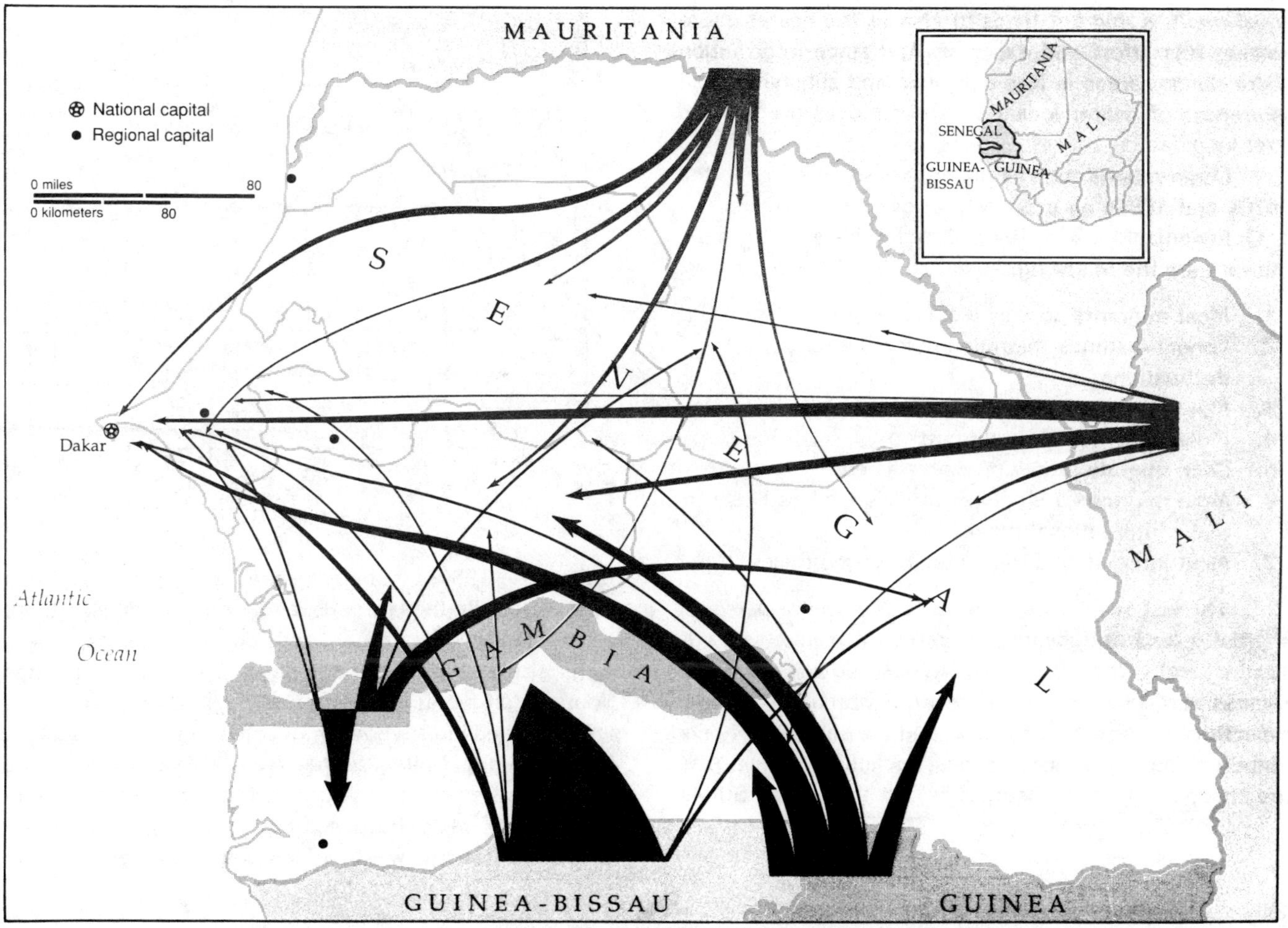

(b)

Voluntary migration is responsive to other controls that influence all forms of spatial interaction. Push-pull factors may be equated with *complementarity*. Costs (emotional and financial) of a residence relocation are expressions of *transferability*. Other things being equal, large cities exert a stronger migrant pull than do small towns, a reflection of the impact of the *gravity model*.

The role of *intervening opportunity* in shaping migration flows is less clear. Determined migrants with specific place utility decisions made, are—research indicates—unlikely to be deflected from their destination goal. Lower-status groups, for whom push factors are more determining than specific locational pull considerations, may limit their migration distance in response to encountered apparent opportunities. From 1940 to 1970, millions of black Americans and Appalachian whites moved by trains running from the rural South up the Atlantic coast and parallel to the Mississippi River to known or hoped-for job opportunities in the industrial North and Northeast. Many were deflected from their original big-city goals, stopping at smaller eastern and midwestern towns on the railroads in response to opportunities perceived there. *Intervening obstacles*—mountain ranges or water bodies, for example, or hostile populations or governments—may also be interrupters of the migrant stream.

Finally, the *distance decay* effect has often been noted in migration studies (Figure 3.35). It has a number of expressions; the impact of intervening opportunities is only one. Movers seek to minimize the *friction of distance*. In selecting between two potential destinations of equal merit, a migrant tends to choose the nearer as involving less effort and expense. And since information about distant areas is less complete and satisfying than awareness of nearer localities, short moves are favored over long.

Observations such as these were summarized in the 1870s and 1880s as a series of "laws of migration" by E. G. Ravenstein (1834–1913). Among those that remain relevant are the following:

1. Most migrants go only a short distance.
2. Longer-distance migration favors big-city destinations.
3. Most migration proceeds step-by-step.
4. Most migration is rural to urban.
5. Each migration flow produces a counterflow.
6. Most migrants are adults; families are less likely to make international moves.
7. Most international migrants are young males.

The last two "laws" introduce the role of personal attributes (and attitudes) of migrants: their age, sex, education, and economic status. Migrants do not represent a cross section of the populace from which they come. Selectivity of movers is evident, and the selection shows some regional differences. In most societies, young adults are the most mobile (Figure 3.36). In the United States during the 1980s, mobility peaked among those in their 20s, especially the later 20s, and tended to decline thereafter. Among West African cross-border migrants, a World Bank study reveals, the age-group 15–39 predominated. Young adult males are particularly prominent in economically pushed international movement. Migrants, legal and illegal, to the United States from Mexico and Central American are primarily young men, as are first-generation "guest workers" in European cities. Males predominate in migration flows to West African countries (Table 3.2).

Figure 3.35

Distance between old and new residences in the Asby area of Sweden. Notice how the number of movers decreased with increasing distance.

Source: From T. Hägerstrand, *Innovation Diffusion as a Spatial Process*. Copyright © 1967 The University of Chicago Press, Chicago, IL. Reprinted by permission.

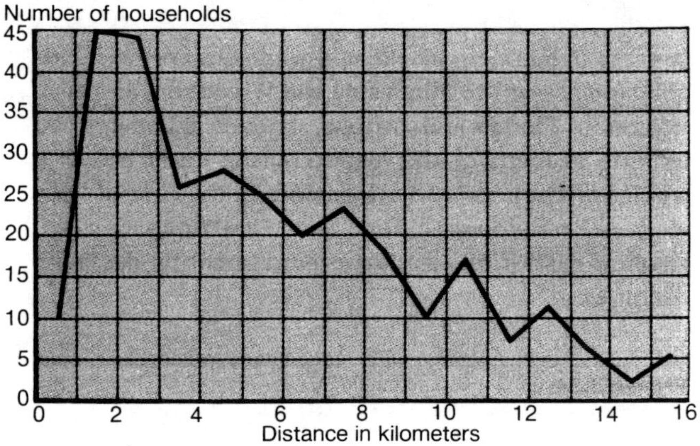

Figure 3.36

Percentage of 1970 population more than 5 years of age with different residence in 1965. Young adults figured most prominently in both short- and long-distance moves in the United States between 1965 and 1970. The same age-related pattern of mobility is recorded in movement data through the middle 1980s.

Source: Reproduced by permission from *Association of American Geographers Resource Paper 77–2*, C. C. Roseman, page 5, Association of American Geographers, 1977.

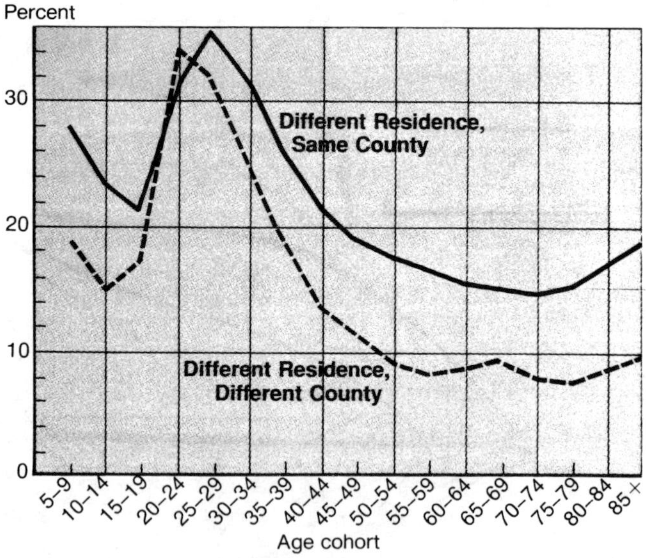

Table 3.2 Males per 100 Females, West African Immigrants, mid-1970s

Country	Males
Gambia	181
Ghana	166
Ivory Coast	177
Liberia	164
Senegal	119
Sierra Leone	187
Togo	88
Burkina Faso	79
All countries	156

Source: K. C. Zachariah and Julien Condé, *Migration in West Africa: Demographic Aspects* (New York: Oxford University Press for the World Bank, 1981), Table 30, p. 46.

For modern Americans in interregional moves, the decisions to migrate appear to involve (l) changes in life cycle (e.g., getting married, having children, getting a divorce); (2) changes in the career cycle (getting a first job or a promotion, receiving a career transfer, retiring); (3) changes of residence associated with individual personality. Work-related relocations are most important in U.S. interstate migrations. Some people simply seem to move often for no discernible reason, whereas others, *stayers,* settle into a community permanently. Of course, for a nation such as the Soviet Union, with its limitations on emigration, a tightly controlled system of internal passports, restrictions on job changes, and severe housing shortages, a different set of migration factors is present.

Summary

Spatial interaction is the dynamic evidence of the areal differentiation of the earth's surface and of the interdependence between geographic locations. The term refers to the movement of goods, information, people, ideas—indeed of every facet of economy and society—between one place and another. It includes the daily spatial activities of individuals and the collective patterns of their short- and long-distance behavior in space. The principles and constraints that unite, define, and control spatial behavior in this sense constitute an essential organizing focus for the study of human geographic patterns of the earth.

We have seen that whatever the type of spatial behavior or flow, a limited number of recurring mechanisms of guidance and control are encountered. Three underlying bases for spatial interaction are *complementarity,* which encourages flows between areas by balancing supply with demand or satisfying need with opportunity; *transferability,* which affects movement decisions by introducing cost, effort, and time considerations; and *intervening opportunity,* which suggests costs of overcoming distance may be reduced by finding closer alternate points where needs can be satisfied. The flows of commodities, ideas, or people governed by these interaction factors are interdependent and additive. Flows of commodities establish and reinforce traffic patterns, for example, and also channelize the movement of information and people.

Those flows and interactions may further be understood by the application of uniform models to all forms of spatial interaction from interregional commodity exchanges to an individual's daily pattern of movement. Distance decay tells us of the inevitable decline of interaction with increasing distance. The gravity model suggests that major centers of activity can exert interaction pulls that partly compensate for distance decay. Recognition of movement biases explains why spatial interaction in the objective world may deviate from that proposed by abstract models.

Humans in their individual and collective short- and long-distance movements are responsive to these impersonal spatial controls. Their spatial behaviors are also influenced by their separate circumstances. Each has an activity and awareness space reflective of individual socioeconomic and life-cycle conditions. Each differs in mobility. Each has unique wants and needs and perceptions of their satisfaction. Human response to distance decay is expressed in a controlling critical distance beyond which the frequency of interaction quickly declines. That decline is partly conditioned by unfamiliarity with distant points outside of normal activity space. Perceptions of home and distant territory therefore color interaction flows and space evaluations. In turn, those perceptions, well or poorly based, underlie travel and migration decisions, part of the continuing spatial diffusion and interaction of people. It is to people and their patterns of distribution and regional growth and change that we turn our attention in the following chapter.

Key Words

activity space 67
complementarity 59
distance decay 61
friction of distance 61
gravity model 62
intervening opportunity 60
migration 80
movement bias 64
network 65
personal space 65
place perception 75
place utility 84
spatial interaction 58
territoriality 65
transferability 59

For Review

1. What is *spatial interaction?* What are the three fundamental conditions governing all forms of spatial interaction? What is the distinctive impact or importance of each condition?
2. What variations in *distance decay* curves might you expect if you were to plot shipments of ready-mixed concrete, potato chips, and computer parts? What do these respective curves tell us about transferability?
3. What is *activity space?* What factors affect the areal extent of an individual's activity space?
4. On a piece of paper, and following the model of Figure 3.11, plot your *space-time path* for your movements on a typical class day. What alterations in your established movement habits might be necessary (or become possible) if (a) instead of walking, you rode a bike, (b) instead of biking, you drove a car, and (c) instead of driving, you had to use the bus or go by bike or afoot?
5. What does the thought that transportation and communication are *space adjusting* imply? In what ways has technology affected the "space adjustment" in commodity flows? In information flows?
6. Recall the places you have visited in the past week. In your movements, were the rules of *distance decay* and *critical distance* operative? What variables affect *your* critical distances?
7. What considerations appear to influence the decision to migrate? How do perceptions of *place utility* induce or inhibit migration?
8. What is a *migration field?* Some migration fields show a *channelized* flow of people. Select a particular channelized migration flow (such as the movement of Scandinavians to Michigan, Wisconsin, and Minnesota, or people from the Great Plains to California, or southern blacks to the North) and speculate why a channelized flow developed.

Suggested Readings

Abler, Ronald. "The Telephone and the Evolution of the American Metropolitan System." In *The Social Impact of the Telephone,* edited by Ithiel de Sola Pool. Cambridge, Mass.: MIT Press, 1981.

Allen, James P. "Changes in the American Propensity to Migrate." *Annals of the Association of American Geographers* 67 (1977): 577–587.

Biggar, Jeanne. "The Sunning of America: Migration to the Sunbelt." *Population Bulletin* 34, no. 2. Washington, D.C.: Population Reference Bureau, 1979.

Bouvier, Leon F., and Robert W. Gardner. "Immigration to the U.S.: The Unfinished Story." *Population Bulletin* 41, no. 4. Washington, D.C.: Population Reference Bureau, 1986.

Burton, Ian, Robert W. Kates, and Gilbert F. White, *The Environment as Hazard.* New York: Oxford University Press, 1978.

Clark, Colin, and G. H. Peters. "The 'Intervening Opportunities' Method of Traffic Analysis." *Traffic Quarterly* 19 (1965): 101–119.

Clark, Gordon L. *Interregional Migration, National Policy, and Social Justice.* Totowa, N.J.: Rowman & Allanheld, 1983.

Clark, W. A. V. *Human Migration.* Vol. 7 of *Scientific Geography Series.* Newbury Park, Calif.: Sage, 1986.

Constantinou, Stavros T., and Nicholas D. Diamantides. "Modeling International Migration: Determinants of Emigration from Greece to the United States, 1820–1980." *Annals of the Association of American Geographers* 75 (1985): 352–369.

Cox, Kevin R., and Reginald G. Golledge. *Behavioral Problems in Geography Revisited.* New York: Methuen, 1982.

Demko, George J., and William B. Wood. "International Refugees: A Geographical Perspective." *Journal of Geography* 86 (1987): 225–228.

Dowty, Alan. *Closed Borders: The Contemporary Assault on Freedom of Movement.* New Haven, Conn.: Yale University Press, 1987.

Eliot Hurst, Michael E. *Transportation Geography.* New York: McGraw-Hill, 1974.

Fligstein, Neil. *Going North: Migration of Whites and Blacks from the South, 1900–1950.* New York: Academic Press, 1981.

Gold, John R. *An Introduction to Behavioural Geography.* Oxford: Oxford University Press, 1980.

Golledge, Reginald G., and Robert J. Stimson. *Analytical Behavioral Geography.* London: Croom Helm, 1987.

Gould, Peter R. *Spatial Diffusion.* Association of American Geographers, Commission on College Geography. *Resource Paper* No. 4. Washington, D.C.: Association of American Geographers, 1969.

Hägerstrand, Torsten. *Innovation Diffusion as a Spatial Process.* Chicago: University of Chicago Press, 1967.

Hay, Alan. "The Geographical Explanation of Commodity Flow." *Progress in Human Geography* 3 (1979): 1–12.

Jakle, John A., Stanley Brunn, and Curtis C. Roseman. *Human Spatial Behavior: A Social Geography.* North Scituate, Mass.: Duxbury Press, 1976. Reprint. Prospect Heights, Ill.: Waveland Press, 1985.

Jones, Richard C., ed. *Patterns of Undocumented Migration: Mexico and the United States.* Totowa, N.J.: Rowman & Allanheld, 1984.

King, Russel, ed. *Return Migration and Regional Economic Problems.* London: Croom Helm, 1986.

Kosinski, Leszek A., and K. Maudood Elahi, eds. *Population Redistribution and Development in South Asia.* Dordrecht, Holland: D. Reidel Publishing Company, 1985.

Lee, Everett S. "A Theory of Migration." *Demography* 3 (1966): 47–57.

Lewis, G. J. *Human Migration: A Geographical Perspective.* London: Croom Helm, 1982.

Lindsay, Beverly, ed. *African Migration and National Development.* University Park: Pennsylvania State University Press, 1985.

Long, Larry H., and Celia G. Boertlein. *The Geographical Mobility of Americans: An International Comparison.* Bureau of the Census, Current Population Reports. *Special Studies Series* P-23, No. 64. Washington, D.C.: U.S. Government Printing Office, 1976.

Long, Larry H., and Kristin A. Hansen. *Reasons for Interstate Migration.* Bureau of the Census, Current Population Reports. *Special Studies Series* P-23, No. 81. Washington, D.C.: U.S. Government Printing Office, 1979.

Lowe, John C., and S. Moryadas. *The Geography of Movement.* Boston: Houghton Mifflin, 1975.

Morrill, Richard L. *The Spatial Organization of Society.* 2d ed. North Scituate, Mass.: Duxbury Press, 1974.

Murdie, Robert A. "Cultural Differences in Consumer Travel." *Economic Geography* 41 (1965): 211–233.

Newland, Kathleen. *International Migration: The Search for Work.* Worldwatch Paper 33. Washington, D.C.: Worldwatch Institute, 1979.

Newland, Kathleen. *Refugees:The New International Politics of Displacement.* Worldwatch Paper 43. Washington, D.C.: Worldwatch Institute, 1981.

Ogden, Philip E. *Migration and Geographical Change.* Cambridge Topics in Geography: Second Series. Cambridge: Cambridge University Press, 1984.

Olsson, Gunnar. *Distance and Human Interaction: A Review and Bibliography.* Regional Science Research Institute. *Bibliography Series* No. 2. Philadelphia: Regional Science Research Institute, 1965.

Parkes, Don, and Nigel Thrift. *Times, Spaces, and Places: A Chronogeographic Perspective.* Chichester, England: John Wiley & Sons, Ltd., 1980.

Porteous, J. Douglas. *Environment & Behavior.* Reading, Mass.: Addison-Wesley, 1977.

Ravenstein, E. G. "The Laws of Migration." *Journal of the Royal Statistical Society* 48 (1885): 167–227; 52 (1889): 241–301.

Rogge, John R., ed. *Refugees: A Third World Dilemma.* Totowa, N.J.: Rowman & Littlefield, 1987.

Roseman, Curtis C. "Channelization of Migration Flows from the Rural South to the Industrial Midwest." *Proceedings of the Association of American Geographers* 3 (1971): 140–170.

Saarinen, Thomas F. *Perception of Environment.* Association of American Geographers, Commission on College Geography. *Resource Paper* No. 5. Washington, D.C.: Association of American Geographers, 1969.

Saarinen, Thomas F. *Environmental Planning: Perception and Behavior.* Boston: Houghton Mifflin, 1976.

Simon, Rita James, and Caroline B. Brettell, eds. *International Migration: The Female Experience.* Totowa, N.J.: Rowman & Allanheld, 1986.

Sommer, Robert. *Personal Space: The Behavioral Basis of Design.* Englewood Cliffs, N.J.: Prentice-Hall, 1969.

Stouffer, Samuel. "Intervening Opportunities: A Theory Relating Mobility and Distance." *American Sociological Review* 5 (1940): 845–867.

Stutz, Frederick P. "Distance and Network Effects on Urban Social Travel Fields." *Economic Geography* 49 (1973): 134–144.

Stutz, Frederick P. *Social Aspects of Interaction and Transportation.* Association of American Geographers, Commission on College Geography. *Resource Paper* No. 76-2. Washington, D.C.: Association of American Geographers, 1976.

Svart, Larry M. "Environmental Preference Migration: A Review." *Geographical Review* 66 (1976): 314–330.

Szalai, Alexander, ed. *The Use of Time: Daily Activities of Urban and Suburban Populations in Twelve Countries.* The Hague: Mouton, 1972.

Tocalis, Thomas R. "Changing Theoretical Foundations of the Gravity Concept of Human Interaction." In *The Nature of Change in Geographical Ideas,* edited by Brian J. L. Berry. Perspectives in Geography 3, pp. 66–124. DeKalb: Northern Illinois University Press, 1978.

Ullman, Edward L. "Amenities as a Factor in Regional Growth." *Geographical Review* 44 (1954): 119–132.

Ullman, Edward L. "The Role of Transportation and the Basis for Interaction." In *Man's Role in Changing the Face of the Earth,* edited by William E. Thomas, Jr., pp. 862–880. Chicago: University of Chicago Press, 1956.

White, Paul, and Robert Woods. *The Geographical Impact of Migration.* London: Longman, 1980.

Zachariah, K. C., and Julien Condé. *Migration in West Africa: Demographic Aspects.* New York: Oxford University Press for the World Bank, 1981.

Zelinsky, Wilbur. "The Hypothesis of the Mobility Transition." *Geographical Review* 61 (1971): 219–249.

Chapter 12

The Political Ordering of Space

They met together in the cabin of the little ship on the day of the landfall. The journey from England had been long and stormy. Provisions ran out, a man had died, a boy had been born. Although they were grateful to have been delivered to the calm waters off Cape Cod that November day of 1620, their gathering in the cramped cabin was not to offer prayers of thanksgiving but to create a political structure to govern the settlement they were now to establish (they had originally been bound for Virginia). The *Mayflower Compact* was an agreement among themselves to "covenant and combine our selves togeather into a civill Body Politick . . . to enacte, constitute, and frame such just and equall Lawes, Ordinances, Acts, Constitutions, and Offices . . . convenient for ye Generall good of ye Colonie." They elected one of their company governor, and only after those political acts did they launch a boat and put a party ashore.

The land they were to colonize had for more than 100 years been claimed by the England they had left. The New World voyage of John Cabot in 1497 had invested their sovereign with title to all of the land of North America and a recognized legal right to govern his subjects dwelling there. That right was delegated by royal patent to colonizers and their sponsors, conferring upon them title to a defined tract and the right to govern it. Although the *Mayflower* settlers were originally without a charter or patent, they recognized themselves as part of an established political system. They chose their governor and his executive department annually by vote of the General Court, a legislature composed of all freemen of the settlement.

As the population grew, new towns were established too distant for their voters to attend the General Court. By 1636 the larger towns were sending representatives to cooperate with the executive branch in making laws. Each town became a legal entity, with election of local officials and enactment of local ordinances the prime purpose of the town meetings that are still common in New England today.

The *Mayflower Compact,* signed by 41 freemen as their initial act in the New World, was the first step in a continuing journey of political development for the settlement and for the larger territory of which it became a part. From company patent to Crown Colony to rebellious commonwealth under the Continental Congress to state in a new nation, Massachusetts (and Plimouth Plantation) was part of a continuing process of the political organization of space.

That process, in one form or another, is as old as human history. From clans to kingdoms, human groups have laid claim to territory and have organized themselves and administered their affairs within it. Indeed, the political organizations of society are as fundamental an expression of culture and cultural differences as are forms of economy or religious beliefs. Geographers are interested in that political structuring because it is an expression of the human organization of space and is also closely related to other spatial evidences of culture, such as religion, language, and ethnicity.

Political geography is the study of political phenomena in their spatial context and of the impact of political actions on society and culture. As a branch of human geography, political geography traditionally has had a primary interest in country units, or *states* (Figure 12.1). Of central concern have been spatial patterns that reflect the exercise of central governmental control, such as questions of boundary delimitation and effect, and the economic and organizational viability of countries. Increasingly, however, attention has shifted both upward and downward on the political scale. International alliances, regional compacts, and producer cartels—some requiring the surrender of at least a portion of national sovereignty—have increased in global and regional prominence since World War II, representing new forms of spatial interaction and control. At the local level, voting patterns and voting strength, constituency boundaries and districting rules, and political fragmentation and governmental consolidation have directed public and academic interest to the significance of area in the domestic political process.

Figure 12.1

This line of flags, symbols of separate member countries, graces the front of the United Nations building in New York City. Although central to political geographic interest, nation-states are only one level of the political organization of space.

Definitions and Roots

The terminology of political geographic discussion is sometimes confusing. Words used in a restrictive sense academically may have a broader or a different meaning in ordinary speech. For example, on the international scene the term **state** refers to an independent political unit occupying a defined, permanently populated territory; it has full sovereign control over its internal and foreign affairs. Within particular countries, however, *state* more likely means a subdivision of national territory, not its totality. The separate "states" comprising the United States of America, Mexico, Brazil, or West Germany are familiar cases.

The more limited definition of the word *state* tells us that not all recognized territorial entities are states in the international, technical sense. Antarctica, for example, has neither established government nor permanent population; it is, therefore, not a state. Nor are *colonies* or *protectorates* recognized as states. Although they have defined extent, permanent inhabitants, and—usually—some degree of separate governmental structure, they lack full control over all of their internal and external affairs. **Country** is sometimes used as a synonym for the territorial and political concept of state, but without clear implication of sovereign status. Although in ordinary speech we may use the words *nation* and *country* as synonyms, more technically **nation** refers to people, not to political structure. It is a *cultural* concept and implies a group of people occupying a particular region or area, bound together by a strong sense of unity arising from shared beliefs and customs. Language and religion may be unifying elements, but even more important are an emotional conviction of cultural distinctiveness and a sense of the ethnocentrism discussed in Chapter 6. Later in this chapter we will encounter the concept of the *nation-state,* where territory and commitment are coupled.

Whether we think in terms of nation or state, the concept and practice of political organization of space and people arose independently in many parts of the world. Certainly, one of the distinguishing characteristics of very early culture hearths—including those shown on Figure 2.15—was the political organization of their peoples and areas. The larger and more complex the economic structures they developed, the more sophisticated became their mechanisms of political control and territorial administration. Along with other cultural introductions, these mechanisms diffused outward from their origin points.

Our Western orientations and biases may incline us to trace ideas of spatial political organization through their Near Eastern, Mediterranean, and western European expressions. Mesopotamian and classical Greek city-states, the Roman Empire, and European colonizing kingdoms and warring principalities were, however, neither unique nor particularly unusual. Southern, southeastern and eastern Asia had their counterparts, as did black Africa and the Western Hemisphere. It is true that western European models and colonization strongly influenced the forms and structures of modern states in both hemispheres, but the cultural roots of statehood run deeper and reach farther back in many parts of the world than European example alone suggests.

The concept of state in all of its places and times of origin had its beginnings in the control of small areas by small groups. Many of the fortified cities of Mesopotamia, each controlling immediately surrounding agricultural lands, were within easy view of each other. The Greek *polis*—the rural and urban community that was the basis of the city-state of that region—centered in a small area of lowland, surrounded and protected by hills. The expansion of such small territorial units might result from a number of causes: an extension of power and domain in pursuit of economic advantage, the association of neighboring communities for protection or environmental control (irrigation, for example), the projection of power farther afield to control resources or acquire territory to support a growing population. Whatever the ultimate size, power, or population of the state, however, and whatever degree of international association might ultimately be developed, the original purpose of the political organization of space was local. It is at the local level that we still find our most intimate and immediate contact with government and with the administration of our affairs.

In this chapter we consider some of the problems of defining political jurisdiction, examine the spatial nature of political processes, look into the connections between the political and other cultural attributes of societies, and seek the elements that lend cohesion to a political entity. We begin where political control began—with local administrative organization—and end with international political systems. Our initial focus will be the North American scene simply because its local political geography is familiar to most of us. We should remember, however, the American structure of municipal governments, minor civil divisions, and special-purpose districts has counterparts in other regions of the world (Figure 12.2).

Organizational Systems

The ways in which we live our lives, conduct our daily affairs, and make plans for our future are affected by decisions made by public agencies. Where children go to school, how old one must be to vote or purchase alcoholic beverages, and how much we pay in taxes are examples of such decisions. In the United States context, the controlling determinations are made by many different organizations, ranging from the local school board through state and federal agencies to the Congress of the United States. Each organizational structure or system operates within a carefully defined geographic area. Congress can make laws for the entire country, but a school board has the power to operate only within its school district.

Even these limited examples suggest that populations are associated with—or are effectively controlled by—two different classes of organizational entities. On the one hand are those, such as Congress (or Parliament), state (or provincial) legislatures, county governments, and city councils, that establish general laws or objectives that apply to the legally bounded territories under their control. Legislatively enacted general bodies of law are converted into specific programs and detailed actions by the second class of control: elected or appointed administrative and regulatory agencies, such as the U.S. Interstate Commerce Commission, state departments of conservation, and local sanitary districts.

Political organizational systems operate within specific geographic areas; they are *spatial systems.* The limits of those areas usually are not evident on the landscape. The boundaries of the wards in a city, the municipal limits, and even most state or provincial boundaries are not physical entities that people can see or touch, and yet we have emotional feelings about many territorial units. In North America, most people identify strongly with their own town and state or province. Most people have a feeling that they are in a "foreign," or at least a different, area when they cross from their own state into another (Figure 12.3).

That sense of belonging is, of course, most important to larger political entities—provinces, states, and countries—where broad areas are joined and different populations must consider themselves as one. At the local level, efficiency of spatial organization and performance of governmental functions is of greater moment than is emotional involvement.

Figure 12.2

The Rathaus—city hall—of Munich, Germany. Before the rise of strong central governments, citizens of wealth and power in medieval and renaissance Europe focused their loyalties upon their home cities and created within them municipal buildings and institutions that would reflect their pride and substance. The city hall was frequently the grandest public building of the community, prominently located in the center of town.

Local Political Structure: The Districting Problem

There are more than 85,000 local governmental units in the United States. Slightly more than half of these are municipalities, townships, and counties. The remainder are school districts, water-control districts, airport authorities, sanitary districts, and other special-purpose bodies. Around each of these districts, boundaries have been drawn. Although the number of districts does not change greatly from year to year, many boundary lines are redrawn in any single year (see "Too Many Governments").

For example, the ruling of the U.S. Supreme Court in 1954 in *Brown v. Board of Education of Topeka, Kansas* led to the redrawing of thousands of attendance

Figure 12.3

The Four Corners Monument, marking the meeting of Utah, Colorado, Arizona, and New Mexico. Within countries, jurisdictional boundaries may be precisely located but are usually not highly visible in the landscape. In the United States, at least, those boundaries may at the same time be very significant in citizens' personal affairs and in the conduct of economic activities. Differing personal and business tax systems and rates, differing controls on firearms or the sale and consumption of alcohol and tobacco, variant administrative systems for the provision of public services and different levels of expenditure for them, and the like may sharply distinguish adjacent states of the United States.

boundaries of school districts. Likewise, the court's "one person, one vote" ruling in *Baker v. Carr* (1962) signified the end of overrepresentation of sparsely populated rural districts in state legislatures and led to the frequent adjustment of electoral districts within states and cities so that they contain roughly equal numbers of voters. Such *redistricting* or *reapportionment* is made necessary by shifts in population, as areas gain or lose people. It may also be mandated when an electoral district is seen to be **gerrymandered,** that is, structured in such a way as to give one political party an unfair advantage in elections, to fragment voting blocs, or to achieve other nondemocratic objectives. However, the drawing of boundaries is a tremendously complicated exercise in spatial decision making, and it is not always clear what is "unfair."

How boundaries are delimited depends on the desired ends. The power of a group of people may be maximized, minimized, or effectively nullified by the way in which the lines are drawn. Assume that *X* and *Y* represent two groups with different policy preferences. In Figure 12.4*a,* the *X*s are concentrated in one district and have one representative of four; in Figure 12.4*b,* the *X*s are dispersed, are not a majority in any district, and stand the chance of not electing any representative at all. The power of the *X*s is maximized in Figure 12.4*c,* where they control two of the four districts. Which of the districting arrangements is "fair" and which "unfair" may be taken to the courts to decide.

Figure 12.4 depicts an idealized district, square in shape with a uniform population distribution and only two groups competing for representation. In actuality, American municipal voting districts may be oddly shaped to consider such factors as the city limits, current population distribution, historic settlement patterns, and transportation routes. In any large city, many groups vie for power. A U.S. Justice Department lawsuit filed successfully against the Los Angeles City Council claimed that the council's 1982 district boundary lines, shown in Figure 12.5, violated the Voting Rights Act because they diluted the voting strength of the city's Hispanic residents by dispersing them among several council districts. The council agreed to draft a redistricting plan that took effect in 1987, although another ethnic political conflict is likely when the council redistricts itself after the 1990 census. Each electoral interest group promotes its version of "fairness" in the way boundaries are delimited. Minorities seek representation in proportion to their numbers so that they will be able to elect representatives who are concerned about and responsive to ethnic community issues. But ethnically homogeneous districts are impossible when—as in the case of Los Angeles—there are only 15 council members representing 3 million people and over 80 ethnic groups.

In Canada, the possibilities for gerrymandering in municipal elections have been reduced since the 1960s by assigning responsibility for redrawing constituency boundaries to independent boundary commissions. That task is made easier or unnecessary in the many Canadian

Figure 12.4

The problem of drawing legislative boundaries. *X*s and *Y*s might represent Republicans and Democrats, liberals and conservatives, urban and rural voters, blacks and whites, or any other distinctive groups.

Source: From Edward F. Bergman, *Modern Political Geography.* Copyright © 1975 Wm. C. Brown Publishers, Dubuque, Iowa. All Rights Reserved. Reprinted by permission.

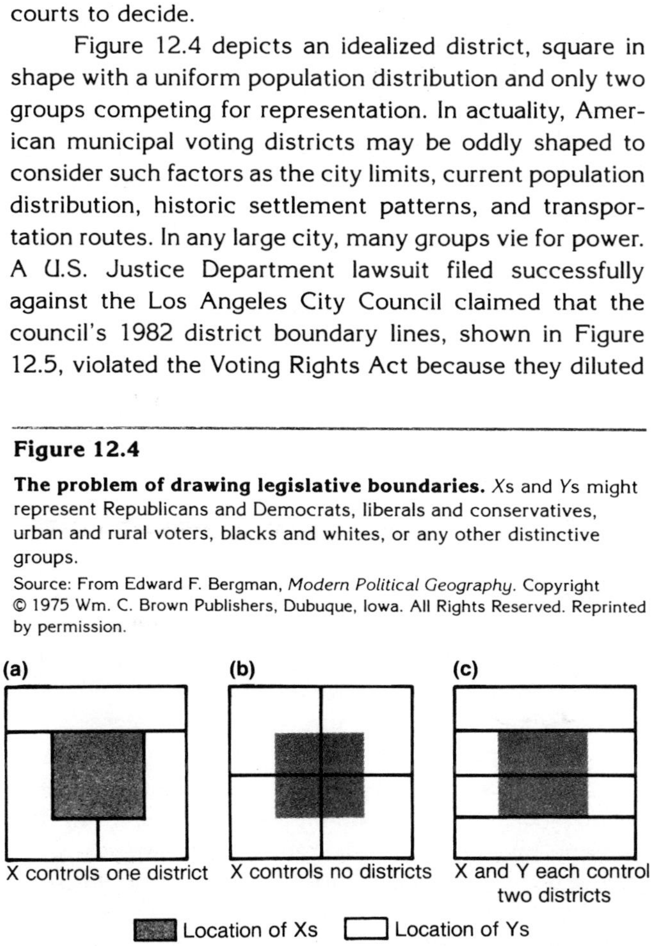

Figure 12.5

City council districts in Los Angeles created by the 1982 redistricting plan had to be redrawn when it was ruled that they violated the Voting Rights Act by splitting the city's Hispanic population. Although Hispanics constituted more than 30% of the city's population, they were a majority—and elected a council member—in only one district, the 14th. In 1986 the city council agreed to redraw the district boundary lines to create a second predominantly Hispanic council district and a third district in which Hispanics are likely to be a majority in a few years.

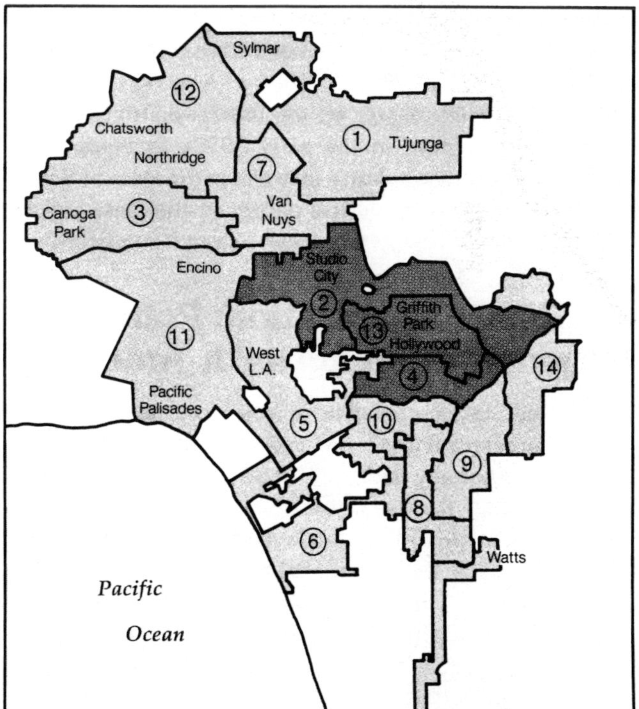

cities that do not observe neighborhood or interest-group election districts ("wards"). Many smaller cities have at-large elections, with the municipality as a whole the single constituency. In the United States, at-large election has been successfully challenged as a device to dilute or prevent minority representation. Other Canadian cities have adopted division into "strip" wards that are unrelated to natural or community boundary lines. Some cities, however, including Montreal, Toronto, and Winnipeg, do retain wards based on city blocks to respect distinctive neighborhood units.

Boundary drawing at any electoral level is never easy, particularly when political groups want to maximize their representation and minimize that of opposition groups. Furthermore, the boundaries that we may want for one set of districts may *not* be those that we want for another. For example, sewage districts must take natural drainage features into account, whereas police districts may be based on the distribution of the population or the number of miles of street to be patrolled, and school attendance zones must consider the numbers of school-aged children and the capacities of individual schools.

Ideally, political boundary making involves the practical application of geographic principles of regionalization, and geographers have been professionally involved in resolving delimitation problems. They were important contributing members of President Woodrow Wilson's staff after World War I, giving the American representatives at the Paris Peace Conference advice on boundary making in eastern Europe, where new nations were created from the defeated Austro-Hungarian Empire. A geographer was appointed special master by the federal district court in Seattle, Washington, in the early 1970s to reapportion congressional and legislative districts in the state of Washington when the two political parties in the state legislature were unable to agree on new district boundaries.

Regional Organization: Political Fragmentation in North America

The United States is subdivided into great numbers of political administrative units whose areas of control and influence are spatially limited. The 50 states are partitioned into more than 3000 counties ("parishes" in Louisiana), most of which are further subdivided into townships, each with a still lower level of spatial influence and governing power. This political fragmentation is further increased by the existence, within both urban and rural areas, of nearly innumerable special-purpose districts whose boundaries rarely coincide with the standard major and minor civil divisions of the nation, or even with each other (Figure 12.6). Each district represents a form of political allocation of territory to achieve a specific aim of local need or legislative intent (see "Too Many Governments"). In addition, the United States administers a number of outlying trusts and territories associated with it, such as the U.S. Virgin Islands, the Trust Territory of the Pacific Islands (Micronesia), and the self-governing Commonwealth of Puerto Rico.

Canada, a federation of 10 provinces and two territories (Northwest Territories and Yukon), has a similar pattern of political subdivision and multiplication. Each of the provinces contains minor civil divisions—municipalities—under provincial control, and all (cities, towns, villages, and rural municipalities) are governed by elected councils. Ontario and Quebec also have counties that group smaller municipal units for certain purposes. In general, municipalities are responsible for police and fire protection, local jails, roads and hospitals, water supply and sanitation, and schools, duties that are discharged either by elected agencies or appointed commissions.

Most North Americans live in large and small cities. In the United States these, too, are subdivided, not only into wards or precincts for voting purposes but also into special districts for such functions as fire and police protection, water and electricity supply, education, recreation, and sanitation. These districts almost never coincide with one another, and the larger the urban area, the greater the proliferation of small, special-purpose governing and taxing units. Although no Canadian community has quite the multiplication of governmental entities as, for example, Chicago, Illinois, with its more than 1000 special- and general-purpose governments, major Canadian cities may find themselves with complex and growing systems of similar nature. Metropolitan Toronto, for example, has more than 100 identified authorities that can be classified as "local governments."

The existence of such a great number of districts in metropolitan areas may cause inefficiency in public services and, perhaps, hinder the orderly use of space. Zoning ordinances, for example, are determined by each municipality. They are intended to allow the citizens to decide how land is to be used and, thus, are a clear example of the effect of political decisions on the division and development of space. Zoning policies dictate the areas where industries may and may not be located and whether they will be light or heavy industries; the sites of parks and other recreational areas; the location of business districts; and what kind of housing is allowed and where it can be built. Unfortunately, in large urban areas, the efforts of one community may be hindered by the practices of neighboring communities. Thus land zoned for an industrial park or shopping mall in one city may abut land zoned for single-family residences in an adjoining municipality. Each community pursues its own interests, which may not coincide with those of its neighbors or of the larger region.

Inefficiency and duplication of effort characterize not just zoning but many of the services local governments provide. The efforts of one community to avert air and water pollution may be and often are counteracted by the rules and practices of other towns in the region. However, state and national environmental protection standards are now reducing such potential conflicts. Social as well as

Figure 12.6

Political fragmentation in Champaign County, Illinois. A few of the independent administrative agencies with separate jurisdictions, responsibilities, and taxing powers within a portion of a single Illinois county are shown on this map. Still others, including—but not limited to—Champaign County itself, a forest preserve district, a public health district, and a community college district are also part of the fragmented political landscape.

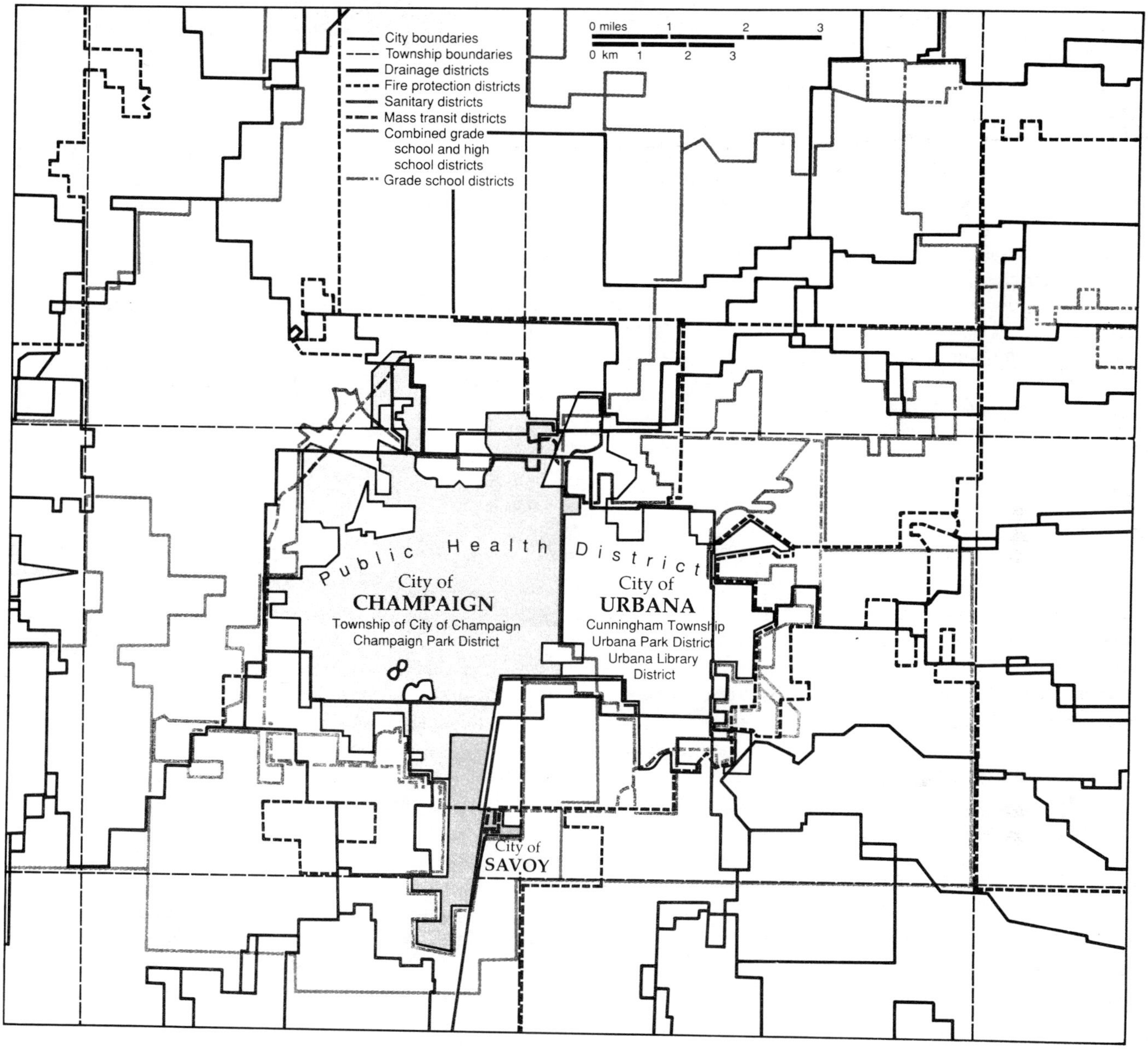

physical problems spread beyond city boundaries. Thus nearby suburban communities are affected when a central city lacks the resources to maintain high-quality schools or to attack social ills. The provision of health-care facilities, electricity and water, transportation, and recreational space affects the whole region and, many professionals think, should be under the control of a single unified metropolitan government.

In counterargument, supporters of special-purpose agencies note that the frequency of their creation indicates they are being formed by public consent to address at the local level particular problems that are not being treated by other units of government. Such agencies provide local control over local concerns and fit the scale of governmental operation to the particular function to be performed. While some see special districts as wasteful fragmentation, others view them as efficient concentrations of attention.

Metropolitan Government

The growth in number and size of metropolitan areas has increased awareness of the problems of their administrative fragmentation. In response, new approaches to the

Too Many Governments

If you are a property owner in Wheeling Township in the city of Arlington Heights in Cook County, Illinois, here's who divvies up your taxes: the city, the county, the township, an elementary school district, a high school district, a junior college district, a fire protection district, a park district, a sanitary district, a forest preserve district, a library district, a tuberculosis sanitarium district, and a mosquito abatement district.

Lest you attribute this to population density or the Byzantine ways of Cook County politics, it's not that much different elsewhere in Illinois—home to more governmental units than any other state in the United Sates. According to late-1980s figures from the U.S. Bureau of the Census, there were 6626 local government units in Illinois. Second-place Pennsylvania had 4956 governments—1670 fewer than Illinois—and the average for all states was 1663.

Along with its 102 counties, Illinois has nearly 1300 municipalities, more than 1400 townships, and over 1000 school districts; but the biggest factor in the governmental unit total is single-function special districts. These were up from 2600 in 1982 to nearly 2800 in 1987, with no end to their increase in sight. Special districts range from Chicago's Metropolitan Sanitary District to the Caseyville Township Street Lighting District. Most of these governments have property-taxing power. Some also impose sales or utility taxes.

This proliferation is in part a historical by-product of good intentions. The framers of the state's 1870 constitution, wanting to prevent overtaxation, limited the borrowing and taxing power of local governments to 5% of the assessed value of properties in their jurisdictions. When this limit was reached and the need for government services continued to grow with population, voters and officials circumvented the constitutional proscription by creating new taxing bodies—special districts. Illinois' special districts grew because they could get around municipality debt limitations and because they could be fitted to service users without regard to city or county boundaries.

Critics say all these governments result in duplication of effort, inefficiencies, higher costs, and higher taxes. Supporters of special districts, townships, and small school districts argue that such units fulfill the ideal of a government close and responsive to its constituents.

integration of those areas have been proposed and adopted. The aims of all plans of metropolitan government are to create or preserve coherence in the management of area-wide concerns and to assure that both the problems and the benefits of growth are shared without regard to their jurisdictional locations. Two approaches—*unified government* and *predevelopment annexation*—have been followed to secure the efficient administration of extensive urbanized areas.

The modern concept of metropolitan government was developed early in the 20th century in the United States as a means of coping with urban growth. The applied North American model of **unified government**—sometimes called *Unigov* or *Metro*—is found, however, in the Province of Ontario, Canada, and, particularly, in metropolitan Toronto. The provincial government undertook the reorganization of Ontario's 900-plus cities, towns, and villages into some 30 two-tiered systems of metropolitan government. The upper tier provided broad-scale direction and fiscal control; the lower tier preserved some measure of local autonomy. The Metropolitan Government of Toronto, established in 1954, was the prototype. The province federated the city of Toronto and its 12 suburbs as a metropolitan corporation that permitted the separate existence of the municipalities but assigned to the 240-square-mile (622-km^2) metropolitan unit the responsibility for public transportation, water and sewerage, centralized police protection, unified capital borrowing, and centralized land use and capital planning. To preserve a sense of identity and to free Metro government from concern with problems best handled locally, the municipalities retained their separate names and local officials and continued to perform such housekeeping tasks as maintaining local roads and sidewalks, providing fire protection, and collecting garbage. In recognition of the growing role and duties of the metropolitan corporation, the number of local municipalities was reduced to 6 in a comprehensive 1967 reorganization. Metro Toronto served as a model for other Unigov introductions in Canada, including that of Greater Winnipeg. The Montreal, Calgary, and Vancouver regions (among others) have metro governments, though not all are structured identically to that of Toronto.

Unigov has won many adherents and some successes in the United States as well, where a variety of approaches to regionwide administration and planning have been tried. Nashville–Davidson County, Tennessee, and Indianapolis–Marion County, Indiana, are among the areas where city-county consolidation has occurred. A different system operates in Metropolitan Dade County, Florida, which includes Miami and 26 other municipalities. There, local governments have retained some powers (such as police and fire services), but others (such as land use planning and tax collection) have been transferred to the county. Other metro government operations are found in

Jacksonville, Florida; Portland, Oregon; Seattle, Washington, and elsewhere. Still another approach to regionalism is represented by the Metropolitan Council of the Twin Cities area of Minneapolis/St. Paul. Here, Minnesota's two largest cities, some 130 smaller municipalities, seven counties, and great numbers of special-purpose agencies—over an area of 2800 square miles (7250 km²) containing about half the state's population (Figure 12.7)—were partially subordinated to a metropolitan council with functions limited to those few activities that demand centralized, areawide administration. These include sewerage, water supply, airport location, highway routes, and the preservation of open space. Local communities retain their own forms of government and the control of fire and police services, schools, street maintenance, and other local functions.

The preservation of the tax base and the retention of expansion room for the central city are the driving forces behind the **predevelopment annexations** permitted under the laws of some, particularly western, states. Cities such as Oklahoma City (650 square miles, or 1680 km²) and Houston (510 square miles, or 1320 km²) have taken advantage of such permissive regulation to assure that they will not be constricted by a ring of incorporated suburbs that would syphon off growth, population, jobs, and income generated by the presence of the central city itself. Since all incorporated communities of a specific minimum size possess such annexation rights over contiguous unincorporated areas, rival communities may initiate unseemly "land grabs" to the detriment of both undeveloped rural areas and the rational control of logically single metropolitan areas.

Problems of local governmental control are not unique to North America. In every metropolitan area, decisions must be made about what services are to be provided, how and where they will be provided, and who is to do the providing. In most countries, however, these decisions are not left to the smallest local units. Central authorities have the power to decide when a reorganization of the political system is necessary and to redraw the boundaries of regions.

In Great Britain, a royal commission was appointed in 1974 to study the problem of the relationship of the more than 1000 local districts that existed. Each district was directly responsible to the central government. The commission redrew county lines and gave control over certain functions—such as education, planning, and transportation—to the counties. Within the counties, district councils were established. These councils were responsible for providing other services, including housing and garbage collection. Believing that certain functions demanded a higher level of organization, the commission created a limited number of regional authorities with control over health services and water supply. The commission's plan attempted to strike a balance between centralization and decentralization of responsibility and control, based on the recognition that some functions should be controlled and provided locally, and others areawide.

Figure 12.7

Twin Cities "metro." The Metropolitan Council of the Twin Cities Area, created by the Minnesota state legislature in 1967, is responsible for the orderly development of a seven-county area. One of its powers is that of review of all land use plans of individual governments with the option of delaying their implementation for 60 days while efforts to coordinate them with regional goals are made.

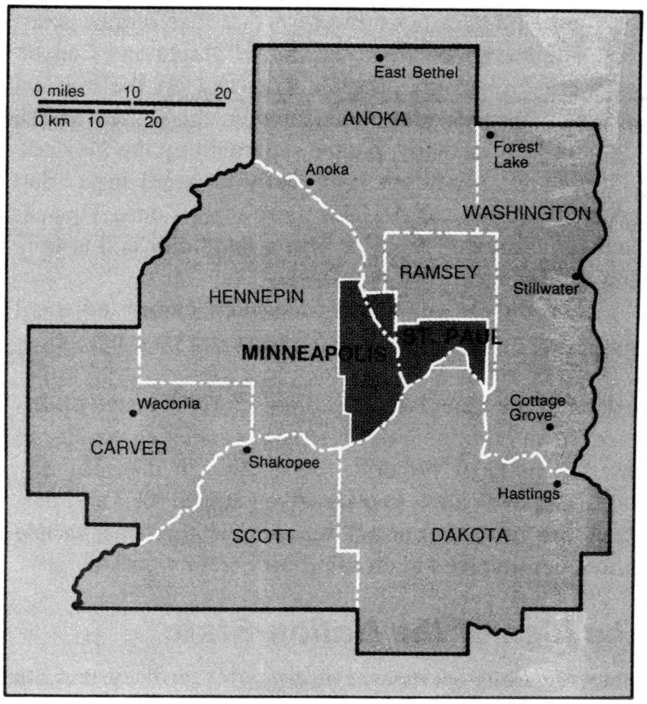

Regional arrangements including metropolitan government are one approach to solving some of today's regional political concerns, equalizing both the tax burden and the delivery of services. A second type of solution is to decentralize political authority to allow the citizens greater control over or input into governmental decisions that affect them and to bring services and government closer to the people. New York City, Chicago, and Los Angeles are among the cities that have experimented with decentralization of schools and with community involvement in decision making. A number of large cities—New York is again an example—have established "little" or neighborhood city halls to improve communication with the citizens, to increase the effectiveness of services, and to reduce neighborhood tensions. A third type of decentralization is the citizens' advisory commission or review board. Usually appointed rather than elected, the members of such boards may represent various segments of the community or specific groups (minorities, young people, environmentalists, and so on). By reflecting the community's wishes, an advisory board may effectively influence policy and its areal application.

The State and the Nation-State

Local control of government and local variations in ordinances and regulations may differently affect citizens' behaviors and personal or business decisions in neighboring communities within the same country. The most profound contrasts in societies, cultures, and political responses, however, tend to occur between rather than within countries. The border between the United States and Canada has more impact on people's lives and on the cultural landscape than do state boundaries in the United States (Figure 12.8). Similarly, people separated by the Mexico–United States boundary lead vastly different lives. San Diego resembles Los Angeles more than it does Tijuana, Mexico, although Tijuana is also a large city and is only 15 miles (24 km) from San Diego.

The differences that divide such closely adjacent peoples as those of the United States and their neighbors to north or south are more than just the territorial implications of the boundary lines themselves. A sense of distinction, an emotional tie to separate societies, loyalty to different national symbols, recollection of different histories, and dedication to distinctive national roles and purposes are more important separations than the simple formalities involved in crossing the border would suggest.

The Idea of the Nation-State

The word *state,* we have seen, connotes territory, and the word *nation* implies a distinct people. In a sense, the former is a legal entity, and the latter is a condition of culture and commitment. The *state* means formal government, laws, taxes, citizenship status, banking systems, cities, roads, and all the other trappings of development and control that distinguish the separate portions of the earth's surface claimed by different countries. The *nation*—in the sense of an ethnically distinct population with its own traditions, language, and sense of identity—may exist within or extend beyond the administrative bounds of the political state. The constitution of Somalia—a state in East Africa—recognizes a Somali population living outside of the country's territorial limits in Ethiopia and Kenya. It commits the state to the reunification of the politically divided "nation." Here, a people's sense of nationality exceeds the areal limits of the state. Conversely, in the constitutional structure of the Soviet Union before the creation of a single-chamber Congress of People's Deputies in December 1988, one division of the legislative branch of the government was termed the *Soviet of Nationalities.* It was composed of prescribed numbers of representatives from civil divisions of the Soviet Union populated by groups of officially recognized "nations": Ukrainians, Kazakhs, Estonians, and others. The current 2250-member Congress of People's Deputies retains 750 representatives of the 15 ethnically based Soviet republics and autonomous territories and regions that are official homelands of more than 100 ethnic groups. In the Soviet instance, the concept of nationality is territorially less than and subservient to the extent and control of the state.

The composite term **nation-state** properly refers to a state whose territorial extent coincides with that occupied by a distinct nation or people or, at least, whose population shares a general sense of cohesion and adherence to a set of common values. Although all countries strive for consensus values and loyalty to the state, few

Figure 12.8

Canadian–United States railroad discontinuity. Canada and the United States developed independent railway systems connecting their respective prairie regions with their separate national cores. Despite extensive rail construction during the 19th and early 20th centuries, the pattern that emerged even before recent track abandonment was one of discontinuity at the border. Note how the political boundary restricted the ease of spatial interaction between adjacent territories. Many branch lines approached the border, but only eight crossed it. In fact, for over 300 miles (480 km), no railway bridged the boundary line. The international boundary inhibits other expected degrees of interaction. Telephone calls between Canadian and U.S. cities, for example, are far less frequent than one would expect if distance alone were the controlling factor.

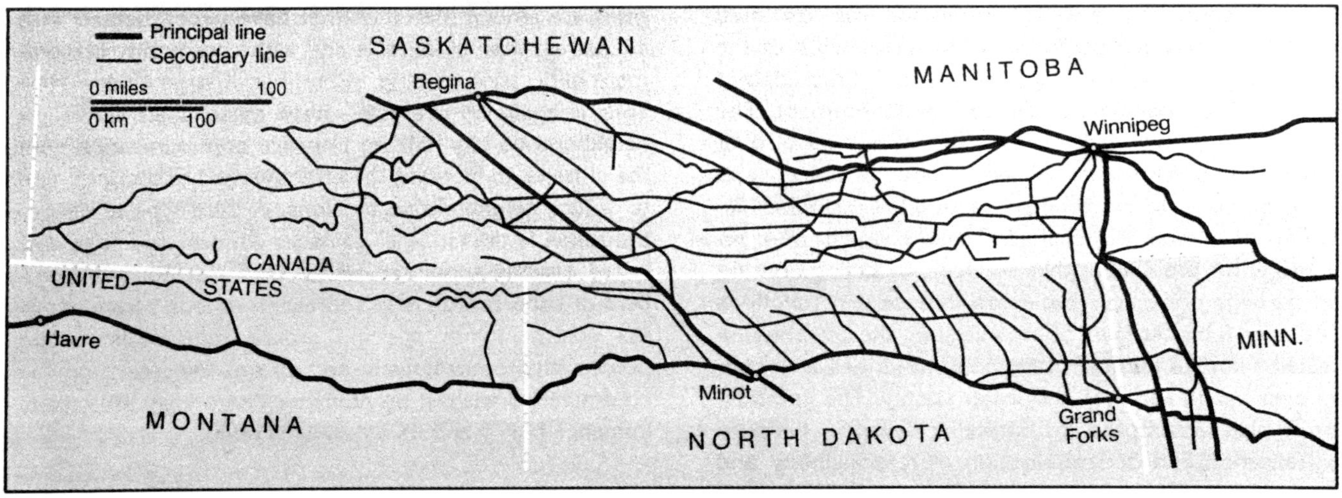

can claim to be ethnic nation-states. France, Iceland, and Poland are often cited as acceptable European examples, though France has dissident minorities within its borders. Japan is an Asian illustration.

For the nation-state, **nationalism**—the concept of allegiance to a single country—is the emotion uniting its inhabitants and providing a sense of identity and loyalty. It is a feeling of adherence and commitment that may be developed within a multicultural society through the integration of different ethnic groups into a unified population. That kind of consensus nationalism has emerged in countries such as the United States and Canada, where different culture groups, few or none with defined North American homelands, have joined together to create political entities commanding the loyalties of all their citizens. It has also been created in other countries, where the state may embrace more than one distinctive national (cultural) group, each expected to place commitment to the state and its government above the nationalism that springs from ethnic distinction and territorial identity.

One of the most familiar cultural geographic realities we have come to accept is the division of the surface of the earth into over 200 separate political entities—states or dependent territories. Most of the nearly 170 independent countries in that political mosaic (shown within the front cover of this book) are not fully coextensive with a single nation or people. Most have within their borders more than one ethnic, linguistic, or other distinctive group that considers itself separate in some important way from other citizens. Even France, often cited as a model of homogeneity and nation-state identity, contains indigenous and sometimes contentious minorities occupying recognized homeland areas (Figure 12.9).

Clearly, an important attribute of the state is the development among all its people of a sense of membership in a single political entity. Not everyone, of course, will feel the same degree of commitment or loyalty; the important consideration is that the majority of a nation's population accept its ideologies, adhere to its laws, and participate in its effective operation. For many countries, such acceptance and adherence have come only recently and partially; in some, they are frail and endangered.

Although nation-states have existed since the days of early Egypt, Mesopotamia, and other culture hearths, only in the last century has the world been almost completely divided into independent governing entities. The idea that people now everywhere accept as normal—the state and its claim to sovereignty within its borders—was developed by European political philosophers in the 18th century. Their views gave rise to the concept that individuals owe allegiance to a state and the people it represents rather than to its leader, such as a king or feudal lord. The new concept coincided in France with the French Revolution and spread over western Europe, to England, Spain, and Germany. It was based on the feeling of *nationhood,* loyalty and attachment to a country and a willingness to abide by its established rules rather than fealty to an individual and the standards of conduct that person demanded. The concept also implies a communality of culture, history, purpose, and territory that transcends purely local associations.

The creation of the nation-states of Europe is still an on-going process. New boundaries and new countries were created after World War I out of the wreckage of the German and the Austro-Hungarian empires; both temporary and more permanent independent states were created from the western border areas of the empire of the Russian czars. Other boundary adjustments followed World War II, and a divided Germany had to create two nation-states from one. Such composite societies as Yugoslavia are still seeking full nation-state acceptance by their own citizens.

Many present-day states are the result of European expansion during the 17th, 18th, and 19th centuries, when much of Africa, Asia, and the Americas was divided into colonies. Usually these colonial claims were given fixed and described boundaries where none had earlier been formally defined. Of course, precolonial native populations had established home areas of control claimed by themselves (and acknowledged by neighboring groups), within

Figure 12.9

Dialect and language regions of France. Until 1970, the spoken regional languages and dialects of France, in some instances the bases for strong feelings of regionalism and separatism, were ignored and denied recognition by the state. Even though the ban on teaching regional tongues was dropped in that year, there are still protests over the imposition of the French language throughout the country. Some linguistically based ethnic groups such as the Bretons have taken direct and occasionally violent steps to assert their separate national identity, and each of the language regions houses elements that seek a degree of regional autonomy.

which there was recognized dominance and border defense and from which there were, perhaps, raids of plunder or conquest of neighboring "foreign" territories. Beyond understood tribal territories, great empires arose, again with recognized outer limits of influence or control: Mogul and Chinese; Benin and Zulu; Incan and Aztec. Upon them, where they still existed, and upon the less formally organized spatial patterns of effective tribal control, European colonizers imposed their arbitrary new administrative divisions of the land. In fact, tribes that had little in common were often joined in the same colony. The new divisions, therefore, were not usually based on meaningful cultural or physical lines. Instead, their boundaries simply represented the limits of the colonizing empire's power.

As these former colonies have gained political independence, they have retained the idea of the state. They have generally accepted the borders established by their former European rulers. In the case of Africa, this was a conscious decision to avoid precolonial territorial or ethnic claims that could lead to war (Figure 12.10). The problem

Figure 12.10

Africa—from colonies to states. (*a*) Africa in 1939 was a patchwork of foreign claims and alien rule, some dating from the 19th century, others of more recent vintage. For example, Germany lost its claim to South West Africa, Tanganyika, Togo, and the Cameroons after World War I, and Italy asserted control over Ethiopia during the 1930s. (*b*) Africa in 1989 was a mosaic of separate states. Their dates of independence are indicated on the map. French West Africa and French Equatorial Africa have been extensively subdivided, and Ethiopia and Somaliland emerged from Italian control. Most of the current countries retain the boundaries of their former colonial existence.

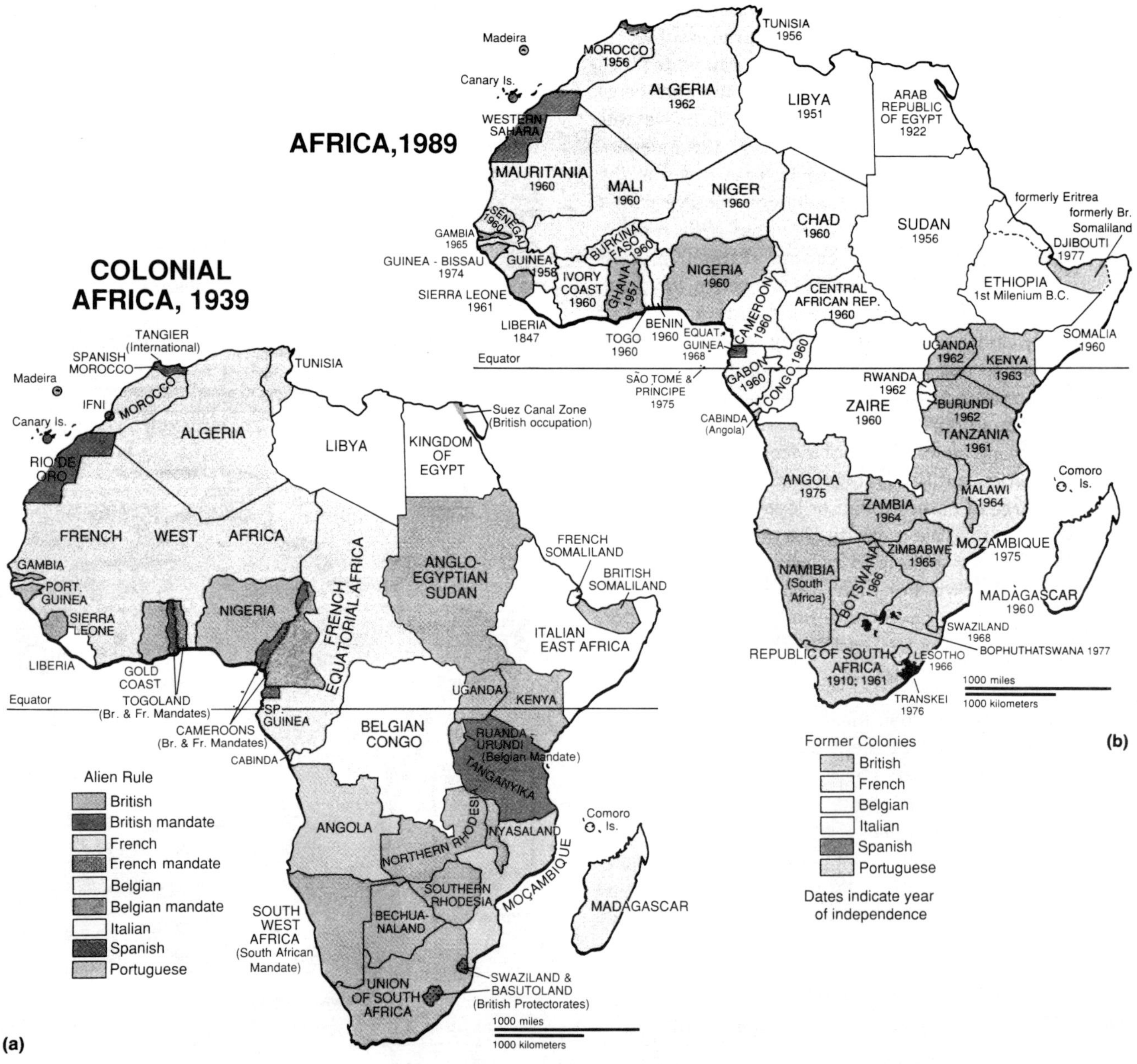

that many of the new countries face is to develop feelings of nationalism and loyalty to the state among their arbitrarily associated citizens. Zaire, the former Belgian Congo, contains more than 250 frequently antagonistic tribes. Only if past tribal animosities can be converted into an overriding spirit of national cohesion will countries such as Zaire truly be consensus nation-states.

State Cohesiveness

The sense of commitment and cohesiveness essential to the state is developed through a number of *centripetal* (uniting and bonding) forces. They help promote and maintain those necessary emotions of unity that bind the people of a state together and counter the *centrifugal* (dividing and divisive) influences present in most societies. Important among the centripetal forces are national symbols, unifying institutions, an efficient and effective organization of the state, and an evenhanded and supportive administration of government.

National Symbols

Iconography is the study of the symbols that bind a people together. National anthems and other patriotic songs; flags, national sports teams, and officially designated or easily identified flowers and animals; and rituals and holidays are all developed as symbols of a state in order to attract allegiance (Figure 12.11). They ensure that all citizens, no matter how diverse the population may be, will have at least these symbols in common. They impart a sense of belonging to a political entity called, for example, Thailand or Canada. In some countries, certain documents, such as the Magna Charta in England or the Declaration of Independence in the United States, serve the same purpose. Royalty may fill the need: in Sweden, Japan, and Great Britain, the monarchy functions as the symbolic focus of allegiance, while real power lies elsewhere.

National legends and heroes perform the same function. Immigrant children in the United States, whether from China or Central Europe, are taught about General Washington at Valley Forge and Betsy Ross and the national flag. Symbols such as these are significant insofar as ideologies and beliefs are an important aspect of culture. When a culture is very heterogeneous, composed of people with different customs, religions, and languages, belief in the national unit can help weld them together.

Institutions That Promote Nationalism

Schools, particularly elementary schools, are among the most important of the institutions that promote national feelings. Children learn the history of their own country and relatively little about other countries. Schools are expected to inculcate the society's goals, values, and traditions; allegiance to the state is accepted as the norm. Schools promote nationalism in another way, by teaching students a common language. Linguistic diversity tends to hamper national cohesiveness; linguistic homogeneity promoted in the schools fosters it.

Other institutions that advance nationalism are the armed forces, political parties, and sometimes, a state church. The military organization fulfills a primary state goal: the provision of security, both internal and external. A high percentage of most states' budgets is spent to secure such protection. The armed forces are of necessity taught to identify with the state. They see themselves as protecting the state's welfare from what are perceived to be its enemies.

Political parties are a less obvious but still significant force for promoting cohesion in a state. In one-party nations, the single official political party serves as the apparent channel through which popular support of governmental programs is expressed. In multiparty states, the political party is an instrument through which separate and perhaps quite divergent economic and social interests are presented for mass support. In their appeals

Figure 12.11

The ritual of the pledge of allegiance is just one way in which schools in the United States seek to instill a sense of national identity in students.

for adherents, political parties strive to promote national unity under their banners. Whether successful or not in advancing their programs, political parties serve to increase participation in the political process and encourage identification with it.

In some countries, the religion of the majority of the people may be designated a state church. In such cases the church may be a force for solidarity, helping to unify the population. This is true of the Roman Catholic church in the Republic of Ireland, Islam in Pakistan and Iran, and Judaism in Israel. In countries such as these, the religion and the church are so identified with the state that belief in one is transferred to allegiance to the other.

The schools, the armed forces, political parties, and the church are just some of the institutions that teach people what it is like to be members of a nation-state. These institutions operate primarily on the level of the *ideological subsystem* of culture. But by themselves, they are not enough to give cohesion, and thus strength, to a state.

Organization and Administration

A further bonding force is public confidence in the effective organization of the state. Can it provide security from external aggression and internal conflict? Are its resources distributed and allocated in such a way as to be perceived to promote the economic welfare of all its citizens? Are there institutions that encourage consultation and the peaceful settlement of disputes? How firmly established are the rule of law and the power of the courts? Is the system of decision making responsive to the people's needs? (See Figure 12.12.) These are not questions of democracy or dictatorship but of citizen perception of the propriety and legitimacy of governmental control.

The answers to such questions and the relative importance of the answers will vary from country to country, but they and similar ones are implicit in the expectation that the state will—in the words of the Constitution of the United States—"establish justice, insure domestic tranquility, provide for the common defence, [and] promote the general welfare." If those expectations are not fulfilled, the loyalties promoted by national symbols and unifying institutions may be weakened or lost.

Ethnic Separatism and Intrastate Conflict

The intergroup cohesion that the spirit of nationalism implies is not easily achieved or, once gained, invariably retained. Since 1945–the end of World War II and the start of movements for nationhood by colonial territories around the world—a number of newly independent states have had to cope with the problem of uniting in a single country disparate groups often more motivated by enmity than by affinity and more responsive to *centrifugal* than to *centripetal* influences. Even among long-established states, the sense of nationalism that overrides small-group consciousness has suffered erosion and been subject to compromise. In recent years **regionalism**—ethnic group self-awareness and identification with a region rather than with the state—has created currents of unrest and division within many countries, threatening established patterns of national cohesion.

In Western Europe alone, five countries (the United Kingdom, France, Belgium, Italy, and Spain) have been affected by **ethnic separatism**; that is, each has had recent separatist political movements demanding regional autonomy (Figure 12.13). Of course, ethnoregional loyalties were present long before 1945. They were, indeed, one of the bases over time for the development of the separate nation-states on the continent. In the form of more localized regionalism, however, those loyalties have presented problems of national unity for almost all European states at one time or another. Czechoslovakia, Greece, Norway, Finland, and many of the Balkan countries (such as Bulgaria, Romania, and Yugoslavia) housed autonomist movements that were successful in achieving independence, either permanently or temporarily.

The two preconditions common to all regional autonomist movements are *territory* and *nationality*. First, the group must be concentrated in a core region that it claims as a national homeland. Second, certain cultural characteristics must provide a basis for the group's perception of separateness and cultural unity. These might be language, religion, or distinctive group customs and institutions. Normally, these cultural differences have persisted over several generations and have survived despite

Figure 12.12

The Supreme Court of Canada building, Ottawa. Continuity, impartiality, and efficiency are attributes of government important in securing public confidence and loyalty.

strong pressures toward assimilation. As Chapter 5 makes clear, both language and religion promote feelings of group identity at the same time that they foster exclusivity. Although language is often thought of as an essential basis for nationhood, it is not necessarily so; language is of little consequence to Scottish nationalists, for example. The regional tongue may be spoken by a majority, as in Catalonia, or by a minority, as in Wales (less than 20%) and Brittany (10%). In the latter two cases, saving the language is a priority of the separatists. In addition to language and religion, separate institutions or the memory of them may form a cultural rallying point for the group. Scotland's distinctive legal system, Catalonia's parliament, and the Basques' ancient privileges of tax collecting are examples of such systems.

Other characteristics common to many separatists movements are a *peripheral location* and *social and economic inequality.* As Figure 12.13 indicates, the troubled regions in Western Europe tend to be peripheral within their countries, and their location away from the seat of central government engenders feelings of alienation and exclusion. They perhaps sense what has been called the *law of peripheral neglect,* which observes that the concern of the capital for its controlled political space decreases with the square of the distance from it. Second, the dominant culture group is often seen as an exploiting class that has suppressed the local language, controlled access to the civil service, and taken more than its share of wealth and power. Poorer regions such as Brittany, Wales, and Corsica complain that they have lower incomes and greater unemployment than prevail in the remainder of the state and that "outsiders" control key resources and industry. Not all the areas shown in Figure 12.13 are poor, however. The Basque country, Catalonia, and Scotland are relatively rich regions where the separatists believe that they could exploit their resources for themselves and do better economically without the constraints imposed by the central state.

Figure 12.13

Regions in Western Europe demanding autonomy. Each was once an independent entity, incorporated within a larger state; each has recently sought or is currently seeking a degree of separatism that recognizes its separate identity.

Analysts have suggested a number of factors to account for the resurgence of separatism in Western Europe. First, because the impact of industrialization and modernization tends to be uneven, areas that are not favored use nationalist demands to protest their position. Second, decolonization has reduced the prestige and the power of many European states since 1945, inadvertently providing regions with successful examples of anticolonial nationalism. Finally, recent years have been marked by a general widespread reaction against the growth of centralized government, which many see as intruding into too many aspects of daily life.

One response to separatist and regionalist movements has been *devolution,* the process by which the central government transfers selected powers to recognized political subunits, granting them a measure of political autonomy. In Spain, the central government has actively pursued devolution policies granting Basques, Catalans, and Andalusians a considerable measure of self-governance. In Belgium, a sweeping measure of devolution was passed in 1981 to regional executives and elected assemblies in Flanders, Wallonia, and the German-speaking eastern part of the country. The different linguistic regions were afforded a wide range of economic and social autonomy. In the United Kingdom, however, an offer by the central government of separate regional assemblies for Wales and Scotland was rejected by referenda in 1979.

Ethnic separatist movements affect many countries outside Western Europe. Scarcely a week goes by without minority agitation in some part of the world. The Basques of Spain and the Bretons of France have their counterparts in the Palestinians in Israel; the Sikhs in India; the Tigre in Ethiopia; Kurds in Turkey, Iraq, and Iran; Moros in the southern Philippines, and many others. One of the most divisive and bitter of the ethnic separatist conflicts of the 1980s wracked the island country of Sri Lanka, pitting a minority Tamil community against a government dominated by the majority Sinhalese population. With a population of about 12 million, the Sinhalese are predominantly Buddhist, while the approximately 3 million Tamils are mostly Hindu or Christian. The goal of the Tamil guerillas has been to create an independent state in the northeastern section of the country. More moderate Tamils proposed autonomy for a Tamil province in the north and some sort of power sharing in the ethnically mixed eastern areas (Figure 12.14). In this instance, the violence of the struggle resulted in the Sri Lankan government agreeing to military intervention by the army of India, whose nearby state of Tamil Nadu itself is the home of over 50 million Tamils and has received some 100,000 Sri Lankan Tamil refugees.

Figure 12.14

Traditional Tamil areas in Sri Lanka. Although Tamils are not a majority in the eastern province, militant Tamils demand a "homeland" embracing the northern and eastern provinces. These were merged into a single administrative unit in late 1988 by executive order of the Sri Lankan president, meeting a major Tamil demand.

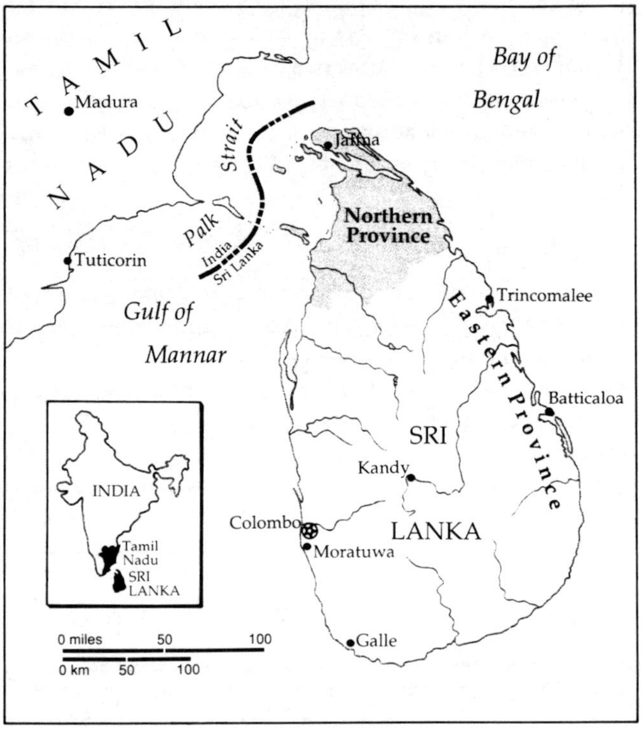

The Physical Endowments of States

Unity and nationalism are important to the stability and vitality of states. Equally important are natural conditions, including the physical makeup and location of the nation, over which governments have little or no control. Another glance at the world political map inside the front cover of this book will make clear the enormous differences between states in their *sizes* and *shapes.* The significance of their *relative location* is not so quickly grasped but is nonetheless real.

Size

The sizes of countries vary tremendously. At 8.6 million square miles (22.4 million km^2) the USSR, the world's largest state, covers nearly one-sixth of the earth's land surface—larger by far than the whole continent of South America. At the other end of the scale are the **microstates** or **ministates** found nearly worldwide. Some of them—for example, the principality of Monaco, at about 400 acres (162 ha)—are smaller than most U.S. or Canadian farms (see "The Microstates").

An easy assumption would be that the larger a state's area, the greater is the chance that its territory will include the ores, energy supplies, and agricultural resources upon which national strength and development potential are based. In general, that assumption is valid, but much depends on accidents of location. Mineral resources, for example, are unevenly distributed and related to geological structure. Size alone does not mean their assured presence or absence within the boundaries of a state. Although small in area, Kuwait and the United Arab Emirates in the Persian Gulf region lie atop tremendous petroleum reserves. The Soviet Union, Australia, and Canada, though large in landmass, have relatively small areas with the types of climates and soils that promise a productive agriculture.

Great size, in fact, may be a disadvantage rather than a benefit to a nation. As we shall see later, most nations have defined "core" areas—concentrated districts of urbanization, industrialization, development, and control. If the core is small in relation to total national territory or is distantly isolated from much of it, the economic efficiency and national security of the state as a whole may be weakened. Areas remote from the center of power may be difficult to govern, defend, or integrate into the mainstream of economy and society. If the size of a state is great relative to its population numbers, resources of money and labor may be insufficient to guarantee full development. Thus *size* alone is not critical in determining a country's stability and strength.

Shape

A country's *shape* may affect its patterns of organization and its problems of development and control. Another quick look at the world political map might suggest that no reasonable classification of country shapes is possible, but a more careful comparison does suggest a tentative typology. A number of states, for example, are easily characterized as **compact** in form. Presumably, the most efficient national shape would be a circle (assuming no major topographic barriers) with the capital located in the center. In such a country, all places could be reached from the center in a minimal amount of time and with the least expenditure for roads, railway lines, and so on. It would also have the shortest possible borders to defend. A perfect circle is an unlikely possibility, but some countries do have roughly circular, *compact* shapes that promise, though do not necessarily deliver, these advantages. Uruguay, Zimbabwe, and Poland are examples. (See Figure 12.15.)

The Microstates

Totally or partially autonomous political units that are small in area and population pose some intriguing questions. Should size be a criterion for statehood? What is the potential of ministates to cause friction among the major powers? Under what conditions are they entitled to representation in international assemblies such as the United Nations?

About half the world's independent countries contain fewer than 5 million people; of these, 37 have fewer than 1 million, the population size adopted by the United Nations as the upper limit defining "small states" (though not too small to be members of that organization). Nauru has about 8000 inhabitants on its 8.2 square miles (21 km²). Other areally small states such as Singapore (224 square miles, 580 km²) have populations (2.6 million) well above the UN criterion. Many are island territories located in the West Indies and the Pacific Ocean (such as Grenada and Tonga), but Europe (Vatican City and Andorra), Asia (Bahrain and Macao), and Africa (Djibouti and Equatorial Guinea) have their share.

Many statelets are vestiges of colonial systems that no longer exist. Some of the small countries of West Africa and on the Arabian peninsula fall into this category. Others, such as Mauritius, served primarily as refueling stops on transoceanic voyages. However, some occupy strategic locations (such as Bahrain, Malta, and Singapore), and others contain valuable minerals (Kuwait, Nauru, and Trinidad). The possibility of claiming 200-nautical-mile (370-km) zones of adjacent seas (see "Specks and Spoils," p. 410) adds to the attraction of still others.

The strategic or economic value of small islands and territories can expose them to unwanted attention from larger neighbors. The 1982 war between Britain and Argentina over the Falkland Islands (claimed as the Islas Malvinas by Argentina) demonstrated the ability of such areas to bring major powers into conflict and to receive world attention out of all proportion to their size and population.

The proliferation of tiny countries raises the question of their representation and their voting weight in international assemblies. Should a minimum size be necessary for participation in such bodies? Should nations receive a vote proportional to their population? The influence of the United States and other major powers in the United Nations has already been eroded by the small nations. Although the United States pays 25% of the UN budget and has about one and one-half times the population of all the small nations combined, its vote can be balanced by that of any of them. The fact that as many as 50 additional territories may gain independence in the next few years underscores the international interest in them.

A much less efficient shape is represented by countries such as Norway, Vietnam, and Chile, which are long and narrow. In such **elongated** states, the parts of the country far from the capital are likely to be isolated because great expenditures are required to link them to the core. These countries are also likely to encompass more diversity of climate, resources, and peoples than compact states, perhaps to the detriment of national cohesion or, perhaps, to the promotion of economic strength. Intermediate between compact and elongated national territories are the **prorupt** states, those nearly compact but possessing one or sometimes two narrow extensions of territory. Burma and Thailand have such shapes. Proruption may simply reflect peninsular elongations of land area, or it may record a state's past history of international negotiation to secure access to river or sea routes or to satisfy some other objective. Whatever their source, proruptions tend to isolate a portion of a state.

A fourth class of shapes is called **fragmented.** This class includes countries composed entirely of islands (e.g., the Philippines, Japan, New Zealand, and Indonesia), countries that are partly on islands and partly on the mainland (Italy and Malaysia), and those that are chiefly on the mainland but with that mainland territory separated by another state (the United States). Incidentally, the United States is an example of fragmentation not only because Alaska is separated overland from it by Canada but also because of the separation of Hawaii from the mainland by international waters. Pakistan was a fragmented country until 1971, when the eastern part broke away from the west and declared itself the independent state of Bangladesh. Fragmentation and isolation can weaken centralized control of state territory and increase the *regionalism* that may lead to separatist movements. That was the case with Bangladesh. It also is a problem in the Philippines and Indonesia, the latter made up of over 13,000 islands.

Much less common is the case of the **perforated** state, where one country is completely surrounded by the territory of another and "perforates" it. South Africa shows this pattern, perforated by its enclosure of Lesotho, an independent state deep in its interior. San Marino perforates Italy, and West Berlin as an **exclave**—an outlier—of West Germany perforates the national territory of East Germany. In fact, Europe has many such outlying bits of one country inside another. Kleinwalsertal, for example, is a patch of Austria accessible only from West Germany. Baarle-Hertog is a fragment of Belgium inside Holland. Campione d'Italia is an Italian outlier in Switzerland, and Büsingen is a West German one. Llivia is a totally Spanish (Catalan) town of 930 residents 3 miles (5 km) inside France.

Figure 12.15

Shapes of states. The national territories are not drawn to scale.

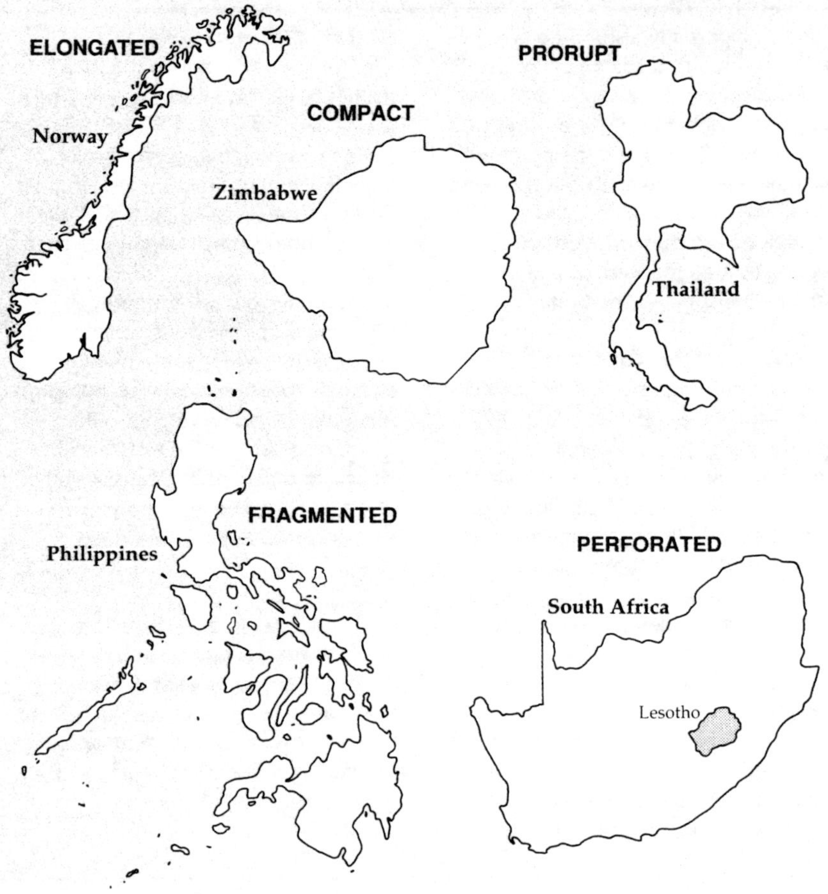

Location

The *location* of a state may add to or modify the significance of size and shape as a factor in national cohesion, stability, or development. By implication we saw that the location of Canada and the Soviet Union in upper-middle latitudes severely reduces their size advantages when the matter of agricultural potential is considered. A **landlocked** location, surrounded by other states and separated from the world's oceans, can restrict the developmental prospects of countries, reduce their roles in the world's commerce, and deprive them of access to maritime trade and also to the resources associated with coastal waters and submerged lands. Bolivia gained 300 miles (480 km) of sea frontier along with its independence in 1825, but it lost its ocean frontage by conquest to Chile in 1879. Its annual Day of the Sea ceremony reminds Bolivians of their loss and of continuing diplomatic efforts to secure an alternate outlet. Other states have never had such access but seek through international law or individual compact to secure passage rights through foreign countries and to share in the world's maritime resources (Figure 12.16).

In a few instances, a favorable relative location constitutes the primary resource of a state. Singapore, a city-state of only 224 square miles (580 km^2) and 2.6 million population that seceded from fragmented Malaysia in 1965, is located at a crossroads of world shipping and commerce. On the basis of its port and commercial activities and buttressed by its more recent industrial development, Singapore has become a notable Southeast Asian economic success.

Transportation and Communication

The importance of access to the world's trade routes is only one expression of the role of transportation and communication in the economic and political well-being of states. A well-developed domestic transportation network fosters internal political integration by promoting interaction between different national areas and by joining them economically and socially into a unified whole. The role of a transportation network in uniting a country has been recognized since ancient times. The saying that all roads lead to Rome had its origin in the impressive system of roads that linked that capital city to the rest of its empire. Centuries later, a similar network was built in France, linking Paris to the various departments of the country.

Figure 12.16

Landlocked countries feel themselves to be at a commercial and strategic disadvantage. When land routes are poorly developed, transit privileges are denied, or ocean access is controlled by unfriendly states, those disadvantages are severely restrictive. Most recently, landlocked nations have felt shut off by coastal states from the potential wealth of marine mineral resources.

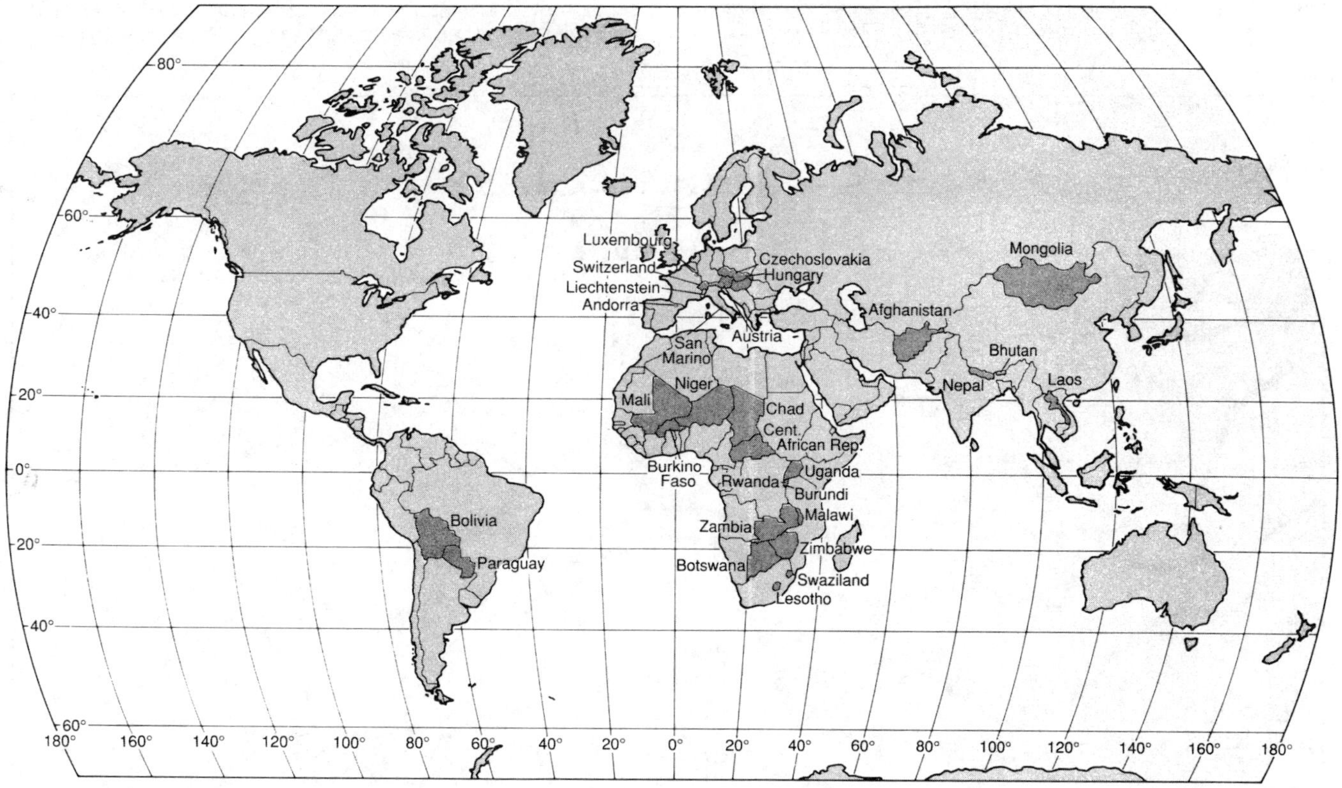

Often the capital city is better connected to other cities than the outlying cities are to one another. In France, for example, it can take less time to travel from one city to another by way of Paris than by direct route.

Roads and railroads and, earlier, such water routes as the Erie Canal have played a historically significant role in promoting political integration. In the United States and Canada, the railroads not only opened up new areas for settlement but increased interaction between rural and urban districts. The Soviet Union anticipates comparable benefits of economic, social, and political interaction from the recently completed Baikal-Amur Mainline Railroad (BAM) through Siberia and the development of its associated territorial production complexes (see p. 291). The cost, some $38 billion, of the 2000-mile (3200-km) project (Figure 12.17) is expected to yield great returns by tapping a development area of 600,000 square miles (1.5 million km²) with some of the world's richest deposits of minerals and timber, by providing access to 13 planned major new industrial centers, and by establishing a militarily secure link between the western and extreme far-eastern sections of the USSR. Similar benefits are anticipated from the planned, but by no means assured, construction of the Amur-Yakutsk Mainline (AYAM) northward from the BAM to Yakutia, the most richly mineralized section of Siberia.

Because transportation systems play a major role in a state's economic development, it follows that the more economically advanced a country is, the more extensive its transport network is likely to be (Figure 8.9). By fostering interdependence between regions a transport network keeps separate parts of the state from having to be self-sufficient. Capitalizing on its *comparative advantage,* each can specialize in the production of the goods and services for which it is best suited. *Complementarity* and *spatial interaction* will tend to reinforce regional identification with the total national economy and to foster state cohesion. At the same time, the higher the level of development, the more money there is to be spent on building transport routes and lines of communication, further strengthening national integration.

Transportation and communication, while encouraged within a state, are frequently curtailed or at least controlled between countries as a conscious device for promoting state cohesion and advancement through limitation on external spatial interaction. The mechanisms of control include restrictions on trade through tariffs or embargoes, legal barriers to immigration and emigration, and limitations on travel through passports and visa requirements. A pointed illustration is the Australian railroad

Figure 12.17

The Baikal-Amur Mainline Railroad is expected to be a vital link in the longer-term development of Siberian resources undertaken according to Soviet developmental plans. A ceremonial golden spike marked the road's completion on October 1, 1984; work still continues on tunnels and structures to make it operational to planned capacity. Other projects, including the Amur-Yakutsk Mainline, leading northward from the southern rail lines into interior Siberia, are also planned or under construction.

Source: Reproduced by permission from *The Professional Geographer of the Association of American Geographers*, Volume 39, 1987, p. 16, fig. 1, p. 21, fig. 2, V. L. Mote.

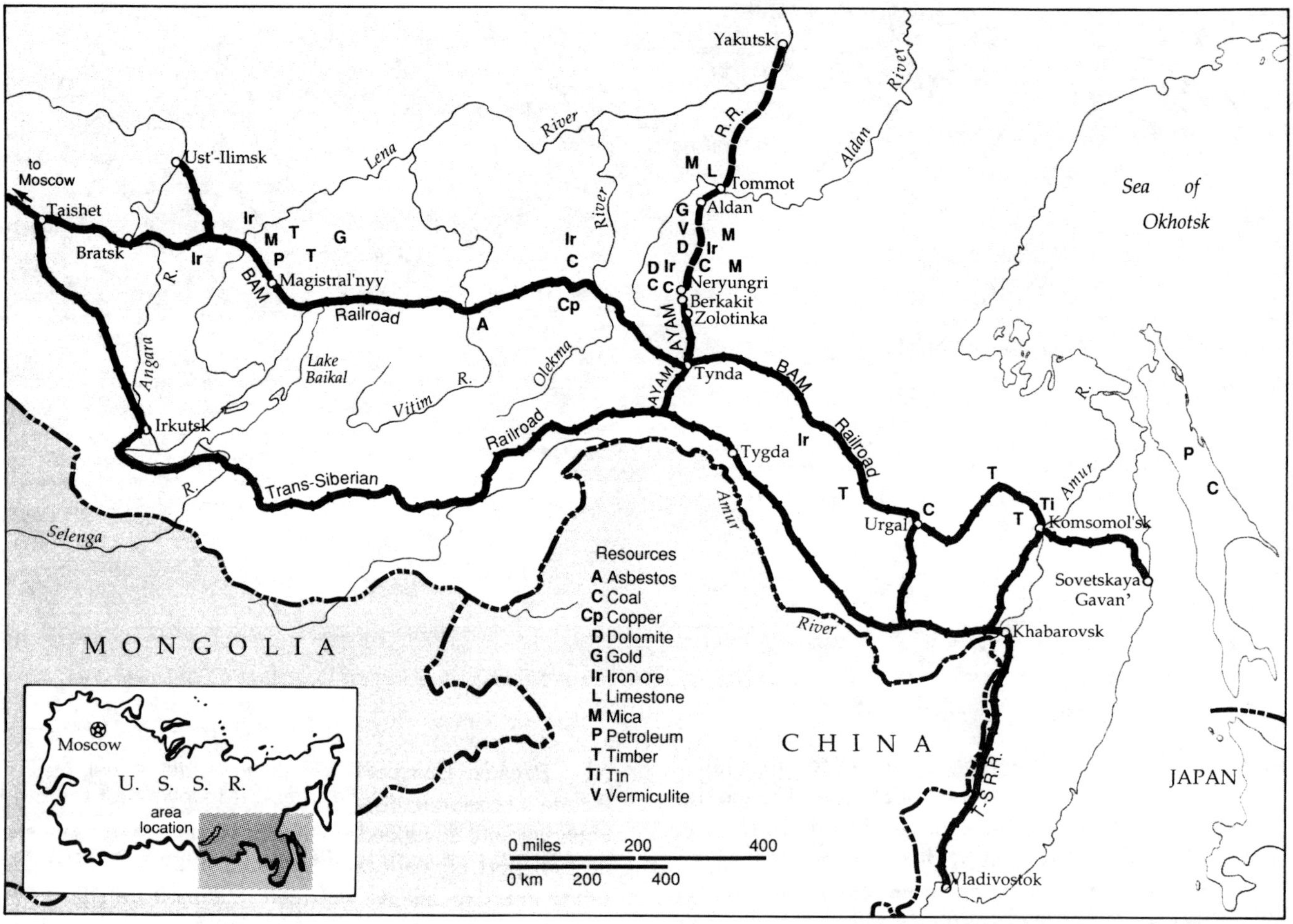

system (Figure 12.18). Until their federation in 1901, the states of Australia were independent of one another. As competitors for British markets, they had no desire to foster mutual interaction. Different rail gauges rather than a single, standard gauge were used, so that people and goods traveling between states had to change trains at the border. Similar changes in rail gauge interrupt freight and passenger movement and force carrier changes at USSR–Eastern Europe border crossings.

Cores and Capitals

States' **core areas** are presumably their heartlands, containing their most developed economies, densest and most urbanized populations, best-articulated transportation systems and communication networks, and—at least formerly if no longer—the resource base that sustained them. Away from national cores all of these characteristics are eroded. Transportation networks thin, urbanization ratios and city sizes decline, and economic development is less intensive. The outlying resource base may be rich, but generally it is of more recent exploitation, with product and benefit tending to flow to the established and distant heartlands. The developed cores of states, then, can be contrasted to their subordinate peripheries just as we saw the core-periphery idea applicable in an international developmental context in Chapter 10.

The concept of a national core area is widely accepted and cited. The Paris Basin of France, London and southeastern England, Moscow and the major cities of the European USSR, the northeastern United States, and southeastern Canada, the Buenos Aires megalopolis in Argentina, and others quickly come to mind as examples of national cores easily recognized and unmistakably dominant. Not all countries have such clearly defined cores—Chad, Mongolia, or Saudi Arabia, for example—and some may have two or more rival core areas.

Figure 12.18

Railway gauges in eastern Australia. Gauges of different widths were established to hinder interaction between the separate states of Australia, which until federation in 1901 owed primary allegiance to Great Britain rather than to one another.

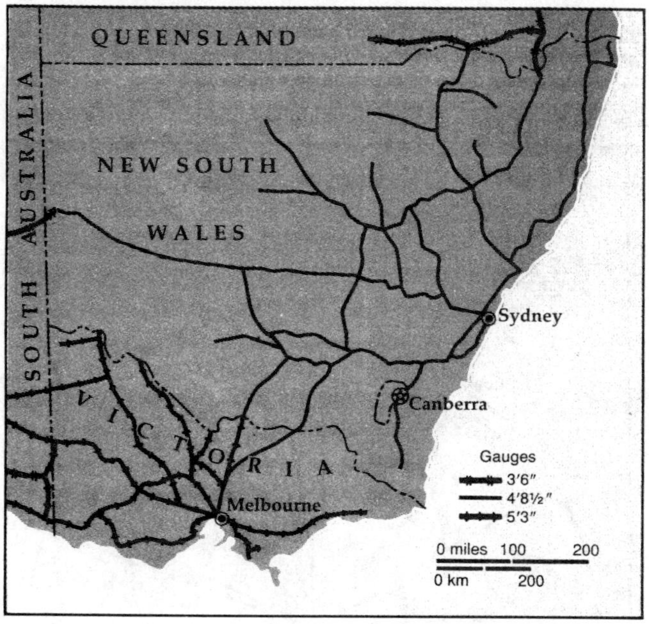

The capital city of a state is usually within its core region and, frequently, is the very focus of it, dominant not only in its political powers but by size and concentration of economic functions as well. That is, in many countries the capital city is also the *primate city,* dominating the structure of the entire country. Paris in France, London in the United Kingdom, and Mexico City are examples of that kind of political and economic primacy (see p. 360 and Figure 11.18). This association of capital with core is common in what have been called the **unitary states,** countries with highly centralized governments, relatively few internal cultural contrasts or centrifugal forces, with a strong sense of national identity, and with borders that are clearly cultural as well as political boundaries. Most European cores and capitals are of this type. It is also usual in many newly independent countries whose former colonial occupiers established a primary center of exploitation and administration and developed a functioning core in a region that lacked an urban structure or organized government. With independence, the new states retained the established infrastructure, added new functions to the capital, and through lavish expenditures on governmental, public, and commercial buildings, sought to create a prestigious symbol of nationhood.

In **federal states**—associations of more or less equal provincial or (local) state entities and with strong regional governmental responsibilities—the capital city may have been newly created or selected to serve primarily or exclusively as the administrative center. Although part of a generalized core region of the state it was not its largest city and acquired few of the additional functions to make it so. Ottawa, Canada; Washington, D.C.; and Canberra, Australia, are examples.

All other things being equal, a capital located in the center of the country provides equal access to the government, facilitates communication to and from the political hub, and enables the government to exert its authority easily. Many capital cities, including Washington, D.C., were centrally located when they were designated as seats of government but lost their centrality as the state expanded. Some capital cities have been relocated outside peripheral national core regions at least in part to achieve the presumed advantages of centrality. A **forward-thrust capital** city is one that has been deliberately sited in a state's frontier zone, signaling the country's interest in encouraging more uniform development and the government's awareness of regions away from an off-center core. In Turkey, the capital was moved to Ankara from Istanbul; in Brazil, to Brasília from Rio de Janeiro. Nigeria has since the late 1970s been building its new capital of Abuja near the geographic center of that West African country, with government offices and foreign embassies scheduled to be relocated there in the early 1990s. In 1986 Argentina's president proposed moving the capital from Buenos Aires to a more central location (Figure 12.19; see also "Transferring the National Capital").

It is important to note that the location of capital cities has been a concern of political units at other than the national level. Capital cities in compromise locations exist all over the United States. In Pennsylvania, Ohio, Michigan, and Illinois, the capitals were moved from previous sites to more central locations. For over 20 years, Alaskans have debated proposals to move the capital from isolated and often inaccessible Juneau closer to Anchorage, the center of population and commerce.

Boundaries: The Limits of the State

No portion of the earth's land surface is outside the claimed control of a national unit. Even uninhabited Antarctica has had territorial claims imposed upon it (Figure 12.20). Each country of the world is separated from its neighbors by **international boundaries,** or lines that establish the limit of each state's jurisdiction and authority. Boundaries indicate where the sovereignty of one state ends and that of another begins. Within its own bounded territory, a state administers laws, collects taxes, provides for defense, and performs other such governmental functions. Thus, the location of the boundary determines the kind of money people in a given area use, the legal code to which they are subject, the army they may be called upon to join, and the language and perhaps the religion children are taught in school. These examples suggest how boundaries serve as powerful reinforcers of cultural variation over the earth's surface.

Figure 12.19

The relocation of Argentina's capital city from Buenos Aires to Viedma was proposed in 1986 as a way of reducing the primacy of Buenos Aires and of directing national attention to the less developed sections of the nation.

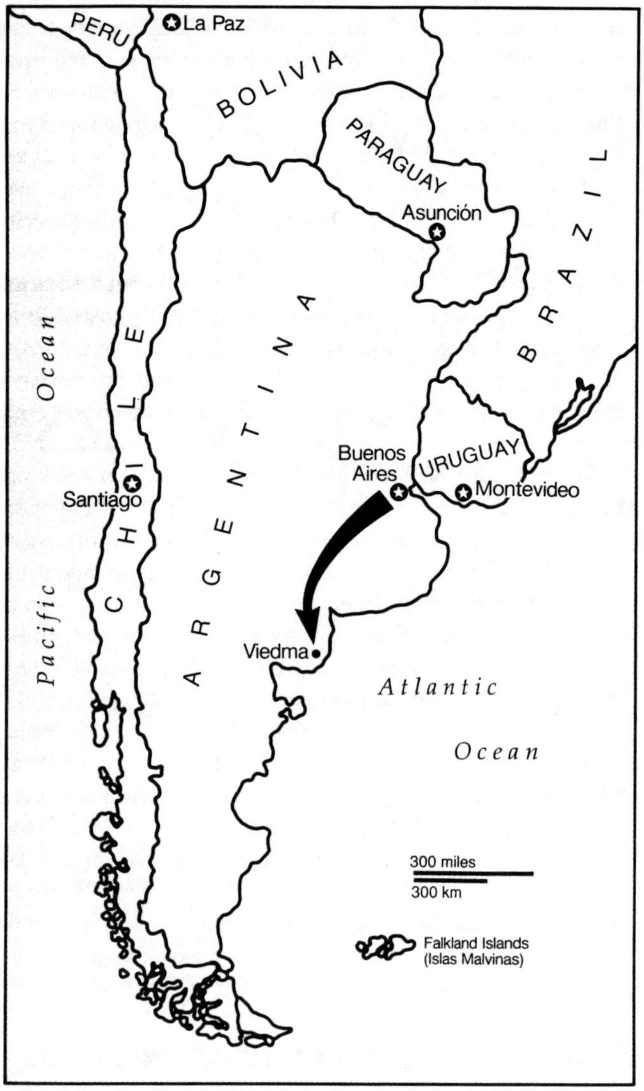

Figure 12.20

Territorial claims in Antarctica. Although 7 countries claim sovereignty over portions of Antarctica and 3 of the claims overlap, the continent has no permanent inhabitants or established local government. In 1961, 12 countries, including the 7 claimants, signed the Antarctic Treaty, which freezes the territorial disputes, bans military use of the continent, and designates it as a theater of scientific research. The discovery of mineral deposits and offshore food and fuel resources prompted increasing exploratory testing for petroleum and serious overfishing of coastal waters. In 1988, representatives of 33 nations agreed on a treaty to govern development of all the natural wealth on and under Antarctica.

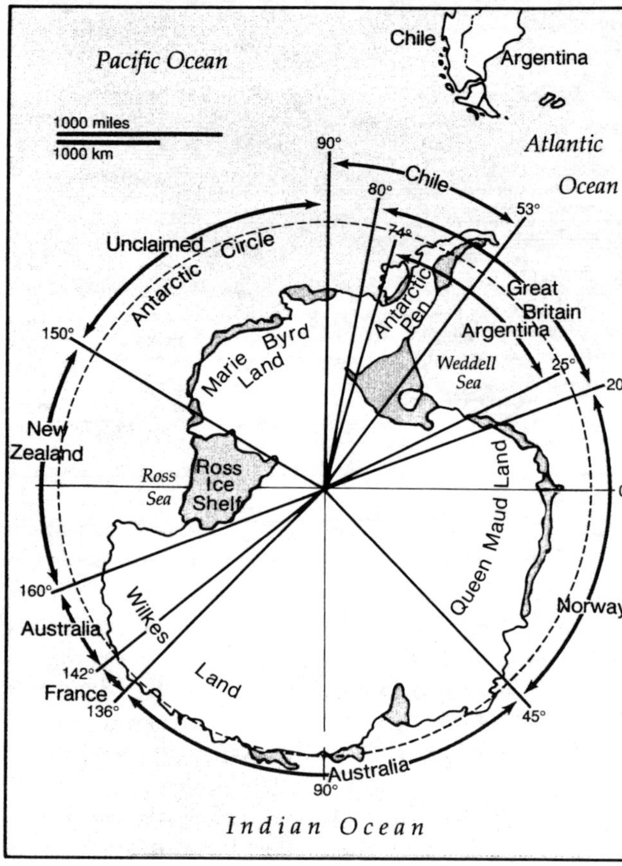

Territorial claims of sovereignty, we should note, are three-dimensional. International boundaries mark not only the outer limits of a state's claim to land (or water) surface; they are also projected downward to the center of the earth in accordance with international consensus allocating rights to subterranean resources. States also project their sovereignty upward, but with less certainty because of a lack of agreement on the upper limits of territorial airspace. The 1983 shooting down of a Korean passenger plane off course in Soviet airspace indicates the seriousness of airspace claims, however. Properly viewed, then, an international boundary is a line without breadth; it is a vertical interface between adjacent state sovereignties.

Before boundaries were delimited, nations or empires were likely to be separated by **frontier zones,** ill-defined and fluctuating areas marking the effective end of a state's authority. Such zones were often uninhabited or only sparsely populated and were liable to change with shifting settlement patterns. Many present-day international boundaries lie in former frontier zones, and in that sense the boundary line has replaced the broader frontier as a marker of a state's authority.

Classification of Boundaries

Geographers have traditionally distinguished between "natural" and "artificial" boundaries. **Natural** (or **physical**) **boundaries** are those based on recognizable physiographic features such as mountains, rivers, and lakes. Although they might seem to be attractive as borders because they actually exist in the landscape and are visible dividing elements, many natural boundaries have proved to be unsatisfactory. That is, they do not effectively separate states.

Transferring the National Capital

President Raúl Alfonsín of Argentina announced in 1986 a plan to move the country's capital from Buenos Aires to the small city of Viedma, about 500 miles to the south . . . .The proposed transfer highlights the recurrent puzzle of why governments throughout history have felt compelled to uproot themselves and move elsewhere.

Argentina's national government is presently entrenched in Buenos Aires, an urban agglomeration of over 10 million people. Viedma, in contrast, has a population of about 34,000 (less than the number of government employees in just one ministry). . . . The move to Viedma, according to Alfonsín, will create a more even national distribution of resources and government attention.

Alfonsín's proposal has a number of historical precedents, including capital transfers in Brazil, Australia, Pakistan, and the United States. Each of the precedents sheds some light on Alfonsín's decision. . . .

Thirty years ago President Kubitschek of Brazil acted upon a constitutional provision for relocating the capital to the country's uninhabited interior and away from the urban sprawl of Rio de Janeiro and São Paulo. Brasília . . . is now the center of national government activity and . . . has established itself as an urban center for Brazil's vast interior.

Australia's capital was established at its present site as a result of rivalry between the country's two largest cities, Sydney and Melbourne. Canberra, located roughly halfway between these two cities, was intended as a demonstration of the national government's commitment to unify Australia's independent-minded states.

Pakistan shifted its capital from Karachi to Islamabad in 1965. . . . Islamabad . . . was to be the capital of a new nation formed from continuous Muslim-majority districts. . . . The capital transfer away from the hot, humid climate of the coast and to the country's cultural heartland also served to emphasize Pakistan's interest in the disputed northern region.

In the early years of the United States, the founding fathers decided to move their capital from New York City to a more centralized location to serve better all of the 13 states and to separate the capital from the administration of individual state governments. . . .

A general look at motives for capital transfers reveals several common themes. First, a capital transfer represents a "fresh start" for a national government concerned with establishing its own mark on a country's future rather than simply continuing previous institutions. For some developing countries, a new, interior capital represents a physical break from a colonial heritage which was centered around the port cities (where local resources were exported to the "Mother Country"). The capital is often moved to a more centralized hinterland location to demonstrate the government's commitment to develop all parts of the country equally.

Abridged from William B. Wood, "Transferring the National Capital," *Focus*, Summer 1986, pp. 31–32.

Many international boundaries lie along mountain ranges, for example in the Alps, Himalayas, and Andes. But while some have proved to be stable, others have not. Mountains are rarely total barriers to interaction. Although they do not invite movement, they are crossed by passes, roads, and tunnels. High pastures may be used for seasonal grazing, and the mountain region may be the source of water for hydroelectric power. Similarly, the definition of a boundary along a mountain range is not a simple matter. Should it follow the crests of the mountains or the *water divide* (the line dividing two drainage areas)? The two are not always the same. Border disputes between China and India are in part the result of the failure of mountain crests and headwaters of major streams to coincide.

Rivers can be even less satisfactory as boundaries. In contrast to mountains, rivers foster interaction. River valleys are likely to be agriculturally or industrially productive and to be densely populated. For example, for hundreds of miles the Rhine River serves as international boundary in Western Europe. It is, as well, a primary traffic route lined by chemical plants, factories, blast furnaces, and power stations, and dotted by the castles and cathedrals that make it one of Europe's major tourist attractions (Figure 9.24). It is more a common intensively used resource than a barrier in the lives of the nations it borders.

It is not precisely clear where a river boundary line should lie: along the right or left bank, along the center of the river, or perhaps along the middle of the navigable channel? Soviet insistence that its sovereignty extended to the Manchurian (Dongbei) bank of the Amur and Ussuri rivers was a long-standing matter of dispute and border conflict between the USSR and the People's Republic of China (resolved with Russian agreement in 1987 that the frontier should pass along the "main channel" of the rivers). Any decision about a river boundary tends to complicate the use of the waterway by people of the states involved. Even an agreement in accordance with international custom–that the boundary be drawn along the main channel–may be impermanent if or when the river changes its course, floods, or dries up.

The alternate to natural boundaries is **geometric** or **artificial boundaries.** Frequently delimited as sections of parallels of latitude or meridians of longitude, they are found chiefly in Africa, Asia, and the Americas. The western portion of the United States–Canada border, which follows the 49th parallel, is an example of a geometric boundary (Figure 12.21). Many such boundaries were established when the areas in question were colonies, the land was only sparsely settled, and detailed geographic knowledge of the frontier region was lacking.

Boundaries can also be classified according to whether they were laid out before or after the principal features of the cultural landscape were developed. An **antecedent boundary** is one drawn across an area before it is well populated, that is, before most of the cultural landscape features were put in place. To continue our earlier example, the western portion of the United States–Canada boundary is such an antecedent line. Boundaries drawn after the development of the cultural landscape are termed **subsequent.**

One variant of the subsequent boundary is **consequent** (also called **ethnographic**), a border drawn to accommodate religious, linguistic, ethnic, or economic differences already developed. Good examples include the partitioning of the Indian subcontinent between Muslim Pakistan and dominantly Hindu India and the boundary drawn between Northern Ireland and Eire (Ireland). Subsequent boundaries may also be **superimposed** on existing cultural landscapes, forced upon a country or a people by a superior conquering or colonizing power or alliance that is unconcerned about already developed ethnic, tribal, religious, or preexisting political patterns. The colonial powers in 19th-century Africa superimposed boundaries on established African cultures, the USSR adjusted unilaterally its borders with Eastern European nations after World War II, and a divided Germany reflected the victors' decisions, not the German cultural landscape.

Stages in the Development of Boundaries

There are three distinct stages in boundary line development. Any individual boundary need not proceed through all three stages, however, and decades or even centuries can elapse between the stages. **Boundary definition** is a general agreement between two states about the allocation of territory, a verbal description of the boundary and the area through which it passes. If the area is only sparsely settled or otherwise lacks value, there is usually little pressure to proceed to the second stage, **delimitation** of the boundary. This involves actually plotting, as precisely as possible, the boundary line on maps and aerial photographs. In Africa and Asia, delimitation became important when former colonies became independent countries. Finally, the boundary may actually be marked on the ground in a process of **demarcation.** The markers may be intermittent, such as poles or pillars, or be continuous fences or walls. Most international boundaries are not demarcated—for example, that between the United States and Canada for most of its length. The "Iron Curtain" separating Soviet-dominated Eastern Europe from Western European countries, in contrast, is marked along much of its length by fences, watchtowers, and mine fields (Figure 12.22).

If a former boundary line that no longer functions as such is still marked by some landscape features or differences on the two sides, it is termed a **relict boundary.** The abandoned castles dotting the former frontier zone

Figure 12.21

The international Peace Arch at Blaine, Washington. Its position is determined by the globe grid rather than by topography or drainage systems. Geometric boundaries such as that between most of the United States and Canada are unrelated to landscape features.

Figure 12.22

Like its historical predecessors, Hadrian's Wall in the north of England and the Great Wall of China, the Berlin Wall is a demarcated boundary. Unlike them, it cuts across a large city and disrupts established cultural patterns. The Berlin Wall, therefore, is a *subsequent, superimposed, demarcated* boundary.

between Wales and England are examples of a relict boundary. They are also evidence of the disputes that sometimes attend the process of boundary making.

Boundary Disputes

Boundaries create many possibilities and provocations for conflict. Since World War II, almost half of the world's sovereign states have been involved in border disputes with neighboring countries. Just like householders, states are far more likely to have disputes with their neighbors than with more distant parties. It follows that the more neighbors a state has, the greater the likelihood of conflict. Although the causes of boundary disputes and open conflict are many and varied, we can reasonably place them into four categories. *Positional disputes* occur when states disagree about the interpretation of documents that define a boundary and/or the way the boundary was delimited. Such disputes typically arise when the boundary is antecedent, preceding effective human settlement in the border region. Once the area becomes populated and gains value, the exact location of the boundary becomes important.

The boundary between Argentina and Chile, originally defined during Spanish colonial rule, was to follow the highest peaks of the southern Andes and the watershed divides between east- and west-flowing rivers. Because the terrain had not been adequately explored, it was not apparent that the two do not always coincide. In some places, the water divide is many miles east of the highest peaks, leaving a long, narrow area of several hundred square miles in dispute (Figure 12.23). During the late 1970s, Argentina and Chile nearly went to war over the disputed territory, whose significance had been increased by the discovery of oil and natural gas deposits. Argentina has also contested its border with Uruguay in the Rio de la Plata estuary. Argentina claims that the international boundary is in the deepest part of the river, while Uruguay contends that it lies in the middle of the estuary.

Territorial disputes over the ownership of a region commonly arise when a boundary that has been superimposed on the landscape divides an ethnically homogeneous population. Each of the two states then has some justification for claiming the territory inhabited by the

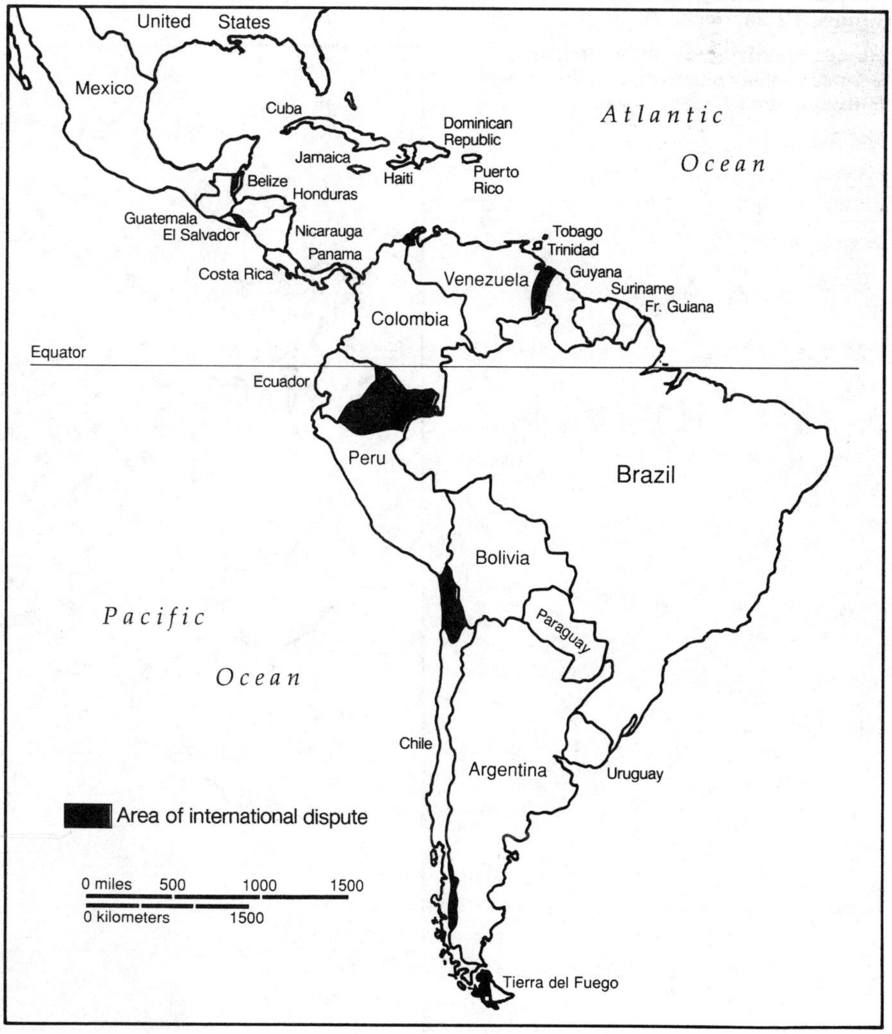

Figure 12.23

Areas of international dispute in Latin America. Among the countries disputing the precise location of their boundaries are Argentina and Chile, Argentina and Uruguay, Venezuela and Guyana, and Honduras and El Salvador.

Ernst Griffin, "Latin America," in D. Gordon Bennett *Tension Areas of the World.* Champaign, Ill.: Park Press, 1982. Reprinted by permission.

ethnic group in question. The Balkan countries of Eastern Europe offer numerous examples of such territorial disputes. Regional tensions provided the sparks that helped ignite both world wars, and although the area is less in the news today, it is far from stable. Ethnic minority problems fueled by historic enmities affect all the Balkan countries (Figure 12.24).

Yugoslavia's six federated republics and two autonomous provinces contain five major ethnic groups in addition to smaller ethnic minorities, three major religions, and three official national languages. Nearly 2 million ethnic Albanians live in Yugoslavia. They form a majority in the Kosovo region, and the minority Serbs and Montenegrins have charged that Albania is fomenting agitation in the province in the hope of detaching it from Yugoslavia. To the north lies Vojvodina, populated largely by former Hungarians, and in the south of the country is Macedonia. From 1941 to 1944 it was occupied by Bulgaria, which has not renounced all claims to it.

Bulgaria itself has disputes with Turkey, which charges that 1 million Turks in southeastern Bulgaria are being brutally assimilated, and with Romania, which claims the Romanians in the Dobruja region of Bulgaria—as well as those in the Moldavian Soviet Socialist Republic of the Soviet Union—have been separated from their homeland. Romania's relations with Hungary have long been troubled by Romania's 1920 annexation of Transylvania. As Figure 12.24 indicates, other ethnic minority areas exist in Austria, Albania, and the islands in the Aegean Sea.

Closely related to territorial conflicts are *resource disputes.* Neighboring states are likely to covet the resources—whether they be valuable mineral deposits, fertile farmland, or rich fishing grounds—lying in border areas and to disagree over their use. In recent years, the United States has been involved in disputes with both its immediate neighbors, Mexico and Canada, over the shared resources of the Colorado River and Gulf of Mexico in the south and the Georges Bank fishing grounds in the northeast.

Functional disputes arise when neighboring states disagree over policies to be applied along a boundary. Such policies may concern immigration, the movement of traditionally nomadic groups, customs regulations, or land use. In Central America, relations between Honduras and

Figure 12.24

Ethnic minority areas in the Balkans. Separatist influences are destabilizing forces in that Eastern European region.

El Salvador, two countries that had long disputed their common boundary, worsened in the late 1970s when Honduras expelled Salvadoran farmers who had illegally occupied parts of western Honduras. More recently, U.S. relations with Mexico have been affected by the increasing number of illegal aliens entering the United States from Mexico.

The Seas: Open or Closed?

Water covers about two-thirds of the earth's surface, and increasingly countries have been projecting their sovereignty seaward to claim as their own maritime areas and resources traditionally outside national control. A basic question of increasing international concern involves the right of states to make such claims of control. The inland waters of a country, such as rivers and lakes, have, of course, traditionally been considered within its sovereignty. Oceans, however, are not within any country's borders. Should they, then, be open to all states to use, or may a single country claim sovereignty and limit access and use by others?

National Claims

For most of human history, the open oceans remained effectively outside individual national control or international jurisdiction. The seas were a common highway for those daring enough to venture on them, an inexhaustible larder for fishermen, and a vast refuse pit for the muck of civilization. By the end of the 19th century, however, most coastal countries claimed sovereignty over a continuous belt 3 or 4 nautical miles wide (a nautical mile equals 1.15 statute miles, or 1.85 km). Such sovereignty, though recognizing the rights of others to innocent passage, permitted national protection of coastal fisheries, allowed the enforcement of quarantine and customs regulations, and made claims of neutrality effective during other people's wars. The primary considerations were security and unrestricted commerce. No separately codified laws of the sea existed, however, and none seemed to be needed until after World War I.

A League of Nations Conference for the Codification of International Law, convened in 1930, inconclusively discussed maritime legal matters and served to identify areas of concern that were to become increasingly pressing after World War II. Important among these was an emerging shift from interest in commerce and national security to a preoccupation with the resources of the seas, an interest fanned by the Truman Proclamation of 1945. Motivated by a desire to exploit offshore oil deposits and to prevent individual U.S. states such as Texas from claiming and taxing such deposits, the federal government under this doctrine laid claim to all resources on the continental shelf contiguous to its coasts.

In many respects the Truman Proclamation set off the great sea rush of the 20th century as other countries, many claiming even broader areas of control, hurried to annex marine resources. Within a few years, a quarter of the earth's surface—four times the total area of Africa—was appropriated by individual states, largely without protest. That annexation was encouraged by the first United Nations conference on the law of the sea which, in 1958, agreed on the "continental shelf" convention enabling countries around the North Sea to divide and exploit petroleum resources. A more encompassing third UN conference from 1973 to 1982 produced an expanded **Convention on the Law of the Sea** that elaborated the continental shelf doctrine of adjacent state control and introduced a new concept of the 200-nautical-mile (370-km) "**exclusive economic zone**" (**EEZ**) within which coastal states would have sovereign rights with respect to natural resources and certain economic activities. In addition, coastal countries were accorded exclusive rights to the resources lying within the continental shelf when this extends farther, up to 350 nautical miles (650 km) beyond their coasts. Although the 1982 convention had not yet been ratified by the requisite number of countries, by the end of the decade some 60 coastal states had used its provisions to proclaim and reciprocally recognize jurisdiction over the 200-mile zones, allocating between themselves nearly a third of all the world's oceans (Figure 12.25). The United States was one of them, proclaiming in 1983 a 200-mile-wide coastal economic zone in which it will exercise exclusive rights over all resources, including minerals and fisheries. This reaffirmation of the Truman Proclamation carried with it no acceptance of the international convention as a whole.

These extended claims have aggravated old disputes and created new ones, particularly among important fishing nations (Figure 12.26). Britain first challenged Iceland's claim to exclusive fishing rights in a 50-nautical-mile (93-km) zone by sending warships to protect British fishing boats and then retaliated by establishing a 200-nautical-mile (370-km) fishing limit around the British Isles. Unlicensed United States fishermen have been arrested for fishing within the 200-mile limit claimed by Peru. An international boundary dispute resulted from overlapping United States and Canadian claims in the Gulf of Maine, one of the richest fishing grounds off the East Coast.

In the years to come, competing claims to the mineral resources on the ocean floor may occasion more disputes than fishing now does. Petroleum and natural gas, in increasing demand as energy supplies, are known resources of continental shelves, and seabed nonfuel mineral deposits are attracting increasing attention. Many countries are competing to secure sovereignty rights over the continental shelf, where fishery and mineral resources

Figure 12.25

(*a*) **Territorial claims permitted by the 1982** United Nations Convention on the Law of the Sea **(UNCLOS)**. Coastal states may exercise full sovereignty over their *territorial sea* up to 12 nautical miles (22 km) in breadth. The *exclusive economic zone,* over which coastal states have sovereign rights of resource exploitation, extends 200 nautical miles (370 km) seaward. The convention reserves the *international seabed area* to the control of an International Sea-Bed Authority, which would be authorized to conduct mining operations or to assign mining rights to private ventures. The resources of the deep ocean areas beyond the guaranteed 200-mile (370-km) *continental shelf* (see p. 263) of any state would be managed as a "common heritage of mankind." (*b*) **EEZ claims by coastal states.** Overlapping claims may take decades to negotiate.

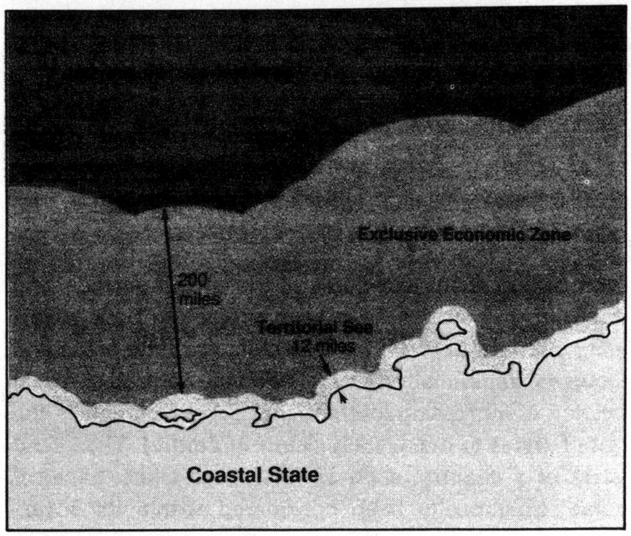

(a)

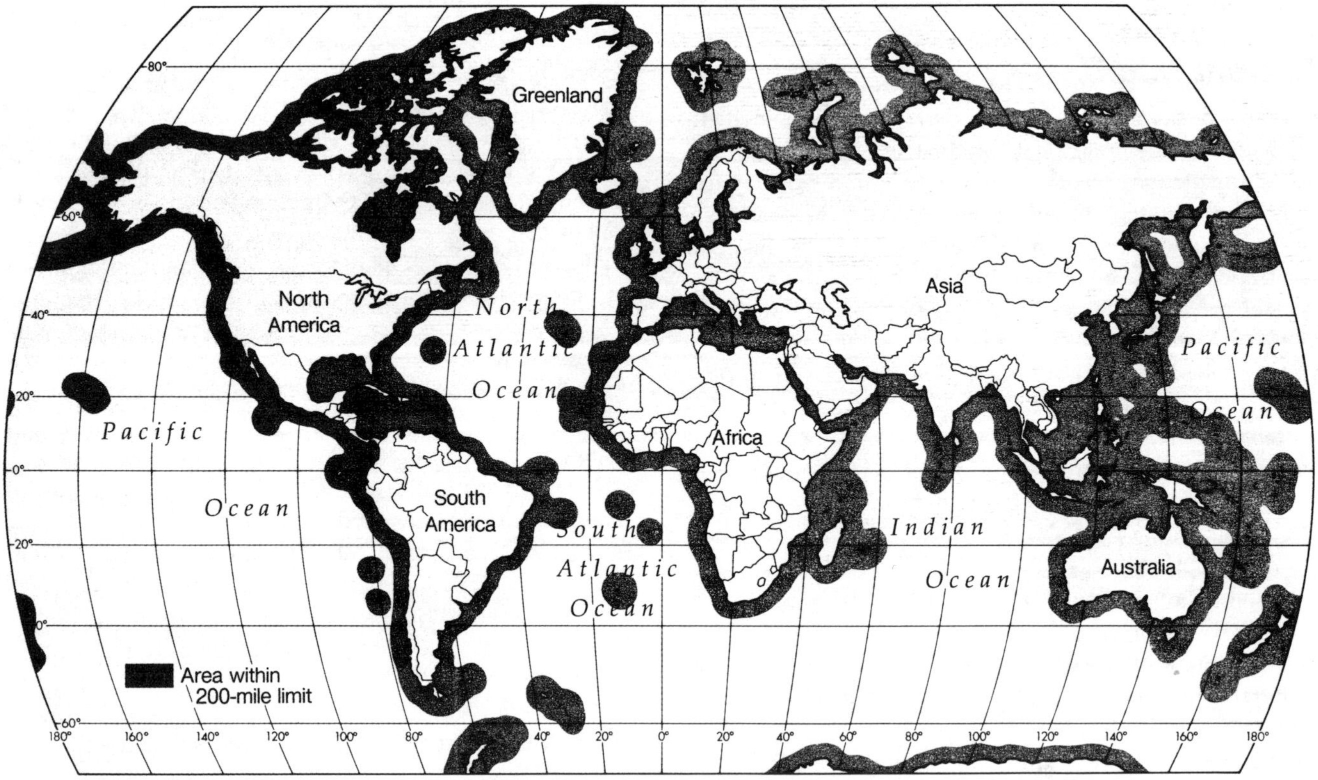

(b)

are known to exist (see "Specks and Spoils"). However, who should have the right to exploit the resources of the ocean floor: internationally licensed companies or countries contiguous to those waters? The answer is as yet undetermined.

The single most controversial provision of the Law of the Sea Convention is that involving *seabed mining* beyond the limits of national jurisdiction. At stake is control of a vast mineral treasure in the form of *manganese nodules* scattered over thousands of square miles of ocean bottom. Potato-sized, crumbly accretions of blackish ore containing up to 27 separate elements, the nodules have four minerals of particular interest: manganese (a crucial ingredient in steel making), copper, cobalt, and nickel. Their potential value has been conservatively estimated at more than $3 trillion. Most deposits lie in international waters outside the jurisdiction of individual countries (Figure 12.27).

Figure 12.26

A U.S. Coast Guard cutter escorts a foreign fishing vessel to port after seizing it for violation of the Fisheries Conservation and Management Act. In 1976 the United States extended its claim of jurisdiction over fishery resources from 12 to 200 nautical miles (22 to 370 km) from its shoreline. The United States has not ratified the Convention on the Law of the Sea and enforces domestic rather than international law in its claimed territorial waters.

Under the convention, the deep seabed and its resources are viewed as the common heritage of humankind, to be managed for the benefit of all the peoples of the earth. Specific clauses, championed by about 110 developing countries, provide that exploitation of these resources be managed by an international agency that will pick the sites to be mined, with industrial countries providing the technology and guaranteeing the profitability of the mining operations. A large portion of the benefits would be used to aid the economic development of the poorer states. The United States and a few other industrial countries that possess the technology and the capital necessary to mine the deep seabeds want to be able to do so through private investment or national corporations. They fear that the developing countries would, by their numbers, dominate an international mining agency. Should countries such as the United States, the USSR, and the United Kingdom fail to ratify the treaty, it is unclear how effective the new United Nations Convention on the Law of the Sea can be.

Aspects of International Political Systems

One of the most significant trends in recent years has been the rapid increase in the number of sovereign states. At the beginning of World War II, in 1939, there were about 70 independent countries. By 1970, the number had more than doubled, and at the end of the 1980s, 159 independent states had membership in the United Nations. Ten or more others were not UN members.

The Dissolution of Empires

Most of the newly formed states were once colonies of one or another of the European empires. Expanding political consciousness in Asia, Africa, and South America brought a demand for change and, subsequently, an end to the colonial era. Faced with strong nationalistic feelings on the part of the inhabitants, and sometimes recognizing the legitimacy of those feelings, convinced at other times that

Specks and Spoils

The Convention on the Law of the Sea gives to owners of islands claims over immense areas of the surrounding sea and—of course—to the fisheries and mineral resources in and under them. Tiny specks of land formerly too insignificant in size or distant in location to arouse the emotions of any nation now are avidly sought and fervently claimed. Remote Rockall, a British islet far west of Scotland, was used by Britain in 1976 to justify extending its fishing rights claim farther into the North Atlantic than otherwise was possible. Argentina nearly went to war with Chile in 1978 over three islands at the tip of South America at the Atlantic end of the Beagle Channel. Chile had lodged its claim of ownership hoping to gain access to known South Atlantic fish resources and hoped-for petroleum deposits. In 1982, Argentina seized the Falkland Islands from Britain, ostensibly to reclaim the Malvinas (their Spanish name) as national territory, but with an underlying economic motive as well. British forces retook the islands and subsequently used sovereignty over them to claim a sea area three times as large as Britain. Japan has encased a disappearing islet in concrete to maintain territorial claims endangered through erosion of the speck of land supporting them.

The Paracel and Spratly islands, straddling trade routes in the South China Sea, have attracted more

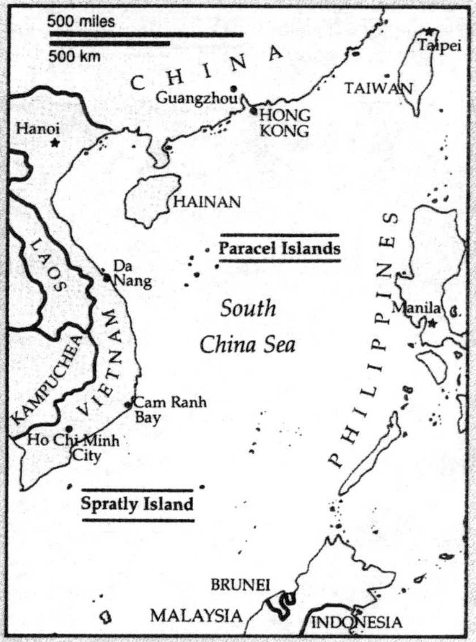

attention and claimants than most island groups, thanks to presumed large reserves of oil and gas in their vicinities. The Japanese seized the Paracels from China during World War II and at its end surrendered them to Nationalist Chinese forces that soon retreated to Taiwan. South Vietnam took them over until 1974, when they were forcibly ejected by the mainland Chinese. In 1979, a united Vietnam reasserted its claims, basing them on 17th- and 18th-century maps. China countered with reference to 3rd-century explorations by its geographers and maintained its control.

The location of the Paracels to the north, near China, in the South China Sea place them in a different and less difficult status than that of the Spratlys, whose nearest neighbors are the Philippines and Malaysia. Mere dots in the sea, the largest of the Spratlys is about 100 acres (40 ha)—no more than one-eighth the size of New York's Central Park. But under the Convention on the Law of the Sea, possession of the island group would confer rights to the resources (oil, it is hoped) found beneath about 150,000 square miles (390,000 km²) of sea. That lure has made rivals of five governments and posed the possibility of conflict. Until early in 1988, Vietnam, the Philippines, Malaysia, and Taiwan had all maintained in peaceful coexistence garrisons on separate islets in the Spratly group. Then China landed troops on islands near the Vietnamese holdings, sank two Vietnamese naval ships, and accused Vietnam of seizing "Chinese" territory on the pretext of searching for their missing sailors. While the Philippines and Malaysia agreed to discuss and negotiate claims, China rejected the idea.

Claims of past discovery, previous or present occupation, proximity, and simple wishful thinking have all served as the basis for the proliferating claims to seas and seabeds. The world's oceans, once open and freely accessible, are increasingly being closed by the lure of specks of land and the spoils of wealth they command.

the colonies were no longer assets either economically or politically, or responding to the pressures of world opinion or military defeat, the European countries relinquished their control.

From the former British Empire and Commonwealth, there have come, just in the period since World War II, the independent countries of India, Pakistan, Bangladesh, Malaysia, and Singapore in Asia, and Ghana, Nigeria, Kenya, Uganda, Tanzania, Malawi, Botswana, Zimbabwe, and Zambia in Africa. Even this extensive list is not complete. These and other now-independent former British colonies are shown in Figure 12.28. A similar process has occurred in most of the former overseas possessions of the Netherlands, Spain, Portugal, and to a lesser extent, France.

Empires have been created—and have fallen—throughout history: the Roman, the Ottoman, and the Mongol, to name only three. Each was, for a while, very strong; each eventually dissolved. What, then, is significant about the dissolution of empires in this century?

The answer lies in the speed of the process, in the areal extent over which it has occurred, and in what has happened to the former colonies. The modern European empires, comprising vast parts of the globe, dissolved almost simultaneously in a few years' span after World War II. The colonies of those empires have now become independent states rather than being absorbed by another empire. This was not a consequence when previous empires broke apart before the concept of nationalism became dominant. Nationalism is now a truly global phenomenon. There will soon be very few people left who are not citizens of independent states.

Figure 12.27

Known manganese nodule deposits. The largest field yet discovered stretches from the coast of California to Hawaii. Most deposits lie outside the proposed exclusive economic zones of coastal countries.

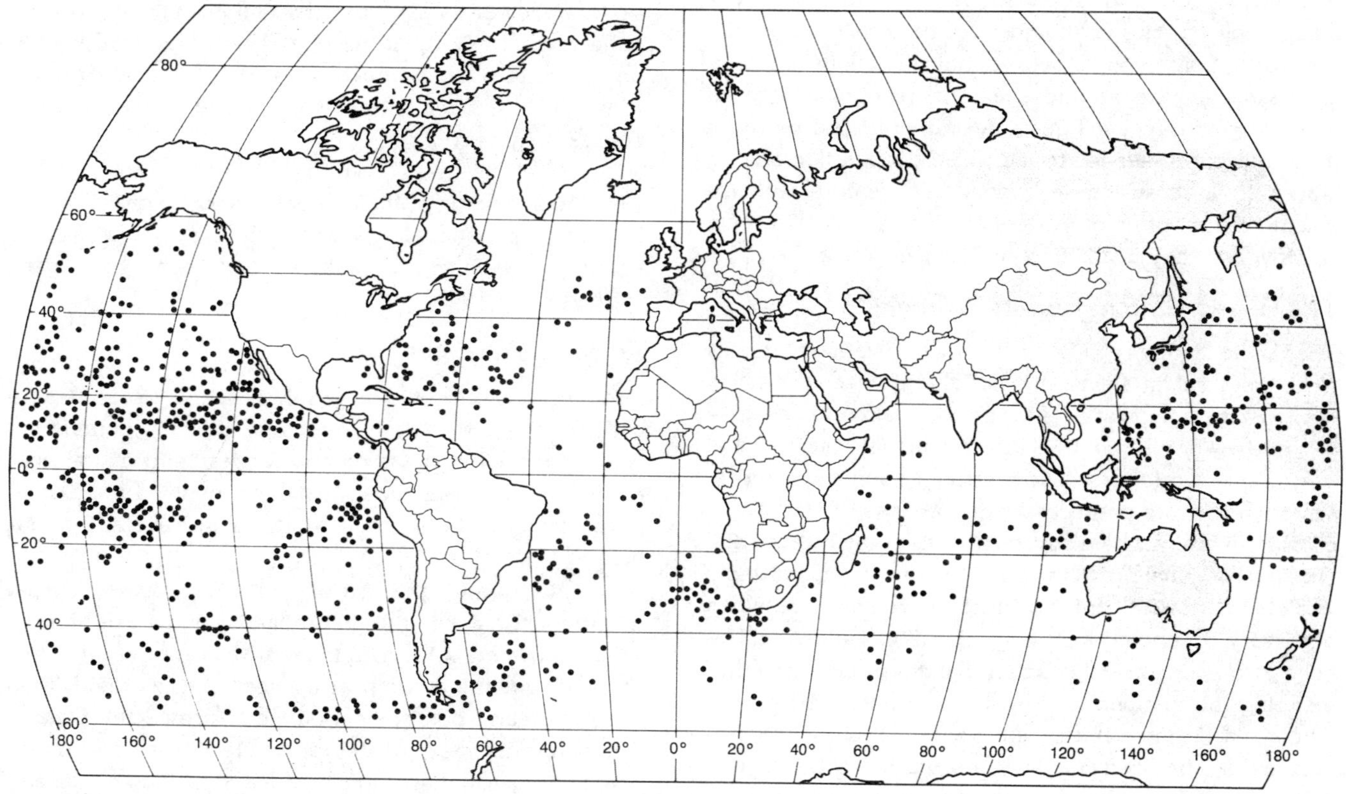

Figure 12.28

The end of empire. Some 54 former possessions, colonies, dominions, and protectorates of Great Britain have achieved independent statehood since 1945. This maps shows how large and far-flung was the territory controlled, well into the 20th century, by a single small European state. At its greatest extent, the British Empire covered almost a quarter of the globe and contained nearly a quarter of the world's population.

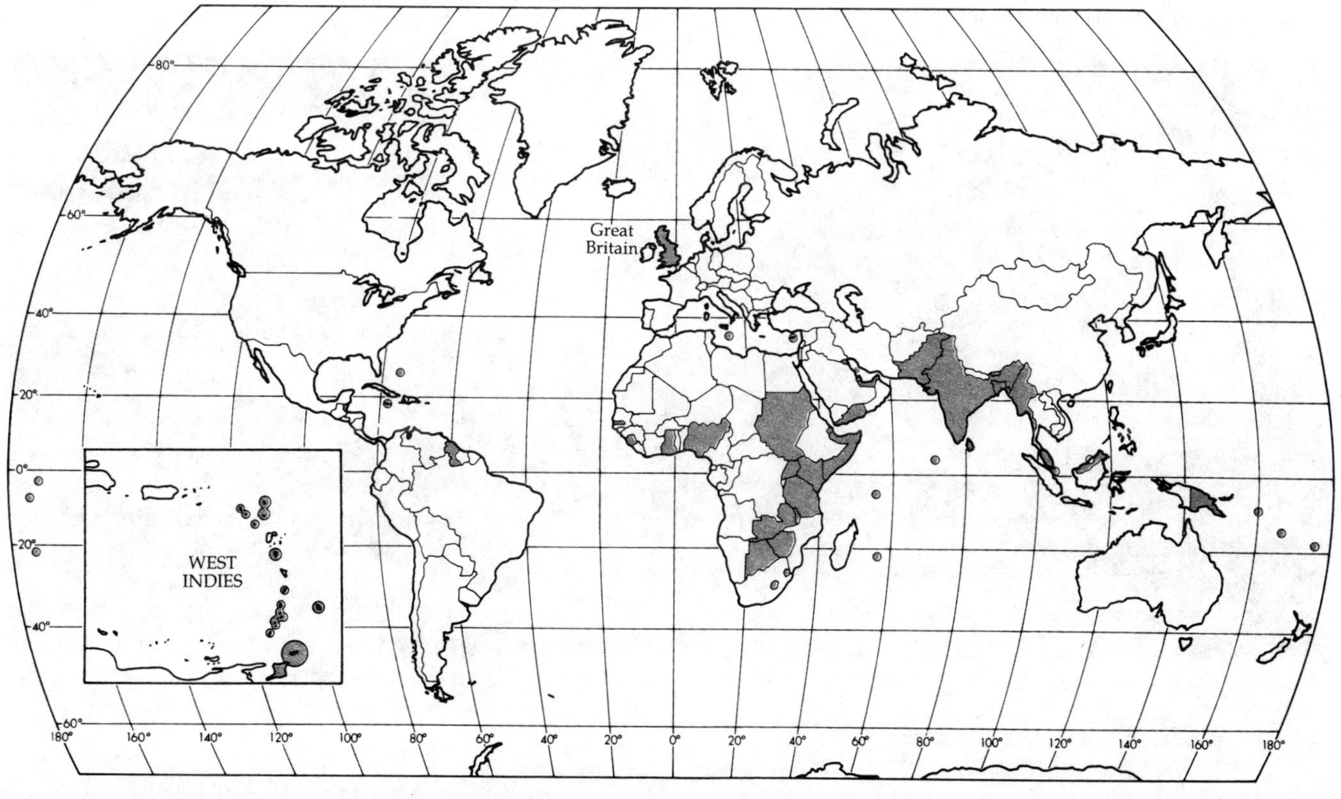

New International Systems

In many ways, countries are now weaker than ever before. Despite their independence, many are economically frail, others are politically unstable, and some are both. Strategically, no country is safe from military attack, for technology now enables us to shoot weapons halfway around the world. Some people believe that no national security is possible in the atomic age. The recognition that a country cannot by itself guarantee either its prosperity or its own security has led to increased cooperation among states. In a sense, these cooperative ventures are replacing the empires of yesterday. They are proliferating quickly, and they involve countries everywhere.

The United Nations and Its Agencies

The United Nations (UN) is the only organization that tries to be universal, and even it is not all-inclusive. The UN is the most ambitious attempt ever undertaken to bring together the world's countries in international assembly and to promote world peace. It is stronger and more representative than its predecessor, the League of Nations. It provides a forum where sovereign states may meet to discuss international problems and a mechanism, admittedly weak but still significant, for forestalling disputes or, when necessary, for ending wars (Figure 12.29). The United Nations also sponsors 40 programs and agencies aimed at fostering international cooperation with respect to specific goals. Among these are the World Health Organization (WHO), the Food and Agriculture Organization (FAO), and the United Nations Educational, Scientific, and Cultural Organization (UNESCO).

Member nations have not surrendered sovereignty to the UN, and the world body is legally and effectively unable to make or enforce a world law. Nor is there a world police force. Although there is recognized international law adjudicated by the International Court of Justice, rulings by this body are sought only by nations agreeing beforehand to abide by its arbitration. The United Nations has no authority over the military forces of individual countries, and it cannot take prompt and effective action to counter aggression.

Countries have shown themselves to be more willing to relinquish some of their independence to participate in smaller multinational systems. These systems may be economic, military, or political, and many have been formed since 1945.

Regional Alliances

Cooperation in the economic sphere seems to come more easily to states than does political or military cooperation. Among the most powerful and far-reaching of the economic alliances are those that have evolved in Europe, particularly the Common Market and its several forerunners.

Shortly after the end of World War II, the Benelux countries (Belgium, the Netherlands, and Luxembourg) formed an economic union to create a common set of tariffs and to eliminate import licenses and quotas. During the same early postwar period were formed the Organization for European Cooperation (1948), which coordinated the distribution and use of Marshall Plan funds, and the European Coal and Steel Community (1952), which integrated the development of that industry in the member

Figure 12.29

Under the auspices of the United Nations, soldiers from many different countries have staffed peacekeeping forces and military observer groups in an effort to halt or mitigate conflicts. Among the areas where they have served are Indonesia, Korea, Cyprus, and the Dominican Republic. Shown here are soldiers on duty in Lebanon.

countries. A few years later, in 1957, the **European Economic Community (EEC),** or Common Market, was created. At first it was composed of only six states: France, Italy, West Germany, and the Benelux countries.

To counteract these Inner Six, as they were called, other countries joined in the European Free Trade Association (EFTA). Known as the Outer Seven, they were the United Kingdom, Norway, Denmark, Sweden, Switzerland, Austria, and Portugal (Figure 12.30). In 1973 the United Kingdom, Denmark, and Ireland applied for and were granted membership in the Common Market. They were joined by Greece in 1981; Spain and Portugal entered in 1986. After 23 years of associate status, Turkey applied for full membership in 1987 but will not achieve it for some years.

The European Community (EC) is the collective name given to three separate organizations with common membership: the European Economic Community (Common Market), the European Coal and Steel Community, and the European Atomic Energy Community. The executive offices of the three communities were merged in 1967. The 12 member states have taken many steps to integrate their economies and coordinate their policies in such areas as transportation and agriculture. A Council of Ministers, a Commission, a European Parliament, and a Court of Justice give the EC supranational institutions with effective ability to make and enforce laws. The EC's objective is to abolish—on December 31, 1992—all barriers to trade, migration, and capital and labor movement among its members (Figure 12.31). Political, military, and cultural integration are scheduled to follow, with eventual political union of the member states the ultimate goal.

Figure 12.30

The original Inner Six and Outer Seven of Europe.

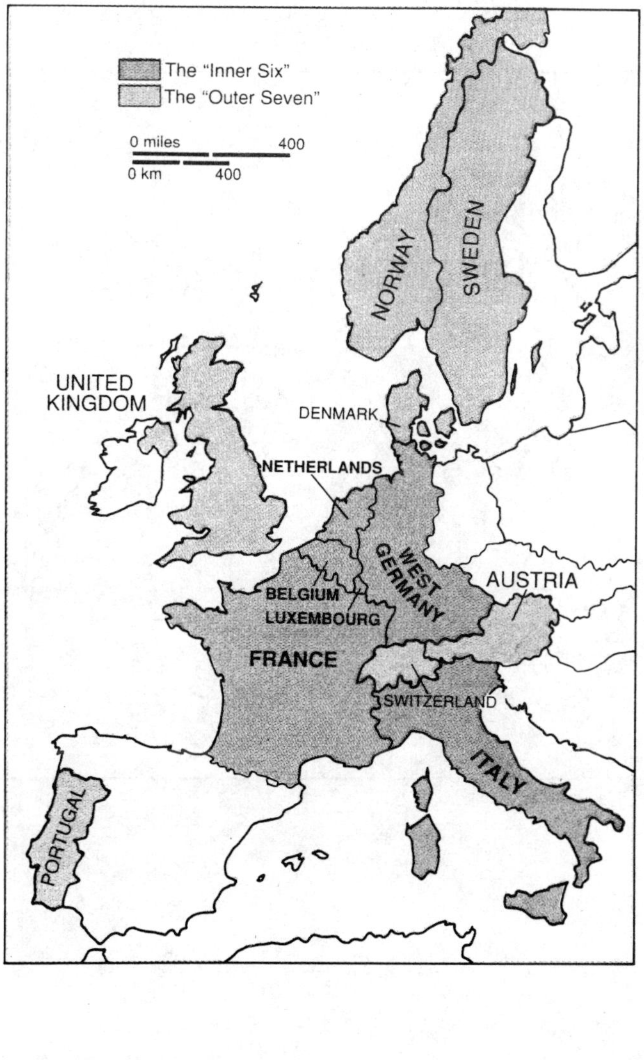

We have traced this European development process to illustrate the fluid process by which regional alliances are made, not because it is important to remember all the forerunners or the present structure of the Common Market. Countries come together in an association, some drop out, and others join. New treaties are made, and new coalitions emerge. It seems safe to predict that although the alliances themselves will change, the idea of economic associations has been permanently added to that of political and military leagues, which are as old as nation-states themselves.

Three further points about economic unions are worth noting. The first, which applies to military and political alliances as well, is that the formation of a coalition in one area often stimulates the creation of another alliance by countries left out of the first. Thus, the union of the Inner Six gave rise to the treaty among the Outer Seven. Similarly, a counterpart of the Common Market is the Council of Mutual Economic Assistance (CMEA), also known as Comecon, which links the Communist countries of Eastern Europe and the USSR through trade agreements.

Second, the new economic unions tend to be composed of voluntarily associated contiguous states (Figure 12.32). This was not the case with the recently dissolved empires, which included far-flung territories (though some, like the Russian Empire, are spatially continuous). Contiguity facilitates the movement of people and goods. Communication and transportation are simpler and more effective among adjoining countries than among those far removed from one another.

Finally, as far as joining economic unions is concerned, it does not seem to matter whether countries are alike or distinctly different in their economies. There are examples of both. If the countries are dissimilar, they may complement each other. This was one basis for the European Common Market. Dairy products and furniture from Denmark are sold in France, freeing that country to specialize in the production of machinery and clothing, although the Common Market is faced with expensive gluts of, for example, butter and wine, which are produced in more than one country (including France) and are basic products of the politically potent national farm groups (see "Policy, Subsidy, and Glut," p. 252). On the other hand, countries that produce the same raw materials hope that by joining together in an economic alliance, they might be

Figure 12.31

The European Economic Community (EEC). At the end of 1992, the 12 nations of the European Economic Community will formally dismantle any remnant trade barriers between them, and the world's largest free-trade region will be a reality. Its unified market of more than 325 million will increasingly be controlled from EEC headquarters in Brussels, Belgium, as individual countries relinquish their sovereign control over industrial standards, residency laws, labor agreements, and the like. In addition to the 12 community members, some 60 states in Africa, the Caribbean, and the Pacific are currently affiliated with the EEC by the Lomé Convention, which provides for development aid and access to European markets.

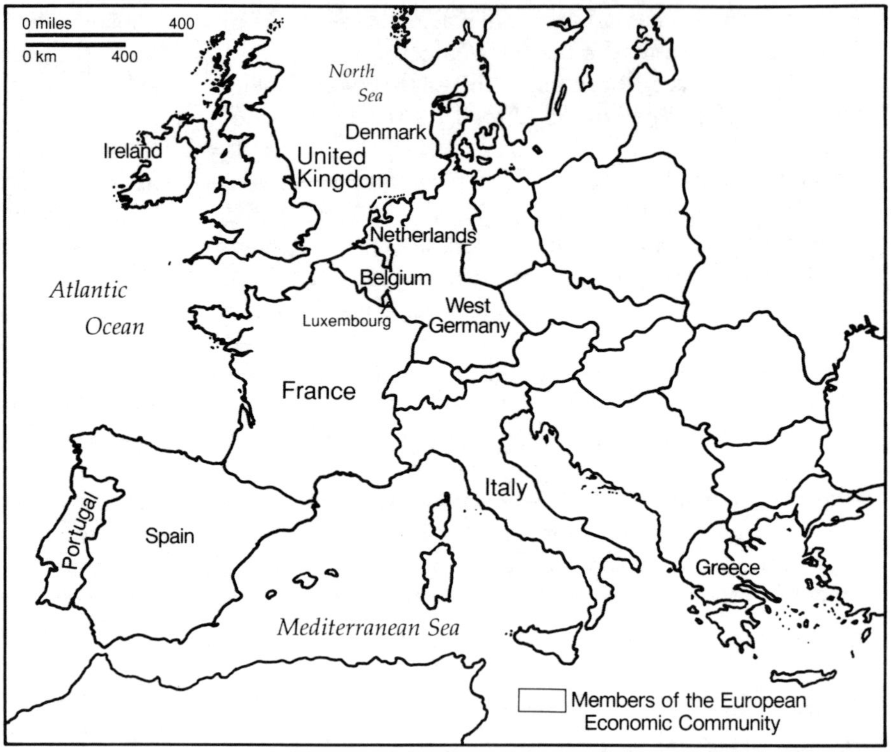

Figure 12.32

Members of the South Asian Association for Regional Cooperation. Formed in December 1985 to promote cooperation in such areas as agriculture and rural development, transportation, and telecommunications, the association is composed of seven neighboring states that contain about one-fifth of the world's population. Until the formation of the association, South Asia was the only major part of the world with no organization for regional cooperation.

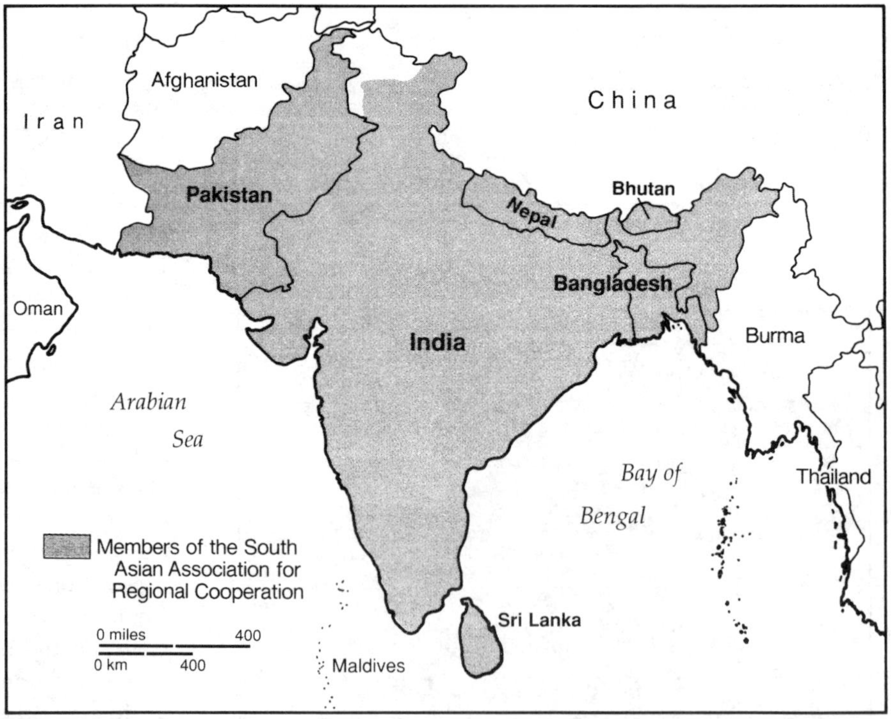

able to enhance their control of markets and prices for their products (Figure 12.33). The Organization of Petroleum Exporting Countries (OPEC), discussed in Chapter 8, is a case in point. Less successful attempts to form commodity cartels and price agreements between producing and consuming nations are represented by the International Tin Agreement, the International Coffee Agreement, and others.

Countries form alliances for other than economic reasons. Strategic, political, and cultural considerations may also foster cooperation. Military alliances are based on the principle that in unity there is strength. Such pacts usually provide for mutual assistance in the case of aggression. Once again, action breeds reaction when such an association is created. The formation of the North Atlantic Treaty Organization (NATO), a defensive alliance of many European countries and the United States, was countered by the establishment of the Warsaw Treaty Organization, which joins the USSR and its satellite countries of Eastern Europe. Both pacts allow the member states to base armed forces in one another's territories—a relinquishment of a certain degree of sovereignty unique to this century.

All international alliances recognize communities of interest. In economic and military associations, common objectives are clearly seen and described, and joint actions are agreed on with respect to the achievement of those objectives. More generalized common concerns or appeals to historical interest may be the basis for alliances that are primarily political. Such associations tend to be rather loose, not requiring their members to yield much power to the union. Examples are the Commonwealth of Nations (formerly the British Commonwealth), composed of many former British colonies and dominions, and the Organization of American States, both of which offer economic as well as political benefits. Many abortive political unions have foundered precisely because the individual countries could not agree on questions of policy and were unwilling to subordinate individual interests to make the union succeed. The United Arab Republic, the Central African Federation, the Federation of Malaysia and Singapore, and the Federation of the West Indies fall within this category.

Although many such political associations have failed, observers of the world scene speculate about the possibility that "superstates" will emerge from one or more of the international alliances that now exist. Will a "United States of Europe," for example, under a single common government be the logical outcome of the successes of the European Community? No one really knows, but as long as the individual state is regarded as the highest form of political and social organization (as it is now) and as the body in which sovereignty rests, such total unification presents difficulties.

Figure 12.33

Members of the Union of Banana Exporting Countries (UPEB). Founded in 1974, the union represents producers of about half of the world's banana exports. The cartel has established a multinational trading corporation in an attempt to challenge the control of, primarily, three United States companies that dominate the banana trade. It has had no more success than many similar producers' cartels seeking to control prices for such commodities as potash, lead, zinc, copper, nickel, sugar, cotton, timber, jute, bauxite, and coffee.

Summary

The political subdivision of the world into sovereign states constitutes an expression of cultural separation and identity as pervasive as that inherent in language, religion, or ethnicity. A broadening array of political entities affects us individually and collectively. The state remains central to the political hierarchy, but multiplying numbers of local administrative units and special-purpose agencies guide our daily lives, while global and regional supranational alliances have assumed some sovereign powers surrendered by countries. In North America, expanding urban areas and governmental responsibilities have increased the number of special-purpose agencies at the local level. Metropolitan or unified government is one form of administration designed to reduce or rationalize them.

The state and the nation-state are the dominant entities in the political subdivision of the world. A product of 18th-century political philosophy, the idea of nation-state was diffused globally by colonizing European powers. The colonial boundaries they established have in many instances been retained as international borders by newly independent countries. Within both new and established states, however, ethnically based nationalism and separatist movements are frequent concerns. State cohesion, on the other hand, is promoted by national symbols, a variety of institutions, and confidence in the organs and aims of government.

The physical endowments of states vary greatly and are important in national development and the retention of stability and viability. Size, shape, and relative location may influence countries' economies and international roles. Transportation and communication connections are important in their economic and political integration, as well as in uniting national cores and peripheries. Boundaries, of course, are the legal definition of a state's size and shape and determine the three-dimensional limits of its sovereignty. Their classification and definition are matters of geographic interest and practical political concern. Boundaries may or may not reflect preexisting cultural landscapes and may or may not be visible presences. Whatever their nature, however, boundaries are at the root of many international territorial and resource disputes. Maritime boundary claims add a new dimension to traditional assertions of territorial sovereignty and have resulted in proposed Law of the Sea conventions.

The end of colonial empires after World War II resulted in a great increase in the number of independent states and in the number and variety of international, regional, and special-interest alliances and affiliations now accorded semisovereign status. These alliances as well as the individual countries behind them seek to address some of the emerging spatial problems of human use of the earth that will concern us in our next and final chapter.

Key Words

antecedent boundary 404
Convention on the Law of the Sea 407
core area 400
exclusive economic zone (EEZ) 407
fragmented state 397
geometric (artificial) boundary 403
gerrymander 385
nation 383
nation-state 390
natural (physical) boundary 402
perforated state 397
prorupt state 397
state 383
subsequent boundary 404
superimposed boundary 404
unified government 388

For Review

1. What reasons can you suggest for the great political fragmentation of the United States? As few other countries have such a multiplicity of local governing and administrative units, is such subdivision—in *your* opinion—an evidence of internal strength or weakness? Support you answer.
2. What is *Unigov?* What would appear to be its advantages? Its disadvantages?
3. What are some of the devices by which countries achieve cohesion and identity? What internal cultural characteristics might serve to disrupt or destroy the cohesiveness that states attempt to foster?
4. What factors account for the wide variety of regionalist movements seeking some or more autonomy within their established state? What characteristics are common to all or most regional autonomist movements? Where are some of these tension areas in Western Europe? Why do they tend to be on the periphery rather than at the national core?
5. How may boundaries be classified? How do they create opportunities for conflict? Describe and give examples of three types of border disputes.
6. What are some of the pressing issues facing the states attempting to codify an international law of the sea? What are some of the provisions of the United Nations Convention on the Law of the Sea? Which provision is the most controversial? Why? Do you feel that the interests of your country would be best served by tight international control of maritime resources? Why or why not?
7. What made the post-1945 dissolution of empires unique in the political history of the world?
8. What types of international organizations and alliances can you name? For what purposes were they established?

Suggested Readings

Archer, J. Clark, and Fred M. Shelley. *American Electoral Mosaics.* Washington, D.C.: Association of American Geographers, 1986.

Bennett, D. Gordon. *Tension Areas of the World.* Delray Beach, Fla.: Park Press, 1982.

Boateng, E. A. *A Political Geography of Africa.* Cambridge, England: Cambridge University Press, 1978.

Boyd, Andrew. *An Atlas of World Affairs.* 8th ed. New York: Methuen, 1987.

Brass, Paul, ed. *Ethnic Groups and the State.* London: Croom Helm, 1985.

Burnett, Alan D., and Peter J. Taylor, eds. *Political Studies from Spatial Perspectives.* Chichester, England: John Wiley & Sons, 1981.

Christopher, A. J. "Continuity and Change of African Capitals." *Geographical Review* 75 (1985): 44–57.

Clarke, C., D. Ley, and C. Peach, eds. *Geography and Ethnic Pluralism.* London: Allen & Unwin, 1984.

Glassner, M. I., and Harm J. de Blij. *Systematic Political Geography.* 4th ed. New York: John Wiley & Sons, 1989.

Hartshorne, Richard. "The Functional Approach in Political Geography." *Annals of the Association of American Geographers* 40 (1950): 95–130.

Johnston, Douglas M., and Phillip M. Saunders, eds. *Ocean Boundary Making: Regional Issues and Developments.* London and New York: Croom Helm, 1988.

Johnston, R. J. *Geography and the State: An Essay in Political Geography.* New York: St. Martin's Press, 1982.

Johnston, R. J., and P. J. Taylor, eds. *A World in Crisis: Geographical Perspectives.* Oxford: Basil Blackwell, 1986.

Kasperson, Roger E., and Julian V. Minghi, eds. *The Structure of Political Geography.* Chicago: Aldine, 1969.

Mikesell, W. Marvin. "The Myth of the Nation State." *Journal of Geography* 82 (1983): 257–260.

Morrill, Richard L. *Political Redistricting and Geographic Theory.* Washington, D.C.: Association of American Geographers, 1981.

O'Loughlin, John. "The Identification and Evaluation of Racial Gerrymandering." *Annals of the Association of American Geographers* 72 (1982): 165–184.

O'Loughlin, John. "Political Geography: Tilling the Fallow Field." *Progress in Human Geography* 10 (1986): 69–83.

Pacione, Michael. *Progress in Political Geography.* London: Croom Helm, 1985.

Paddison, Ronan. *The Fragmented State: The Political Geography of Power.* New York: St. Martin's Press, 1983.

Prescott, J. R. V. *The Maritime Political Boundaries of the World.* London: Methuen, 1985.

Prescott, J. R. V. *Political Frontiers and Boundaries.* London: Allen & Unwin, 1987.

Rokkan, Stein, and Derek W. Urwin, eds. *The Politics of Territorial Identity: Studies in European Regionalism.* Newbury Park, Calif.: Sage, 1982.

Sanger, Clyde. *Ordering the Oceans: The Making of the Law of the Sea.* Toronto: University of Toronto Press, 1987.

Smith, Anthony D. *The Ethnic Origins of Nations.* Oxford: Basil Blackwell, 1986.

Taylor, Peter J. *Political Geography: World-Economy, Nation-State and Locality.* London and New York: Longman, 1986.

Williams, C. H., ed. *National Separatism.* Cardiff: University of Wales Press, 1982.

Appendix

1989 World Population Data

Region or Country	Population Estimate Mid-1989 (Millions)	Crude Birth Rate	Crude Death Rate	Natural Increase (Annual, %)	Population "Doubling Time" in Years (at Current Rate)	Population Projected to 2000 (Millions)	Population Projected to 2020 (Millions)	Infant Mortality Rate[a]	Total Fertility Rate[b]	% Population under Age 15/65+	Life Expectancy at Birth (years)	Urban Population (%)	Population with Access to Safe Water Supply (%, 1983)	Energy Consumption Per Capita, 1984 (Gigajoules)/ Percent Change Since 1970	Per Capita GNP, 1987 (U.S.$)
World	**5,234**	**28**	**10**	**1.8**	**39**	**6,323**	**8,330**	**75**	**3.6**	**33/6**	**63**	**41**	**—**	**55/6**	**$ 3,330**
More Developed	**1,206**	**15**	**9**	**0.6**	**122**	**1,268**	**1,339**	**15**	**1.9**	**22/12**	**73**	**73**	**—**	**—/—**	**12,070**
Less Developed	**4,028**	**31**	**10**	**2.1**	**32**	**5,055**	**6,991**	**84**	**4.1**	**37/4**	**60**	**32**	**—**	**—/—**	**670**
Less Developed (Excl. China)	**2,924**	**35**	**11**	**2.4**	**29**	**3,763**	**5,468**	**93**	**4.7**	**40/4**	**58**	**36**	**51**	**—/—**	**820**
Africa	**646**	**45**	**15**	**2.9**	**24**	**898**	**1,523**	**113**	**6.3**	**45/3**	**51**	**30**	**41**	**12/41**	**610**
Northern Africa	**142**	**39**	**11**	**2.8**	**24**	**188**	**278**	**93**	**5.6**	**42/4**	**59**	**41**	**73**	**—/—**	**1,150**
Algeria	24.9	42	10	3.2	22	33.7	49.4	81	6.4	46/4	62	43	90	21/95	2,760
Egypt	54.8	38	9	2.8	24	71.2	103.0	93	5.3	40/4	59	45	75	19/142	710
Libya	4.1	39	8	3.1	22	5.6	8.4	74	5.6	46/2	65	76	98	111/463	5,500
Morocco	25.6	36	10	2.6	27	33.3	48.5	90	4.9	42/4	61	43	—	9/69	620
Sudan	24.5	45	17	2.8	24	33.6	54.6	113	6.5	45/3	49	20	48	2/–46	330
Tunisia	7.9	32	7	2.4	29	9.9	13.6	77	4.3	40/4	64	53	—	20/137	1,210
Western Sahara	0.2	49	23	2.5	28	0.2	0.4	—	—	—/—	—	—	—	—	—
Western Africa	**200**	**47**	**18**	**2.9**	**24**	**280**	**480**	**124**	**6.6**	**45/2**	**47**	**28**	**34**	**—/—**	**380**
Benin	4.7	51	20	3.0	23	7.1	14.2	115	7.0	47/3	45	39	20	1/–14	300
Burkina Faso	8.7	47	19	2.8	25	11.8	20.6	144	6.5	45/3	46	8	30	1/125	170
Cape Verde	0.4	36	11	2.5	28	0.5	0.8	71	5.2	46/5	60	27	—	6/–	500
Côte d'Ivoire	12.1	51	15	3.6	19	18.5	35.4	101	7.4	46/3	52	43	20	6/5	750
Gambia	0.8	47	22	2.5	28	1.1	1.7	148	6.4	44/4	42	18	—	3/49	220
Ghana	14.6	44	13	3.1	22	20.4	33.9	90	6.4	47/3	53	32	43	2/–55	390
Guinea	7.1	47	23	2.4	29	9.2	14.4	153	6.2	43/3	41	22	17	2/–28	—
Guinea-Bissau	1.0	41	22	1.9	37	1.2	2.0	138	5.4	41/4	44	27	33	1/–40	170
Liberia	2.5	45	13	3.2	22	3.5	6.5	87	6.6	47/3	54	42	40	10/–26	440
Mali	8.9	50	22	2.9	24	12.3	21.1	175	6.7	46/3	43	18	14	1/60	200
Mauritania	2.0	46	20	2.7	26	2.7	4.5	132	6.5	46/3	45	35	—	5/2	440
Niger	7.4	51	22	2.9	24	10.6	18.9	140	7.1	47/3	44	16	34	2/133	280
Nigeria	115.3	46	17	2.9	24	160.9	273.6	122	6.6	45/2	47	28	37	7/383	370
Senegal	7.2	46	20	2.6	27	9.7	15.2	135	6.4	44/3	45	36	44	5/32	510
Sierra Leone	4.1	48	24	2.4	29	5.4	8.9	160	6.5	41/3	40	28	23	2/–44	300
Togo	3.4	47	13	3.4	20	4.9	8.9	113	6.6	45/3	54	22	37	2/6	300

Region or Country	Population Estimate Mid-1989 (Millions)	Crude Birth Rate	Crude Death Rate	Natural Increase (Annual, %)	Population "Doubling Time" in Years (at Current Rate)	Population Projected to 2000 (Millions)	Population Projected to 2020 (Millions)	Infant Mortality Rate[a]	Total Fertility Rate[b]	% Population under Age 15/65+	Life Expectancy at Birth (years)	Urban Population (%)	Population with Access to Safe Water Supply (%, 1983)	Energy Consumption Per Capita, 1984 (Gigajoules)/ Percent Change Since 1970	Per Capita GNP, 1987 (U.S.$)
Eastern Africa	**194**	**48**	**17**	**3.1**	**23**	**281**	**511**	**118**	**6.9**	**47/3**	**49**	**17**	**—**	**—/—**	**240**
Burundi	5.5	48	15	3.3	21	7.8	13.8	114	6.9	45/3	52	5	26	z/51	240
Comoros	0.4	47	14	3.3	21	0.7	1.3	96	7.0	47/4	55	23	—	2/-	380
Djibouti	0.4	48	18	3.0	23	0.6	1.0	127	6.6	46/3	46	74	—	8/34	—
Ethiopia	49.8	44	23	2.1	33	71.1	127.6	155	6.1	46/4	41	11	—	z/-34	120
Kenya	24.1	54	13	4.1	17	37.6	69.6	76	8.1	51/2	57	19	28	3/-8	340
Madagascar	11.6	46	15	3.1	22	16.6	29.6	110	6.6	44/3	53	22	23	2/-20	200
Malawi	8.7	51	18	3.3	21	11.6	21.5	134	7.7	46/4	48	13	51	1/-10	160
Mauritius	1.1	19	7	1.3	54	1.3	1.6	25.2	2.3	31/5	68	41	95	8/-18	1,470
Mozambique	15.2	45	19	2.6	27	20.4	31.9	147	6.4	47/3	46	20	13	3/-31	150
Reunion	0.6	22	6	1.7	41	0.7	0.8	14	2.4	33/5	71	98	—	—/—	—
Rwanda	7.0	52	18	3.4	20	10.4	19.7	127	8.5	49/3	48	6	60	1/280	310
Seychelles	0.1	25	8	1.8	39	0.1	0.1	18.4	3.2	36/6	70	52	—	—/—	3,180
Somalia	8.2	48	21	2.6	26	10.4	18.6	137	7.4	45/3	44	33	31	4/228	290
Tanzania	26.3	50	15	3.6	19	39.6	75.4	111	7.1	48/3	52	19	46	1/-24	220
Uganda	17.0	50	16	3.4	20	24.7	46.4	108	7.3	48/3	50	9	16	1/-64	260
Zambia	8.1	51	14	3.7	19	12.2	22.7	84	7.2	50/3	52	43	—	10/-11	240
Zimbabwe	10.1	47	11	3.6	19	15.2	29.5	76	6.5	48/3	57	24	52	13/-46	590
Middle Africa	**66**	**45**	**17**	**2.8**	**25**	**91**	**156**	**117**	**6.0**	**45/3**	**49**	**37**	**22**	**—/—**	**400**
Angola	8.5	47	21	2.6	27	11.5	19.1	143	6.4	45/3	43	25	28	4/-25	—
Cameroon	10.8	43	16	2.6	26	14.5	23.5	128	5.9	44/3	50	42	26	13/600	960
Central African Republic	2.8	44	19	2.5	28	3.7	5.9	146	5.6	42/3	46	35	16	1/-26	330
Chad	4.9	43	23	2.0	35	6.3	9.5	143	5.3	44/2	39	27	26	1/12	150
Congo	2.2	47	13	3.4	21	3.2	5.9	112	6.9	45/3	56	48	29	3/-29	880
Equatorial Guinea	0.4	38	20	1.9	37	0.4	0.7	130	5.6	40/5	45	60	—	3/-24	—
Gabon	1.1	36	17	1.9	36	1.6	2.6	108	4.8	35/6	50	41	—	33/139	2,750
Sao Tome and Principe	0.1	36	9	2.7	25	0.2	0.3	61.7	5.4	46/5	67	38	—	—/—	280
Zaire	34.9	45	15	3.1	23	49.3	88.9	103	6.1	47/3	51	40	19	2/2	160
Southern Africa	**44**	**36**	**10**	**2.7**	**26**	**59**	**97**	**71**	**4.7**	**36/5**	**62**	**53**	**—**	**88/20**	**1,780**
Botswana	1.2	47	13	3.4	20	1.8	3.1	72	6.3	48/4	57	16	65	—/—	1,030
Lesotho	1.7	41	13	2.8	25	2.4	3.9	106	5.8	42/4	55	17	14	—/—	360
Namibia	1.8	44	13	3.1	22	2.6	4.5	111	6.1	44/5	55	51	—	—/—	—
South Africa	38.5	35	9	2.6	27	51.5	83.5	65	4.5	35/5	63	56	—	—/—	1,890
Swaziland	0.8	47	13	3.4	21	1.1	2.0	124	6.5	49/2	54	26	—	—/—	700
Asia	**3,061**	**28**	**9**	**1.9**	**36**	**3,743**	**4,883**	**78**	**3.6**	**34/5**	**62**	**29**	**—**	**20/55**	**1,170**
Asia (Excl. China)	**1,957**	**32**	**10**	**2.2**	**32**	**2,452**	**3,360**	**90**	**4.3**	**38/4**	**60**	**33**	**50**	**—/—**	**1,730**
Western Asia	**129**	**36**	**8**	**2.8**	**25**	**174**	**279**	**74**	**5.3**	**41/4**	**64**	**58**	**70**	**—/—**	**2,890**
Bahrain	0.5	31	4	2.8	25	0.7	0.9	26	4.6	32/2	67	81	—	388/402	—
Cyprus	0.7	19	9	1.0	71	0.8	0.9	11	2.3	25/10	76	62	—	56/56	5,210
Gaza	0.6	48	7	4.1	17	0.8	1.4	56	7.2	48/3	62	—	—	—/—	—
Iraq	18.1	45	8	3.8	18	27.2	51.0	69	7.1	45/3	66	68	73	15/-13	—
Israel	4.5	23	7	1.6	43	5.4	6.9	10.7	3.1	32/9	75	90	—	69/10	6,810
Jordan	4.0	41	6	3.5	20	5.7	9.6	54	6.2	46/2	69	69	89	29/274	1,540
Kuwait	2.1	30	3	2.8	25	2.8	4.1	15.6	4.5	36/1	73	94	89	195/37	14,870
Lebanon	3.3	28	7	2.1	33	4.1	5.7	50	3.8	41/5	68	80	92	25/26	—
Oman	1.4	46	13	3.3	21	2.1	3.8	100	7.2	44/3	55	9	—	391/8,434	5,780
Qatar	0.4	30	2	2.7	25	0.6	1.1	31	4.5	28/1	69	88	—	637/65	12,360
Saudi Arabia	14.7	42	8	3.4	20	22.4	42.9	71	7.2	45/3	63	73	93	104/667	6,930
Syria	12.1	44	6	3.8	18	17.6	29.5	48	6.8	49/4	65	50	71	27/141	1,820
Turkey	55.4	30	8	2.2	32	68.6	92.2	81	3.7	37/4	64	53	63	26/84	1,200
United Arab Emirates	1.7	27	4	2.3	30	2.1	2.8	32	5.2	30/1	69	81	93	223/210	15,680
Yemen, North	6.9	52	18	3.5	20	10.0	19.5	132	7.7	49/3	48	20	31	6/1,243	580
Yemen, South	2.5	48	15	3.4	21	3.6	6.3	113	7.1	48/3	51	40	50	25/241	420

Region or Country	Population Estimate Mid-1989 (Millions)	Crude Birth Rate	Crude Death Rate	Natural Increase (Annual, %)	Population "Doubling Time" in Years (at Current Rate)	Population Projected to 2000 (Millions)	Population Projected to 2020 (Millions)	Infant Mortality Rate[a]	Total Fertility Rate[b]	% Population under Age 15/65+	Life Expectancy at Birth (years)	Urban Population (%)	Population with Access to Safe Water Supply (%, 1983)	Energy Consumption Per Capita, 1984 (Gigajoules)/ Percent Change Since 1970	Per Capita GNP, 1987 (U.S.$)
Southern Asia	**1,166**	**35**	**12**	**2.3**	**30**	**1,489**	**2,080**	**105**	**4.8**	**39/4**	**57**	**26**	**50**	**—/—**	**290**
Afghanistan	14.8	49	23	2.6	26	26.6	38.4	172	6.9	46/4	42	13	10	2/24	—
Bangladesh	114.7	43	15	2.8	25	153.4	230.2	138	5.8	43/3	52	13	42	2/—	160
Bhutan	1.5	38	17	2.1	32	1.9	2.7	128	5.5	39/4	48	5	17	z/—	150
India	835.0	33	11	2.2	32	1,042.5	1,374.5	96	4.3	37/4	58	26	54	7/71	300
Iran	53.9	44	10	3.4	20	75.7	130.3	93	6.2	45/3	62	54	—	39/40	—
Maldives	0.2	47	10	3.7	19	0.3	0.6	80	7.0	45/2	60	26	—	—/—	300
Nepal	18.7	42	17	2.5	28	24.4	37.4	112	6.0	42/3	52	7	15	z/14	160
Pakistan	110.4	43	14	2.9	24	145.3	242.2	120	6.5	43/4	54	28	39	6/30	350
Sri Lanka	16.9	22	6	1.6	43	19.4	24.0	22.5	2.5	35/4	70	22	36	4/-1	400
Southeast Asia	**445**	**31**	**9**	**2.2**	**31**	**549**	**740**	**68**	**3.8**	**39/4**	**61**	**26**	**43**	**—/—**	**680**
Brunel	0.3	31	3	2.7	25	0.3	0.4	11	3.6	37/3	71	59	—	—/—	15,390
Burma	40.8	34	11	2.3	30	51.1	68.7	80	4.6	39/4	58	24	25	2/40	—
East Timor	0.7	44	22	2.2	31	0.9	1.1	166	5.4	35/3	43	12	—	—/—	—
Indonesia	184.6	29	9	2.0	35	222.0	287.4	83	3.5	39/3	58	26	33	8/126	450
Kampuchea	6.8	40	17	2.3	31	8.5	12.0	134	4.6	35/3	48	11	—	z/-92	—
Laos	3.9	41	16	2.5	28	5.0	6.9	110	5.7	43/3	49	16	21	1/-72	—
Malaysia	17.4	31	6	2.5	28	20.9	26.6	31	3.9	37/4	67	35	80	26/57	1,800
Philippines	64.9	35	7	2.8	25	85.5	130.7	48	4.6	41/3	66	41	54	9/17	590
Singapore	2.7	17	5	1.2	59	2.9	3.1	7.4	1.6	23/5	73	100	100	196/429	7,940
Thailand	55.6	24	7	1.7	41	65.5	81.6	50	2.7	34/4	65	17	65	12/125	840
Vietnam	66.8	34	8	2.6	27	86.0	121.2	53	4.4	40/4	64	19	—	4/-59	—
East Asia	**1,321**	**20**	**6**	**1.3**	**52**	**1,531**	**1,785**	**41**	**2.3**	**28/6**	**67**	**29**	**—**	**—/—**	**1,910**
China	1,103.9	21	7	1.4	49	1,291.6	1,523.2	44	2.4	29/6	66	21	—	19/87	300
Hong Kong	5.7	12	5	0.8	91	6.3	6.7	7.7	1.3	23/8	76	93	99	—/—	8,260
Japan	123.2	11	6	0.5	141	130.0	131.7	4.9	1.7	20/11	78	77	—	111/18	15,770
Korea, North	22.5	29	5	2.4	29	28.4	38.5	33	3.5	38/4	70	64	—	78/29	—
Korea, South	43.1	19	6	1.3	53	48.8	54.7	30	2.1	30/4	68	65	—	44/131	2,690
Macao	0.4	23	6	1.7	41	0.5	0.6	12	—	23/8	68	97	—	—/—	—
Mongolia	2.1	37	9	2.9	24	2.8	4.4	49	5.4	42/4	65	52	—	47/85	—
Taiwan	20.0	16	5	1.1	62	22.4	24.9	6.9	1.7	29/5	73	71	—	—/—	—
North America	**275**	**16**	**9**	**0.7**	**97**	**297**	**325**	**10**	**1.9**	**21/12**	**75**	**74**	**—**	**—/—**	**18,110**
Canada	26.3	15	7	0.8	92	28.4	30.2	7.9	1.7	21/11	76	77	—	288/11	15,080
United States	248.8	16	9	0.7	98	268.3	294.4	9.9	1.9	21/12	75	74	—	280/-12	18,430
Latin America	**438**	**29**	**7**	**2.1**	**33**	**535**	**705**	**55**	**3.6**	**38/5**	**66**	**68**	**70**	**—/—**	**1,820**
Central America	**115**	**32**	**6**	**2.5**	**27**	**144**	**196**	**52**	**4.1**	**42/4**	**67**	**61**	**70**	**—/—**	**1,630**
Belize	0.2	36	6	3.0	23	0.2	0.3	36	5.1	45/5	69	50	—	—/—	1,250
Costa Rica	3.0	29	4	2.5	28	3.8	5.3	17.4	3.5	37/5	76	45	93	13/22	1,590
El Salvador	5.1	35	9	2.7	26	6.3	8.8	62	4.4	46/4	63	43	51	5/-10	850
Guatemala	8.9	40	9	3.1	23	11.8	17.6	59	5.6	46/3	62	40	51	6/6	940
Honduras	5.0	40	8	3.2	22	6.8	10.6	69	5.5	47/3	64	41	69	7/-3	780
Mexico	86.7	30	6	2.4	29	107.2	142.1	50	3.8	42/4	68	66	74	50/65	1,820
Nicaragua	3.5	43	8	3.5	20	5.1	7.7	69	5.7	47/3	62	57	53	8/-11	830
Panama	2.4	27	5	2.2	31	2.9	3.7	25	3.3	37/5	72	52	62	22/-6	2,240
Caribbean	**33**	**25**	**8**	**1.7**	**41**	**38**	**47**	**56**	**3.0**	**33/7**	**68**	**57**	**—**	**—/—**	**—**
Antigua and Barbuda	0.1	15	5	1.0	71	0.1	0.1	11	1.7	28/7	72	58	—	—/—	2,570
Bahamas	0.2	17	5	1.2	60	0.3	0.4	26.4	1.9	34/5	71	75	—	—/—	10,320
Barbados	0.3	16	9	0.7	94	0.3	0.3	12.6	1.9	30/11	75	32	—	36/42	5,330
Cuba	10.5	17	6	1.1	62	11.4	12.5	13.3	1.8	25/8	74	72	—	43/47	—

Region or Country	Population Estimate Mid-1989 (Millions)	Crude Birth Rate	Crude Death Rate	Natural Increase (Annual, %)	Population "Doubling Time" in Years (at Current Rate)	Population Projected to 2000 (Millions)	Population Projected to 2020 (Millions)	Infant Mortality Rate[a]	Total Fertility Rate[b]	% Population under Age 15/65+	Life Expectancy at Birth (years)	Urban Population (%)	Population with Access to Safe Water Supply (%, 1983)	Energy Consumption Per Capita, 1984 (Gigajoules)/ Percent Change Since 1970	Per Capita GNP, 1987 (U.S
Caribbean *(continued)*															
Dominica	0.1	26	5	2.1	33	0.1	0.1	14	2.7	34/7	76	—	—	—/—	1,440
Dominican Republic	7.0	31	7	2.5	28	8.6	11.0	65	3.8	39/3	65	56	60	14/55	730
Grenada	0.1	37	7	3.0	23	0.1	0.1	30	4.0	39/7	72	—	—	—/—	1,340
Guadeloupe	0.3	20	7	1.4	51	0.4	0.5	18	2.3	31/7	73	90	—	—/—	—
Haiti	6.4	35	13	2.2	32	7.8	10.8	117	4.9	40/5	55	26	33	2/43	360
Jamaica	2.5	22	5	1.7	41	2.9	4.1	20	2.8	37/6	74	49	86	35/-2	960
Martinique	0.3	19	6	1.3	55	0.4	0.4	11	2.1	30/7	74	82	—	—/—	—
Netherlands Antilles	0.2	19	6	1.4	51	0.2	0.2	10	2.1	30/7	76	53	—	—/—	—
Puerto Rico	3.3	19	7	1.2	56	3.4	4.0	14.9	2.2	28/10	74	67	—	—/—	5,520
St. Kitts-Nevis	0.04	23	11	1.3	55	0.04	0.1	39.7	2.7	34/9	68	45	—	—/—	1,700
Saint Lucia	0.2	28	6	2.2	32	0.2	0.3	21.5	4.1	44/6	71	44	—	—/—	1,370
St. Vincent and the Grenadines	0.1	25	6	1.9	37	0.1	0.2	24.7	2.9	44/6	70	18	—	—/—	1,070
Trinidad and Tobago	1.2	27	7	2.0	34	1.5	2.1	13.7	3.1	34/6	70	64	99	148/15	4,220
Tropical South America	**242**	**29**	**8**	**2.1**	**33**	**298**	**395**	**61**	**3.6**	**37/4**	**65**	**69**	**72**	**—/—**	**1,850**
Bolivia	7.1	40	14	2.6	27	9.3	13.4	110	5.5	43/3	53	49	43	10/47	570
Brazil	147.4	28	8	2.0	34	179.5	233.8	63	3.4	36/4	65	71	76	19/51	2,020
Colombia	31.2	28	7	2.0	34	38.0	49.3	46	3.4	36/4	64	67	81	25/41	1,220
Ecuador	10.5	33	8	2.6	27	13.6	19.5	63	4.3	42/4	65	54	59	20/140	1,040
Guyana	0.8	26	7	1.9	36	0.8	1.1	44	2.9	38/4	66	32	80	18/-41	380
Paraguay	4.2	36	7	2.9	24	5.5	8.4	45	4.8	41/4	67	43	25	7/54	1,000
Peru	21.4	29	8	2.1	33	26.4	35.1	69	4.4	41/4	61	69	52	19/5	1,430
Suriname	0.4	27	7	2.1	33	0.5	0.6	33	3.3	37/4	69	66	—	49/-28	2,360
Venezuela	19.1	28	4	2.4	29	24.1	33.3	36	3.4	39/4	70	83	—	96/46	3,230
Temperate South America	**48**	**22**	**8**	**1.4**	**50**	**55**	**66**	**27**	**2.9**	**31/8**	**70**	**84**	**70**	**—/—**	**2,070**
Argentina	31.9	22	9	1.4	51	36.2	43.8	29.7	3.1	31/9	70	85	63	50/9	2,370
Chile	13.0	22	6	1.6	43	15.3	19.0	19.1	2.4	31/6	71	83	85	27/-20	1,310
Uruguay	3.0	18	10	0.8	83	3.2	3.5	27.9	2.3	26/11	71	84	79	18/-29	2,180
Europe	**499**	**13**	**10**	**0.3**	**269**	**507**	**502**	**12**	**1.7**	**20/13**	**74**	**74**	**—**	**124/13**	**9,700**
Northern Europe	**84**	**13**	**11**	**0.2**	**285**	**84**	**83**	**8**	**1.8**	**19/15**	**75**	**84**	**—**	**—/—**	**11,640**
Denmark	5.1	11	11	0.0	(—)	5.2	4.8	8.4	1.5	18/15	75	84	—	133/-16	15,010
Finland	5.0	12	10	0.3	277	5.0	4.8	5.8	1.6	19/13	75	62	—	146/23	14,370
Iceland	0.3	17	7	1.0	69	0.3	0.3	3.4	2.0	25/10	78	90	—	150/22	16,670
Ireland	3.5	17	9	0.8	89	3.7	4.3	7.4	2.3	29/11	73	56	—	96/27	6,030
Norway	4.2	13	11	0.2	315	4.3	4.3	7.8	1.8	19/16	76	60	—	193/44	17,110
Sweden	8.5	13	11	0.2	367	8.3	8.0	5.7	1.9	17/18	77	83	—	137/-19	15,690
United Kingdom	57.3	14	11	0.2	289	57.3	56.6	9.1	1.8	19/15	75	90	—	138/-3	10,430
Western Europe	**157**	**12**	**10**	**0.2**	**365**	**157**	**148**	**8**	**1.6**	**18/14**	**75**	**83**	**—**	**—/—**	**13,610**
Austria	7.6	12	11	0.1	1,155	7.6	7.1	8.3	1.4	18/14	75	55	—	117/22	11,970
Belgium	9.9	12	11	0.1	577	9.7	9.1	9.7	1.5	19/14	75	95	—	145/-12	11,360
France	56.1	14	10	0.4	161	57.9	58.7	7.6	1.8	20/14	76	73	—	114/1	12,860
Germany, West	61.5	11	11	-0.1	(—)	59.6	51.5	8.3	1.4	15/15	75	94	—	163/8	14,460
Luxembourg	0.4	11	11	0.1	1,155	0.4	0.4	9.4	1.4	17/13	74	78	—	328/-34	15,860
Netherlands	14.9	13	8	0.4	158	15.2	14.7	7.6	1.6	19/12	77	88	—	172/28	11,860
Switzerland	6.6	12	9	0.3	267	6.6	6.1	6.8	1.5	17/15	77	61	—	108/14	21,250

Region or Country	Population Estimate Mid-1989 (Millions)	Crude Birth Rate	Crude Death Rate	Natural Increase (Annual, %)	Population "Doubling Time" in Years (at Current Rate)	Population Projected to 2000 (Millions)	Population Projected to 2020 (Millions)	Infant Mortality Rate[a]	Total Fertility Rate[b]	% Population under Age 15/65+	Life Expectancy at Birth (years)	Urban Population (%)	Population with Access to Safe Water Supply (%, 1983)	Energy Consumption Per Capita, 1984 (Gigajoules)/ Percent Change Since 1970	Per Capita GNP, 1987 (U.S.$)
Eastern Europe	**113**	**15**	**11**	**0.3**	**203**	**117**	**120**	**17**	**2.1**	**24/11**	**71**	**63**	**—**	**—/—**	**—**
Bulgaria	9.0	13	12	0.1	770	9.0	8.9	15.0	2.0	21/11	71	66	—	166/50	—
Czechoslovakia	15.6	14	12	0.2	301	16.2	17.1	13.1	2.0	24/11	71	75	—	183/16	—
Germany, East	16.6	14	13	0.1	990	16.5	16.3	8.7	1.7	19/14	73	77	—	223/24	—
Hungary	10.6	12	13	−0.2	(—)	10.5	10.3	17.3	1.8	21/13	70	59	—	112/36	2,240
Poland	38.2	16	10	0.6	114	39.9	42.4	17.5	2.2	26/9	71	61	—	133/27	1,920
Romania	23.2	16	11	0.5	141	24.4	25.4	25.6	2.3	25/10	70	51	—	134/52	—
Southern Europe	**145**	**12**	**9**	**0.3**	**253**	**149**	**150**	**15**	**1.6**	**21/12**	**74**	**68**	**—**	**—/—**	**6,860**
Albania	3.2	25	6	2.0	35	3.8	4.8	42	3.2	35/5	70	35	—	39/117	—
Greece	10.0	11	10	0.1	630	10.2	10.1	12.6	1.6	21/13	74	58	—	66/103	4,350
Italy	57.6	10	9	0.0	2,310	58.2	55.9	10.1	1.3	19/13	74	72	—	90/16	10,420
Malta	0.4	15	8	0.7	99	0.4	0.4	7.3	2.0	24/10	75	85	—	45/62	4,010
Portugal	10.4	12	9	0.3	257	10.5	10.5	14.2	1.6	23/12	74	30	—	38/74	2,890
Spain	39.2	11	8	0.3	210	40.8	42.4	8.8	1.5	22/12	76	91	—	65/50	6,010
Yugoslavia	23.7	15	9	0.6	114	25.0	26.2	25.4	2.0	24/9	71	46	—	73/77	2,480
USSR	**289**	**20**	**10**	**1.0**	**70**	**312**	**355**	**25**	**2.5**	**25/9**	**69**	**66**	**—**	**176/45**	**8,375**
Oceania	**26**	**20**	**8**	**1.2**	**56**	**30**	**38**	**26**	**2.7**	**28/9**	**72**	**70**	**—**	**135/25**	**8,440**
Australia	16.8	15	7	0.8	88	18.7	21.9	8.8	1.9	23/10	76	86	—	180/27	10,900
Fiji	0.8	28	5	2.3	31	0.8	1.0	21	3.1	38/3	67	39	—	12/−12	1,510
French Polynesia	0.2	28	4	2.4	29	0.3	0.3	23	3.5	36/3	71	62	—	—/—	—
New Caledonia	0.2	24	6	1.8	39	0.2	0.2	35	3.0	34/5	69	58	—	—/—	—
New Zealand	3.4	17	8	0.8	83	3.6	4.0	10.0	2.0	24/10	74	84	—	133/38	8,230
Pacific Islands[c]	0.2	34	5	2.9	24	0.2	0.3	33	4.9	47/4	71	29	—	—	—
Papua-New Guinea	3.9	39	12	2.7	26	5.1	7.9	59	5.7	41/2	54	13	16	9/93	730
Solomon Islands	0.3	41	5	3.6	19	0.5	0.8	41	6.4	48/3	69	9	—	8/26	420
Vanuatu	0.2	38	5	3.2	21	0.2	0.3	37	5.6	45/3	69	18	—	—/—	—
Western Samoa	0.2	34	7	2.8	25	0.2	0.3	48	4.8	44/3	66	21	—	—/—	560

[a]Infant deaths per 1000 live births

[b]Average number of children born to a woman during her lifetime

[c]Comprising the Federated States of Micronesia, Palau, and the Marshall and N. Mariana Islands

A dash (—) indicates data unavailable or inapplicable

A (z) signifies an amount that rounds to zero

Table modified from the *1989 World Population Data Sheet* of the Population Reference Bureau. Data for safe water supply and energy consumption were taken from the 1987 edition of the *World Population Data Sheet.*

Glossary

Terms in italics identify related glossary entries.

A

absolute direction
Direction with respect to cardinal east, west, north, and south reference points.

absolute distance
(*syn*: geodesic distance) The shortest-path separation between two places measured on a standard unit of length (miles or kilometers, usually); also called real distance.

absolute location
(*syn*: mathematical location) The exact position of an object or place stated in spatial coordinates of a grid system designed for locational purposes. In geography, the reference system is the *globe grid* of parallels of *latitude* north or south of the *equator* and of meridians of *longitude* east or west of a *prime meridian.* Absolute globe locations are cited in degrees, minutes, and (for greater precision) seconds of latitude and longitude north or south and east or west of the equatorial and prime meridian base lines.

absorbing barrier
A condition that blocks the *diffusion* of an *innovation* or prevents its adoption.

accessibility
The relative ease with which a destination may be reached from other locations; the relative opportunity for *spatial interaction.* This may be measured in geometric, social, or economic terms.

acculturation
Cultural modification or change that results when one *culture* group or individual adopts traits of a dominant or *host society*; cultural development or change through "borrowing."

acid rain
Precipitation that is unusually acidic; created when oxides of sulfur and nitrogen change chemically as they dissolve in water vapor in the *atmosphere* and return to earth as acidic rain, snow, or fog.

activity space
The area within which people move freely on their rounds of regular activity.

adaptation
A presumed modification of heritable traits through response to environmental stimuli.

agglomeration
The spatial grouping of people or activities for mutual benefit; in *economic geography,* the concentration of productive enterprises for collective or cooperative use of *infrastructure* and sharing of labor resources and market access.

agglomeration economies
(*syn*: external economies) The savings to an individual enterprise derived from locational association with a cluster of other similar economic activities, such as other factories or retail stores.

agriculture
The science and practice of farming, including the cultivation of the soil and the rearing of livestock.

amalgamation
In *ethnic geography,* the concept that multiethnic societies become a merger of the *culture traits* of their member groups.

anecumene
See *nonecumene.*

animism
A belief that natural objects may be the abode of dead people, spirits, or gods who occasionally give the objects the appearance of life.

antecedent boundary
A *boundary* line established before the area in question is well populated.

aquaculture
Production and harvesting of fish and shellfish in land-based ponds.

arithmetic density
See *crude density.*

artifacts
The material manifestations of *culture,* including tools, housing, systems of land use, clothing, and the like. Elements in the *technological subsystem* of culture.

artificial boundary
See *geometric boundary.*

assimilation
A two-part *behavioral* and *structural* process by which a minority population reduces or completely loses its identifying cultural characteristics and blends into the *host society.*

atmosphere
The air or mixture of gases surrounding the earth.

awareness space
Locations or places about which an individual has knowledge even without visiting all of them; includes *activity space* and additional areas newly encountered or about which one acquires information.

B

basic sector
Those products or services of an *urban* economy that are exported outside the city itself, earning income for the community.

behavioral assimilation
(*syn*: cultural assimilation) The process of integration into a common cultural life through acquisition of the sentiments, attitudes, and experiences of other groups.

beneficiation
The enrichment of low-grade ores through concentration and other processes to reduce their waste content and increase their *transferability.*

bilingualism
A society's use of two *official languages.*

biomass
The total dry weight of all living organisms within a unit area; plant and animal matter that can in any way be used as a source of energy.

biome
A major ecological community, including plants and animals, occupying an extensive earth area.

biosphere
(*syn*: ecosphere) The thin film of air, water, and earth within which we live, including the *atmosphere,* surrounding and subsurface waters, and the upper reaches of the earth's crust.

birth rate
The ratio of the number of live births during 1 year to the total population, usually at the midpoint of the same year, expressed as the number of births per year per 1000 population.

boundary
A line separating one political unit from another; see *international boundary.*

boundary definition
A general agreement between two *states* about the allocation of territory between them.

boundary delimitation
The plotting of a *boundary* line on maps or aerial photographs.

boundary demarcation
The actual marking of a *boundary* line on the ground; the final stage in boundary development.

Brandt report
Entitled *North-South: A Program for Survival,* a report of the Independent Commission on International Development Issues, published in 1980 and named for the commission chairman, Willy Brandt.

break-of-bulk point
A location where goods are transferred from one type of carrier to another (e.g., from barge to railroad).

Buddhism
A *universalizing religion,* primarily of eastern and central Asia, based on teachings of Siddartha Gautama, the Buddha, that suffering is inherent in all life but can be relieved by mental and moral self-purification.

built environment
That part of the *physical landscape* that represents *material culture*; the buildings, roads, bridges, and similar structures large and small of the *cultural landscape.*

carrying capacity
The maximum population numbers that an area can support on a continuing basis without experiencing unacceptable deterioration; for humans, the numbers supportable by an area's known and used resources—usually agricultural ones.

cascade diffusion
A form of *expansion diffusion* in which dispersal progresses rapidly from larger to smaller centers or from higher to lower levels of a hierarchy.

caste
One of the hereditary social classes in *Hinduism* that determine one's occupation and position in society.

central business district (CBD)
The nucleus or "downtown" of a city, where retail stores, offices, and cultural activities are concentrated, mass transit systems converge, and land values and building densities are high. See also *core-frame concept.*

central city
That part of a *metropolitan area* contained within the boundaries of the main city around which suburbs have developed.

central place
An *urban* or other settlement node whose primary function is to provide goods and services to the consuming population of its *hinterland, complementary region,* or trade area.

central place theory
A deductive theory formulated by Walter Christaller (1893–1969) to explain the size and distribution of settlements through reference to competitive supply of goods and services to dispersed rural populations.

chain migration
The process by which *migration* movements from a common home area to a specific destination are sustained by links of friendship or kinship between first movers and later followers.

channelized migration
The tendency for *migration* to flow between areas that are socially and economically allied by past migration patterns, economic and trade connections, or some other affinity.

charter group
In plural societies, the early arriving ethnic group that created the *first effective settlement* and established the recognized cultural norms to which other, later groups are expected to conform.

choropleth map
A *thematic map* presenting spatial data as average values per unit area.

Christianity
A *monotheistic, universalizing religion* based on the teachings of Jesus Christ and of the Bible as sacred scripture.

circular and cumulative causation
A process through which tendencies for economic growth are self-reinforcing; an expression of the *multiplier effect,* it tends to favor major cities and *core* regions over less-advantaged *peripheral* regions.

climate
A summary of weather conditions in a place or region over a period of time.

cluster migration
A pattern of movement and settlement resulting from the collective action of a distinctive social or *ethnic group.*

cognitive map
See *mental map.*

cohort
A group of people who are unified by a specific common characteristic, such as age, and are subsequently treated as a statistical unit during their lifetimes.

collective farm
In the Soviet *planned economy,* the cooperative operation of an agricultural enterprise under state control of production and market, but without full status or support as a state enterprise.

colony
In *ethnic geography,* an urban ethnic area serving as point of entry and temporary *acculturation* zone for a specific immigrant group.

commercial economy
A system of production of goods and services for exchange in competitive markets where price and availability are determined by supply and demand forces.

compact state
A *state* whose territory is nearly circular.

comparative advantage
The principle that an area produces the items for which it has the greatest ratio of advantage or the least ratio of disadvantage in comparison to other areas, assuming free trade exists.

complementarity
The actual or potential relationship of two places or regions that each produce different goods or services for which the other has an effective demand, resulting in an exchange between the locales.

complementary region
The area served by a *central place.*

concentration
In *spatial distributions,* the clustering of a phenomenon around a central location.

concentric zone model
A model describing urban land uses as a series of circular belts or rings around a core *central business district,* each ring housing a distinct type of land use.

Confucianism
A Chinese *value system* and *ethnic religion* emphasizing ethics, social morality, tradition, and ancestor worship.

connectivity
The directness of routes linking pairs of places; an indication of the degree of internal connection in a transport *network.* More generally, all of the tangible and intangible means of connection and communication between places.

consequent boundary
(*syn*: ethnographic boundary) A *boundary* line that coincides with some cultural divide, such as religion or language.

contagious diffusion
A form of *expansion diffusion* that depends on direct contact. The process of dispersion is centrifugal, strongly influenced by distance, and dependent on interaction between actual and potential adopters of the innovation. Its name derives from the pattern of spread of contagious diseases.

continental shelf
A gently sloping seaward extension of the landmass found off the coasts of many continents; its outer margin is marked by a transition to the ocean depths at about 660 feet (200 m).

conurbation
A continuous, extended *urban* area formed by the growing together of several formerly separate, expanding cities.

Convention on the Law of the Sea
A code of maritime law approved by the United Nations in 1982 that authorizes, among other provisions, territorial waters extending 12 nautical miles (22 km) from shore and 200-nautical-mile-wide (370-km-wide) *exclusive economic zones.*

core area
1: In *economic geography,* as "core region," the national or world districts of concentrated economic power, wealth, innovation, and advanced technology. **2:** In *political geography,* the heartland or nucleus of a *state,* containing its most developed area, greatest wealth, densest populations, and clearest national identity.

core-frame concept
An observation that the traditional *central business district* has two distinct spatial elements: (1) a core of tall buildings on high-value land, with large daytime population and concentrated, strongly linked *tertiary* and *quaternary* activities; and (2) a surrounding less intensively developed frame with lower land values and dispersed activities (including wholesaling, light manufacturing, warehousing, and parking) that are not closely linked functionally.

core-periphery model
A model of the spatial structure of an economic system in which underdeveloped or declining peripheral areas are defined with respect to their dependence on a dominating developed *core region.*

core region
See *core area* **1.**

counter migration
(*syn*: return migration) The return of migrants to the regions from which they earlier emigrated.

country
See *state.*

creole
A *language* developed from a pidgin to become the native tongue of a society.

critical distance
The distance beyond which cost, effort, and/or means play a determining role in the willingness of people to travel.

crop rotation
The annual alteration of crops that make differential demands on or contributions to soil fertility.

crude birth rate (CBR)
See *birth rate.*

crude death rate (CDR)
See *death rate.*

crude density
(*syn*: arithmetic density) The number of people per unit area of land.

cultural assimilation
See *behavioral assimilation.*

cultural convergence
The tendency for *cultures* to become more alike as they increasingly share *technology* and organizational structures in a modern world united by improved transportation and communication.

cultural divergence
The likelihood or tendency for *cultures* to become increasingly dissimilar with the passage of time.

cultural ecology
The study of the interactions between societies and the natural *environments* they occupy.

cultural geography
A branch of *systematic geography* that focuses on culturally determined human activities, the impact of *material* and *nonmaterial* human *culture* on the environment, and the human organization of space.

cultural integration
The interconnectedness of all aspects of a *culture*; no part can be altered without creating an impact on other components of the culture.

cultural lag
The retention of established *culture traits* despite changing circumstances rendering them inappropriate.

cultural landscape
The *natural landscape* as modified by human activities and bearing the imprint of a *culture* group or society; the *built environment.*

culture
1: A society's collective beliefs, symbols, values, forms of behavior, and social organizations, together with its tools, structures, and artifacts created according to the group's conditions of life; transmitted as a heritage to succeeding generations, and undergoing adoptions, modifications, and changes in the process. **2:** A collective term for a group displaying uniform cultural characteristics.

culture complex
A related set of *culture traits* descriptive of one aspect of a society's behavior or activity. Culture complexes may be as basic as those associated with food preparation, serving, and consumption or as involved as those associated with religious beliefs or business practices.

culture hearth
A nuclear area within which an advanced and distinctive set of *culture traits,* ideas, and *technologies* develops and from which there is *diffusion* of those characteristics and the *cultural landscape* features they imply.

culture realm
A collective of *culture regions* sharing related culture systems; a major world area having sufficient distinctiveness to be perceived as set apart from other realms in terms of cultural characteristics and complexes.

culture rebound
The readoption by later generations of *culture traits* and identities associated with immigrant forebears or ancestral homelands.

culture region
A *formal* or *functional region* within which common cultural characteristics prevail. It may be based on single *culture traits,* on *culture complexes,* or on political, social, or economic integration.

culture trait
A single distinguishing feature of regular occurrence within a *culture,* such as the use of chopsticks or the observance of a particular caste system. A single element of learned behavior.

cumulative causation
See *circular and cumulative causation.*

custom
The body of traditional practices, usages, and conventions that regulate social life.

D

Daoism
See *Taoism.*

death rate
(*syn*: mortality rate) A mortality index usually calculated as the number of deaths per year per 1000 population.

deforestation
The clearing of land through total removal of forest cover.

deglomeration
The process of deconcentration; the location of industrial or other activities away from established *agglomerations* in response to growing costs of congestion, competition, and regulation.

deindustrialization
The cumulative and sustained decline in the contribution of manufacturing to a national economy.

demographic cycle
A model of the effect of economic development on population growth. A first stage involves stable numbers with both high *birth rates* and *death rates*; the second displays high birth rates, falling death rates, and population increases. Stage three shows reduction in population growth as birth rates decline to the level of death rates. The fourth, and final, stage again shows a population stable in size, with low birth and death rates. An idealized summary of the population history of industrializing Europe, its application to underdeveloped countries is questioned. See also *demographic transition.*

demographic equation
A mathematical expression that summarizes the contribution of different demographic processes to the population change of a given area during a specified time period: $P_2 = P_1 + B_{1-2} - D_{1-2} + IM_{1-2} - OM_{1-2}$, where P_2 is population at time 2, P_1 is population at beginning date; B_{1-2} is the number of births between times 1 and 2, D_{1-2} is the number of deaths during that period, IM_{1-2} is the number of in-migrants and OM_{1-2} the number of out-migrants between times 1 and 2.

demographic momentum
(*syn*: population momentum) The tendency for population growth to continue despite stringent family planning programs because of a relatively high concentration of people in the childbearing years.

demographic transition
The change in *birth rates* and *death rates* from high to low levels in a population passing through the *demographic cycle.* Since decline in death rates precedes falling birth rates, population growth tends to be rapid during the period of transition.

demography
The scientific study of population, with particular emphasis on quantitative aspects.

density
The quantity of anything (people, buildings, animals, traffic, etc.) per unit area.

dependency ratio
The number of children (usually, 0–14 years of age) and of old people (usually, 65 and over) in a population that each 100 people in the economically productive years must on average support.

desertification
Extension of desertlike landscapes as a result of overgrazing, destruction of the forests, or other human-induced changes, usually in semiarid regions.

development
The process of growth, expansion, or realization of potential; bringing regional resources into full productive use.

devolution
The transfer of certain powers from the *state* central government to separate political subdivisions within the state's territory.

dialect
A *language* variant marked by vocabulary, grammar, and pronunciation differences from other variants of the same common language. When those variations are spatial or regional, they are called *geographic dialects;* when they are indicative of socioeconomic or educational levels, they are called *social dialects.*

dialect geography
See *linguistic geography.*

dibble
Any small hand tool or stick used to make a hole for planting.

diffusion
The spread or movement of a phenomenon over space or through time. The dispersion of a *culture trait* or characteristic or new ideas and practices from an origin area (e.g., *language,* plant *domestication,* new industrial *technology*). Recognized types include *relocation, expansion, contagious, cascade,* and *hierarchical* diffusion.

diffusion barrier
Any condition that hinders the flow of information, the movement of people, or the spread of an *innovation.*

direction bias
A statement of *movement bias* observing that among all possible directions of movement or flow, one or only a very few are favored and dominant.

dispersion
In *spatial distributions,* a statement of the amount of spread of a phenomenon over area or around a central location. Dispersion in this sense represents a continuum from clustered, concentrated, or agglomerated (at one end) to dispersed or scattered (at the other).

distance bias
A statement of *movement bias* observing that short journeys or interchanges are favored over more distant ones.

distance decay
The declining intensity of any activity, process, or function with increasing distance from its point of origin.

domestication
The successful transformation of plant or animal species from a wild state to a condition of dependency upon human management, usually with distinct physical change from wild forebears.

doubling time
The time period required for any beginning total experiencing a compounding growth to double in size.

E

economic base
The manufacturing and service activities performed by the *basic sector* of a city's labor force; functions of a city performed to satisfy demands external to the city itself and, in that performance, earning income to support the urban population.

economic geography
The branch of *systematic geography* concerned with how people support themselves, with the spatial patterns of production, distribution, and consumption of goods and services, and with the areal variation of economic activities over the surface of the earth.

economic rent
The net surplus yielded by any factor of production (frequently land) after all costs, including interest and all income from possible alternative uses, have been subtracted.

ecosphere
See *biosphere.*

ecosystem
A population of organisms existing together in a small, relatively homogeneous area (pond, forest, small island), along with the energy, air, water, soil, and chemicals upon which it depends.

ecumene
That part of the earth's surface physically suitable for permanent human settlement; the permanently inhabited areas of the earth.

elongated state
A *state* whose territory is long and narrow.

environment
Surroundings; the totality of things that in any way may affect an organism, including both physical and cultural conditions; a region characterized by a certain set of physical conditions.

environmental determinism
The view that the physical *environment,* particularly *climate,* controls human action, molds human behavior, and conditions cultural development.

environmental perception
The concept that people of different *cultures* will differently observe and interpret their *environment* and make different decisions about its nature, potentialities, and use.

environmental pollution
See *pollution.*

epidemiologic transition
The reduction of periodically high mortality rates from epidemic diseases as those diseases become essentially continual within a population that develops partial immunity to them.

equator
An imaginary east-west line that encircles the globe halfway between the North and South poles.

erosion
The wearing away and removal of rock and soil particles from exposed surfaces by agents such as moving water, wind, or ice.

ethnic geography
The study of spatial distributions and interactions of *ethnic groups* and the cultural characteristics on which they are based.

ethnic group
People sharing a distinctive *culture,* frequently based on common national origin, *religion, language,* or *race.*

ethnic island
A small rural area settled by a single, distinctive *ethnic group* that placed its imprint on the landscape.

ethnic province
A large territory, urban and rural, dominated by or closely associated with a single *ethnic group.*

ethnic religion
A *religion* identified with a particular *ethnic group* and largely exclusive to it. Such a religion does not seek converts.

ethnic separatism
Desired *regional autonomy* expressed by a culturally distinctive group within a larger, politically dominant *culture.*

ethnicity
Ethnic quality; affiliation with a group whose racial, cultural, religious, or linguistic characteristics or national origins distinguish it from a larger population within which it is found.

ethnocentrism
Conviction of the evident superiority of one's own *ethnic group.*

ethnographic boundary
See *consequent boundary.*

European Economic Community (EEC)
An economic association established in 1957 by a number of countries in Western Europe to promote free trade among its members; often called the Common Market.

evapotranspiration
The return of water from the land to the *atmosphere* through evaporation from the soil surface and transpiration from plants.

exclave
A portion of a *state* that is separated from the main territory and surrounded by another country.

exclusive economic zone (EEZ)
As proposed in the *Convention on the Law of the Sea,* a zone of exploitation extending 200 nautical miles (370 km) seaward from a coastal state and in which that state has exclusive mineral and fishing rights.

expansion diffusion
The spread of ideas, behaviors, or articles through a culture area or from one *culture* to neighboring areas through contact and exchange of information; the dispersion leaves the phenomenon intact or intensified in its area of origin.

extensive agriculture
A crop or livestock system characterized by low inputs of labor per unit area of land. It may be part of either a *subsistence* or a *commercial* economy.

external economies
See *agglomeration economies.*

extractive industries
Primary activities involving the mining and quarrying of *nonrenewable* metallic and nonmetallic mineral resources.

F

fallowing
The practice of allowing plowed or cultivated land to remain (rest) uncropped or only partially cropped for one or more growing seasons.

federal state
A *state* with a two-tier system of government and a clear distinction between the powers vested in the central government and those residing in the governments of the component regional subdivisions.

fertility rate
The average number of live births per 1000 women of childbearing age.

filtering
In *urban geography,* a process whereby individuals of a lower-income group replace, in a portion of an urban area, residents who are of a higher-income group.

first effective settlement
The influence that the characteristics of an early dominant settlement group exert on the later *social* and *cultural geography* of an area.

fixed cost
An activity cost (as of investment in land, plant, and equipment) that must be met without regard to level of output; an input cost that is spatially constant.

fixed costs of transportation
See *terminal costs.*

folk culture
The body of institutions, customs, dress, artifacts, collective wisdoms and traditions of a homogeneous, isolated, largely self-sufficient, and relatively static social group.

folklore
Oral traditions of a *folk culture,* including tales, fables, legends, customary observations, and moral teachings.

folkway
A learned manner of thinking and feeling and a prescribed mode of conduct common to a traditional social group.

footloose industry
An industry or firm relatively insensitive to variations in transport costs and showing neither *market* nor *material orientation.*

form utility
A value-increasing change in the form—and therefore in the "utility"—of a raw material or commodity.

formal region
(*syn*: uniform region, homogeneous region, structural region) A *region* distinguished by a uniformity of one or more characteristics that can serve as the basis for areal generalization and of contrast with adjacent areas.

forward-thrust capital
A capital city deliberately sited in a *state's* frontier zone.

fossil fuel
(*syn*: mineral fuel) Any of the fuels derived from decayed organic material converted by earth processes; especially, coal, petroleum, and natural gas, but also including tar sands and oil shales.

fragmented state
A *state* whose territory contains isolated parts, separated and discontinuous.

frame
In *urban geography,* that part of the *central business district* characterized by such low-intensity uses as warehouses, wholesaling operations, and automobile dealerships. See also *core-frame concept.*

freight rate
The charge levied by a transporter for the loading, moving, and unloading of goods; includes *line-haul costs* and *terminal costs.*

friction of distance
A measure of the retarding or restricting effect of distance on *spatial interaction.* Generally, the greater the distance, the greater the "friction" and the less the interaction or exchange, or the greater the cost of achieving the exchange.

frontier
That portion of a country adjacent to its boundaries and fronting another political unit.

frontier zone
A belt lying between two *states* or between settled and uninhabited or sparsely settled areas.

functional dispute
(*syn*: boundary dispute) In *political geography,* a disagreement between neighboring *states* over policies to be applied to their common border; often induced by differing customs regulations, movement of nomadic groups, or illegal immigration or emigration.

functional region
(*syn*: nodal region) A *region* differentiated by what occurs within it rather than by a homogeneity of physical or cultural phenomena; an earth area recognized as an operational unit based on defined organizational criteria. The concept of unity is based on interaction and interdependence between different points within the area.

G

gatekeeper
In information flows, the individual or agency that collects information and selectively transmits a portion of it.

gathering industries
Primary activities involving the *subsistence* or *commercial* harvesting of *renewable* natural resources of land or water. Primitive gathering involves local collection of food and other materials of nature, both plant and animal; commercial gathering usually implies forestry and fishing industries.

genetic drift
A chance modification of gene composition occurring in an isolated population and preserved and accentuated through inbreeding.

gentrification
The movement into the inner portions of American cities of middle- and upper-income people who replace low-income populations, rehabilitate the structures they occupied, and change the social character of neighborhoods.

geodesic distance
See *absolute distance.*

geographic dialect
(*syn*: regional dialect) See *dialect.*

geometric boundary
(*syn*: artificial boundary) A boundary without obvious physical geographic basis; often a section of a *parallel of latitude* or a *meridian of longitude.*

geophagy
The practice of eating earthy substances, usually clays.

gerrymander
To redraw voting district boundaries in such a way as to give one political party maximum electoral advantage and to reduce that of another party, to fragment voting blocks, or to achieve other nondemocratic objectives.

ghetto
A forced or voluntarily segregated residential area housing a racial, ethnic, or religious minority.

globe grid
The set of imaginary lines of *latitude* and *longitude* that intersect at right angles to form a coordinate reference system for locating points on the surface of the earth.

gravity model
A mathematical prediction of the interaction between two bodies (places) as a function of their size and of the distance separating them. Based on Newton's law, the model states that attraction (interaction) is proportional to the product of the masses (population sizes) of two bodies (places) and inversely proportional to the square of the distance between them.

Green Revolution
A term suggesting the great increases in food production, primarily in subtropical areas, accomplished through the introduction of very high yielding grain crops, particularly wheat, maize, and rice.

greenhouse effect
Heating of the earth's surface as shortwave solar energy passes through the *atmosphere,* which is transparent to it but opaque to reradiated long-wave terrestrial energy; also, increasing the opacity of the atmosphere through the addition of increased amounts of carbon dioxide and other gases that trap heat.

grid system
See *globe grid.*

gross national product (GNP)
The total value of goods and services (with some adjustments) produced by the residents of a country during a specified period (usually a year).

groundwater
Subsurface water that accumulates in the pores and cracks of rock and soil.

guest worker
A foreign worker, usually male and frequently under contract, who migrates to secure permanent work in a host country without intention to settle permanently in that country; particularly, workers from North Africa and the eastern, southern, and southeastern parts of Europe employed in industrialized countries of Western Europe.

H

hazardous waste
Discarded solid, liquid, or gaseous material that poses a substantial threat to human health or to the *environment* when improperly disposed of or stored.

hierarchical diffusion
A variant of *cascade diffusion* in which spread of an *innovation* can proceed either upward or downward through a hierarchy.

hierarchical migration
The tendency for individuals to move from small places to larger ones. See also *step migration.*

high-level waste (HLW)
Nuclear waste with a relatively high level of radioactivity.

Hinduism
An ancient and now dominant *value system* and *religion* of India, closely identified with Indian *culture* but without central creed, single doctrine, or religious organization. Dharma (customary duty and divine law) and *caste* are uniting elements.

hinterland
The market area or region served by an *urban* center.

homeostatic plateau
(*syn*: carrying capacity) The application of the concept of homeostasis, or a relatively stable state of equilibrium, to the balance between population numbers and areal resources; the equilibrium level of population that available resources can adequately support.

horticultural farming
See *truck farming.*

host society
The established and dominant society within which immigrant groups seek accommodation.

human geography
One of the two major divisions (the other is *physical geography*) of *systematic geography*; the spatial analysis of human populations, their *cultures,* their activities and behaviors, and their relationship with and impact on the physical landscapes they occupy.

hunting-gathering
An economic and social system based primarily or exclusively on the hunting of wild animals and the gathering of food, fiber, and other materials from uncultivated plants.

hydrologic cycle
The natural system by which water is continuously circulated through the *biosphere* by evaporation, condensation, and precipitation.

hydrosphere
All water at or near the earth's surface that is not chemically bound in rocks, including the oceans, surface waters, *groundwater,* and water held in the *atmosphere.*

I

icebox effect
The tendency for certain kinds of air pollutants to lower temperatures on earth by reflecting incoming sunlight back into space, and thus preventing it from reaching (and heating) the earth.

iconography
In *political geography,* the study of symbols that unite a country.

ideological subsystem
The complex of ideas, beliefs, knowledge, and means of their communication that characterizes a *culture.*

independent invention
(*syn*: parallel invention) *Innovations* developed in two or more unconnected locations by individuals or groups acting independently. See also *multilinear evolution.*

Industrial Revolution
The term applied to the rapid economic and social changes in agriculture and manufacturing that followed the introduction of the factory system to the textile industry of England in the last quarter of the 18th century.

infrastructure
The basic structure of services, installations, and facilities needed to support industrial, agricultural, and other economic development; included are transport and communications, along with water, power, and other public utilities.

innovation
The introduction of new ideas, practices, or objects; usually, an alteration of *custom* or *culture* that originates within the social group itself.

intensive agriculture
Any agricultural system involving the application of large amounts of capital and/or labor per unit of cultivated land; this may be part of either a *subsistence* or a *commercial economy.*

interaction model
See *gravity model.*

international boundary
The outer limit of a *state's* claim to land or water surface, projected downward to the center of the earth and, less certainly, upward to the height the state can effectively control.

International Date Line
By international agreement, the designated line where each new day begins, generally following the 180th *meridian.* The line compensates for accumulated 1-hour time changes for each 15 degrees of *longitude* by adding (from east to west) or subtracting (from west to east) 24 hours for travelers crossing the line.

interrupting barrier
A condition that delays the rate of *diffusion* of an *innovation* or that deflects its path.

intervening opportunity
The concept that closer opportunities will materially reduce the attractiveness of interaction with more distant—even slightly better—alternatives; a closer alternative source of supply between a demand point and the original source of supply.

in-transit privilege
The application of a single-haul *freight rate* from origin to destination even though the shipment is halted for processing en route, after which the journey is completed.

Islam
A *monotheistic, universalizing religion* that includes belief in Allah as the sole deity and in Mohammed as his prophet completing the work of earlier prophets of *Judaism* and *Christianity.*

isochrone
A line connecting points equidistant in travel time from a common origin.

isogloss
A mapped boundary line marking the limits of a particular linguistic feature.

isoline
A map line connecting points of equal value.

isotropic plain
A hypothetical portion of the earth's surface assumed to be an unbounded, uniformly flat plain with uniform and unvarying distribution of population, purchasing power, transport costs, accessibility, and the like.

J

J-curve
A curve shaped like the letter J, depicting exponential or geometric growth (1, 2, 4, 8, 16. . .).

Judaism
A *monotheistic, ethnic religion* first developed among the Hebrew people of the ancient Near East; its determining conditions include descent from Israel (Jacob), the Torah (law and scripture), and tradition.

K

krill
A form of marine *plankton* composed of crustaceans and larvae.

L

landlocked
Describing a *state* that lacks a seacoast.

land race
A genetically diverse, naturally adapted, native food plant.

language
The system of words, their pronunciation, and methods of combination used and mutually understood by a community of individuals.

language family
A group of *languages* thought to have descended from a single, common ancestral tongue.

latitude
Angular distance north or south of the *equator,* measured in degrees, minutes, and seconds. Grid lines marking latitudes are called *parallels.* The equator is 0°, the North Pole is 90° N, the South Pole is 90° S. Low latitudes are considered to fall within the tropics (23° 30′ N and 23° 30′ S); midlatitudes extend from the tropics to the Arctic and Antarctic circles (66° 30′ N and S); high latitudes occur from those circles to the North and South poles.

law of peripheral neglect
The observation that a government's awareness of or concern with regional problems decreases with the square of the distance of an outlying region from the capital city.

Law of the Sea Convention
See *Convention on the Law of the Sea.*

leachate
The contaminated liquid discharged from a *sanitary landfill* to either surface or subsurface land or water.

least-cost theory
(*syn*: Weberian analysis) The view that the optimum location of a manufacturing establishment is at the place where the costs of transport and labor and the advantages of *agglomeration* or *deglomeration* are most favorable.

limiting factor principle
The distribution of an organism or the structure of an *ecosystem* can be explained by the control exerted by the single factor (such as temperature, light, water) that is most deficient, that is, that falls below the levels required.

line-haul costs
(*syn*: over-the-road costs; variable costs of transportation) The costs involved in the actual physical movement of goods (or passengers); costs of haulage (including equipment and routeway costs) excluding *terminal costs.*

lingua franca
Any of various auxiliary *languages* used as common tongues among people of an area where several languages are spoken; literally, "Frankish language."

linguistic geography
(*syn*: dialect geography; dialectology) The study of local variations within a speech area by mapping word choices, pronunciations, or grammatical constructions.

link
In *network* theory, a transportation or communication connection or route within a network.

lithosphere
The earth's solid crust.

locational interdependence
The circumstance under which the locational decision of a particular firm is influenced by the locations chosen by competitors.

locational triangle
A simple graphic model used in *Weberian analysis* to illustrate the derivation of the least-transport-cost location of an industrial establishment.

longitude
Angular distance of a location in degrees, minutes, and seconds measured east or west of a designated *prime meridian* given the value of 0°. By general agreement, the *globe grid* prime meridian passes through the old observatory of Greenwich, England. Distances are measured from 0° to 180° both east and west, with 180° E and W being the same line. For much of its extent the 180° meridian also serves as the *International Date Line.* Because of the period of the earth's axial rotation, 15 degrees of longitude are equivalent to a difference of 1 hour in local time.

long lot
A farm or other property consisting of a long, narrow strip of land extending back from a river or road.

low-level waste (LLW)
Nuclear waste with relatively moderate levels of radioactivity.

malnutrition
Food intake insufficient in quantity or deficient in quality to sustain life at optimal conditions of health.

Malthus
Thomas R. Malthus (1766–1843). English economist, demographer, and cleric who suggested that unless self-control, war, or natural disaster checks population, it will inevitably increase faster than will the food supplies needed to sustain it. This view is known as Malthusianism. See also *neo-Malthusianism.*

map projection
A method of transferring the *globe grid* system from the earth's curved surface to the flat surface of a map. Projection automatically incurs error, but an attempt is usually made to preserve one or more (though never all) of the characteristics of the spherical surface: equal area, correct distance, true direction, proper shape.

map scale
See *scale.*

marginal cost
The additional cost of producing each successive unit of output.

mariculture
The production and harvesting of fish and shellfish in fenced confinement areas along coasts and in estuaries.

market equilibrium
The point of intersection of demand and supply curves of a given commodity; at equilibrium, the market is cleared of the commodity.

market gardening
See *truck farming.*

market orientation
The tendency of an economic activity to locate close to its market; a reflection of large and variable distribution costs.

material culture
The tangible, physical items produced and used by members of a specific culture group and reflective of their traditions, life-styles, and technologies.

material orientation
The tendency of an economic activity to locate near or at its source of raw material; this is experienced when material costs are highly variable spatially and/or represent a significant share of total costs.

mathematical location
See *absolute location.*

maximum sustainable yield
The maximum rate at which a *renewable resource* can be exploited without impairing its ability to be renewed or replenished.

Mediterranean agriculture
An agricultural system based on the mild, moist winters, hot, sunny summers, and rough terrain of the Mediterranean basin. It involves cereals as winter crops, summer tree and vine crops (olives, figs, dates, citrus and other tree fruits, and grapes), and animals (sheep and goats).

megalopolis
1: A large, sprawled *urban* complex with contained open, nonurban land, created through the spread and joining of separate *metropolitan areas*; **2:** When capitalized, the name applied to the continuous functionally urban area of the coastal northeastern United States from Maine to Virginia.

mental map
(*syn*: cognitive map) The maplike image of the world, country, region, city, or neighborhood a person carries in the mind. The representation is therefore subjective; it includes knowledge of actual locations and spatial relationships and is colored by personal perceptions and preferences related to place.

mentifacts
The central, enduring elements of a *culture* that express its values and beliefs, including *language, religion, folklore,* artistic traditions, and the like. Elements in the *ideological subsystem* of culture.

meridian
A north-south line of *longitude*; on the *globe grid,* all meridians are of equal length and converge at the poles.

Mesolithic
Middle Stone Age. The *culture* stage of the early postglacial period, during which the earliest phases of *domestication* of animals and plants occurred, refined and specialized tools were developed, pottery was produced, and semipermanent settlements were established as climate change reduced the game-animal herds earlier followed for food.

metes-and-bounds survey
A system of property description using natural features (streams, rocks, trees, etc.) to trace and define the boundaries of individual parcels.

Metro
See *unified government.*

metropolitan area
In the United States, a large functionally integrated settlement area comprising one or more whole county units and usually containing several *urbanized areas* discontinuously built up but operating as a coherent economic whole.

microdistrict
The basic neighborhood planning unit characteristic of new urban residential construction in the planned "socialist city."

microstate
(*syn*: ministate) An imprecise term for a *state* or territory small in both population and area. An informal definition accepted by the United Nations suggests a maximum of 1 million population combined with a territory of less than 700 km^2 (270 square miles).

migration
The permanent (or relatively permanent) relocation of an individual or group to a new, usually distant, place of residence and employment.

migration field
The area from which a given city or place draws the majority of its in-migrants.

mineral
A natural inorganic substance that has a definite chemical composition and characteristic crystal structure, hardness, and density.

mineral fuel
See *fossil fuel.*

ministate
See *microstate.*

model
An idealized representation, abstraction, or simulation of reality. It is designed to simplify real-world complexity and eliminate extraneous phenomena in order to isolate for detailed study causal factors and interrelationships of *spatial systems.*

monolingualism
A society or country's use of only one *language* of communication for all purposes.

monotheism
The belief that there is but a single God.

mortality rate
See *death rate.*

movement bias
Any aggregate control on or regularity of movement of people, commodities, or communication. Included are *distance bias, direction bias,* and *network bias.*

multilinear evolution
A concept of independent but parallel cultural development advanced by anthropologist Julian Steward (1902–1972) to explain cultural similarities between widely separated peoples existing in similar environments but who could not have benefited from shared experiences, borrowed ideas, or diffused technologies. See *independent invention.*

multilingualism
The common use of two or more *languages* in a society or country.

multiple-nuclei model
The postulate that large cities develop by peripheral spread not from one *central business district* but from several nodes of growth, each of specialized use. The separately expanding use districts eventually coalesce at their margins.

multiplier effect
The direct, indirect, and induced consequences of change in an activity. **1:** In industrial *agglomerations,* the cumulative processes by which a given change (such as a new plant opening) sets in motion a sequence of further industrial employment and infrastructure growth. **2:** In *urban geography,* the expected addition of nonbasic workers and dependents to a city's total employment and population that accompanies new basic employment.

N

nation
A culturally distinctive group of people occupying a specific territory and bound together by a sense of unity arising from shared ethnicity, beliefs, and customs.

nationalism
A sense of unity binding the people of a *state* together; devotion to the interests of a particular country or *nation*; an identification with the state and an acceptance of national goals.

nation-state
A *state* whose territory is identical to the homeland occupied by a particular *ethnic group* or *nation.*

natural boundary
(*syn*: physical boundary) A *boundary* line based on recognizable physiographic features, such as mountains or rivers.

natural hazard
A process or event in the physical environment that has consequences harmful to humans.

natural increase
The growth of a population through an excess of births over deaths, excluding the effects of immigration and emigration.

natural landscape
The physical *environment* unaffected by human activities. The duration and near totality of human occupation of the earth's surface assure that little or no "natural landscape" so defined remains intact. Opposed to *cultural landscape.*

natural resource
A physically occurring item that a population perceives to be necessary and useful to its maintenance and well-being.

neocolonialism
A disparaging reference to economic and political policies by which major developed countries are seen to retain or extend influence over the economies of less developed countries and peoples.

Neolithic
New Stone Age. The *culture* (succeeding that of the *Mesolithic*) of the middle postglacial period, during which polished stone tools were perfected, the economy was solely or largely based on cultivation of crops and *domestication* of animals, and the arts of spinning, weaving, smelting, and metalworking were developed. More formalized societies and *culture complexes* emerged as cities developed and trade routes were established.

neo-Malthusianism
The advocacy of population control programs to preserve and improve general national prosperity and well-being.

net migration
The difference between in-migration and out-migration of an area.

network
The areal pattern of sets of places and the routes (*links*) connecting them along which movement can take place.

network bias
The view that the pattern of *links* in a *network* will affect the likelihood of flows between specific nodes.

nodal region
See *functional region.*

node
In *network* theory, an origin, destination, or intersection in a communication network.

nomadic herding
Migratory but controlled movement of livestock solely dependent on natural forage.

nonbasic sector
(*syn*: service sector) Those economic activities of an urban unit that supply the resident population with goods and services and that have no "export" implication.

nonecumene
(*syn*: anecumene) The portion of the earth's surface that is uninhabited or only temporarily or intermittently inhabited. See also *ecumene.*

nonmaterial culture
The oral traditions, songs, and stories of a *culture* group along with its beliefs and customary behaviors.

nonrenewable resource
A *natural resource* that is not replenished or replaced by natural processes or is used at a rate that exceeds its replacement rate.

North
The general term applied in the *Brandt report* to the developed countries of the Northern Hemisphere plus Australia and New Zealand.

O

official language
A governmentally designated *language* of instruction, of government, of the courts, and of other official public and private communication.

outsourcing
Producing abroad parts or products for domestic use or sale.

overpopulation
A value judgment that the resources of an area are insufficient to sustain adequately its present population numbers.

over-the-road costs
See *line-haul costs.*

ozone
A gas molecule consisting of three atoms of oxygen (O_3) formed when diatomic oxygen (O_2) is exposed to *ultraviolet radiation.* In the upper *atmosphere* it forms a normally continuous, thin layer that blocks ultraviolet light; in the lower atmosphere it constitutes a damaging component of *photochemical smog.*

P

Paleolithic
Old Stone Age. An early stage of human *culture* largely coinciding with the *Pleistocene* glacial period and characterized by *hunting-gathering* economies and the use of fire and simple stone tools, especially those made from flint.

parallel
An east-west line of *latitude* indicating distance north or south of the *equator.*

parallel invention
See *independent invention.*

pattern
The design or arrangement of phenomena in earth space.

peak value intersection
The most accessible and costly parcel of land in the *central business district* and therefore in the entire *urbanized area.*

perception
The acquisition of information about a place or thing through sensory means; the subjective organization and interpretation of acquired information in light of cultural attitudes and individual preferences or experiences. See also *environmental perception.*

perceptual region
A *region* perceived to exist by its inhabitants or the general populace. Also known as a *vernacular region* or popular region, it has reality as an element of *popular culture* or *folk culture* represented in the *mental maps* of average people.

perforated state
A *state* whose territory is interrupted ("perforated") by a separate, independent state totally contained within its borders.

periodic market
A market operating at a particular location (village, city neighborhood) on one or more fixed days per week or month.

periphery/peripheral
The outer regions or boundaries of an area. See also *core-periphery model.*

permeable barrier
An obstacle raised by a culture group or one culture group's reluctance to accept some, but not all, *innovations* diffused from a related but different *culture.* Acceptance or rejection may be conditioned by religious, political, ethnic, or similar considerations of suitablity or compatiblity.

personal communication field
An area defined by the distribution of an individual's short-range informal communications. The size and shape of the field are defined by work, recreation, school, and other regular contacts and are affected by age, sex, employment, and other personal characteristics.

personal space
An invisible, usually irregular area around a person into which he or she does not willingly admit others. The sense (and extent) of personal space is a situational and cultural variable.

photochemical smog
A form of polluted air produced by the interaction of hydrocarbons and oxides of nitrogen in the presence of sunlight.

physical boundary
See *natural boundary.*

physical geography
One of two major divisions (the other is *human geography*) of *systematic geography*; the study of the structures, processes, distributions, and change through time of the natural phenomena of the earth's surface that are significant to human life.

physical landscape
The *natural landscape* plus visible elements of *material culture.*

physiological density
The number of people per unit area of cultivable land.

pidgin
An auxiliary *language* derived, with reduced vocabulary and simplified structure, from other languages. Not a native tongue, it is used for limited communication between people with different languages.

place perception
See *perception.*

place utility
1: In human movement and *migration* studies, a measure of an individual's perceived satisfaction or approval of a place in terms of its social, economic, or environmental attributes. **2:** In *economic geography,* the value achieved by goods transported to places where they are needed but absent.

plankton
Microscopic freely floating plant and animal organisms of lakes and oceans.

planned economy
A system of production of goods and services, usually consumed or distributed by a governmental agency, in quantities, at prices, and in locations determined by governmental program.

plantation
A large agricultural holding, frequently foreign owned, devoted to the production of a single export crop.

Pleistocene
The geological epoch dating from 2 million to 11,000 years ago, during which four stages of continental glaciation occurred.

political geography
A branch of *human geography* concerned with the spatial analysis of political phenomena.

pollution
The introduction into the biosphere of materials that because of their quantity, chemical nature, or temperature have a negative impact on the *ecosystem* or that cannot be readily disposed of by natural recycling processes.

polytheism
Belief in or worship of many gods.

popular culture
The constantly changing mix of material and nonmaterial elements available through mass production and the mass media to an urbanized, heterogeneous, nontraditional society.

popular region
See *vernacular region.*

population density
A measurement of the number of persons per unit area of land within predetermined limits, usually political or census boundaries. See also *physiological density.*

population geography
A branch of *human geography* concerned with spatial variations in distribution, composition, growth, and movements of population and the relationship of those concerns with the geographic character of areas.

population momentum
See *demographic momentum.*

population projection
A statement of a population's future size, age, and sex composition based on the application of stated assumptions to current data.

population pyramid
A bar graph in pyramid form showing the age and sex composition of a population, usually a national one.

port of entry
Any place, not necessarily a port, where aliens may enter a country or where incoming foreign goods pass through customs.

positional dispute
(*syn*: boundary dispute) In *political geography,* disagreement about the actual location of a *boundary.*

possibilism
The philosophical viewpoint that the physical *environment* offers human beings a set of opportunities from which (within limits) people may choose according to their cultural needs and technological awareness. The emphasis is on a freedom of choice and action not allowed under *environmental determinism.*

postindustrial
A stage of economic development in which service activities become relatively more important than goods production, professional and technical employment supersedes employment in agriculture and manufacturing, and level of living is defined by the quality of services and amenities rather than by the quantity of goods available.

potential model
A measurement of the total interaction opportunities available under *gravity model* assumptions to a center in a multicenter system.

precipitation
All moisture—solid and liquid—that falls to the earth's surface from the *atmosphere.*

predevelopment annexation
The inclusion within the *central city* of nonurban peripheral areas for the purpose of securing to the city itself the benefits of their eventual development.

primary activities
Those parts of the economy involved in making *natural resources* available for use or further processing; included are mining, *agriculture,* forestry, fishing and hunting, and grazing.

primate city
A country's leading city, disproportionately larger and functionally more complex than any other; a city dominating an urban hierarchy composed of a base of small towns and an absence of intermediate-sized cities.

prime meridian
An imaginary line passing through the Royal Observatory at Greenwich, England, serving by agreement as the 0° line of *longitude.*

principle of first choice
The generalization that when a suitable area for production of a commodity in high demand is limited, that commodity will have first claim on any area capable of its production.

private plot
In Soviet and Eastern bloc economies, a small garden parcel allotted to collective farmers and urban workers.

projection
See *map projection.*

prorupt state
A *state* of basically *compact* form but with one or more narrow extensions of territory.

protolanguage
An assumed, reconstructed, or recorded ancestral *language.*

pull factors
Characteristics of a locale that act as attractive forces, drawing migrants from other regions.

push factors
Unfavorable characteristics of a locale that contribute to the dissatisfaction of its residents and impel their emigration.

quaternary activities
Those parts of the economy concerned with research, with the gathering and dissemination of information, and with administration, including administration of the other economic activity levels; often considered a specialized subdivision of *tertiary activities.*

quinary activities
A sometimes separately recognized subsection of *tertiary activity* management functions involving the highest level decision making in all types of large organizations. Also deemed the most advanced form of the *quarternary* subsector.

race
A subset of human population whose members share certain distinctive, inherited biological characteristics.

rank-size rule
An observed regularity in the city-size distribution of some countries. In a rank-size hierarchy, the population of any given town will be inversely proportional to its rank in the hierarchy; that is, the nth-ranked city will be $1/n$ the size of the largest city.

rate
The frequency of an event's occurrence during a specified time period.

reapportionment
The process and outcome of a reallocation of electoral seats to defined territories, such as congressional seats to states of the United States.

recycling
The reuse of disposed materials after they have passed through some form of treatment (e.g., melting down glass bottles to produce new bottles).

redistricting
The drawing of new electoral district boundary lines in response to changing patterns of population or changing legal requirements.

region
Any earth area with distinctive and unifying physical or cultural characteristics that set it off and make it substantially different from surrounding areas. A region may be defined on the basis of its homogeneity or its functional integration as a single organizational unit. Regions and their boundaries are devices of areal generalization, intellectual concepts rather than visible landscape entities.

regional autonomy
A measure of self-governance afforded a subdivision of a *state.*

regional concept
The view that physical and cultural phenomena on the surface of the earth are rationally arranged by complex, diverse, but comprehensible interrelated spatial processes.

regional dialect
(*syn*: geographic dialect) See *dialect.*

regional geography
The study of geographic *regions*; the study of areal differentiation.

regionalism
In *political geography,* group—frequently ethnic group—identification with a particular region of a *state* rather than with the state as a whole.

relational direction
See *relative direction.*

relative direction
(*syn*: relational direction) A culturally based locational reference, as the Far West, the Old South, or the Middle East.

relative distance
A transformation of *absolute distance* into such relative measures as time or monetary costs. Such measures yield different explanations of human spatial behavior than do linear distances alone. Distances between places are constant by absolute terms, but relative distances may vary with improvements in transportation or communication technology or with different psychological perceptions of space.

relative location
The position of a place or activity in relation to other places or activities. Relative location implies spatial relationships and usually suggests the relative advantages or disadvantages of a location with respect to all competing locations.

relict boundary
A former *boundary* line that is still discernable and marked by some *cultural landscape* feature.

religion
A personal or institutionalized system of worship and of faith in the sacred and divine.

relocation diffusion
The transfer of ideas, behaviors, or articles from one place to another through the *migration* of those possessing the feature transported; also, spatial relocation in which a phenomenon leaves an area of origin as it is transported to a new location.

renewable resource
A *natural resource* that is potentially inexhaustible either because it is constantly (as solar radiation) or periodically (as *biomass*) replenished as long as its use does not exceed its *maximum sustainable yield.*

resource
See *natural resource.*

resource dispute
In *political geography,* disagreement over the control or use of shared resources, such as boundary rivers or jointly claimed fishing grounds.

return migration
See *counter migration.*

rotation
See *crop rotation.*

roundwood
Timber as it is harvested, before squaring, sawing, or pulping.

S-curve
The horizontal bending, or leveling, of an exponential *J-curve.*

Sahel
The semiarid zone between the Sahara and the grassland areas to the south in West Africa; a district of recurring drought, famine, and environmental degradation and *desertification.*

salinization
The process by which *soil* becomes saturated with salt, rendering the land unsuitable for *agriculture.* This occurs when land that has poor drainage is improperly irrigated.

sanitary landfill
Disposal of solid wastes by spreading them in layers covered with enough soil to control odors, rodents, and flies; sited to minimize water pollution from runoff and *leachate.*

satisficing location
A less than ideal best location, but one providing an acceptable level of utility or satisfaction.

scale
1: In cartography, the ratio between the size of area on a map and the actual size of that same area on the earth's surface. **2:** In more general terms, the size of the area studied, from local to global.

secondary activities
Those parts of the economy involved in the processing of raw materials derived from *primary activities* and in altering or combining materials to produce commodities of enhanced utility and value; included are manufacturing, construction, and power generation.

sector model
A description of urban land uses as wedge-shaped sectors radiating outward from the *central business district* along transportation corridors. The radial access routes attract particular uses to certain sectors, with high-status residential uses occupying the most desirable wedges.

secularism
A rejection of or indifference to *religion* and religious practice.

segregation
A measure of the degree to which members of a minority group are not uniformly distributed among the total population.

separatism
See *ethnic separatism.*

service sector
See *nonbasic sector.*

shamanism
A form of *tribal religion* based on belief in a hidden world of gods, ancestral spirits, and demons responsive only to a shaman, or interceding priest.

shifting cultivation
(*syn*: slash-and-burn agriculture, swidden agriculture) Crop production on tropical forest clearings kept in cultivation until their quickly declining fertility is lost. Cleared plots are then abandoned and new sites are prepared.

Shinto
The *polytheistic, ethnic religion* of Japan that includes reverence of deities of natural forces and veneration of the emperor as descendent of the sun-goddess.

site
The *absolute location* of a place or activity described by local relief, landform, and other physical (or sometimes cultural) characteristics.

situation
The *relative location* of a place or activity in relation to the physical and cultural characteristics of the larger regional or *spatial system* of which it is a part. Situation implies spatial interconnection and interdependence.

slash-and-burn cultivation
See *shifting cultivation.*

social area
An area identified by homogeneity of the social indices (age group, socioeconomic status, ethnicity) of its population.

social dialect
See *dialect.*

social distance
A measure of the perceived degree of social separation between individuals, ethnic groups, neighborhoods, or other groupings; the voluntary or enforced *segregation* of two or more distinct social groups for most activities.

social geography
The branch of *cultural geography* that studies *social areas* and the social use of space, especially urban space; the study of the spatial distribution of social groups and the processes underlying that distribution.

sociofacts
The institutions and links between individuals and groups that unite a *culture,* including family structure and political, educational, and religious institutions. Components of the *sociological subsystem* of culture.

sociolinguistics
The study of the relationship between *language* and society.

sociological subsystem
The totality of expected and accepted patterns of interpersonal relations common to a *culture* or subculture.

soil
The complex mixture of loose material including minerals, organic and inorganic compounds, living organisms, air, and water found at the earth's surface and capable of supporting plant life.

South
The general term applied in the *Brandt report* to the poor, developing countries of the world, generally (but not totally) located in the Southern Hemisphere.

space-time convergence
An expression of the extent to which improvements in transportation and communication have reduced distance barriers.

space-time prism
A diagram of the volume of space and the length of time within which our activities are confined by constraints of our bodily needs (eating, resting) and the means of mobility at our command.

spatial
Of or pertaining to space on the earth's surface. Often a synonym for *geographical* and used as an adjective to describe specific geographic concepts or processes, such as *spatial interaction* or *diffusion.*

spatial diffusion
See *diffusion.*

spatial distribution
The arrangement of things on the earth's surface; the descriptive elements of spatial distribution are *density, dispersion,* and *pattern.*

spatial interaction
The movement (e.g., of people, goods, information) between different places; an indication of interdependence between different geographic locations or areas.

spatial margin of profitability
The set of points delimiting the area within which a firm's profitable operation is possible.

spatial system
The arrangement and integrated operation of phenomena produced by or responding to spatial processes on the earth's surface.

spatially fixed cost
An input cost in manufacturing that remains constant wherever production is located.

spatially variable cost
An input cost in manufacturing that changes significantly from place to place in its amount and its relative share of total costs.

speech community
A group of people having common characteristic patterns of vocabulary, word arrangement, and pronunciation.

standard language
A *language* substantially uniform with respect to spelling, grammar, pronunciation, and vocabulary and representing the approved community norm of the tongue.

state
(*syn*: country) An independent political unit occupying a defined, permanently populated territory and having full sovereign control over its internal and foreign affairs.

state farm
In the Soviet Union (and other planned economies), a government agricultural enterprise operated with paid employees.

step (stepwise) migration
A *migration* in which an eventual long-distance relocation is undertaken in stages as, for example, from farm to village to small town to city. See also *hierarchical migration.*

structural assimilation
The distribution of immigrant ethnics among the groups and social strata of a *host society,* but without their full *behavioral assimilation* into it.

subsequent boundary
A *boundary* line that is established after the area in question has been settled and that considers the cultural characteristics of the bounded area.

subsistence agriculture
Any of several farm economies in which most crops are grown for food nearly exclusively for local or family consumption.

subsistence economy
An economic system of relatively simple technology in which people produce most or all of the goods to satisfy their own and their family's needs; little or no exchange occurs outside of the immediate or extended family.

substitution principle
In industry, the tendency to substitute one factor of production for another in order to achieve optimum plant location.

suburb
A functionally specialized segment of a large *urban* complex located outside the boundaries of the *central city*; usually, a relatively homogeneous residential community, separately incorporated and administered.

superimposed boundary
A *boundary* line placed over and ignoring an existing cultural pattern.

swidden
See *shifting cultivation.*

syncretism
The development of a new form of *culture trait* by the fusion of two or more distinct parental elements.

systematic geography
A division of geography that selects a particular aspect of the physical or cultural environment for detailed study of its areal differentiation and interrelationships. Branches of systematic geography are labeled according to the topic studied (e.g., *recreational geography*) or the related science with which the branch is associated (e.g., *economic geography*).

systems analysis
An approach to the study of large systems through (1) segregation of the entire system into its component parts, (2) investigation of the interactions between system elements, and (3) study of inputs, outputs, flows, interactions, and boundaries within the system.

T

Taoism
(*syn*: Daoism) A Chinese *value system* and *ethnic religion* emphasizing conformity to Tao (Way), the creative reality ordering the universe.

tapering principle
A *distance decay* observation of the diminution or tapering of costs of transportation with increasing distance from the point of origin of the shipment because of the averaging of *fixed costs* over a greater number of miles of travel.

technological subsystem
The complex of material objects together with the techniques of their use by means of which people carry out their productive activities.

technology
The integrated system of knowledge, skills, tools, and methods developed within or used by a *culture* to successfully carry out purposeful and productive tasks.

technology gap
The contrast between the *technology* available in developed *core* regions and that present in *peripheral* areas of *underdevelopment.*

technology transfer
The *diffusion* to or acquisition by one *culture* or *region* of the *technology* possessed by another, usually more developed, society.

terminal costs
(*syn*: fixed costs of transportation) The costs incurred and charged for loading and unloading freight at origin and destination points; costs charged each shipment for terminal facility use and unrelated to distance of movement or *line-haul costs.*

terracing
The practice of planting crops on steep slopes that have been converted into a series of steplike level plots (terraces) at right angles to the natural slope of the land.

territorial dispute
(*syn*: boundary dispute) In *political geography,* disagreement between *states* over the control of surface area.

territorial production complex
In Soviet economic planning, a design for large regional industrial, mining, and agricultural development leading to regional self-sufficiency, diversification, and the creation of specialized production for a larger national market.

territoriality
An individual or group attempt to identify and establish control over a clearly defined territory considered partially or wholly an exclusive domain; the behavior associated with the defense of the home territory.

tertiary activities
Those parts of the economy that fulfill the exchange function, that provide market availability of commodities, and that bring together consumers and providers of services; included are wholesale and retail trade, associated transportational and governmental services, and personal and professional services of all kinds.

thematic map
A map depicting a specific *spatial distribution* or statistical variation of abstract objects (e.g., unemployment) in space.

Third World
Originally (1950s), designating countries uncommitted to either the "First World" Western capitalist bloc or the Eastern "Second World" Communist bloc; subsequently, a term applied to countries considered not yet fully developed or in a state of *underdevelopment* in economic and social terms.

threshold
In *economic geography* and *central place theory,* the minimum market needed to support the supply of a product or service.

tipping point
The degree of neighborhood racial or ethnic mixing that induces the former majority group to move out rapidly.

toponym
A place name.

toponymy
The place names of a region or, especially, the study of place names.

total fertility rate
The average number of children that would be born to each woman if during her childbearing years she bore children at the current year's rate for women that age.

toxic waste
Discarded chemical substances that can cause serious illness or death.

traditional religion
See *tribal religion.*

tragedy of the commons
The observation that in the absence of collective control over the use of a resource available to all, it is to the advantage of all users to maximize their separate shares even though their collective pressures may diminish total yield or destroy the resource altogether.

transferability
Acceptable costs of a spatial exchange; the cost of moving a commodity relative to the ability of the commodity to bear that cost.

tribal religion
(*syn*: traditional religion) An *ethnic religion* specific to a small, localized, preindustrial culture group.

tropical rain forest
Tree cover composed of tall, high-crowned evergreen deciduous species, associated with the continuously wet tropical lowlands.

truck farming
(*syn*: horticultural farming, market gardening) The intensive production of fruits and vegetables for market rather than for processing or canning.

U

ubiquitous industry
A *market-oriented* industry whose establishments are distributed in direct proportion to the distribution of population.

ultraviolet (UV) radiation
Electromagnetic radiation from the sun with wavelengths shorter than the violet end of visible light and longer than X rays.

underdevelopment
A level of economic and social achievement below what could be reached—given the natural and human resources of an area—were necessary capital and technology available.

underpopulation
A value statement reflecting the view that an area has too few people in relation to its resources and population-supporting capacity.

unified government
(*syn*: Unigov, Metro) Any of several devices federating or consolidating governments within a *metropolitan* region.

uniform region
See *formal region.*

Unigov
See *unified government.*

unitary state
A *state* in which the central government dictates the degree of local or *regional autonomy* and the nature of local governmental units; a country with few cultural conflicts and a strong sense of national identity.

universalizing religion
A *religion* that claims global truth and applicability and seeks the conversion of all humankind.

urban
Characteristic of, belonging to, or related to a city or town; the opposite of rural. An agglomerated settlement whose inhabitants are primarily engaged in nonagricultural occupations.

urban geography
The geographical study of cities; the branch of *human geography* concerned with the spatial aspects of (1) the locations, functional structures, size hierarchies, and intercity relationships of national or regional systems of cities, and (2) the *site,* evolution, *economic base,* internal land use, and social geographic patterns of individual cities.

urban hierarchy
A ranking of cities based on their size and functional complexity.

urbanization
The transformation of a population from rural to *urban* status; the process of city formation and expansion.

urbanized area
A continuously built up *urban* landscape defined by building and population densities with no reference to the political boundaries of the city; it may contain a *central city* and many contiguous towns, *suburbs,* and unincorporated areas.

V

value system
Mentifacts of the *ideological subsystem* of a *culture* summarizing its common beliefs, understandings, expectations, and controls.

variable cost
A cost of enterprise operation that varies either by output level or by location of the activity.

variable costs of transportation
See *line-haul costs.*

vernacular
1: The nonstandard indigenous *language* or *dialect* of a locality. **2:** Of or related to indigenous arts and architecture, such as a *vernacular house.* **3:** Of or related to the perceptions and understandings of the general population, such as a *vernacular region.*

vernacular house
An indigenous style of building constructed of native materials to traditional plan, without formal drawings.

vernacular region
A region perceived and defined by its inhabitants, usually with a popularly given or accepted nickname.

von Thünen rings
The concentric zonal pattern of agricultural land use around a single market center proposed in the model developed by Johann H. von Thünen (1783–1850).

W

wattle and daub
A building technique featuring walls of interwoven twigs, branches, or poles (wattles) plastered (daubed) with clay and mud.

Weberian analysis
See *least-cost theory.*

Z

zero population growth (ZPG)
A term suggesting a population in equilibrium, fully stable in numbers with births (plus immigration) equaling deaths (plus emigration).

zoning
Designating by ordinance areas in a municipality for particular types of land use.

CREDITS

Photo Credits

Part Openers and Table of Contents

Part 1, p. iii: © J. Robert Stottlemeyer/Biological Photo Service; **Part 2,** p. iv: © Robert Frerck/Odyssey Productions; **Part 3,** p. v: © B. Templeman/Valan Photos; **Part 4,** p. vi: © Cameramann International

Chapter 1

Opener: © Robert Frerck/Odyssey Productions; **1.1**: NOAA; **1.5**: © Walter Frerck/Odyssey Productions; **1.6**: NASA; **1.7**: U.S. Geological Survey; **1.9**: Chicago Area Transportation Study, "Final Report," vol. I, p. 44, 1959

Chapter 2

Opener: © Maptec International LTD/Science Photo Library/Photo Researchers, Inc.; **p. 36**: © Elizabeth Hovinen; **2.27**: NASA; **2.28**: Exxon Corporation and American Petroleum Institute; **2.30**: Courtesy of Intergraph Corporation

Chapter 3

Opener: © Robert Frerck/Odyssey Productions; **p. 52**: NASA; **p. 74**: L. A. Yehle/USGS; **3.10b**: © B. F. Molnia/Terraphotographics/BPS; **3.13b**: © N. Myers/Bruce Coleman, Inc.; **3.14a**: © Kevin Shafer/Tom Stack and Associates; **3.15a**: Wide World Photos; **3.15b**: M. Celebi/USGS; **3.17a**: Dr. Mullineux/USGS; **3.17b**: Austin Post/USGS; **3.18a**: © B. F. Molnia/Terraphotographics/BPS; **3.20a**: © W. D. Kleck/Terraphotographics/BPS; **3.20b**: © John Cunningham/Visuals Unlimited; **3.21**: © William E. Ferguson; **3.22**: NASA; **3.23a**: © John Cunningham/Visuals Unlimited; **3.23b**: © B. F. Molnia/Terraphotographics/BPS; **3.24**: © W. D. Kleck/Terraphotographics/BPS; **3.25**: © Arthur Getis; **3.27b**: NASA; **3.30**: © Dr. Carla Montgomery; **3.31**: © Steve McCutcheon/Alaska Photo; **3.32b**: © Dr. Carla Montgomery; **3.34**: © Carl May/Terraphotographics/BPS; **3.35**: © Arthur Getis

Chapter 4

Opener: © Carl Purcell/Photo Researchers, Inc.; **4.11b**: © David Baird/Tom Stack and Associates; **4.19a**: © Stephen Trimble; **4.19b**: © William E. Ferguson; **4.19c**: © Steve McCutcheon/Alaska Photo; **4.27**: © H. Oscar/Visuals Unlimited; **4.32**: © Lawrence Naylor/Photo Researchers, Inc.; **4.33a**: © Aubrey Lang/Valan Photos; **4.33b**: © Arthur Getis; **4.36**: © Ann Duncan/Tom Stack and Associates; **4.39**: © William E. Ferguson; **4.44**: © John Cunningham/Visuals Unlimited; **4.47, 4.50**: © William E. Ferguson

Chapter 5

Opener: © J. A. Wilkinson/Valan Photos; **5.1**: © Joe Sohm/The Image Works; **p. 136**: © Robert Eckert/EKM-Nepenthe; **p. 147, top**: © Gene Alexander/Soil Conservation Service, USDA; **bottom**: © Robert Frerck/Odyssey Productions; **5.8**: © Cameramann International; **5.9**: © William E. Ferguson; **5.12**: © Stephen Trimble; **5.15**: © Will McIntyre/Photo Researchers, Inc.; **5.18**: NASA; **5.20a**: Courtesy of Kennecott; **5.20b**: © Stephen Trimble; **5.21**: © Lawrence E. Stager/Anthro-Photo; **5.22**: Courtesy of the California Department of Water Resources; **5.23**: © Kirschenbaum/Stock Boston; **5.24**: © F. Botts/United Nations; **5.27**: Australian Information Service; **5.28**: © Mark Boulvon/Photo Researchers, Inc.; **5.34**: © ISP/PSI Fotofile

Chapter 6

Opener: © Cameramann International; **6.2**: © Bob Daemmrich/Stock Boston; **6.11**: © Patrick Ward/Stock Boston; **6.17**: Historical Pictures Service, Inc.; **6.18**: © William E. Ferguson; **6.22**: © Robert Frerck/Odyssey Productions; **6.26**: © Mark Antman/The Image Works; **6.27**: United Nations; **6.29**: © Ben Mathes; **6.30**: © Christine Osborne/Valan Photos

Chapter 7

Opener: © Eugene Gordon; **7.1a**: © Robert Frerck/Odyssey; **7.1b**: © Robert Eckert/EKM-Nepenthe; **p. 198**: © Joe Sohm/The Image Works; **7.2**: © J. Popp/Anthro-Photo; **7.3**: © Bob Daemmrich/The Image Works; **7.4a**: © Robert Frerck/Odyssey Productions; **7.4b**: © R. Aguirre and G. Switkes, AMAZONIA/Terraphotographics/BPS; **7.8a**: © Warren Garst/Tom Stack and Associates; **7.8b**: © Robert Frerck/Odyssey Productions; **7.8c**: © Cary Wolinsky/Stock Boston; **7.8d**: © Harriet Gans/The Image Works; **7.9**: © Aubrey Lang/Valan Photos; **7.10a**: © Mark Antman/The Image Works; **7.10b**: © Marc Bernsau/The Image Works; **7.10c**: © Owen Franken/Stock Boston; **7.14a**: © Robert Frerck/Odyssey Productions; **7.14b**: © William H. Allen, Jr.; **7.15**: © David Austen/Stock Boston; **7.16**: © Jean Fellmann; **7.17**: © Mark Antman/The Image Works; **7.24**: © Charles Gateman/The Image Works; **7.30a**: © David Burney; **7.30b**: © Robert Frerck/Odyssey; **7.32**: © T. Joyce/Valan Photos; **7.33**: © Fred Bruemmer/Valan Photos; **7.35, 7.36**: © Wolfgang Kaehler

Chapter 8

Opener: © Bob Daemmrich/The Image Works; **8.1**: © Joe Sohm/The Image Works; **8.8**: Australian Information Service; **8.10**: © AP/Wide World Photos; **8.16**: © Michael Siluk; **8.22**: AP/Wide World Photos; **8.28**: © Cameramann International

Chapter 9

Opener: © Genzo Sugino/Tom Stack; **9.1**: © Bob Daemmrich/The Image Works; **9.6**: AP/Wide World; **9.15**: © Robert Frerck/Odyssey Productions; **9.20**: © J. K. Isaac/United Nations

Chapter 10

Opener: © Robert E. Ford/Terraphotographics/BPS; **10.1**: © John Maher/EKM-Nepenthe; **10.3**: © Pierre R. Chabot/Valan Photos; **10.5**: © Kurt Thorson/EKM-Nepenthe; **10.7**: © Philip Jon Bailey/Stock Boston; **10.8**: © V. Crowl/Visuals Unlimited; **10.10**: © Jean Fellmann; **10.15**: © Brian Brake/Photo Researchers, Inc.; **10.16**: © Jonathan Wright/Bruce Coleman, Inc.; **10.18**: © Thomas Kitchin/Valan Photos; **10.19**: © Cameramann International; **10.20**: United Nations; **10.25**: © Cameramann International; **10.26**: Courtesy of John Deere; **10.30**: © Aubrey Diem/Valan Photos; **10.33**: Culver Pictures, Inc.; **10.34**: Courtesy Fairfax County Economic Development Authority

Chapter 11

Opener: © David M. Doody/Tom Stack and Associates; **11.1**: © Grant Heilman/Grant Heilman; **p. 356**: © American Petroleum Institute; **11.9**: © Visuals Unlimited; **11.12**: © ISP/PSI Fotofile; **11.13**: © American Petroleum Institute, Tosco Corporation; **11.14**: © American Petroleum Institute, Gulf Oil Corporation; **11.17**: © Harold W. Scott; **11.18**: © Christine Osborne/Valan Photos; **11.20**: © Cameramann International; **11.21**: © American Petroleum Institute and Union Oil Company of California; **11.22**: Courtesy Southern California Edison Company and the American Petroleum Institute; **11.28**: © Christine Osborne/Valan Photos; **11.31**: © John Maher/Stock Boston; **11.32**: © Cameramann International; **11.34**: © John Bova/Photo Researchers, Inc.; **11.36a**: © R. Moller/Valan Photos; **11.36b**: © Halle Flygare Photos/Valan Photos; **11.39b**: Goddard Space Flight Center/NASA; **11.41**: © Joe Sohm/The Image Works

Chapter 12

Opener: © Louis Goldman/Photo Researchers, Inc.; **p. 405**: © Arthur Getis; **12.5a**: © B. F. Molnia/Terraphotographics/BPS; **12.5b**: © Carl May/Biological Photo Service; **12.21**: © John Jakle; **12.30**: © Ellis Herwig/Stock Boston; **12.32**: © Eugene Gordon; **12.33**: © The Pulitzer Publishing Company/St. Louis Post-Dispatch; **12.34**: Courtesy of The Chicago Housing Authority; **12.35**: © David Woo/Stock Boston; **12.36**: © Robin L. Ferguson/William E. Ferguson; **12.37**: © Robert Frerck/Odyssey Productions; **12.39**: © IPA/The Image Works; **12.40, 12.43**: © Carl Purcell; **12.44**: © Aubrey Diem/Valan Photos

Chapter 13

Opener: © Robert E. Ford/Terraphotographics/BPS; **13.5**: © B. F. Molnia/Terraphotographics/BPS; **13.8**: © William E. Ferguson; **13.13**: © J. Y. Rabeuf/The Image Works; **13.17, 13.19**: © Elizabeth L. Hovinen

Illustration Credits

Chapter 1

Fig. 1.8 Source: Redrawn by permission from Howard J. Nelson and William A. V. Clark, *The Los Angeles Metropolitan Experience,* p. 49. Association of American Geographers, 1976. **Fig. 1.12** Source: Reproduced by permission from the *Annals of the Association of American Geographers,* John R. Borchert, Vol. 62, p. 358, Association of American Geographers, 1972. **Fig. 1.10** Source: From Thomas J. Mason, et al., *An Atlas of Mortality from Selected Diseases.* NIH Publication #81–2397, May 1981. National Institutes of Health, Bethesda, Maryland.

Chapter 2

Fig. 2.9b Source: Copyright 1977, Brooks-Roberts; with permission. **Fig. 2.11** Copyright The University of Chicago Department of Geography. **Fig. 2.12a** Source: Copyright—Florence Thierfeldt, Milwaukee, Wisconsin. **Fig. 2.14** Source: U.S. Geological Survey. **Fig. 2.15** Source: U.S. Geological Survey. **Fig. 2.18a** From George O. Carney, "From Down Home to Uptown" in Journal of Geography 76: 107, 1977. Copyright © 1977 National Council for Geographic Education, Indiana, PA. Reprinted by permission. **Fig. 2.18b** From Daniel B. Arreola, "Urban Mexican Americans" in Focus. Vol. 34, No. 3, January/February, 1984. Copyright © 1984 American Geographical Society. **Fig. 2.21** Source: U.S. Census Bureau. **Fig. 2.22** Source: Reprinted with permission from *The Economist,* London. *The Economist* The Economist Building, 25 St. James' Street, London SW1A 1HG ENGLAND. **Fig. 2.23** From Gunther Glebe and John O'Loughlin, editors, Foreign Minorities in Continental European Cities, Vol. 84 of Erkundliches Wissen. Copyright © 1987. Franz Steiner Verlag, Weisbaden, West Germany. Reprinted by permission. **Fig. 2.24** Source: U.S. Geological Survey. **Fig. 2.31** From Stephen C. Guptill and Lowell E. Starr, "Making Maps with Computers" in American Scientist, Vol. 76, March-April, 1988 Sigma XI, the Scientific Research.

Chapter 3

Fig. 3.1 Source: After U.S. Geological Survey publication, "Geologic Time." **Fig. 3.3** Source: Exxon Corporation, 1983. **Fig. 3.4** Source: American Petroleum Institute. **Fig. 3.6** Source: From Barazangi and Dorman, *Bulletin of Seismological Society of America,* 1969. Reprinted by permission. **Fig. 3.12b** Copyright © 1986 by the New York Times Company. **Fig. 3.19** From Carla W. Montgomery, *Physical Geology,* 2d edition. Copyright © 1990 Wm. C. Brown Publishers, Dubuque, Iowa. All rights reserved. Reprinted by permission. **Fig. 3.36** From Charles B. Hunt, Geology of Soils: Their Evolution, Classification, and Uses. Copyright © W.H. Freeman and Company.

Chapter 4

Fig. 4.23 From T. McKnight, Physical Geography: A Landscape Approach, 2d ed. Copyright © 1987 Prentice-Hall, Inc., Englewood Cliffs, NJ. Reprinted by permission. **Fig. 4.26** Source: After U.S. Navy Oceanographic Office. **Illustration, Page 103** From T. L. McKnight, Physical Geography: A Landscape Appreciation, 2nd ed. Copyright © 1987 Prentice-Hall, Inc., Englewood Cliffs, NJ. Reprinted by permission.

Chapter 5

Fig. 5.6 Source: Redrawn from "Channelization: Shortcut to Nowhere" by Raymond V. Corning, in *Virginia Wildlife,* February 1975. Copyright *Virginia Wildlife.* Used with permission. **Table 5.1** Reproduced from *Environmental Issues* by Lawrence G. Hines, by permission of W. W. Norton & Co., Inc., copyright 1973 by W. W. Norton & Co., Inc. **Fig. 5.10a,b** Sources: National Air Pollution Control Administration; U.S. Department of Health and Human Services. **Fig. 5.16** Source: Environmental Protection Agency. **Fig. 5.25** Sources: Based upon H. E. Dregne, *Desertification of Arid Lands,* figure 1.2. Copyright 1983 Harwood Academic Publishers; and *A World Map of Desertification.* Unescol FAO. **Fig. 5.26** Source: Based on data from 1934 Reconnaissance Erosion Survey of the United States and other soil conservation surveys by the Soil Conservation Service—U.S. Soil Conservation Service. **Fig. 5.30** Charles J. Krebs, "Ecology," *The Economist,* April 26, 1986, p. 93. *The Economist* The Economist Building 25 St. James' Street London SW1A 1HG ENGLAND. **Fig. 5.35** Artwork by Richard Farrell. Copyright © 1983, Richard Farrell.

Chapter 6

Fig. 6.3 Source: Data from Population Reference Bureau. **Fig. 6.4** Source: Data from United Nations and U.S. Bureau of the Census. **Fig. 6.5** Source: Data from Population Reference Bureau. **Fig. 6.6** Source: Data from Population Reference Bureau. **Fig. 6.7** Source: Data from the U.S. Bureau of the Census and Population Reference Bureau. **Fig. 6.8** Sources: Data from the United Nations, *Demographic Yearbook*; Population Reference Bureau; and the censuses of Mexico, the United States, and West Germany. **Fig. 6.9** Sources: Carl Haub and Mary Mederios Kent, *1988 World Population Data Sheet* (Washington D.C.: Population Reference Bureau, 1988). **Fig. 6.10** Source: Data from Population Reference Bureau. **Fig. 6.12** Sources: U.S. Bureau of the Census and "The Population Future of West Germany: Three Possibilities," *Population Today,* Vol. 13, No. 10 (Washington, D.C.: Population Reference Bureau, October 1985) p. 5. **Fig. 6.13** Source: Data from Population Reference Bureau. **Fig. 6.23** Sources: Data from World Bank and FAO. **Fig. 6.24** Source: World Bank, *Population Growth and Policies in sub-Saharan Africa* (A World Bank Policy Study), 1986. Redrawn with permission. **Fig. 6.25** Source: Data from Population Reference Bureau. **Table 6.1** Sources: Thomas W. Merrick, with PRB staff, "World Population in Transition," *Population Bulletin* Vol 41, no. 2 (Washington, D.C.: Population Reference Bureau, 1986); and *1989 World Population Data Sheet,* Population Reference Bureau. **Table 6.5** Source: The World Bank, *World Development Report 1984.* **Table 6.6** Brinley Thomas, *Migration and Economic Growth* (2nd ed.) Cambridge: Cambridge University Press, 1973. p. 131. **Table 6.7** K. C. Zachariah and Julien Conde, *Migration in West Africa: The Demographic Aspects,* Table 30, p. 46. World Bank Research Publication Service. **Table 6.8** Sources: *1989 World Population Data Sheet* (Population Reference Bureau) and data derived from UN Food and Agricultural Organization (FAO) and United States Department of State.

Chapter 7

Fig. 7.5 Source: Data from World Bank. **Fig. 7.22** Redrawn from *A Word Geography of the Eastern United States* by Hans Kurath. Copyright © 1949 by The University of Michigan. By permission of the University of Michigan Press. **Fig. 7.29** Redrawn from Isma'il al-Farugi and David E. Soper (eds.), *Historical Atlas of the Religions of the World.* New York: Macmillan, 1974, p. 235. **Fig. 7.38** Redrawn from St. Clair Drake and Horace R. Cayton, *Black Metropolis.* New York: Harper & Row, 1962. **Table 7.2** Sources: Based upon *Britannica Book of the Year* and other sources; 1989 population data from Population Reference Bureau, Inc.

Chapter 8

Fig. 8.4 From *World Regional Geography: A Question of Place* by Paul Ward English, with James Andrew Miller. **Fig. 8.11** Ernst Griffin, "Latin America" in D. Gordon Bennett *Tension Areas of the World.* Champaign, IL, Park Press, 1982. **Fig. 8.17** Reproduced by permission from The Professional Geographer of the Association of American Geographers, Vol. 39, 1987, p. 16, Fig. 1, p. 21, Fig. 2, V. L. Mote. **Fig. 8.18** Source: Redrawn from *Oxford Atlas for New Zealand,* 1960. Reproduced by permission of Oxford University Press. **Fig. 8.29** Source: From Edward F. Bergman, *Modern Political Geography,* Copyright 1975 Wm. C. Brown Publishers, Dubuque, Iowa. All Rights Reserved. Redrawn by permission of the author.

Chapter 9

Fig. 9.3 Redrawn with permission from Herbert A. Whitney, "Preferred Locations in North America: Canadians Clues and Conjectures." in *Journal of Geography,* 83, No. 5, 1984. Copyright © 1984 National Council for Geographic Education, Indiana, PA. **Fig. 9.4** From Los Angeles City Planning Department, The Visual Environment of Los Angeles, 1971. Used by permission. **Fig. 9.9** Source: Data from Chicago Area Transportation Study, *1970 Travel Characteristics.* **Fig. 9.8** Source: Redrawn with permission from Robert A. Murdie, "Cultural Differences in Consumer Travel" in *Economic Geography,* Vol 41, No. 3, July 1965. Copyright 1965 *Economic Geography.* **Fig. 9.10** Source: Metropolitan Toronto and Region Transportation Study, by Maurice Yeates. The Queen's Printer, Toronto, 1966, fig. 42. Used with permission. **Fig. 9.12** Reproduced by permision from Resource Publications for College Geography, Spatial Diffusion, Peter R. Gould, Page 4, Association of American Geographers, 1969. **Fig. 9.14** Source: Redrawn with permission from Paul F. Mattingly, "Disadoption of Milk Cows by Farmers in Illinois: 1930–1978" *The Professional Geographer,* Vol. 36, No. 1, p. 45. Association of American Geographers, Washington, D.C., 1984.

Fig. 9.17 Source: Redrawn with permission from Yoshio Sugiura, "Diffusion of Rotary Clubs in Japan, 1920–1940: A Case of Non-Profit Motivated Innovation Diffusion Under A Decentralized Decision-Making Structure" in *Economic Geography,* Vol. 62, No. 2, 1986. Copyright 1986 *Economic Geography.* **Fig. 9.18** Source: Redrawn with permission from Yoshio Sugiura, "Diffusion of Rotary Clubs in Japan, 1920–1940: A Case of Non-Profit Motivated Innovation Diffusion Under A Decentralized Decision-Making Structure" in *Economic Geography,* Vol. 62, No. 2, 1986. Copyright 1986 *Economic Geography.* **Fig. 9.22** Source: Redrawn with permission from C. C. Roseman, *Association of American Geographers Resource Paper* 77.2, p. 5. Association of American Geographers, Washington, D.C., 1977. **Table 9.1** Source: Ian Burton and Robert Kates, "The Perception of Natural Hazards in Resource Management." Reprinted with permission from *3 Nat. Res. J.* 435 (1964), published by the University of New Mexico School of Law, Albuquerque, N.M. **Fig. 9.23** Data from T. Hägerstrand, *Innovation Diffusion as a Spatial Process.* Copyright © 1967 The University of Chicago Press, Chicago, IL. **Fig. 9.25** Source: Redrawn with permission from C. C. Roseman, "Channelization of Migration Flows From the Rural South to the Industrial Midwest," *Proceedings of the Association of American Geographer,* Vol. 3, p. 142. Association of American Geographers, Washington, D.C., 1971. **Fig. 9.24** Source: J. A. Jakle, S. Brunn, and C. C.Roseman, *Human Spatial Behavior: A Social Geography* (1976), reissued 1985 by Waveland Press, Inc., P.O. Box 400, Prospect Heights, Illinois 60070, p. 167. Reprinted with permission. **Fig 9.26** Redrawn by permission from Annals of the Association of American Geographers, Vol. 76, 1986, J. O. Huff. **Part 3 Opener** Source: From *Partick Geddes: Spokesman for Man and the Environment,* ed, by Marshall Stalley. New Brunswick, N.J.: Rutgers University Press, 1972, p. 123.

Chapter 10

Fig. 10.4 Source: Copyright permission: Hammond Incorporated. Maplewood, New Jersey 07040. **Fig. 10.9** Source: Redrawn from Robert E. Huke, "The Green Revolution," *Journal of Geography,* Vol. 84, No. 6, 1985, Fig. 1, p. 248. With permission of the National Council for Geographic Education. **Fig. 10.11b** Modified from Bernd Andreae, Farming, Development and Space, (Berlin: De Gruyter, 1981), Figure 63. **Fig. 10.13** Sources: U.S. Bureau of Agricultural Economics; Agriculture Canada; Mexico. Secretaria de Agricultura y Recursos Hidraulicos. **Fig. 10.23** Redrawn from Truman A. Harshorn and John W. Alexander, Economic Geography, 3rd ed. Englewood Cliffs, NJ: Prentice-Hall, 1988, Fig. 13–2, P. 198. **Fig. 10.27** Redrawn from: Charles Whynne-Hammond, *Elements of Human Geography* (2nd ed.). London: George Allen and Unwin, 1985. Figure 9.7, p. 124.

Chapter 11

Fig. 11.2 Adapted from Macmillan Publishing Company, a Division of Macmillan, Inc., from the United States Energy Atlas, 2nd edition, by David J. Cuff and William J. Young. Copyright © 1985 by Macmillan Publishing Company. **Fig. 11.3** Source: From World Bank. *World Development Report, 1986,* published by Oxford University Press. Used with permission. **Fig. 11.5** Source: Data from British Petroleum Company, *BP Statistical Review of World Energy,* 1989. **Fig. 11.6** Source: British Petroleum Company, *BP Statistical Review of World Energy,* 1989. **Fig. 11.7** British Petroleum Co. BP Statistical Review of World Energy, 1989, Page 3. **Fig. 11.10** Sources: Courtesy of U.S. Department of Energy and American Petroleum Institute. **Fig. 11.11** British Petroleum Co. BP Statistical Review of World Energy, 1989, Page 3. **Fig. 11.16** Source: U.S. Council for Energy Awareness. **Table 11.1** Source: Data from *Minerals Handbook, 1984–85.* Phillip Crowson, ed. Macmillan, London, 1984. **Fig. 11.25** Source: Data from U.S. Bureau of Mines, *Mineral Commodity Summaries,* 1983. **Fig. 11.27** J. W. Mallor and R. H. Adams, Jr. "Feeding the Underdeveloped World," Chemical and Engineering News, 62(17), 1984. Pg. 32–39. **Fig. 11.38** Source: U.S. Forest Service.

Chapter 12

Fig. 12.3 Source: List of metropolitan areas prepared by Richard L. Forstall for Rand McNally and Company's *The International Atlas,* 1989. Used by permission. **Fig. 12.2** Source: Redrawn with permission from Elaine M. Murphy, *World Population Toward the Next Century,* Population Reference Bureau, Inc., May, 1985. **Fig 12.10** Source: Redrawn by permission from R. B. Vance and S. Smith, "Metropolitan Dominance and Integration" in *The Urban South,* R. B. Vance and N. S. Demeruth, Eds. Copyright 1954 University of North Carolina Press, Chapel Hill, North Carolina. **Fig. 12.11** Source: From A. Getis and J. Getis, "Christaller's Central Place Theory," *Journal of Geography,* 1966. With permission of the National Council for Geographic Education. **Fig. 12.15** Source: Data from U.S. Bureau of the Census. **Fig. 12.19** Source: Redrawn with permission from P. J. Smith, "Calgary: A Study in Urban Patterns" in *Economic Geography,* Vol. 38, 1962. Copyright 1962 *Economic Geography.* **Fig. 12.20** Redrawn from B. J. L. Berry, Commercial Structure and Commercial Blight. Research Paper 85, Department of Geography. Research Series, The University of Chicago, 1963. **Fig. 12.23** Redrawn from Arthur Getis, "Urban Population Spacing Analysis" in Urban Geography, Vol. 6, 1985. Copyright © 1985 V. H. Winston and Son, Inc. Siver Srings, MD. **Fig. 12.24** Redrawn with permission from "The Nature of Cities" by Chauncy D. Harris and Edward L. Ullman in Volume CCXLII of the Annals of the American Academy of Political and Social Science. Copyright © 1945 The American Academy of Political and Social Science, Philadelphia, PA. **Fig. 12.25** Source: Redrawn from Philip Rees, "The Factorial Ecology of Metropolitan Chicago" M.A. Thesis, University of Chicago, 1968. **Fig 12.26** Source: Redrawn from E. F. Bergman and T. W. Pohl, *A Geography of the New York Metropolitan Region.* **Fig 12.28** Source: Redrawn with permission from B. J. L. Berry, Chicago: *Transformation of an Urban System.* **Fig. 12.29** Figure from The North American City by Maurice Yeates and Barry Garner. Copyright © 1980 by Maurice Yeates and Barry J. Garner. Reprinted by permission of Harper & Row, Publishers, Inc. **Table 12.3** Source: U.S. Bureau of the Census. **Fig. 12.38** Paul White, *The West Europe City: A Social Geography.* Copyright © 1984, Longman Group U K Limited, Essex England. Reprinted by permission. **Fig. 12.41** Redrawn with permission from Ernst Griffin and Larry Ford, "A Model of Latin American City Structure" in *Geographical Review,* 70, p. 406, 1980. Copyright © 1980 The American Geographical Society, New York, NY. **Fig. 12.42** Figure from Cities of the World. World Regional Development, by Stanley D. Brunn and Jack F. Williams. Copyright © 1983 by Harper & Row, Publishers, Inc. Reprinted by permission.

Chapter 13

Fig. 13.9 From Preston E. James. Introduction to Latin America. Copyright © 1964 Odyssey Press. Macmillan Publishing Co., Inc. **Fig 13.12** Source: Redrawn with permission from T. T. Waterman, "Yorok Geography," *University of California Publications in American Archaeology and Ethnology,* Vol. 16, 1920.

Credits

Photo Credits

Part Openers

Part One: © Robert Frerck/Odyssey Productions; **Part Two:** © Wolfgang Kaehler; **Part Three:** © Robert Frerck/Odyssey Productions; **Part Four:** © Eastcott/Momatiuk/The Image Works, Inc.; **Part Five:** © Val Wilkinson/Valan Photos.

Chapter 1

Opener: Courtesy of NASA; **Fig. 1.1:** © Chris Malazdrewicz/Valan Photos; **Fig. 1.9:** © Robert Frerck/Odyssey Productions; **Fig. 1.10:** © Courtesy of NASA; **Fig. 1.11:** U.S. Geological Survey; **Fig. 1.12:** Prepared by the Chicago Area Transportation Study, Final Report, vol. 1, pg. 44, 1959.

Chapter 2

Opener: © Harriet Gans/The Image Works, Inc.; **Fig. 2.1:** © Robert Eckert/EKM-Nepenthe; **Fig. 2.2a:** © Harriet Gans/The Image Works, Inc.; **Fig. 2.2b:** © Genzo Sugino/Tom Stack & Associates; **Fig. 2.3:** © Harriet Gans/The Image Works, Inc.; **Fig. 2.5:** © Bob Daemmrich/The Image Works, Inc.; **Fig. 2.6:** © Aubrey Lang/Valan Photos; **Fig. 2.7:** © Wolfgang Kaehler; **Fig. 2.12:** © Aubrey Lang/Valan Photos; **Fig. 2.14:** © Val & Alan Wilkinson/Valan Photos; **Fig. 2.16:** © Robert Frerck/Odyssey Productions; **Fig. 2.17a:** © Robert Frerck/Odyssey Productions; **Fig. 2.17b:** © R. Aguirre & G. Switkes, Amazonia/Terraphotographics/Biological Photo Service; **Fig. 2.18a:** © Warren Garst/Tom Stack & Associates; **Fig. 2.18b:** © Robert Frerck/Odyssey Productions; **Fig. 2.18c:** © Cary Wolinsky/Stock Boston; **Fig. 2.18d:** © Harriet Gans/The Image Works, Inc.; **Fig. 2.19a:** © Mark Antman/The Image Works, Inc.; **Fig. 2.19b:** © Marc Bernsau/The Image Works, Inc.; **Fig. 2.19c:** © Owen Franken/Stock Boston; **Fig. 2.21:** © Robert Frerck/Odyssey Productions; **Fig. 2.24a:** © Robert Frerck/Odyssey Productions; **Fig. 2.24b:** © William H. Allen, Jr.; **Fig. 2.25:** © Robert Frerck/Odyssey Productions; **Fig. 2.26:** © Rick Grosse/EKM-Nepenthe; **pg. 37:** © Joe Sohm/The Image Works, Inc.

Chapter 3

Opener: © Robert Frerck/Odyssey Productions; **Fig. 3.1:** © Bob Daemmrich/The Image Works, Inc.; **Fig. 3.8:** © Robert Frerck/Odyssey Productions; **Fig. 3.17:** © William E. Ferguson; **Fig. 3.27:** The Bettmann Archive; **Fig. 3.28:** © Alexandrea Sassaki/Gamma-Liaison; **Fig. 3.33:** AP/Wide World Photos, Inc.

Chapter 4

Opener: © Bob Daemmrich/Stock Boston; **Fig. 4.2:** © Bob Daemmrich/Stock Boston; **Fig. 4.11:** © Patrick Ward/Stock Boston; **Fig. 4.16:** Historical Picture Service; **Fig. 4.17:** © William E. Ferguson; **Fig. 4.21:** © Robert Frerck/Odyssey Productions; **Fig. 4.22:** © William E. Ferguson; **Fig. 4.25:** © Robert Frerck/Odyssey Productions; **Fig. 4.26:** United Nations; **Fig. 4.28:** © Ben Mathes.

Chapter 5

Opener: © Robert Frerck/Odyssey Productions; **Fig. 5.7:** © Mark Antman/The Image Works, Inc.; **Fig. 5.10:** © Andrea Barrist Stern/The Image Works, Inc.; **Fig. 5.19:** © Charles Gatewood/The Image Works, Inc.; **Fig. 5.26:** © David Burney; **Fig. 5.29b:** © Robert Frerck/Odyssey Productions; **Fig. 5.31:** © Robert Frerck/Odyssey Productions; **Fig. 5.32:** © Fred Bruemmer/Valan Photos; **Fig. 5.34:** © Wolfgang Kaehler; **Fig. 5.35:** © Wolfgang Kaehler.

Chapter 6

Opener: © Margot Granitsas/The Image Works, Inc.; **Fig. 6.1:** © Philippe Gontier/The Image Works, Inc.; **Fig. 6.2:** © J. Messerschmidt/Bruce Coleman, Inc.; **Fig. 6.3:** Culver Pictures, Inc.; **Fig. 6.6:** © Joe Sohm/The Image Works, Inc.; **Fig. 6.7:** © Wolfgang Kaehler; **Fig. 6.15:** © Willy Spiller/Bruce Coleman, Inc.; **Fig. 6.16a:** © Joseph Pearce/Valan Photos; **Fig. 6.16b:** © Harold V. Green/Valan Photos; **Fig. 6.18:** © Larry Kolvoord/The Image Works, Inc.; **Fig. 6.20:** © Margot Granitsas/The Image Works, Inc.; **Fig. 6.25:** © David Wells/The Image Works, Inc.

Chapter 7

Opener: © Wolfgang Kaehler; **Fig. 7.1:** Culver Pictures, Inc.; **Fig. 7.3:** Courtesy of the Plimoth Plantation, Plymouth, Massachusetts; **Fig. 7.6a:** © Ralph Gates; **Fig. 7.6b:** © Pam E. Hickman/Valan Photos; **Fig. 7.7d:** © Wolfgang Kaehler; **Fig. 7.7e:** © Val & Alan Wilkinson/Valan Photos; **Fig. 7.7f:** © Wolfgang Kaehler; **Fig. 7.9a:** © Peter Menzel/Stock Boston; **Fig. 7.15a:** © Nancy Anne Dawe; **Fig. 7.16a:** © J. R. Page/Valan Photos; **Fig. 7.16c:** © Margot Granitsas/The Image Works, Inc.; **Fig. 7.18a:** Courtesy of the Bureau of Alcohol, Tobacco, & Fire Arms; **Fig. 7.21b:** © David Burney; **Fig. 7.26:** © Robert Frerck/Odyssey Productions; **Fig. 7.27:** © Ralph Gates; **Fig. 7.30:** © Bob Coyle; **Fig. 7.31:** © Val Wilkinson/Valan Photos.

Chapter 8

Opener: © Wolfgang Kaehler; **Fig. 8.1:** © John Maher/EKM-Nepenthe; **Fig. 8.3:** © Pierre R. Chabot/Valan Photos; **Fig. 8.4:** © Kurt Thorson/EKM-Nepenthe; **Fig. 8.7:** © Wedigo Ferchland/Bruce Coleman, Inc.; **Fig. 8.8:** © Thomas J. Bassett; **Fig. 8.10:** © V. Crowl/Visuals Unlimited; **Fig. 8.11:** © V. Crowl/Visuals Unlimited; **Fig. 8.19:** © John D. Cunningham/Visuals Unlimited; **Fig. 8.22:** © Brian Brake/Photo Researchers, Inc.; **Fig. 8.23:** © Jonathan Wright/Bruce Coleman, Inc.; **Fig. 8.25:** © Wolfgang Kaehler; **Fig. 8.26:** © Grant Heilman/Grant Heilman Photography; **Fig. 8.28:** © Dr. A. Farquhar/Valan Photos; **Fig. 8.30:** © Larry Tackett/Tom Stack & Associates; **Fig. 8.34:** © Sygma.

Chapter 9

Opener: © Bob Daemmrich; **Fig. 9.1:** AP/Wide World Photos, Inc.; **Fig. 9.2:** © Paolo Koch/Photo Researchers, Inc.; **Fig. 9.9:** © BLM/Visuals Unlimited; **Fig. 9.17:** © Cammeramann International, LTD.; **Fig. 9.18:** Courtesy of John Deere; **Fig. 9.21:** © David Burney; **Fig. 9.22:** © Bob Daemmrich; **Fig. 9.24:** © Aubrey Diem/Valan Photos; **Fig. 9.25:** © Joachim Messerschmidt/Bruce Coleman, Inc.; **Fig. 9.29:** © J. P. Laffont/Sygma; **Fig. 9.30:** Culver Pictures, Inc.; **Fig. 9.31:** © David M. Doody/Tom Stack & Associates.

Chapter 10

Opener: © Byron Augustin/Tom Stack & Associates; **Fig. 10.1:** © Dilip Mehta/Contact Press Images; **Fig. 10.2a:** © Cammeramann International, LTD.; **Fig. 10.2b:** © Jason Clay/Anthro Photo; **Fig. 10.4:** © Wolfgang Kaehler; **Fig. 10.5:** © William E. Ferguson; **Fig. 10.6:** United Nations; **Fig. 10.7:** The Bettmann Archive; **Fig. 10.9:** © Byron Augustin/Tom Stack & Associates; **Fig. 10.14:** © D. Willetts/Gamma-Liaison; **Fig. 10.18:** © Bob Daemmrich/The Image Works, Inc.; **Fig. 10.23:** © Ben Mathes; **Fig. 10.26:** © Topham/The Image Works, Inc.

Chapter 11

Opener: © Joe Sohm/The Image Works, Inc.; **Fig. 11.1:** © Elizabeth L. Hovinen; **Fig. 11.7a:** © William E. Ferguson; **Fig. 11.7c:** © Wolfgang Kaehler; **Fig. 11.9a:** © B. F. Molnia/Terraphotographics/Biological Photo Service; **Fig. 11.9b:** © Carl May/Terraphotographics/Biological Photo Service; **Fig. 11.19:** © Norman Owen Tomalin/Bruce Coleman, Inc.; **Fig. 11.20:** © Joe Sohm/The Image Works, Inc.; **Fig. 11.32:** Courtesy of the Fairfax County Economic Development Authority; **Fig. 11.33:** © AP/Wide World Photos, Inc.; **Fig. 11.35:** © IPA/The Image Works, Inc.; **Fig. 11.37a:** © Aubrey Diem/Valan Photos; **Fig. 11.37b:** © Eastcott/Momatiuk/The Image Works, Inc.; **Fig. 11.38:** © Robert Frerck/Odyssey Productions; **Fig. 11.39:** © Robert Frerck/Odyssey Productions; **Fig. 11.40:** © Mark Antman/The Image Works, Inc.; **pg. 393:** United Nations.

Chapter 12

Opener: © Michael J. Johnson/Valan Photos; **Fig. 12.1:** © Karen D. Rooney/Valan Photos; **Fig. 12.2:** © Keystone/The Image Works, Inc.; **Fig. 12.3:** © Cammeramann International, LTD.; **Fig. 12.11:** © Michael Siluk; **Fig. 12.12:** © Michael J. Johnson/Valan Photos; **Fig. 12.21:** © Jeannie R. Kemp/Valan Photos; **Fig. 12.22:** AP/Wide World Photos, Inc.; **Fig. 12.26:** Courtesy of the United States Coast Guard; **Fig. 12.29:** AP/Wide World Photos, Inc.

Chapter 13

Opener: © David Burney; **Fig. 13.1:** © John Elk III/Bruce Coleman, Inc.; **Fig. 13.2:** Courtesy of Kennecott; **Fig. 13.7:** © R. Maiman/Sygma; **Fig. 13.8:** © Robert Eckert/EKM-Nepenthe; **Fig. 13.9:** © William E. Ferguson; **Fig. 13.11a:** Courtesy of the Exxon Corporation and the American Petroleum Institute; **Fig. 13.11b:** © David Burney; **Fig. 13.12:** © C. M. Pringle, University of California, Berkeley/Biological Photo Service; **Fig. 13.13:** © Y. R. Tymstra/Valan Photos; **Fig. 13.15:** © Wolfgang Kaehler; **Fig. 13.21:** © Robert Frerck/Odyssey Productions; **Fig. 13.22:** © Stephen Trimble; **Fig. 13.23:** © Lawrence E. Stager/Anthro Photo; **Fig. 13.25:** © The Image Works Archives; **Fig. 13.26a:** © John Griffin/The Image Works, Inc.; **Fig. 13.26b:** © Roger A. Clark/Photo Researchers, Inc.; **pg. 446:** © Gamma-Liaison; **Fig. 13.27:** © Illustrator's Stock Photos.